THE ECONOMICS OF SALMON AQUACULTURE

The Economics of Salmon Aquaculture

Editor:

Bruno Augusto Amato Borges

Delve Publishing

www.delvepublishing.com

The Economics of Salmon Aquaculture

Editor: **Bruno Augusto Amato Borges**

708 3rd Avenue, 6th Floor
New York, NY 10017
United States of America
www.delvepublishing.com

ISBN: 978-1-68095-854-6

Library of Congress Control Number: 2016951607

About the Editor

Bruno Augusto Amato Borges

Aquaculture Engineer Florianópolis, Brazil

Bruno Augusto is Aquaculture Engineer from Federal University of Santa Catarina. His focus is on writing projects and development of new technologies in aquaculture. He is currently project service provider, consultant and develops research in the field of Biofloc Technology (BFT) for freshwater fish farming.

Contents

List of Contributors

Bjørn Hersoug
The Norwegian College of Fishery Science, University of Tromsøthe Arctic University of Norway

Haakan T Sandersen
Faculty of Social Science, University of Nordland

Ingrid Kvalvik
Nofima—Norwegian Institute of Food, Fisheries and Aquaculture Research

Henry Clifford
From 5th Congress of the Brazilian Biotechnology Society (SBBIOTEC) Florianópolis, Brazil.

Shuichi Kitada
Graduate School of Marine BiosciencesTokyo University of Marine Science and Technology

Hsin Yuan TsaiEmai
The Roslin Institute and Royal (Dick) School of Veterinary Studies, The University of Edinburgh

Alastair Hamilton
Landcatch Natural Selection Ltd., 15 Beta Centre, Stirling University Innovation Park

Derrick R Guy
Landcatch Natural Selection Ltd., 15 Beta Centre, Stirling University Innovation Park

Alan E Tinch
Landcatch Natural Selection Ltd., 15 Beta Centre, Stirling University Innovation Park

Stephen C Bishop
The Roslin Institute and Royal (Dick) School of Veterinary Studies, The University of Edinburgh

Ross D Houston
The Roslin Institute and Royal (Dick) School of Veterinary Studies, The University of Edinburgh

Jim Thorsen
Department of Basic Sciences and Aquatic Medicine, Section of Genetics, Norwegian School of Veterinary Science, Oslo, Norway,
BACPAC Resources, Children's Hospital Oakland Research Institute, Oakland, CA, USA,

Baoli Zhu
BACPAC Resources, Children's Hospital Oakland Research Institute, Oakland, CA, USA,

Eirik Frengen
Biotechnology Centre of Oslo, University of Oslo, Oslo, Norway,
Institute of Medical Genetics, University of Oslo, Blindern, N-0315 Oslo, Norway,

Kazutoyo Osoegawa
BACPAC Resources, Children's Hospital Oakland Research Institute, Oakland, CA, USA,

Pieter J de Jong
BACPAC Resources, Children's Hospital Oakland Research Institute, Oakland, CA, USA,

Ben F Koop
Biotechnology Centre of Oslo, University of Oslo, Oslo, Norway,
Institute of Medical Genetics, University of Oslo, Blindern, N-0315 Oslo, Norway,

William S Davidson
Department of Molecular Biology and Biochemistry, Simon Fraser University, Burnaby, BC, V5A 1S6, Canada

Bjørn Høyheim
Department of Basic Sciences and Aquatic Medicine, Section of Genetics, Norwegian School of Veterinary Science, Norway,
Department of Dermatology, National Hospital, Rikshospitalet, N- 0027 Oslo, Norway

Gary S Burr
ARS National Cold Water Marine Aquaculture Center, USDA, 25 Salmon Farm Road, Franklin, ME 04634, USA

William R Wolters
ARS National Cold Water Marine Aquaculture Center, USDA, 25 Salmon Farm Road, Franklin, ME 04634, USA

Frederic T Barrows
Agricultural Research Service, Fish Technology Center, USDA, Bozeman, MT 59715, USA.

Alan W Donkin
Northeast Nutrition Inc, 494 Willow St, Truro, Nova Scotia B2N 6X8, Canada.

Elżbieta Król
Institute of Biological and Environmental Sciences, University of Aberdeen, Aberdeen AB24 2TZ, UK.

Alex Douglas
Institute of Biological and Environmental Sciences, University of Aberdeen, Aberdeen AB24 2TZ, UK.
Douglas R. Tocher
Institute of Aquaculture, University of Stirling, Stirling FK9 4LA, UK.
Viv O. Crampton
EWOS Innovation AS, Dirdal, Norway.

John R. Speakman
Institute of Biological and Environmental Sciences, University of Aberdeen, Aberdeen AB24 2TZ, UK.

Christopher J. Secombes
Institute of Biological and Environmental Sciences, University of Aberdeen, Aberdeen AB24 2TZ, UK.

Samuel A. M. Martin
Institute of Biological and Environmental Sciences, University of Aberdeen, Aberdeen AB24 2TZ, UK.

Luca Tacchi
Institute of Biological and Environmental Sciences, University of Aberdeen, Tillydrone Avenue, Aberdeen AB24 2TZ, UK.

Christopher J Secombes
Institute of Biological and Environmental Sciences, University of Aberdeen, Tillydrone Avenue, Aberdeen AB24 2TZ, UK.

Ralph Bickerdike
BioMar Ltd, Grangemouth Docks, Grangemouth FK3 8UL, UK.

Michael A Adler
BioMar Ltd, Grangemouth Docks, Grangemouth FK3 8UL, UK.

Claudia Venegas
AVS Chile S.A., Casilla 300, Puerto Varas, Chile.

Harald Takle
AVS Chile S.A., Casilla 300, Puerto Varas, Chile.
Nofima, Aas, Norway.

Samuel AM Martin
Institute of Biological and Environmental Sciences, University of Aberdeen, Tillydrone Avenue, Aberdeen AB24 2TZ, UK.

Torfinn Moldal
Norwegian Veterinary Institute, Sentrum, 0106 Oslo, Norway.

Guro Løkka
Norwegian School of Veterinary Science, Dep, 0033 Oslo,Norway.

Jannicke Wiik-Nielsen
Norwegian Veterinary Institute, Sentrum, 0106 Oslo, Norway.

Lars Austbø
Norwegian School of Veterinary Science, Dep, 0033 Oslo,Norway.

Bente E Torstensen
National Institute of Nutrition and Seafood Research, Nordnes , 5817 Bergen, Norway.

Grethe Rosenlund
Skretting ARC, Stavanger, Norway.

Ole Bendik Dale
Norwegian Veterinary Institute, Sentrum, 0106 Oslo, Norway.

Magne Kaldhusdal
Norwegian Veterinary Institute, Sentrum, 0106 Oslo, Norway.

Erling Olaf Koppang
Norwegian School of Veterinary Science, Dep, 0033 Oslo,Norway.

Sofia Morais
Institute of Aquaculture, University of Stirling, Stirling FK9 4LA, Scotland, UK.

Tomé Silva
Aquaculture Research Group, Centre of Marine Sciences (CCMAR), University of Algarve, 8005-139 Faro, Portugal.

Odete Cordeiro
Aquaculture Research Group, Centre of Marine Sciences (CCMAR), University of Algarve, 8005-139 Faro, Portugal.

Pedro Rodrigues
Aquaculture Research Group, Centre of Marine Sciences (CCMAR), University of Algarve, 8005-139 Faro, Portugal.

Derrick R Guy
Landcatch Natural Selection Ltd., Cooperage Way Business Centre, Alloa FK10 3LP, Scotland, UK.

James E Bron
Institute of Aquaculture, University of Stirling, Stirling FK9 4LA, Scotland, UK.

John B Taggart
Institute of Aquaculture, University of Stirling, Stirling FK9 4LA, Scotland, UK.

J Gordon Bell
Institute of Aquaculture, University of Stirling, Stirling FK9 4LA, Scotland, UK.

Douglas R Tocher
Institute of Aquaculture, University of Stirling, Stirling FK9 4LA, Scotland, UK.

Tyler D. Brown
Department of Ocean Sciences, Memorial University of Newfoundland, 1 Marine Lab Road, St. John's, NL A1C 5S7, Canada

Tiago S. Hori
Department of Ocean Sciences, Memorial University of Newfoundland, 1 Marine Lab Road, St. John's, NL A1C 5S7, Canada

Xi Xue
Department of Ocean Sciences, Memorial University of Newfoundland, 1 Marine Lab Road, St. John's, NL A1C 5S7, Canada

Chang Lin Ye
Department of Plant and Animal Sciences, Faculty of Agriculture, Dalhousie University, Truro, NS, Canada B2N 5E3

Derek M. Anderson
Department of Plant and Animal Sciences, Faculty of Agriculture, Dalhousie University, Truro, NS, Canada B2N 5E3

Matthew L. Rise
Department of Ocean Sciences, Memorial University of Newfoundland, 1 Marine Lab Road, St. John's, NL A1C 5S7, Canada

HsinYuan Tsai
The Roslin Institute and Royal (Dick) School of Veterinary Studies, The Univer sity of Edinburgh, Midlothian EH25 9RG, UK.

Alastair Hamilton
Landcatch Natural Selection Ltd., 15 Beta Centre, Stirling University Innovation Park, Stirling FK9 4NF, UK.

Alan E. Tinch
Landcatch Natural Selection Ltd., 15 Beta Centre, Stirling University Innovation Park, Stirling FK9 4NF, UK.

Derrick R. Guy
Landcatch Natural Selection Ltd., 15 Beta Centre, Stirling University Innovation Park, Stirling FK9 4NF, UK.

James E. Bron
Insti tute of Aquaculture, University of Stirling, Stirling FK9 4LA, UK.

John B. Taggart
Insti tute of Aquaculture, University of Stirling, Stirling FK9 4LA, UK.

Karim Gharbi
Edinburgh Genomics, Ashworth Laboratories, King's Buildings, University of Edinburgh, Edinburgh EH9 3JT, UK.

Michael Stear
Institute of Biodiversity, Animal Health and Compara tive Medicine, University of Glasgow, Bearsden Road, Glasgow G61 1QH, UK.

Oswald Matika
The Roslin Institute and Royal (Dick) School of Veterinary Studies, The Univer sity of Edinburgh, Midlothian EH25 9RG, UK.

Ricardo PongWong
The Roslin Institute and Royal (Dick) School of Veterinary Studies, The Univer sity of Edinburgh, Midlothian EH25 9RG, UK.

Steve C. Bishop
The Roslin Institute and Royal (Dick) School of Veterinary Studies, The Univer sity of Edinburgh, Midlothian EH25 9RG, UK.

Ross D. Houston
The Roslin Institute and Royal (Dick) School of Veterinary Studies, The Univer sity of Edinburgh, Midlothian EH25 9RG, UK.

Anna K Sonesson
Nofima AS, 1431 Ås, Norway.

Jørgen Ødegård
Nofima AS, 1431 Ås, Norway.

Lars Rönnegård
Dalarna University, 791 88 Falun, Sweden.

Sofia Morais
Institute of Aquaculture, University of Stirling, Stirling, FK9 4LA Scotland, UK.

John B Taggart
Institute of Aquaculture, University of Stirling, Stirling, FK9 4LA Scotland, UK.

Derrick R Guy
Landcatch Natural Selection Ltd, Cooperage Way, Alloa, FK10 3LP Scotland, UK

J Gordon Bell
Institute of Aquaculture, University of Stirling, Stirling, FK9 4LA Scotland, UK.

Douglas R Tocher
Institute of Aquaculture, University of Stirling, Stirling, FK9 4LA Scotland, UK.

Marcos G Godoy
Biovac SA, Puerto Montt, Chile,

Alejandra Aedo
Biovac SA, Puerto Montt, Chile,

Molly JT Kibenge
Department of Pathology and Microbiology, OIE Reference Laboratory for ISA, Atlantic Veterinary College, University of Prince Edward Island, Charlottetown, PE, Canada,

David B Groman
Aquatic Diagnostic Services, Atlantic Veterinary College, University of Prince Edward Island, Charlottetown, PE, Canada,

Carmencita V Yason
Regional Diagnostic Virology Laboratory, Atlantic Veterinary College, University of Prince Edward Island, Charlottetown, PE, Canada

Horts Grothusen
Biovac SA, Puerto Montt, Chile,

Angelica Lisperguer
Regional Diagnostic Virology Laboratory, Atlantic Veterinary College, University of Prince Edward Island, Charlottetown, PE, Canada

Marlene Calbucura
Biovac SA, Puerto Montt, Chile,

Fernando Avendaño
Marine Harvest S.A., Puerto Montt, Chile

Marcelo Imilán
Marine Harvest S.A., Puerto Montt, Chile

Miguel Jarpa
Marine Harvest S.A., Puerto Montt, Chile

Frederick SB Kibenge
Department of Pathology and Microbiology, OIE Reference Laboratory for ISA, Atlantic Veterinary College, University of Prince Edward Island, Charlottetown, PE, Canada,

Helena Hauge
Norwegian Veterinary Institute, Sentrum, 0106 Oslo, Norway

Maria Dahle
Norwegian Veterinary Institute, Sentrum, 0106 Oslo, Norway

Torfinn Moldal
Norwegian Veterinary Institute, Sentrum, 0106 Oslo, Norway

Even Thoen
Norwegian Veterinary Institute, Sentrum, 0106 Oslo, Norway

Anne-Gerd Gjevre
Norwegian Veterinary Institute, Sentrum, 0106 Oslo, Norway

Simon Weli
Norwegian Veterinary Institute, Sentrum, 0106 Oslo, Norway

Marta Alarcón
Norwegian Veterinary Institute, Sentrum, 0106 Oslo, Norway

Søren Grove
Norwegian Veterinary Institute, Sentrum, 0106 Oslo, Norway

Laura Martinez-Rubio
Institute of Aquaculture, School of Natural Sciences, University of Stirling, Stirling FK9 4LA, Scotland, UK.

Øystein **Evensen**
Norwegian Life Science University, Faculty of Veterinary Medicine, Dep, Oslo N-0033, Norway.

Aleksei Krasnov
Nofima AS, Ås N-1431, Norway.

Sven Martin Jørgensen
Nofima AS, Ås N-1431, Norway.

Simon Wadsworth
EWOS Innovation AS, Dirdal N-4335, Norway

Kari Ruohonen
EWOS Innovation AS, Dirdal N-4335, Norway

Jose LG Vecino
EWOS Innovation AS, Dirdal N-4335, Norway

Douglas R Tocher
Institute of Aquaculture, School of Natural Sciences, University of Stirling, Stirling FK9 4LA, Scotland, UK.

Fabian Grammes
Institute of Animal and Aquaculture Sciences, Norwegian University of Life Sciences, N-1432 As-UMB, Norway.
NOFIMA, N-1432 s, Norway.

Harald Takle
NOFIMA, N-1432 s, Norway.
AVS Chile SA, Casilla 300, Puerto Varas, Chile

Preface

The group of salmonids covers several species currently farmed in aquaculture, being the main Atlantic salmon Salmo salar. Its cultivation began to spread in the United Kingdom in order to repopulate the rivers for sport fishing, through knowledge of the reproduction, hatchery and rebuilds of fingerlings. However, it was in Norway in the middle of the last century that salmon farming began to gain greater proportions from the crops in offshore cages.

Currently, 70% of the salmon consumed in the world comes from aquaculture, being the main producers are Norway, Chile and Canada and represents the third largest economy in aquaculture, behind shrimp and cyprinids. His production in 2014 exceeded 3,000,000 tons (FAO, 2014). Is a high-value fish and very well accepted by the market. Several problems related to the productive chain of the salmon are pressing producers and the scientific community to propose changes in search of better results using resources in a responsible manner. The search for an effective diet that does not use fishmeal is a big problem, as other species are fished to feed farmed salmon. Many ingredients of plant origin are being tested to replace fishmeal, making increasingly sustainable salmon farming. Another issue that affects strongly the salmon farming industry is the increasing occurrence of diseases causing high mortality of production batches and subsequent financial loss. The four infectious diseases that most affect salmon cultivation were, respectively, infectious salmon anaemia (ISA), infectious hematopoietic necrosis (IHN), furunculosis and bacterial kidney disease. The main one, infectious salmon anaemia was first detected in Norway in 1984 and recently sharply affected the economy of Chile's salmon. ISA and IHN are listed as reportable diseases by the World Organization for Animal Health. On the other hand, the search for alternatives and improvements has been going on in several ways, including genetic mapping and the quality of the diet for the welfare and animal nutrition. The most important thing when you want to start a production of any fish, not just the salmon, is to understand the mechanisms of self-regulation of the environment, respecting the support capacity and seeking comfort and proper nutrition of animals. In this way the chances of problems with diseases will be smaller by decreasing the risk of investment.

Editor
Bruno Augusto Amato Borges
Aquaculture Engineer
Florianópolis, Brazil

CHAPTER 1 THE GREENING OF NORWEGIAN SALMON PRODUCTION

Bjørn Hersoug

The Norwegian College of Fishery Science, University of Tromsøthe Arctic University of Norway

ABSTRACT

The Norwegian salmon industry has, in terms of production and economic results, been a great success. However, due to environmental problems such as sea lice and escapes, the industry encounters increasing resistance from many quarters, headed by the wild salmon sport fishers. In order to improve the image of the industry, the fisheries authorities have suggested four new reforms that all contribute to the greening of the industry. At present, the outcome of these reforms is uncertain. The article discusses the pros and cons of these reforms. It sees them as an introduction to the next large battle for space, which is due to take place if the salmon industry is going to expand as projected.

The Norwegian aquaculture industry is at the crossroads. In the last 45 years, the production of salmon has increased from less than 1,000 tons to 1.2 million tons in 2014. The industry has, over the years, become a financial success with significant repercussions in many coastal communities. It is frequently referred to as "a Norwegian answer to IKEA"[1] and is seen as one (of a few) promising sectors in terms of growth when the petroleum industry gradually has to be reduced. The future prospects are enormous, but the industry is also facing increasing opposition from many different quarters in Norwegian society. Currently, the main focus is on salmon escapes and the incidence of sea lice, which are believed to have significant negative effects on wild salmon and trout. Sport fishers, professional fishers, conservationists and recreational groups have demanded full stop and no further increase in production, and they have been heard. Both the current and the former Ministers of Fisheries have publicly declared that further growth is impossible until the problems of sea lice and escapes have been solved or, at least, considerably reduced. Meanwhile, the fish farming companies have shown that they have the capacity and market access for an increase of production. In this situation, both fish farmers and authorities have been searching high and low for acceptable new solutions. Is it possible to increase production while at the same time reducing the ecological footprint of the industry? In this article I shall describe four such initiatives, assess their possible effects and discuss whether these instruments may contribute to a more sustainable aquaculture industry, not only in Norway but also in other salmon producing countries where Norwegian interests are involved. The article is based on several previous projects (Hersoug and Johnsen 2012, Johnsen et al. 2014, Andreassen et al. 2014, Hovland et al. 2014), interviews with key stakeholders, a review of public documents and, not least, 40 years of contact with the aquaculture industry.

Theoretical Perspectives on Aquaculture Regulations

Natural resource governance is based on interventions that regulate human behaviour (Berkes2008). These interventions are defined on the basis of assumptions about how human

Citation: Hersoug, B (2015). The greening of Norwegian salmon production. Maritime Studies, v. 14:16. DOI: 10.1186/s40152-015-0034-9.

actions affect nature (Pálsson 2006). Because nature is not directly accessible, it has to be represented through specific governable objects, which in turn become the foundation for defining interventions (Johnsen et al. 2009, Johnsen 2013). Governable objects are constituted when the components and processes in an ecosystem are represented symbolically. Specific techniques are used to translate and assemble the components and processes into bounded, homogenous objects that can be measured, quantified, or modelled in ways that make it possible to create specific intervention mechanisms for governance, as is done with a fish stock or with the maximum allowable biomass (MTB) in aquaculture.[2]

Aquaculture regulations, much like fisheries regulations, are public regulations. That is, they are a distinct type of public regulation. From the early 1970s, there were two main reasons for regulating access to this industry. The first applied to how much should be produced and by whom. Salmon farming was a risky business and overproduction could create bankruptcies and loss of private and public capital. Hence, volume had to be controlled. The authorities also had a plan of using aquaculture as a means to increase employment in rural coastal areas, facing severe problems as the fisheries and processing industry were contracting. Therefore, the right to do fish farming was directed to certain groups or persons. The measure to be used was licensing, combined with limited production volume. The second reason applies to the environment. Fish farming is regulated because it takes place in what is defined as *common property* (coastal sea areas) and because fish farming involves important *externalities.* What one fish farmer is doing may influence not only other farmers in the same area but also other stakeholders in other sectors and industries. Consequently, regulating aquaculture is done for different reasons, but they all depend on having legitimacy—that is, by being accepted as valid by most stakeholders (Jentoft 2004).

Licenses or concessions have been used in Norway for more than 100 years to regulate, among other sectors, hydroelectric power. They were used in regulating access to the fisheries, and were quite familiar when the Ministry of Fisheries was going to limit the number of fish farmers in the 1970s. As pointed out by Mikalsen (1983), the administrative allocation of rights to participation in aquaculture does not imply the elimination of competition, but that access to competition is regulated and normally limited. Licensing certain operators means that these fish farmers are allowed to enter an exclusive market, a market with limited participation. According to Hernes (1978), such licensing may be seen as foreclosing the decisions of the market. Without intervention, the market would have eliminated many operators and, in the next round, probably opened the arena for *other* operators. Hence, the judgment of the market is not effectuated because the effects may be deemed too costly or unwanted in political terms. What is worth noting is that this transfer of responsibility for access from the market to a public hierarchy makes public authorities responsible for the results of the allocation, while the fish farmers acquire status as clients. Hence, the critical issue is how the aquaculture authorities organize the allocation in terms of goals, criteria, and administrative processes (Hersoug 2005).

The environmental aspects of aquaculture had to be regulated by another management measure, namely the allocation of space, or *locality.* Deciding where to place the aquaculture activities thus became part of marine spatial planning. Marine planning is a process whereby coastal space becomes framed as a governable object (Johnsen et al. 2014). Coastal space as a governable object is not equal to the undefined natural space, but a translation of an ungovernable complexity into a defined, governable object that can serve as a foundation for governance interventions. In a marine setting, this takes place when a variety of biotic and abiotic components in a specific location are translated into specific spatial properties that can be defined on a map representing the space.

When aquaculture licenses became freely transferable from 1991 onwards, this was a step in the direction of using the market to a larger degree than before. The authorities were less interested in *who* did the farming than in *how* they farmed. It was not a question of substituting one system for the other (from plan to market), but how to change the mix of the two steering systems. However, such changes do not happen haphazardly. They are mediated through*institutions.* Here we are not going into detail about the many definitions of institutions, but suffice to say that many theories are based on rational choice theory, where institutions are seen as sets of positive (inducements) and negative (constraints) motivations for individuals, in which the individual utility maximization is acting as the dynamic element in the institutional set-up. Against this instrumental perception of institutions, it is possible to supply a more sociological concept, where management institutions are *embedded* in a larger social structure (Polanyi 1944, Granovetter 1985). These authors argue that markets and other coordinating mechanisms are shaped and are shapers of social systems of production (Hollingworth and Boyer 1997). The interconnectedness of social and economic systems works both ways. The crucial point of this theoretical understanding is that politics is not an exogenous variable that can easily be eliminated. Politics is the very essence of resource management, i.e. of allocating scarce resources (Easton 1953).

As pointed out by Holm (2001), modern fisheries management builds on science and, if we are to understand fisheries management, we need to understand the role of science. The same applies to aquaculture management. Science and scientific knowledge have to be seen as institutions. In this respect, I have found it fruitful to introduce a *social constructivism perspective.* A central element of this perspective is *framing*, which may explain why certain solutions have been chosen while others have been neglected or refuted. According to Holm and Nielsen (2007), a frame is a boundary, and framing is the process of producing this boundary. In our case, framing can be used on two levels: first on how problems are perceived and then on how they can be solved. Framing is central in most science and technology studies (STS), but this is not the place to explicate the various positions (see Barnes 1982, Bloor 1991, Collins1992). Actor-network theory, as developed by Latour (1987, 1993) and other writers in the same tradition, has proved very useful when considering various elements of resource management (Holm and Nielsen 2007, Johnsen 2013).

In our case, the regulation of aquaculture implies the use of two very different sets of management measures; the license regime, which in effect limits the number of competitors, and the locality regulation, which is part of a larger system regulating the use of coastal space. Both systems can be seen as social constructs, but as soon as they are established and fortified by institutions, laws, science, stakeholders and lobby groups they may be difficult to change. The network created by various types of actors seems stabilized and may be taken as a fact, or as the (only) solution. Consequently, establishing a new regulatory system implies the creation of new institutions (or the modification of old), laws, networks of stakeholders and material artefacts. In our case, a new regulatory regime is critically dependent on the production of knowledge—that is, on science—which in turn can be made operational for management interventions.

Aquaculture—a success or a Horror Story?

According to the fisheries authorities and the industry Norwegian salmon farming is, per 2015, a great success story. During the last 45 years, the industry has grown from under 1,000 tons to 1250,000 tons (salmon and trout). The average annual growth has been around ten percent, and aquaculture is the most rapidly increasing sector in the entire Norwegian economy. Norway is the world's largest producer of Atlantic salmon (*salmo salar*). Chile is in second place, with about half of Norway's production. Salmon and trout amounted in 2014 to 70 percent of the total value of seafood exports, and salmon farming has thus in a few years taken over the hegemony within Norwegian seafood production. The industry employs approximately 10,000 persons directly, i.e. in producing brood stock, juveniles, on-growth farms, slaughterhouses and directly value-added processing, while derivative effects in terms of production of intermediate goods and services employ around 14,000 persons. Many of these jobs are located in marginal coastal municipalities, which in the same period have experienced a sharp decline in terms of fishing and fish processing. The seafood industry is currently the third largest export industry, after petroleum and petroleum related input factors, and is considered one of the industries that can take over when petroleum activities have to scale down over the next 30–40 years.

The ambitions for further growth are formidable. The previous government stated in its report to Parliament that Norway would be "the world's leading seafood nation" by inter alia, providing a regulatory framework that increases further growth, based on biological sustainability (Meld. St. 22 (2012–2013)). The current government is even more specific: "The Government will follow up the vision of a quadrupling of GDP in the industry over the next decades ..." (Andreassen et al. 2014). Both governments have largely relied on expert reports from the Royal Norwegian Academy of Sciences (DKNVS) and the Norwegian Technical Academy (NTVA), which in their report outlined a future scenario with a tripling of salmon production in 2030 and increasing fivefold in 2050 (DKNVS/NTVA 2012). Shortly summarized, "The sky is the limit!" Fig. 1

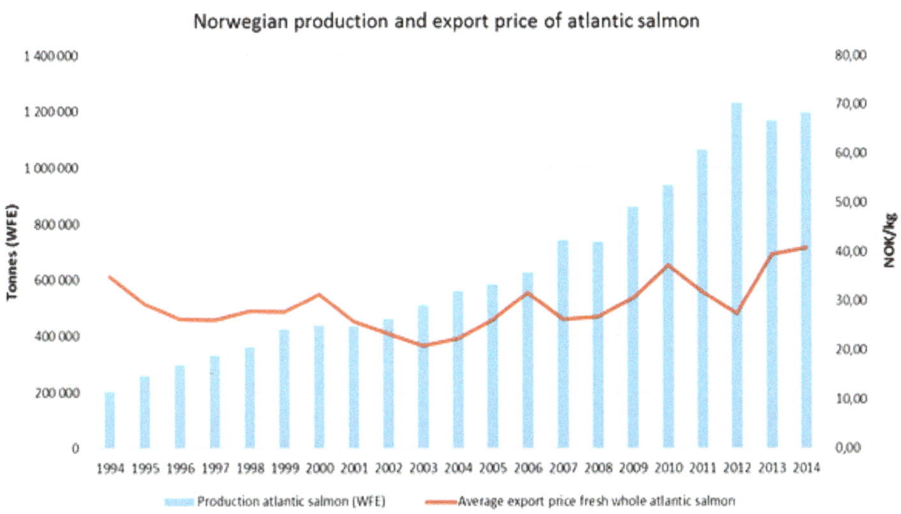

Figure. 1 Quantity produced of Atlantic salmon, nominal export prices. Source: Norwegian Seafood Council

This view is by no means shared by the critics. The most eager critics are the sport fishermen and their organizations, *Norges Jeger-og Fiskerforbund* [3] (organizing the users) and *Norske lakseelver* (organizing the river owners).[4] They frequently accuse the fisheries authorities of managing the salmon industry in an unsustainable manner, to the detriment of the wild stocks. The representatives of the river owners are even more outspoken, demanding that there be no further increase before a new management regime has been established.[5] But opposition is not limited to the recreational sport fishers and river owners. The conservation interest groups have followed suit, with traditional groups like *Norges Naturvernforbund* and *Natur og ungdom*, as well as the newer NGOs like World Wide Fund for Nature (WWF), Bellona and Greenpeace, supporting the recreational fishers.

The professional fishers, organized through *Norges Fiskarlag* and *Norges Kystfiskarlag*, have for years been rather accommodating vis-à-vis the new industry. They recognize that many former fishers are now fish farmers and that the industry creates employment and secures settlements all along the coast. But, with concentration of ownership and a complete reorganization of the salmon industry during the 1990s, the critics became more predominant. The salmon farming companies, especially in the north, were soon owned by large national or multinational companies with limited local connections. Fishers felt that traditional breeding grounds and fishing grounds were threatened, and they often opposed plans suggesting new and expanded areas for the aquaculture industry. One particular issue caught their attention: the claim that spawning cod would not return to traditional fjords in the north due to aquaculture farms located in these fjords. A more recent concern is the massive use of *flubenzuroner* in the delousing of infected salmon, which in the next round affects all bottom-dwelling creatures, including snails, shrimp, crabs and lobster. In sum, fishers are no longer welcoming the unconditional growth of the salmon industry.

Finally, we have the general public, represented by their local councillors all along the coast, which is organized into 275 coastal municipalities. As long as the industry offered new jobs, they welcomed the new salmon industry but, with the more recent developments, many turned to scepticism and, quite often, straightforward resistance against any further expansion. This can be explained by the fact that coastal plans, normally worked out as part of the municipality plan, which is to be revised each fourth year, are costly and cumbersome to make. One issue is that most small, coastal municipalities lack capacity and competence to do planning at sea, and therefore need to hire external consultants or enter into some type of inter-municipality cooperation (Hersoug 2013). The other issue relates to the political costs of manoeuvring a coastal plan with generous allocation to aquaculture, which quite often has to be adopted against the interests of fishers, both professional and recreational, recreational dwellers, conservationists and tourist interests. By 2014, many municipalities said no to further expansion until they are offered a better economic deal.

A Complex Management Regime

While the traditional fisheries are subject to one ministry, aquaculture depends on at least four ministries and their subordinate directorates and supervision. Here we will briefly look at the two basic elements of the system: the use of *licenses* (permits) and the regulation of *localities*(sea space or sites).

The use of licenses has a long tradition in Norwegian fisheries. The Lysø Committee, which had been tasked to investigate how the new fish farming industry should be regulated, also suggested the use of licenses (NOU 1977:39). At the outset, all existing farms were recorded and, from then on, all new facilities were required to apply for a license. Each owner could only have one license and, through irregular allocation rounds, the fisheries authorities could regulate both *where* (geographically) and *who* should be prioritized. Licensing rounds were arranged in 1981, 1984, 1985, 1989, 2002, 2003, 2006 and 2009. In 1991, the ownership restrictions were lifted. It had already been under pressure for years, since large owners operated with minority interests and "straw men", which ensured effective control of several farms (Hovland et al. 2014). With the bankruptcy of the salmon farmers' mandatory sales union (*Fiskeoppdretternes Salgslag*) in 1991, many salmon farmers went bankrupt, which in turn paved the way for a large-scale reorganization of the industry (ibid).

From the start, the licenses were regulated according to net pen volume, originally 8000 m^3, then 3000 m^3, 5000 m^3, 8000 m^3 and, finally, 12,000 m^3. It proved to be a highly approximate production limitation, because the volume was just measured down to five meters depth, and many farmers had already acquired nets that were both 20 and 30 meters deep. Later, the volume restriction was combined with feed quotas as a result of the production limitations following the EU's threats of trade restrictions due to accusations of Norwegian dumping and use of irregular subsidies. Thus, throughout the 1990s, licenses and quotas acted as production control measures, through which the fisheries authorities tried, to the best of their abilities, to limit growth, so that the industry would not be affected by further trade restrictions. In 2005, feed quotas were replaced by a system based on maximum allowable biomass (*MTB*), a system that still persists, despite criticism from those farmers who carry out extensive processing in Norway. MTB for a standard license is 780 tons from Nordland County and southwards, while farmers in Troms and Finnmark have 945 tons as the upper limit. This is to accommodate the northern locations' shorter growing seasons, due to lower seawater temperatures.

The present government has decided that the licensing system should remain unchanged. There are, however, challenges from several quarters, not least by the conservatives who, in opposition, advocated that the system should be discontinued in favour of a system based on technical approval, i.e. on the basis of environmental considerations only. The Senior Think Tank, a grouping of established fish farmers and people with solid experience in all aspects of farming operations, has advocated that the state's role as "production manager" should be abolished (Seniortanken 2014). Their main objection is that the aquaculture industry should be treated like any other industry, where it is up to the actors themselves to determine production volume, within applicable technical and environmental regulations. Another type of criticism has been that licensing rounds act as "beauty contests", with applicants promising gold and green forests, such as further processing, local employment, etc. These promises are seldom checked and tested, and many of the lucky license winners are selling shortly afterwards, normally with windfall gains (Andreassen and Robertsen 2014). Nevertheless, licensing regulation is no longer regarded as a production control tool, but as a necessity in terms of "distribution and scarcity considerations", which refers to the environment and optimum use of the coastal zone (NFD 2014).

But a license is not enough. The farmer must also have a site (locality), and in Norway it is the coastal municipalities that reign over sea space. Planning at sea has, until now, been a voluntary task, but such plans were earlier on regarded as a precondition to be awarded farming licenses. In 1985 coastal municipalities got the opportunity to plan their sea areas to the baselines, i.e. the straight lines connecting the outermost reefs and islands. In 1989, the area was extended by one nautical mile (1852 m). Thus, the planning responsibility should coincide with EU's Water Directive, which Norway has also acceded. In practice, this means that the municipalities now reign over an area of about 100,000 km^2. Here, municipalities allocate space for aquaculture (A-sites) or create multipurpose areas where aquaculture is included as one of several possible activities (FFFNA-areas). But the process toward an approved locality is long and difficult, as many considerations and agencies are involved. Proceedings are outlined in Fig. 2, which states that the Norwegian Food Safety Authority plays a central role as manager of the Animal Welfare Act and Food Act; the Coast Directorate manages the Harbour Act; the county governors act as managers of a number of laws, primarily the Biological Diversity Act, the Pollution Control Act and the Open Air Act; while the Directorate of Fisheries manages the Aquaculture Act and the Marine Resources Act. The county is responsible for coordinating the proceedings, ensuring the quality assurance of the applications, and assigning the localities. Each of the afore-mentioned agencies have, in principle, a veto opportunity in relation to proposals for new farm localities, while the Ministry of Local Government and Modernisation is called on to mediate if the agencies do not reach agreement with the municipality plan.

Figiure. 2 Application procedure for aquaculture site (locality)

No further growth, except

In 2013, the Minister of Fisheries and Coastal Affairs was hard pressed. The salmon farmers had the capacity to produce more, the markets were favourable and profits were excellent. On the other hand, an increasingly sceptical opinion demanded a full stop until the two major threats had been solved, or at least significantly reduced. The Ministry of Fisheries and Coastal Affairs then came up with a new idea: to introduce a new type of license, under much stricter conditions than the present ones (which have a fixed critical limit of a maximum of 0.5 sea lice per fish), and to allocate licenses only to those able to implement new solutions regarding escapes and sea lice. With an open competition, a number of promising new solutions could be tested out on an industrial scale and, within a few years, it was hoped that the industry would be able to cope with both sea lice and escapes. In sum, it was a large-scale industrial experiment that had the support of most salmon farmers and did not encounter much resistance from the traditional opposition groups described above.

One year later, a new conservative/right government had taken over, and it decided to finalize the allocation of green licenses. In addition, the minister wanted to open for immediate growth, partly in order to compensate for the shelving of another growth proposal to use an *average*MTB over the year. She therefore proposed that each existing farmer should be able to increase its MTB volume by five percent under new and very strict regulations regarding sea lice. The minister also wished to initiate a more predictable, long-term growth, under which the growth parameters were fixed and transparent. A new proposal was therefore presented for public hearing during the fall of 2014 (NFD 2014) and as a White Paper in the spring of 2015 (Meld. St. nr. 16 (2014–2015)).

Finally, a new green management policy was developed, largely under the radar of most public debate. The Norwegian Food Safety Authority (*Mattilsynet*) has the task of controlling fish farms and checking not only public health concerns, but also animal welfare, according to the mandate granted in the Animal Welfare Act. Based on a system of counting sea lice in all sea pens at all farms every week, the Norwegian Food Safety Authority is able to keep track of the offenders and perform risk based surveillance. In the more serious instances, the Norwegian Food Safety Authority demanded the slaughter of all or some of the salmon at the farm. By January 2015, the Norwegian Food Safety Authority made public its list of worst offenders, based on counting sea lice in 2013 and 2014. In the next round, these companies will be required to halve their maximum allowable biomass (MTB) on specified locations.[6]

All four measures can be seen as part of the actual effort of *greening the Norwegian salmon industry*, i.e. making the industry more acceptable to the Norwegian public, to the central and local politicians and, not least, to the foreign consumers buying 95 percent of the salmon produced in Norway. In the following we shall have a closer look at these four elements.

Green and Super-Green Licenses

In 2011–2012, it was obvious that a further expansion of aquaculture was not on the agenda. The Ministry of Fisheries and Coastal Affairs had received strong criticism from the National Audit Office (*Riksrevisjonen* 2012) regarding the management of the Norwegian aquaculture industry, and the planned expansion of five percent of MTB across the country had to be cancelled. Only farmers in Troms and Finnmark (where the sea lice problem was relatively small) got the opportunity to increase production. Meanwhile, there was a strong desire to increase production by the farmers; they had the capacity and skills, while there were few problems on the market, with some exceptions for restrictions in China, and prices were high. The solution was the introduction of so-called *green licenses*, with which farmers were given the opportunity to expand production if they adopted new solutions that could lead to a reduction of sea lice and escapes. The scheme was eventually relatively complicated, in that the Ministry simultaneously wanted to embed both regional priorities and maintain a diverse farming structure in terms of company size. Furthermore, the Ministry would want to use both public auction and allocation by fixed price. Finally, the environmental criteria were differentiated, as 35 of the allocated licenses should have an upper limit of 0.25 adult female sea lice per salmon, while ten licenses had to commit themselves to maximum of 0.10 sea lice per salmon. The distribution of the different groups and conditions is illustrated in Fig. 3.

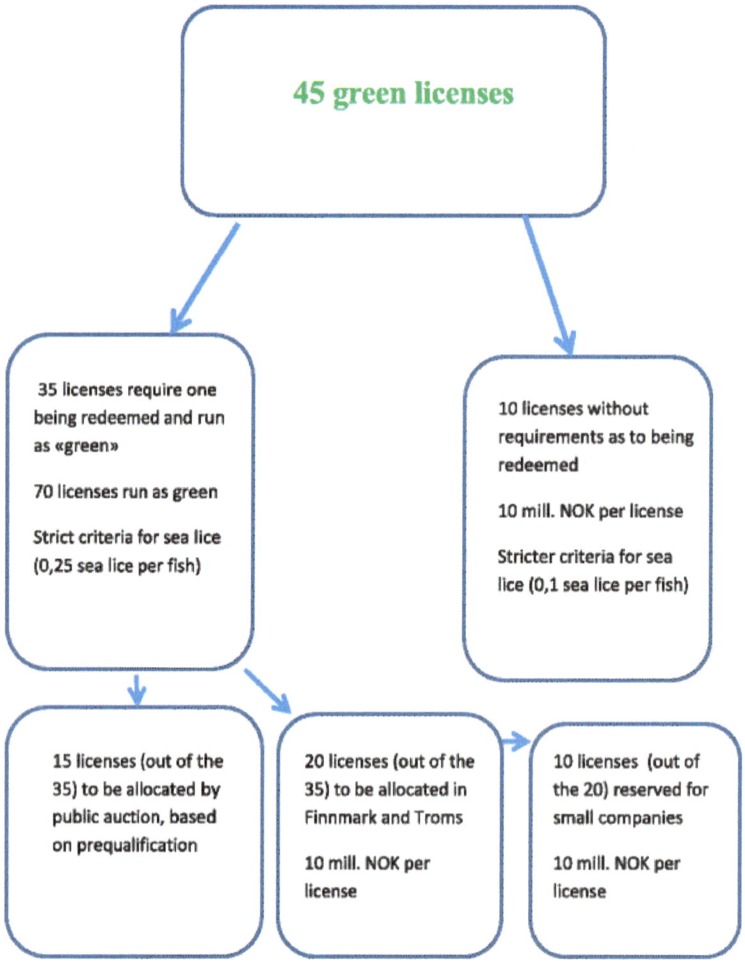

Figugr. 3 The allocation of green licenses in 2014.

The assignment process was organized so that a small group of professionals, headed by a lawyer, accounted for the selection, based on criteria that were prepared in advance. They chose to concentrate on measures that would reduce the incidence of sea lice and the risk of escapes. In practice, this implied a variety of louse defeating devices, which ranged from the use of skirts (outside the net pens) to the use of laser cannons to shoot sea lice. Furthermore, the use of large smolt (up to one kilo, produced on land-based facilities) and triploid salmon was prioritized. Interest was great and 255 applications were received in all. Groups A and B are now finally awarded, while in Group C there are still legal complaints, and final allocation is not expected until the appeals procedure is completed, which will probably occur in the summer of 2015. It is also important to note that, in Groups A and B, companies must commit themselves to redeem an existing license per awarded license, which will also run on green terms. This means that, when the scheme is implemented, there will be 80 licenses that will be run on green terms, of which ten will be run on stringent or *super-green* conditions (0.1 sea lice).

An Incremental Step: Five Percent Increase on Strict Conditions

For a long time, the present MTB-regime regulating production has been heatedly debated. An expert committee, looking into the pros and cons, concluded in 2013 by recommending the use of an average MTB (over the year) instead of the present system of using the given MTB as an absolute ceiling at any time (Guttormsen et al. 2013). However, the farmers were split; while some of the large companies preferred this more flexible approach, others, headed by Marine Harvest, were dead against it. In the end, the Minister of Fisheries decided to drop the proposal and instead introduced a new offer to the fish farmers: the possibility of buying a five percent increase, provided that they were able to keep the sea lice level down to 0.1 during a production cycle with a maximum of two delousing treatments. The offer was presented in the summer of 2014 and could be effective a year later, depending on the reactions of the fish farmers and other stakeholders.

The fish farmers, represented by their generic organization FHL, were in principle in favour of the opportunity to increase production. They also found the price acceptable (1.5 million NOK for a five percent increase), but they found the suggested regime untenable. The suggested level of 0.1 sea lice per license was found unrealistic and the sanctions were seen as risky (economic fines and, ultimately, withdrawal of the increase without compensation). In addition, the farmers found it difficult that the fisheries authorities should operate with three different sea lice regimes, one with 0.1, one with 0.25 and one with 0.5 sea lice. In the spring of 2015, the offer was repeated, in a slightly modified version (a lower price and an extended deadline). The actual number of farmers responding positively to this offer will first be known in 2016. At the outset, it is expected that farmers with small sea lice problems will probably be interested, while others are sceptical and will wait and see, their final decision depending, among other factors, on the possibility of transferring licenses across administrative regions.[7] Among the sceptics, the attitude is fairly uniform; there should be no increase whatsoever until the problems of sea lice and escapes have been solved.

A New Allocation Regime

While the outgoing Labour/centre/left government tried to get around the political limitations on further growth by introducing green licenses, the new conservative/right government proposed a completely new policy regulating growth. The proposal, introduced in 2014, was sent out for public hearing with a view to be presented for the Norwegian Parliament by late spring 2015. The proposal is concerned with the lack of clarity regarding future growth in the salmon sector, and tries to suggest a predictable growth path, thus reducing the uncertainty for salmon farmers. Right from the beginning, the authorities stressed that the future growth of the sector has to take place within a sustainable framework. This means that further growth will be determined by biological sustainability, as defined by various possible parameters. The proposal, which was later developed into a White Paper, contained three alternatives for further growth:

1. The first implies "business as usual", i.e. to continue the present practice of allocating new licenses through license rounds, when the government sees fit. However, the new government is adamant that licenses from now on shall be allocated on "objective criteria", not on shifting political likes and dislikes.
2. The second suggests that growth shall be determined by a fixed annual growth rate, leaving the environmental concerns to other regulatory measures.
3. The third is to make growth dependent on an operational management rule (similar to the Operational Management Procedures we find in fisheries management). Predictability is here secured by a set of environmental indicators that are decisive for growth (or reduction) of production.

The Ministry made no attempt to hide that the third alternative was its preferred solution, and used most of the 60-page document to spell out how such a system could be set up and how it would operate in practice. We do not have the space to elaborate the proposal in detail, but suffice it to say that a number of questions are left open, with the request that stakeholders comment and, if they disagree, to come up with better solutions. The groundwork for the proposal was made by an expert group, the Area Committee, which delivered its report in 2010. The main idea of this report was to divide the coast into different production areas and manage growth according to fish mortality as the central indicator. While the idea of using production areas and put-out zones received wide acclaim, the use of fish mortality as the main indicator met with stern resistance. So when the idea was taken up again in 2014, the main indicator was changed to the frequency of sea lice in a particular area. For the sake of illustration, the coast was divided in 11–13 production areas, where all licenses are connected to a particular area (different from the present seven management areas). Critical values have been suggested, similar to what we find in the fisheries management system, where traffic lights (green, yellow and red) indicate where growth can take place, what areas are under observation, and what areas require reduction in terms of annual production, as measured by allowable biomass (MTB). Other indicators can be connected to the system, such as local and regional pollution, but at this stage, only the frequency of sea lice has been developed to the necessary sophisticated level to be used as an indicator.

The idea behind the proposal is not only to regulate growth, but to make the salmon farmers collectively more responsible for the environmental standard in their production areas. For obvious reasons, the presence of sea lice will vary from one put-out zone to another, thus paving the way for different grading within a production area. But in the end, the different grades will be weighted and the area will be assigned a value or traffic light colour. On the issue of how production capacity shall be allocated, the proposal suggests a Solomonic solution that partly increases capacity via annual or semi-annual license rounds and allows existing farmers to expand by a certain percentage. In terms of payment, the Ministry prefers the public auction of new licenses, while also allowing the use of fixed prices and lottery. For fear of subsidy or dumping allegations (primarily from the EU), the Ministry will not prioritize any particular group in terms of size (small-scale versus large-scale) or in terms of preferred technical solutions. "The Ministry considers biological sustainability the most important requirement for future growth, while the market seems not to be a reason why the authorities shall regulate the production capacity" (NFD 2014: 22). Nevertheless, the Ministry insists on maintaining the license system untouched, although the actual production connected to each license will vary depending on where it is located and how well the farmers in the area are able to reduce the sea lice level.

At this stage, it is important to stress that the new regulatory regime is still on the drawing table and many details are to be decided at a later stage, not least influenced by a rather diversified response from the various stakeholders, who often combine elements from all three solutions to create their preferred options.

A New Regime of Sanctions

While green licenses and any new farming regime will produce results sometime in the future, it is already one regime that is operational. The sanctions' regime builds on the simple logic of reducing the MTB at farm level, based on long time violation of given louse limits (0.5 adult female sea lice per salmon). The Norwegian Food Safety Authority (*Mattilsynet*), which is responsible for the organization of the system, has for a long time had the authority to impose restrictions on farmers and, at worst, require the slaughtering of fish. The Norwegian Food Safety Authority has systematized sea lice data for 2013 and 2014 based on weekly reported counts, which makes it possible to pick out the worst offenders in the first place, i.e. localities that, over a long time, have had sea lice over the allowed limit. In January 2015 these farms have received a warning letter stating that stocking up on these sites must be halved in the next round (production cycle). This does not mean that the company must halve its MTB, but that the company in this case must find other localities where the company's given license MTB can be placed. What is interesting to note, is that company size does not seem to have any significant impact on the frequency of its sea lice-infested localities. As can be seen from Fig. 4, all size categories have red localities (ranging from 21 to 40 percent).

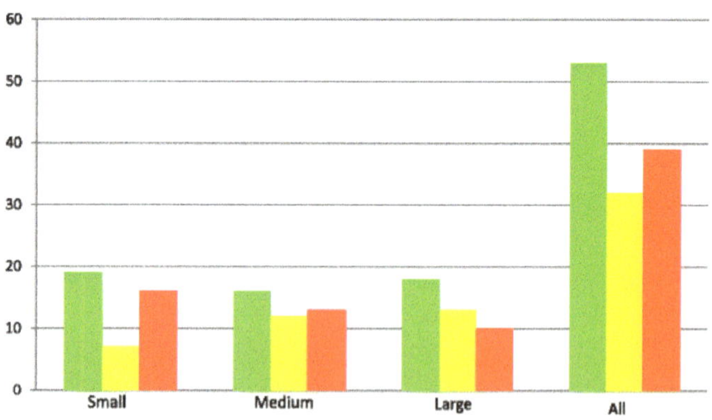

Figure. 4 Distribution of different size groups in terms of having sea lice infested localities. Source: Moen, NFA (2015)

The measure has, as expected, triggered strong reactions. The publication of the companies included on the list and the use of data are highly controversial issues, but the Norwegian Food Safety Authority seems to have strong support in the Ministry of Industry and Fisheries that the measures will be implemented as planned, and that the authorities will take complaints and possible litigations as they come.

DISCUSSION

In the proposal for a new growth regime, it is clearly stated that the industry does not need the state to regulate production according to market needs. Nevertheless, it is equally clear that the government has no intention of changing the existing license regime. This may at the outset seem somewhat paradoxical, but makes sense when one considers the values involved. At the last auction of salmon licenses (in 2014), prices were in the range of 55 to 66 million NOK. If we make a calculation based on the more modest average of 40 million NOK per license, the value of the 1,000 licenses allocated so far will be in the order of 40 billion NOK. No owner would like to see these values scrapped overnight, not least because many have bank loans where the license values are used as collateral.

However, the present license regime has many drawbacks. The most important is that it is very hard for newcomers to enter the industry. They can, of course, buy a license on the open market, but an entry ticket of 40 million NOK is too much for most prospective entrants. This means that the industry is nearly closed to newcomers, which creates the risk of blocking new ideas. In recent years ownership has been heavily concentrated. Although there are still 130 companies involved, with 89 owners, the ten largest are actually producing 80 percent of the quantity (NFD 2014).[8] Over time, the ownership concentration has seemed to increase even further, not necessarily due to economies of scale in the actual production, but due to the advantages of handling all the other operations, including the access to sea space. So far, the ESA (the controlling organ of the European Economic Space agreement) has overruled the previous Norwegian ownership regulation of a maximum of 25 percent of the total number of licenses, and instead set maximum ownership at 40 percent. This has, somewhat hesitantly, been accepted by the previous government, so long as certain requirements regarding further processing, research and the use of labour are fulfilled.[9]

The four management measures described above have one feature in common: they are all based on one simple indicator, namely the frequency of sea lice (adult, female sea lice) as reported by the farmers themselves and controlled by the Norwegian Food Safety Authority. This implies that the question of sustainability has been framed as a question of biological sustainability, i.e. of not affecting the wild salmon and trout negatively. The indicator and critical threshold values have been defined, which in turn have been transformed to a simple system of traffic lights (red, yellow and green). The success of such a regime will therefore depend on whether the authorities are able to *stabilize* the regime, i.e. to get the system accepted and institutionalized. Along this route, there are several challenges. The first relates to the connection between the frequency of sea lice and the status of the wild salmon and trout stocks. While the Institute of Marine Research is proclaiming a definite and clear connection, other researchers and the farmers themselves point to other factors explaining the poor conditions for out-migrating wild salmon and trout (FHL 2015, Bellona 2015). Basing the environmental state on one indicator only may also seem like a risky strategy, although the Ministry is open to adding more indicators (such as pollution) at a later stage.

Secondly, the critical values are still being debated, and while 0.5 sea lice per salmon seems to be accepted as a common denominator, the new limits introduced by the green licenses (0.25 and 0.1) are by many experts considered unrealistic. Many farmers still demand a more scientific explanation for the fixed critical values, although most seem to accept that keeping down the level of sea lice, will, in general, reduce the impact on wild salmon and trout. Even better (for the farmers), this will, if successful, reduce the costs of delousing, which already constitute more than 3 billion NOK per year.

Thirdly, many stakeholders, both inside and outside the industry, doubt the counting regime that has been established, and accuse some farmers of under-reporting. The authorized count by the Norwegian Food Safety Authority is accepted, but the NFA will, in any case, only be able to control a few farms every week. Fourthly, the public sanctions are heatedly debated. Slaughtering part of or an entire batch of salmon will of course have dramatic economic consequences, as will the requirement of halving the put-out for the next production cycle on specific locations, as is now being implemented by the NFA's sanctions' regime. Many farmers and their organizations have already warned that they find the collective punishment inherent in the new growth regime unfair and illegal. All farmers in an area could be forced to reduce their put-out due to the practice of one bad farmer. Fig. 5

License MTB (780 or 945)	**45 Green licenses (1)** **5% MTB increase (2)**	**Increased or reduced *license* MTB for a period** ↑
Locality (site) MTB From 780 to 10 000	**Sanctions' regime(4)**	**New growth regime (3), based on environmental standard in 11-13 new production areas based on sea lice as indicator**

Figure. 5 New management measures according to license MTB and locality MTB

Last but not least, the new proposed growth regime (measure no. 3) seems to confound two different but closely related systems: the license system, which gives a person or a company the right to farm, and the locality (site) system, which allocates space where one might farm. So far, these two systems have been kept isolated, as the right to farm is a prerequisite to apply for a site, but not the other way around. The right to farm has been decided by the state (the Ministry), while the site allocation is, in principle, up to the municipalities, although the county authorities are responsible for the process and the final distribution. By suggesting, as in the new proposal, that companies risk their *license MTB* if they do not comply with the environmental requirements as measured by the sea lice indicator, the authorities mix the two systems, which may not be the most convenient way of obtaining the goal that all partners share (a sustainable salmon industry).

In that respect, the "new" sanctions regime is built upon a different logic, under which specific*sites* are monitored and required to reduce MTB for the next production cycle. Sites are here the central theme, not the licenses and their accompanying MTB. It should also be mentioned that the responsible authority (the Norwegian Food Safety Authority) already has the mandate to reduce production in entire *areas* with a bad environmental standard. The Norwegian Food Safety Authority may reduce the MTB on such localities and also refuse to grant permission to new localities. The possibility that companies' license MTB can be reduced may open a host of legal claims as to the legality of this measure. The comments from the Directorate of Fisheries (2015) to the proposed new growth regime speak for themselves:

"The consultation paper envisages a radical change of management regime for Norwegian aquaculture. This will partly require a significant research work, extensive rulemaking, significant changes in management, restructuring of registers besides changes and strengthening of audit. The production areas must be created in separate regulations, which also contain the necessary procedural rules for changes. The operation of a system of production areas is difficult to realize widath the current fragmented management system. A system of production areas involves a more complicated administration itself, and will be very demanding with the current roles and responsibilities." (Author's transl.).

Hence, the administrative organ most closely involved with the present regime considers the new management regime "a radical change" and "very demanding". While green licenses may be seen as a front-runner of a new regime, the actual proposal of the traffic light system is the full implementation of a more environmentally oriented production system. Whether this system will be viable (fully institutionalized) depends not only on governmental will (political support in the Norwegian parliament) and the acceptance of and cooperation with the major stakeholders (farmers as well as environmental NGOs), but also on scientific backing (solid science to support the indicators, the threshold values and, ultimately, the results in terms of improved survival of wild salmon). Shortly summarized; the green licenses, the five percent increase and the current sanctions' regime operated by the Norwegian Food Safety Authority can all be seen as *modifications* of the existing management regime, although they contain a number of new elements and principles. The suggested new management regime (based on increasing or reducing the *license MTB*) is an entirely new approach, which may require dramatic changes in the institutional set-up of the sector administration. All four initiatives depend critically on the quality of the science involved to prove sustainability, i.e. whether the scientific results are being accepted by the stakeholders and the larger public (including foreign consumers).

The discussion so far has been limited to Norwegian salmon farming, but the reforms under way may also have repercussions for other salmon producing countries. While it may be argued that the policy of diversification into other countries (Chile, Canada, Ireland, and Scotland) by Norwegian salmon producers originally may have postponed stronger regulations at home, the management logic also works the other way around. Stronger demands on safety for escapes and improved treatment of sea lice in Norway will, within relatively short time, be transferred to other countries where the multinational companies are involved (Perkins 2010). Over time, these demands will be incorporated into "best practices" and may also be included in international standards, such as the Aquaculture Stewardship Council's Salmon Standard.

CONCLUSION

Through the new initiative of providing "green growth", the past and present governments have tried to circumvent the impasse created by massive problems connected to sea lice and escapes. We have earlier claimed that Norwegian salmon production is not limited by lack of space or by exceeding ecological limits (Hersoug and Johnsen 2012). The main issue is what is *acceptable*in terms of ecological footprints to Norwegian politicians (acting on behalf of the Norwegian people) and, in a similar manner, what is acceptable to foreign consumers, who are increasingly occupied with sustainability concerns. At present (spring 2015), the government is hard pressed between an industry that would like to expand further, having both the capacity and the markets, and an increasingly sceptical public opinion, which clamours for no further expansion until the main challenges have been solved. In this situation, at least three other public debates are interfering to complicate a political solution. The first is a technical, scientific debate: how can sustainability be measured, monitored and used as the main mechanism for regulating growth? The second applies to how the proceeds from the very profitable salmon sector should be divided, i.e. to what extent should the municipalities managing the near shore sea areas on behalf of the Norwegian people receive an area fee (in addition to a share of the license fees).[10]The third relates to the sale of Norwegian companies, which offers Norwegian sea territories to foreign companies free of charge.[11] This presentation of the greening of the Norwegian salmon industry has not suggested any easy way out of these conflicts. Whether the green licenses are successful in producing the results promised in terms of reduced sea lice and escapes remains to be seen. The large number of technical solutions now being implemented is by all means promising, but the jury is still out as to tangible results. The promise of a five percent increase in production volume (MTB) in return for much stricter limits regarding sea lice has not met with enthusiasm and implementation is pending. How the new proposed growth regime will be implemented is even more uncertain; it depends not only on the technical feasibility but also on gaining the political support and cooperation of salmon farmers. The new sanctions regime has already been introduced, but we have still not seen the results and definitely not the legal reactions of those who are affected (having to reduce their MTB to half on certain "red" localities).

According to a former prime minister (Mrs. Gro Harlem Brundtland), "Everything is dependent on everything".[12] This seems to be an apt description of the current situation in the aquaculture industry. Further growth cannot be obtained before the industry has succeeded in reducing its ecological footprints as measured in terms of frequency of sea lice and escapes. Whether the new management measures (the green licenses, the five percent increase, the growth regime and the sanctions regime) will produce the expected results remains to be seen. Using the Actor Network Theory terminology, it depends on whether all the necessary steps toward a new regime can be stabilized, accepted as legitimate and supported by science. This is by no means secure. The different debates related to the use of sea lice as the main indicator in all the suggested management reforms indicate that the new framework is far from taken for granted. Regarding the question of institutional reform, the crucial question is how to institute a new management regime where future production is dependent on environmental considerations while at the same time presenting predictable growth opportunities to the salmon farmers. In many ways, it is a question of "squaring the circle". Predictable, actual growth cannot possibly be guaranteed if it is dependent on changing environmental conditions. Increased license MTB may be granted, but whether the growth will be realized depends on the environmental conditions on the site or area level, which at present depends on the farmers' ability to reduce sea lice and escapes. Based on risk analysis, we may say that the political risk of operating in Norway will increase. However, if the regime turns out to be successful, the biological risk will be considerably reduced, thus improving not only legitimacy but also profits.

In the short run, the main challenges are connected to reducing sea lice and escapes. In this context, four new initiatives have been presented, all with good intentions but variable degrees of probability that the problems they address will be solved. As soon as these problems have been solved or considerably reduced, the area issue is going to resurface. Larger and better areas are definitely needed if the industry is going to expand as planned (Arealutvalget 2010, Hersoug and Johnsen 2012). The willingness to grant new areas to the aquaculture industry depends in the last instance on legitimacy—on whether the demands of the industry are seen as legitimate and valid. The attempts of greening the industry may, from this perspective, be seen as a preparation for the larger battle—the battle for space, which is bound to be the central issue if the Norwegian industry is going to increase its production five-fold by 2050. It is still an open question whether a greening of the existing management institutions is sufficient or if this dramatic increase would require a wholesale institutional reform. A third possibility is that the present conflicts will be reduced as a result of new technological solutions, such as transferring the production onshore or further offshore. While the first option certainly will reduce the comparative advantage of Norwegian producers, who have access to large areas of sheltered sea waters, the second will require large investments in new technology, and thus drive the industry toward further concentration. In this respect, the greening initiatives may be seen as a compromise that allows for the continuation of the present success of open or semi-open salmon farming in net pens in the fjords while, at the same time, reducing the environmental problems.

ENDNOTES

[1]http://www.ilaks.no/solberg-laks-er-det-norske-ikea/.

[2]Maximum allowable biomass is here indicated by the Norwegian acronym MTB (Maksimalt Tillatt Biomasse).

[3]All Norwegian organizations are presented by their Norwegian names, in italics.

[4]The organization represents rights holders and river associations in 70 salmon rivers. Members are spread across the entire country. Through local chapters Norwegian Salmon

Rivers represents approximately 7000 rights holders. It constitutes about 70 percent of all licensees to salmon rivers in Norway.

[5]http://www.lakseelver.no/Nyheter/2015/januar/Pressemelding%20vekst.pdf.

[6]It should be noted that the MTB should be halved on specific locations, not affecting the MTB guaranteed by the license. In principle the affected companies will then have to find other (and better) localities for the production being reduced. If small companies have only few localities that are all being restricted, this will, other factors equal, reduce the company's total production. This will in the end, be left to the discretion of the Food Authority.

[7]At present licenses are geographically bound to the seven administrative regions of the Fisheries Directorate.

[8]In Chile there are 24 companies controlling 80 percent of the total production.

[9]The present government has suggested that these requirements should be scrapped in 2015, thus leaving the salmon sector on an equal footing with most other industries.

[10]The main income of the municipalities is income tax and economic transfers from the state. Until 1999 part of the company tax was allocated to the municipalities. Today the tax goes directly to the state and is then redistributed through the economic transfer to the municipalities (Isaksen and Mikkelsen 2012).

[11]The former state owned company, Cermaq was recently sold to the Japanese multinational Mitsubishi.

[12]http://www.ordtak.no/index.php?fn=Gro%20Harlem&en=Brundtland.

REFERENCES

1. Andreassen, O, R Robertsen, and JP Johnsen. 2014. Framtidig lokalitetstilgang-Gir krav og dokumentasjon areal og goodwill? Lecture on TEKMAR Conference 2. December, 2014. Tromsø: Nofima.
2. Arealutvalget. 2010. Effektiv og bærekraftig arealbruk i havbruksnæringen–Areal til begjær. Rapport fra et ekspertutvalg oppnevnt av Fiskeri-og kystdepartementet. Oslo: FKD.
3. Barnes, B. 1982. T. S. Kuhn and Social Science. London: Macmillan.
4. Bellona. 2015. Høringssvar: Melding til Stortinget om vekst i norsk lakse-og ørretoppdrett. Oslo: Bellona.
5. Berkes, F. 2008. Sacred Ecology. New York: Routledge Science.
6. Bloor, D. 1991. [1976]: Knowledge and Social Imagery. Chicago: Chicago Press.
7. Collins, HM. 1992. [1985]: Changing order: Replication and Induction in Scientific Practice. Chicago: Chicago Press.
8. Directorate of Fisheries. 2015. Høringssvar–Vekstkriterier havbruk. Bergen: Fiskeridirektoratet.
9. DKNVS/NTVA. 2012. Verdiskaping basert på produktive hav i 2050. Oslo.
10. Easton, D. 1953. The political system: An inquiry into the state of political science. New York: Knopf.
11. FHL. 2015. Høringssvar–Vekstkriterier havbruk. Notat 14.01.2015. Oslo: FHL.
12. Granovetter, M. 1985. Economic action and social structure: The problem of embeddedness. American Journal of Sociology 91(3): 481–510.
13. Guttormsen, A, T Davidsen, K Sæther, GM Knutsen, L Ellingsen, and R Kongsvik Brandvik. 2013. Videreutvikling av MTB-systemet. Oslo, FKD: Utredning.
14. Hernes, G. 1978. Markedet som domstol. In Forhandlingsøkonomi og blandingsadministrasjon, ed. G Hernes. Oslo: Universitetsforlaget.
15. Hersoug, B. 2005. Closing the Commons, 29. Norwegian fisheries from open access to private property. Delft: Eburon.
16. Hersoug, B. 2013. The battle for space–the position of Norwegian aquaculture in integrated coastal zone planning. In Integrated coastal zone management, ed. P Moksnes, E Dahl, and J Støttrup. Oxford: Wiley-Blackwell.
17. Hersoug, B, and JP Johnsen (eds.). 2012. Kampen om plass på kysten. Interesser og utviklingstrekk I kystsoneplanleggingen. Oslo: Universitetsforlaget.
18. Hollingworth, JR, and R Boyer. 1997. Contemporary Capitalism. The embeddedness of institutions. Cambridge: Cambridge University Press.
19. Holm, P. 2001. The invisible revolution. The construction of institutional change in the fisheries. PhD thesis, the Norwegian College of Fishery Science, University of Tromsø.
20. Holm, P, and KN Nielsen. 2007. Framing fish, making markets: the construction of Individual Transferable Quotas (ITQs). In Market Devices, ed. M Callon, Y Millo, and F Muniesa. Oxford: Blackwell Publishing.
21. Hovland, E, D Møller, A Haaland, N Kolle, B Hersoug, and G Nævdal. 2014. Over den leiken ville han rå. Fagbokforlaget: Havbruksnæringens historie. Trondheim.
22. Isaksen, J, and E Mikkelsen. 2012. Økonomer i kystsonen: Kan kunnskap om verdiskaping gi bedre arealforvaltning? In Kampen om plass på kysten. Interesser og utviklingstrekk i kystsoneplanleggingen, ed. B Hersoug and JP Johnsen. Oslo: Universitetsforlaget.
23. Jentoft, S. 2004. Institutions in fisheries: what they are, what they do, and how they change. In Fisheries development: The institutional challenge, ed. B Hersoug, S Jentoft, and P Degnbol. Eburon: Delft.
24. Johnsen, JP. 2013. Is fisheries governance possible? Governability and Governmentality in Norwegian fisheries. Fish and Fisheries. doi:10.1111/faf.12024. 30.
25. Johnsen, JP, P Holm, PS Sinclair, and D Bavington. 2009. The cyborgization of the fisheries. On attempts to make fisheries management possible. Mast 7(2): 9–34.
26. Johnsen, JP, B Hersoug, and AM Solås. 2014. The creation of coastal space–how local ecological knowledge becomes relevant. Maritime Studies 13: 2.
27. Latour, B. 1987. Science in Action: How to follow scientists and engineers through society. Cambridge: Harvard University Press.

28. Latour, B. 1993. We have never been modern. Cambridge (Massachusetts): Harvard University Press.

29. Meld. St. nr. 16. 2014-2015. Forutsigbar og miljømessig bærekraftig vekst i norsk lakse-og ørretoppdrett. Oslo: Nærings-og fiskeridepartementet.

30. Meld. St. nr. 22. 2013-2014. Verdens fremste sjømatnasjon. Oslo: Nærings-og fiskeridepartementet.

31. Mikalsen, K. 1983. Reguleringspolitikk og offentlig styring. Institutt for fiskerifag, Universitetet i Tromsø.

32. Moen, M. 2015. Midlertidig tilbaketrekking–bruk av RUM (Risikobasert utvelgelse av matfiskvirksomheter, laksefisk). Mattilsynet.

33. Nærings og fiskeridepartementet (NFD). 2014. Høringsnotat–melding til Stortinget om vekst i norsk lakse-og ørretoppdrett. Oslo: NFD. NOU. 1977. 39. Fiskeoppdrett.

34. Pallson, G. 2006. Nature and society in the age of postmodernity. In Reimagining political ecology, ed. A Biersack and J Greenberg. Durham, NC: Duke University Press.

35. Perkins, R. 2010. The internationalisation of managerial environmentalism: globalisation, diffusion and territorialisation. Geography Compass 4(8): 1069–1083.

36. Polanyi, K. 1944. The great transformation. New York: Holt, Rinehart.

37. Riksrevisjonen. 2012. Riksrevisjonens undersøkelse av havbruksforvaltningen. Dokument 3: 9 (2011–2012). Oslo: Riksrevisjonen.

38. Seniortanken. 2014. Tanker om veien mot havbrukets Horisont 2050. Notat. Oslo

CHAPTER 2

ACCESS TO AQUACULTURE SITES: A WICKED PROBLEM IN NORWEGIAN AQUACULTURE DEVELOPMENT

Haakan T. Sandersen[1] and Ingrid Kvalvik[2]

[1]Haakan T Sandersen, Faculty of Social Science, University of Nordland

[2]Ingrid Kvalvik, Nofima—Norwegian Institute of Food, Fisheries and Aquaculture Research

ABSTRACT

Increased complexity and interdependencies lead to problems and issues that are hard to solve. Such chronic and enduring problems are often referred to as 'wicked problems'. This article focusses on Norwegian salmon aquaculture and addresses the problems related to the industry's and government's ambitions for growth in aquaculture production and the municipalities' desire for economic compensation, usually termed area rent, from the aquaculture companies that use municipal waters for aquaculture. The government has resisted the introduction of such an area rent due to several complex institutional and distributional considerations. This complex and enduring issue is analysed here as a wicked problem, where we find that the main source of wickedness does not lie within the definition of the problem rather it's related to the available solution as the proposed solution requires changes that are not politically viable.

INTRODUCTION

Development in marine aquaculture is dependent on several factors, including availability of feed, healthy marine environments and suitable locations. A major and enduring debate in Norwegian aquaculture involves the question of access to suitable new production sites (Hersoug 2013; Sandersen and Nikolaisen 2007). Both the government and the aquaculture industry have high ambitions for continued growth of aquaculture production. The country is already by far the world's largest producer of salmon (*salmo salar*), and both government and industry aim to at least double the current production of more than 1 million tonnes, which had an export value of 5 billion euros in 2013 (Seafood Federation] 2013; Seafood Federation] 2012; Report to the Storting Stortingsmelding no. 22 2013). Achieving this goal would require both more and better production sites (Gullestad et al. 2011).

However, there is now increasingly a general understanding that the coastal municipalities are becoming bottlenecks for this growth. The municipalities has since the 1989 revision of the Planning and Building Act (Act of 21. April 1989) had the authority to allocate and designate areas for aquaculture production in their coastal zone. Through municipal coastal zone planning the municipalities decide whether and where to designate areas for salmon production, as well as other economic and non-economic uses, exclusively or in combination. Municipal autonomy in coastal zone planning is limited by sector laws and national guidelines and policy provisions, but it is nevertheless considerable and offers plenty leeway for independent decision-making (Sandersen and Kvalvik 2014; Jentoft and Buanes 2005). This

Citation: Haakan T. Sandersen and Ingrid Kvalvik, Access to aquaculture sites: A wicked problem in Norwegian aquaculture development, DOI: 10.1186/s40152-015-0027-8

local planning autonomy gives the municipalities a gatekeeper role regarding access to the coastal zone that is seen as a growing obstacle to further growth in salmon production. Many municipalities have increasingly become more reluctant and negative to further aquaculture growth in their waters. This reluctance is due to environmental concerns (such as sea lice and genetic interference on wild salmon, and release of faeces and chemicals), competition with other uses of the sea space (mainly fisheries and transport but also other industries, leisure and tourist activities) and the relatively modest local economic benefits (mainly due to restructuring of the industry) (Hersoug, 2013; Isaksen and Mikkelsen 2012; Gullestad et al. 2011).

Both the municipalities and the salmon producers are concerned about this issue, though for different reasons. Many municipalities argue that some kind of economic compensation for the use of their sea space for salmon production will make them more amenable to allocating new areas for the salmon industry, and a "Network of fjord and coastal municipalities (Nettverk for fjord og kystkommuner—NFKK), organizing 56 municipalities, is established to promote the introduction of an area rent (NFKK 2013; Kontali 2011; Isaksen et al. 2012).. On their webpage you can read: "Many of us are proud hosts of the aquaculture industry. But many of us saw the business become centralized. Workplaces disappear and municipalities are left with executive work and cages in the sea. Tax money does not necessarily have the same anchor as fish cages. This is one of the reasons to establish the network." (http://kystnettverk.no/). Most fish farming companies have sponsored various types of local activities to foster good will and a collaborative atmosphere with their host municipalities, and in several municipalities aquaculture companies have established industry funds. The municipalities, however, want a general, permanent and predictable arrangement. One suggestion put forward during the last 15 years has been the introduction of an *area rent*, according to which the municipalities would be compensated for the administrative burden of planning, facilitating and hosting the aquaculture industry (Kontali analyse as 2011). It is assumed that this payment would make municipalities more receptive to allocating new areas for the salmon industry, thereby solving or reducing the access problem (Isaksen et al. 2012; Sandersen and Nikolaisen 2007; (NOU 2004: 19). The idea has been supported by both municipalities and many salmon farmers, but the national government has resisted it due to several complex institutional and distributional interdependencies related to the introduction of such a rent.

We will refer to this policy dilemma, created by the discrepancy between the three factors of (1) national ambitions for industrial growth, (2) local autonomy in allocating necessary sites for production and (3) lack of incentives for municipalities to allocate areas to the salmon industry, as *the wicked problem in Norwegian salmon aquaculture development*. The concept of wicked problems was introduced by Rittel and Webber (1973) but has since been developed in various fields to describe a certain type of highly complex problems. Jentoft and Chuenpagdee (2009) applied the concept to fisheries, stating that coastal management often involves wicked problems. More recently, Berkes (2012) has conceptualised ecosystem-based management in fisheries as a wicked problem, as have Khan and Neis (2010) in their analysis of the rebuilding of fish stocks. The present study attempts to apply the concept to the context of aquaculture and consider its usefulness in understanding the area rent debate.

We certainly acknowledge that there are other and serious challenges related to further aquaculture development, in particular the problems related to environmental impact and ecosystem effects (sea lice and genetic interference on wild salmon, release of faeces and chemicals). Among stakeholders and in the local debates these environmental issues and the area rent issue are certainly intertwined, but strikingly in the national public debate of area rent and environmental concerns are rather two parallel discourses with limited interference with each other. The more general question of area access and environmental concerns are however tightly interconnected in debates. The area rent discussion is taking place in other forum with other actors, including municipalities that do not want aquaculture in their waters or do not want an increase due to environmental impact. Further, competing uses of the coastal zone, like fishing and recreational use, might be affected by the introduction of an area rent. This has received very little attention in the debate, but are still considered to be an important factor in the debate about aquaculture growth.

The aim of this article is to examine the 'wickedness' of Norwegian discussion of area rent for aquaculture. In general problems are wicked when they are difficult to define and delineate from other problems and when they cannot be solved once and for all but tend to reappear (Jentoft and Chuenpagdee 2009; Rittel and Webber 1973). Wicked problems are complex and hard to solve mainly because they are inter-linked with broader social, economic and policy issues present at multiple levels (Weber and Khademian 2008). Our point of departure is that the multi-level character and organisation of the above-described policy dilemma in Norwegian salmon aquaculture and the resulting political stalemate can be described and analysed as a wicked problem. Our research questions are: What are the main considerations that contribute to the barriers regarding introduction of an area rent for salmon farming? Will an area rent solve the Norwegian aquaculture industry's difficulties in gaining access to new suitable production sites? How wicked or solvable is this policy dilemma?

The following discussion is based on literature studies, document analysis, and interviews with municipal, regional and national decision makers in Norway and industry actors. Another source of information has been public documents related to area access for the aquaculture industry and area rent and news articles, in addition to interviews with the interest organisation for the industry, the Norwegian Seafood Federation (Fiskeri- og havbruksnæringens landsforening—FHL/Sjømat Norge) and the Network for fjord and coastal municipalities (Nettverk for fjord og kystkommuner—NFKK) which was established to further an area rent on aquaculture to be allocated to the municipalities, as well as the Ministry of industry and fisheries. We also participated at the 2014 Annual Meeting of the NFKK where political or administrative leaders of the municipal members were present. In addition representatives of the general public and opponents to aquaculture growth in selected municipalities have been interviewed.

In the next section the wicked approach is presented and discussed. This is followed by a description of the development of the Norwegian aquaculture industry and of the municipalities' justifications for a rent on salmon production sites. After that follow an outline of the area rent debate and national policy answers. Based on this we analyse the area rent as a wicked problem and discusses whether an area tax can be expected to solve the problem of area access. Finally, we consider the wickedness of the problem and discuss whether the area rent issue really can be considered a wicked.

Wicked Problems

The term 'wicked problem', originally developed to focus on micro-level issues, is now increasingly used to describe large, fragmented, complex and interconnected societal challenges that span disciplines, knowledge bases and value systems. It has also been applied to a wide range of public problems such as the climate crisis, conservation of forests, obesity, global health, drug use and many others with no technical, simple and definitive solution or quick fix (Weber and Khademian 2008). Such problems share the characteristic of being highly resistant to resolution.

Ritter and Webber introduced the concept of 'wicked problems' in their seminal work on rationality in city planning in 1973. The article is a critique of the linear, rational-technical understanding of planning, decision-making and problem-solving in a societal context, and the approach is developed to analyse complex and interlocking problems that do not fit into this pattern. 'Wicked problems' are contrasted to 'tame problems' that are simpler and can be solved within existing management or problem solving systems (Rittel and Webber 1973).The following four elements of wicked problems are based on Rittel & Webbers (1973) and Conklin (2006) and Weber and Khademian' (2008) summary of their work.

Wicked public problems have a tendency to *defy definition* in the sense that it is hard to point at the factor(s) causing the problem. According to Rittel and Webber (1973) you cannot truly understand the problem until you have developed a solution. Furthermore, as stated by Weber and Khademian (2008), wicked problems constantly 'morph and move'. When a new solution is offered, it exposes new aspects of the problem, again requiring further adjustment of the solution. Setting up and constraining the solution space is thus the wicked part of the problem (Ritter and Webber 1973). Wicked problems are therefore not objectively given, as their formulations and solutions rely on the viewpoint of those presenting them. They also involve complex problem-solving processes as resources, stakeholders, political ramifications and other constraints are constantly changing. Wicked problems are thus relentless, persistent and subject to redefinition in different ways over time (Weber and Khademian 2008; Rittel and Webber 1973).

Since there is no definitive problem, there is also *no definitive, final or correct solution*. There are no objective criteria or identifiable rules to tell us when *the solution* or even *a solution* has been found (Rittel and Webber 1973). However, we cannot avoid making a decision at some point. According to Conklin (2006), the process of solving a wicked problem ends when one runs out of resources and the available solution or compromise is 'good enough', not when a final solution emerges. Solutions to wicked problems are therefore not right or wrong. In a wicked problem situation the context is ultimately ambiguous and resolution depends on elusive political judgments. From this it follows that solutions to such problems are shaped in negotiation among multiple stakeholders who are equally equipped, interested and entitled to judge. Rather than definitively solving wicked problems we have to design more or less useful, effective and efficient compromises based on who participates and how the problem is defined.

Further, *every wicked problem is essentially unique* and novel and every situation is likely to be one of a kind. 'The wicked problem space comprises multiple, overlapping, interconnected subsets of problems that cut across multiple policy domains and levels of government. Wicked problems, in other words, cut across hierarchy and authority structures within and between organisations and across policy domains, political and administrative jurisdictions, and political 'group' interests' (Weber and Khademian 2008: 336). According to Rittel and Webber (1973) there are apparent similarities between wicked problems, but 'one can never be certain that the particularities of a problem do not override its commonalities with other problems already dealt with' (Rittel and Webber 1973: 165). This makes it difficult or impossible to transfer known solutions to new wicked problems. It is not very likely that such a process can follow any pre-determined recipe or model of established practices; rather, solutions must be tailor-made and fitted to the unique situation.

Finally, because wicked problems are multi-causal with connections to many other issues, there is no reliable way to test in advance if a public policy will work; it is necessary to try implementing the proposed solutions and see what happens. Every solution to a wicked problem is therefore a '*one-shot operation'*. Still more problematic, each try may be expensive and may generate lasting unintended traces and consequences that can in turn create new wicked problems, or it may have other undesirable consequences and repercussions that could be difficult or impossible to undo.

Rittel and Webber (1973) seem to contend that nearly all modern societal or public planning problems are inherently wicked (as people are what make problems wicked). The term 'wicked problems' seems to indicate that there is no solution, and to avoid this slippery linguistic trap it is more accurate to talk about the degree of 'wickedness' present in a given situation (Conklin 2006). In line with this understanding, our premise here is that not all societal issues are wicked, but rather that wicked issues display various degrees of wickedness.

The Norwegian Aquaculture Industry and Site Requirements

To produce salmon in Norway one needs both a license and a production site. The licenses are allocated by the national government. They decide when new licenses are to be allocated, the number of licenses and also the volume of fish allowed on each license and each site (i.e. the maximum allowed biomass (MTB) for each license and each site) (Gullestad et al. 2011). Of the 276 coastal municipalities in Norway, 160 currently are hosting salmon farms (Parliamentary bill (Stortingsproposisjon) no. 1 (2013)).

The industry's development has been profound. Back in 1995 the production was 200.000 tonnes of salmon a year, and the industry was composed of 565 small, local and family-run companies holding 725 licenses. Today, due to changes in government regulations, technological developments and a restructuring of the industry, the amount of production has soared to more than 1 million tonnes on 932 licenses. At the same time, the number of producers has been reduced to 90, which own 132 companies (Report to the Storting (Stortingsmelding) no. 22 (2013); Gullestad et al. 2011).

Despite the doubling of licenses and the six-fold increase in production, there are fewer aquaculture sites today than 25 years ago (Report to the Storting (Stortingsmelding) no. 22 (2013).[1] The production on each site has therefore increased considerably. We also see a trend toward even bigger sites, with more applications for so-called super-locations allowing MTB over 5,400 tonnes or more (Isaksen et al. 2012). The total area occupied by the aquaculture industry is therefore increasing. Still, the aquaculture sites, including safety and anchoring area, amounts to only 0.5 per cent of the Norwegian coastal areas inside the baseline (Isaksen et al. 2012).

Salmon farming is taking place virtually all along the coast, but the main areas of production have been extending from the west coast northward. As for production growth, there remains some room for increasing production on existing sites as their capacity is not yet fully utilised (Hersoug et al. 2014)[2], but the more industrialised type of production requires bigger and better-suited areas. Also, veterinary regulations requiring distance to other aquaculture facilities, fallowing and a rotating use of the sites impact the industry's need for suitable production sites (Gullestad et al. 2011). The municipalities are, as outlined above, the authority that designates areas for salmon farming in the coastal zone through their planning (Sandersen and Nikolaisen 2007). It is this willingness to allocate areas for salmon farming that is of concern to the industry.

The Coastal Municipalities: Door Openers and Gatekeepers

The reasons for the municipalities' reluctance towards increased aquaculture production in their coastal zones are, as mentioned, diverse. Some can be ascribed to the perceived or actual environmental impact of the industry and to competition with other uses of the coastal zone (Hersoug 2013; Sandersen and Nikolaisen 2007). Parallel with this and to an increasing extent, concerns about the municipalities' share of the value created by the aquaculture industry have been voiced (Hersoug 2013; Isaksen et al. 2012; Sandersen and Nikolaisen 2007). In the industry's first years, hosting the industry was attractive as it contributed to employment in many small, rural coastal municipalities facing depopulation problems. The jobs were created at the production sites at sea, at the landing and processing plants in the community, and through a general ripple effect on local businesses (Henriksen et al.2012; Robertsen et al. 2012). This economic growth yielded revenue to the municipalities through the income tax and company tax, as well as generating activity in local communities and optimism on behalf of the prospects of rural areas.

The changing structure of the industry during the last two decades has altered some of this dynamic. From small and family-run companies rooted in their local communities, the industry has been restructured into a more large-scale organisation with fewer but bigger companies and a more cost-effective concentration of salmon landing and processing plants (Isaksen et al. 2012; Sandersen and Nikolaisen 2007). Five listed companies with headquarters in main Norwegian cities today own more than 50 per cent of the licenses (Robertsen et al. 2012). The number of landing and processing plants has been reduced from 250 to 50 in a 20-year time period (Norsk Fiskeoppdrett2013). For many municipalities this change has resulted in a loss of jobs and revenue, as well as a reduced local commitment by the salmon producers as the companies have no personal or local connection (Isaksen et al. 2012; Robertsen et al. 2012).

The increased concentration of aquaculture landing and processing plants and of administrative support services, along with the accompanying relocation of economic ripple-effect benefits, makes it relevant to talk today about 'production municipalities' on one hand and 'base and processing municipalities' on the other. The first category consists of municipalities where there is only salmon production at sea, whereas the second group of municipalities also has processing plants, headquarters and other aquaculture-related service functions (Sandersen and Nikolaisen 2007). The production municipalities do not enjoy a comparable number of jobs or amount of ripple effect as the base and processing municipalities, and this asymmetry in the distribution of benefits is one of the drivers in the quest for an area rent. Continued rationalization and concentration processes can also make today's "base- and processing municipalities" tomorrows "production municipalities", creating a situation of uncertainty for many municipalities.

Another driver is dissatisfaction with the allocation of the government's income from the aquaculture industry. The main income of the municipalities is income tax and economic transfers from the state. Until 1999 part of the company tax was allocated to the municipalities. Today the tax goes directly to the state and is then redistributed through the economic transfer to the municipalities (Isaksen et al. 2012).[3] Even if the aquaculture companies are contributing through national taxation like other industries, their contribution to the municipalities is not directly visible, as there are no direct links between local value creation, profit from aquaculture companies and income for the hosting municipalities. Therefore, even though the value creation and profit from the industry have grown considerably (Report to the Storting (Stortingsmelding) no. 22 (2013)), many municipalities no longer experience a corresponding local return or benefit from the very profitable aquaculture industry.

Both the structural changes in the industry and the changes in taxation have modified the incentive structure of the municipalities in relation to the aquaculture industry. Many municipalities along the coast have therefore increasingly asked for compensation or their 'fair share' of the high profits generated by the aquaculture industry from the use of the municipalities' sea spaces (Isaksen et al. 2012; NFKK 2013). The municipal representatives often also refer to the costs and resource expenditures related to coastal zone planning and management and to the handling of aquaculture and site matters as reasons why they should be compensated (Isaksen et al. 2012; NFKK 2013). Hence the municipalities simply want their share of the income from this Klondike business as they provide the production sites, the core natural resource that makes this industry possible. It is also of relevance that aquaculture has been seen as a rather distinct and non-traditional way to exploit coastal and marine resources (Sandersen and Nikolaisen 2007). Whereas the traditional use forms have been transitory and seasonal, aquaculture displays a semi-permanent presence that resembles private ownership on land, and it is often seen as an atypical and illegitimate way of occupying the traditionally free-for-all coastal commons (Sandersen and Nikolaisen 2007).

The municipalities' justifications for a rent on salmon production sites are thus related to compensation for costs, fair share of a profitable endeavour, alleged property rights (i.e. a belief that the municipalities are the 'owner' of the sites and should thus be entitled to a part of the income that the sites are generating) and the somewhat exclusive character of aquaculture's use of coastal spaces.

The Question of Area Rent and National Policy Answers

The question of local compensation and area rent was first put forward in the early 2000s, and it gained momentum in the second half of the decade (White Paper (Odelstingsproposisjon) no. 65 (2000); Report to the Storting (Stortingsmelding) no. 19 (2005) and Report to the Storting (Stortingsmelding) no. 22 (2013); Parliamentary bill (Storingsproposisjon) no. 1 (2007)). The government stated in 2007 that it supported in principle the idea that the municipalities could require compensation for the use of the area in their coastal zone, and it appointed an inter-ministerial working group, headed by the Ministry of Finance, to assess how this could be done (Parliamentary bill (Storingsproposisjon) no. 1 (2007)). After a process involving the organisations of affected industries and recreational users, coastal municipalities, counties, and government-sector agencies, the government concluded that an area rent was 'not a suited instrument' because it would be difficult to determine the base for such a rent and because an area rent on salmon production does not harmonise with the established tax system.[4] Instead the Ministry suggested allowing the municipalities to levy property taxes on aquaculture facilities in the sea (White paper (White Paper (Odelstingsproposisjon) no. 1 (2008); Parliamentary bill (Stortingsproposisjon) no. 1 (2008). This would give the municipalities some direct and annual income from the industries' permanent use of their coastal zone. The suggestion was adopted, and today most aquaculture municipalities had levied this tax. The returns from this tax have been very modest, however, as the maximum tax rate is 7 % and the value set on the installations at sea (not the fish in the net) is quite low (Isaksen et al. 2012).

Another initiative to benefit the municipalities was to allot part of the fees for new aquaculture licenses to the County Councils, to be devoted (among other uses) to supporting municipal coastal zone planning, a task in which many municipalities lack both competence and resources (Hersoug and Johnsen 2012; Robertsen et al. 2014).

The municipalities were clearly not satisfied with this proposal, and in 2009 a 'Network for fjord and coastal municipalities' (NFKK)[5] was established by 20 small coastal municipalities. The network's main (if not its only) agenda is to lobby for the introduction of an area rent (Kontali analyse as 2011).[6] Their work is directed at members of Parliament and the government, as well as at other stakeholders and the media, and according to themselves dedicated efforts by the network and their municipal members have led to an increasing understanding of and support for the demand for an area rent or other form of compensation to the municipalities for the aquaculture industry's use of the coastal waters (personal communication with administrator of network). Today the network has 56 members; almost all of them are small municipalities and all but one has aquaculture sites in their waters (see their homepage http://kystnettverk.no/ for more information).

In the same period, the industry's attitude toward an area rent seems to have changed. In the consultation process related to the inter-ministerial work in 2007–2008 the industry was critical of the idea of paying for use of the coastal zone. However, as both the need for new production sites and general support for the municipalities' demand for compensation have increased, the aquaculture industry has increasingly adapted its approach to the issue, as supportive and facilitating municipalities are the key to getting access to new sites (Seafood Federation] 2013). The roles have therefore changed; previously the municipalities had to make themselves attractive to the aquaculture industry, but now it is the other way around. The aquaculture industry now increasingly supports an area rent, but on the condition that it goes directly to the municipalities and that the total tax burden of the industry does not increase (Seafood Federation] 2012).

The Norwegian Seafood Federation (FHL/Sjømat Norge), the prominent interest organisation of the industry, has also come up with a suggestion of how to compensate the municipalities, according to which the industry would pay an annual fee to the municipalities based on an annual growth in production of 3–5 %. This should be an incentive for the municipalities to allocate more areas to the aquaculture industry. The industry prefers to pay to stimulate further growth than to pay for existing sites. The Norwegian Seafood Federation (FHL) also argues that such a fee would not disturb the established tax system, and hence that this approach would be more able to gain government approval. The municipalities, however, do not find this suggestion very appealing. The main objection is that the payment would compensate only for growth, i.e. not for the production already taking place, and that it therefore would not benefit all municipalities (personal communication NFKK).

The network has more recently launched a new proposal which it claims will not alter the basic principles and standardisation of taxation. This proposal suggests introducing a rent on production sites within the Aquaculture Act, and it calls for coordinating this rent with the company tax so that the total tax burden on the industry do not increase (Lund and Co 2013). By this the NFKK tries to meet both the industry's condition that the tax burden should not increase and the Ministry of Finance's resistance to additional taxes under the existing tax system (personal communication NFKK).

The government did not respond to either of the proposals. It did, however, seek other (perhaps preliminary) ways to meet the municipalities' claims for economic compensation by granting the municipalities the part of the license fee that, in the last license allocation, was given to the County Councils.[7] This change was welcomed by the NFKK and the municipalities, but both pointed out that it involves only a one-time payment and also benefits only the municipalities hosting the new licenses (NFKK 2013). In a subsequent press release, however, the Minister of Fisheries and Coastal Affairs suggested reconsidering the introduction of an area rent and asked whether a property tax is indeed the right way to promote the municipalities' further facilitation for aquaculture (FKD [Ministry of Fisheries and Coastal Affairs] 2013).[8] The new coalition government has so far not responded to this initiative.[9] It has however suggested allowing a 5 per cent increase in the production of salmon and trout on existing licenses, under strict environmental conditions, for a fee of 1.5 mil. NOK (NFD Ministry of industry and fisheries 2014a; NFD Ministry of industry and fisheries 2014b). In the proposed state budget for 2015 it is stated that "it should be worthwhile for the municipalities to facilitate aquaculture", and the compensation to the municipalities should therefore be increased from 40 to 50 per cent of the fee (Parliamentary bill (Stortingsproposisjon), no 1 (2014):98). It therefore seems that the government prefers to pay to stimulate further growth rather than to pay for existing sites. The response from the NFKK is, again, that the compensation to the municipalities should be divided between all municipalities hosting aquaculture sites, not only the ones hosting companies that are allowed to increase production (NFKK 2014). This arrangement will give the municipalities' incentive for allocating new areas to aquaculture production, which is defined as an important obstacle to further growth. The arrangement may also open up for new alliances between individual municipalities and aquaculture companies, where access to new areas will be negotiated and the municipality compensated. It might however weaken the municipal network.

To sum up, the municipalities' reluctance to allocate more area to the salmon industry is considered a main obstacle to the industry's continued growth. An area rent is assumed to solve or reduce the industry's challenge of acquiring new production sites, but this is hindered by resistance, on the part of the national government and the Ministry of Finance, to altering the principles for taxation due to the risk of setting unwanted precedents. Neither of the compensations introduced by the government is regarded as sufficient by the municipalities, and the one-time payment is considered to reinforce rather than remedy the challenges related to asymmetric distribution of benefits from aquaculture production.

Area rent as a Wicked Problem

To reiterate, wicked problems are persisting, recurring societal problems that are difficult to define and delineate from other problems, have competing and changing requirements, and involve many clients and decision makers with conflicting values, and where solutions are difficult to identify because of the complex political, institutional and social considerations that surround the problem (Ludwig 2001; Rittel and Webber 1973).

A first task when analysing the issue of the salmon producers' access to production sites and area rent as a wicked question is, therefore, to examine how the problem is defined. Three main groups of actors are involved in this enduring policy debate: the coastal municipalities, the aquaculture industry, and the national government. The three approach the area rent issue from rather different interests and points of view. For the municipalities voicing the demand for economic compensation for additional aquaculture-related costs and the resource use through an area rent it is framed as a *self-government and income issue*, and an area rent is justified as their just or fair share of the value creation of the industry, and through their institutionally backed 'sense of ownership' of the coastal area. For the industry area rent is an *access and growth issue* as well as a *tax issue*; when industry representatives talk about compensation, it is framed mainly as a bonus to the municipalities for allowing further growth and less as a general compensation for the activities involved in planning for and facilitating aquaculture in general or existing production. One major concern for the industry is to demonstrate and make visible its contribution to the municipalities and its willingness to pay for continued presence and especially for additional growth. The industry would rather see its taxes go directly to the municipalities than to the state, as they need the municipalities' support and good will in order to obtain access to new production sites. The industry would certainly also welcome a change in the veterinary regulation that creates the scarcity or lack of access to production sites, or a reduction of the municipalities' autonomy in the coastal zone. The Norwegian Seafood Federation has suggested introducing a national 'protection of aquaculture' scheme (with reference to the existing national 'protection of farmland' arrangement) to avert the problem of well-suited super-locations for aquaculture being unused, or being used for other purposes, due to municipal and other priorities (Seafood Federation] 2012). But since the Norwegian local government system, which gives considerable autonomy to municipalities, enjoys such strong national support, such efforts to reframe the issue has been highly unfruitful. This conflict of interest is mainly between the industry on the one hand and the municipalities on the other, and where the government supports the industry's wish for growth, but as the same time is a guarantor for municipal autonomy.

Finally, for the national government the question of imposing an area rent on salmon sites to be allocated to the municipalities is not only a *rural and industry development issue*, but also (or even more) a *tax system* and *just tax distribution issue* as well as a *multi-level government issue*. These concerns or conflict of interests are represented by the Ministry of Industry and Fisheries and the Ministry of Finance respectively. The issue of area rent raises concerns over the possible consequences on the overall tax system. An area rent on aquaculture sites does not fit well into the current tax system. Introducing such a new or additional tax on aquaculture production to be allocated to the municipalities will affect not only the distribution of income between government levels, an issue long debated in Norway, but also how other industries such as wind power companies and the mineral industry might meet, or be subjected to, similar tax arrangements. The government is therefore reluctant to make the changes necessary to incorporate such a rent, as this will further complicate the taxation system as well as the income systems for the municipalities. For the government it is also a question of who should pay the municipalities: the government (by a reallocation of some of the income received from the industry) or the aquaculture industry directly through a new and additional tax, hence it is also a question of income distribution between municipalities.

Interdependencies, multiple causes and internally conflicting goals make wicked problems hard to define. In this case the consensus between the parties regarding problem definition—namely, that increased municipal friendliness to aquaculture is required to increase salmon production—is rather high. The problem is here rather more related to the actors' different interests and to different and incompatible concerns regarding the suggested solution of an area rent. The wickedness lies in the solution, that an area rent will solve the area access problem. But this solution is difficult or impossible for the government to accept as it will disturb the tax system and have other unwanted effects. The suggested solution is blocked by the cross-cutting and interlocking character of the issue, and both political and bureaucratic issues and considerations of juridical precedent make this a tough nut to crack. Neither weakening municipal autonomy nor changing the veterinary regulations in order to facilitate the industry's access to the remaining super-locations along the coast seems to be a viable political option, as these actions would challenge other highly prized values like local autonomy and environmental protection.

So far the issue has found *no definitive or final solution*, another characteristic of wicked problems. Hence, the current measures targeting the problem of access have been characterised by a lukewarm compromise approach on behalf of the government. Instead, of addressing the core question of an area rent it has settled for measures like imposing a property tax on aquaculture installations at sea and allocating part of the license fee for new licences or production increase to the hosting municipalities and counties. The pressure from the strong aquaculture industry, in alliance with an increasing number of coastal municipalities, has caused national politicians to propose new measures, but the wicked, interlocking and complex character of the solution has so far made it impossible to settle on an area rent, the only solution that seem to be satisfactory for both the municipalities and the industry. The efforts to solve the problem by levying a property tax and by allocating a portion of the license fees to the municipalities show that, instead of targeting the core problem, the authorities are trying to reduce and 'tame' the problem by introducing measures that resemble a real solution, but without challenging the tax system, governmental structure or more deeply ingrained values as an area rent would.

The government's resistance to introduce an area rent is related to the fear of establishing an unwanted precedent. If area rent allocated directly to the municipalities is introduced for salmon production, this will open up the question of also including other industries that are dependent on coastal resources or that use natural resources more generally, such as wind power, coastal fisheries and coastal tourism. Indeed, this trend is also seen in some municipalities' proposal to tax cruise ships that are entering the Norwegian fjords but not providing much ripple economic effect. Interestingly, the hydroelectric power companies are already paying the hosting municipalities a tax[10] that could be seen as a parallel to an area rent for salmon production, and the suggestion to levy an area rent in aquaculture has been inspired partly by this example.[11] The tax on hydroelectric power has been in place since the early twentieth century. The tax is an important source of income and the term 'power municipalities' has been used to indicate the resulting wealth of these municipalities. There has through the years been some debate about the justness of the system, especially raised by municipalities not hosting hydropower plants. The criticism is that the system results in big differences in the economic situation in municipalities with and without hydropower (Borge et al. 2013). Several of our informants refer to this as a reason for not introducing an area rent on aquaculture production to be allocated directly to the municipalities.

Further, not all types of aquaculture will be able to bear the burden of an additional tax. Other forms of aquaculture, such as cod, shellfish and sea ranching, do not enjoy the same high profit and are often operating on the margins. Also, an area rent on salmon sites could reduce the municipalities' willingness to allocate areas to these less profitable industries that do not produce as much local income. An area rent on salmon production sites will also undermine the very strong tradition of perceiving the coast as an open and common area free for all, based on the*allemannsrett* (literally 'all men's right') codified in 1957. An area rent may therefore not only change the tax system but also impact the deeply rooted coastal regime.

The lack of sufficient political support for an area rent also stems from the fact that it is not clear that the current diagnosis will stand the test of time for very long, as both the situation and its framing can change quickly. So far the question of access has been framed mainly as an economic issue, but at some point it may also be reframed as an environmental issue, placing greater emphasis on environmental concerns related to nutrient wastes, disease outbreaks, escapes and salmon lice. Some of the municipalities who are not taking part in the NFKK network do not see economic compensation as a condition for growth in their waters, while others are critical and will not designate new areas even with economic compensation. Hence, it is not only economic reasons for the municipalities' reluctance to host aquaculture. These issues have however not been central in the discussion on area rent. Nevertheless, area rent as a means to solve the area access challenge of the industry will be influenced by this. It is also possible that competing activities in the coastal zone, such as coast-based tourism or developments of second homes, will be more highly prioritised in the future, making it even harder for the industry to 'buy out' the municipalities. An area rent explicitly tailored to make salmon aquaculture more interesting to the municipalities may also have democratic implications, as this may put non-commercial interest and less affluent actors and stakeholders in a weaker position relative to aquaculture.

How likely, then, is it that an area tax will solve the problem? Is access to new sites really necessary to sustain a strong growth in salmon production? Only the introduction of an area rent will show conclusively whether its use would cause the municipalities to become more positive toward allocating areas to the salmon industry. This would reveal whether the municipalities' reluctance towards the salmon industry is actually the bottleneck for further growth in the industry—and, hence, if the problem has been properly defined. Industrial representatives frequently admit the main obstacle to continued growth, at least in the short run, is not access to new production sites but rather government license politics, MTB limits and environmental regulations. On a longer term, however, new sites will be required. Further, an area rent does not necessarily give the aquaculture industry access to its preferred production sites, either because the municipalities still may not want to allocate more areas to aquaculture even if more generously compensated or because the national government's nature-protection, environmental and veterinary regulations would still limit access to desired areas. Thus an area rent might not be the right cure for the problem. The very existence of the debate over area rent may itself make access to new sites more burdensome and difficult at this time, as municipalities may be reluctant to offer any new sites pending an expected lucrative policy change, as seen in the cases of Alta and Tysnes municipalities. A worst-case scenario for the industry is that the municipalities, once they obtain an area rent, may conclude that they have already allocated sufficient area to the salmon industry from which they would now be receiving an area rent. In this scenario the municipalities would get their money but the industry would not get its desired expansion. It seems quite clear that introducing an area rent is no panacea for the industry's problem of access to new production sites for growth, and that moreover it would create a variety of new and related issues, especially for the government.

Conclusion: How wicked is the problem?

The core of the problem is the industry's and government's ambition for continued growth of the aquaculture industry and the municipalities' status as independent and autonomous political bodies in coastal zone planning. When the municipalities were granted the right to designate their coastal areas for different uses, the national government certainly gave away substantial decision-making power and gave the municipalities a role as gatekeepers regarding the aquaculture industry's access to the coastal zone. To some extent this power can be counterbalanced by government guidelines and so-called 'national expectations', but so far the municipalities' autonomy in the coastal zone has been respected by the government. For an ever-expanding aquaculture industry, this has made site availability an increasingly urgent issue as many municipalities have become more reluctant to granting new areas to salmon production without being economically compensated.

The area rent issue is wicked in the sense that it is inescapably connected to other issues and in that the proposed solution challenges deeply held values. The strong inter-linkages with other issues certainly contribute to making the issue of access to aquaculture sites a wicked context. However, consensus about the core problem seems to be shared among the three main parties. Thus, the problem should not be classified as 'super-wicked'. The national government has resisted the introduction of an area rent due to the complex institutional and distributional interdependencies related to the proposed rent. There could also be more practical, institutional reasons behind the national government's resistance to introducing an area rent on salmon farming. It might be receptive to the idea but unable to find a way to integrate it into the existing tax system without creating precedents for other sectors. The main problem is thus not that the problem defies definition, but more that the suggested solution requires too big changes to be politically viable. The wickedness therefore lies mainly in the problems related to the available solutions, not necessarily in the problem definition. This study therefore suggest that Rittel and Webbers statement that "The process of formulating the problem and of conceiving a solution (or re-solution) are identical, since every specification of the problem is a specification of the direction in which a treatment is considered" (Rittel & Webber1973: 157) in this case should be understood mainly as an encouragement to try to by-pass the "wickedness of the available solutions" by finding new, less wicked and more legitimate and feasible solutions, as the wicked problem lies in high political costs of the available solution, not in the wickedness of the problem per se.

So far the government does not see the problem as serious enough to compromise the other values at stake, even if the alliance between the industry and the municipalities is pushing hard. In this sense the logjam may as well be analysed in terms of political power and legitimacy than as wickedness, and the problem framed more as a 'tragic choice' rather than a wicked problem. In their work on tragic choices Calabresi and Bobbit (1978) deal with the public-policy dilemma of allocating scarce resources, contending that conflicts over deeply held values are tragic both in the sense that some deeply held values lose out to others and in view of the legitimacy problems that may occur when sources of social and political power are revealed and laid bare in such decisions. However, a management system will always favour some interests, type of knowledge and worldviews at the expense of others. In the "wicked problems" tradition such elements are obscured or masked, focusing more on the established definition of a problem and the suggested solution. The wicked approach can contribute in understanding enduring, unsolved goal or means conflicts where the stalemate is due to the intertwined character of the issues. With wicked problems it is often unclear if there is a conflict of goals or conflict of means as the complexity makes these distinctions interwoven and blurred. Analyzing a problem's degree of wickedness contributes to the understanding of why some defined solutions to a problem might be difficult to gain acceptance for due to its interconnectedness with other issues. The wicked problems approach's contribution is to focus on the importance of the framing of a given management problem. The strength of the approach is that it emphasizes that a societal steering or management issue has to be framed in a way that fit the larger management system, and that some problems becomes largely unsolvable if they are framed too differently.

The government's previous measures and policies and its hesitancy to introduce an area rent system can thus be understood as a way to avoid the deeper and much more politically costly conflicts that establishing such a rent system would create. The results of the 'tame' measures already introduced by the government, however, indicate that the range of politically viable, legitimate and practical solutions may not be sufficient to address the defined problem,—and solution. But again, the defined solution of area rent might not solve the underlying problem of area access.

The environmental debate related to aquaculture is in Norway rather loosely linked to the area rent debate, as the opposition groups have so far not been interested in this issue. But if the environmental problems are not rather soon handled through changes in technology, regulations or incentive structures it is likely that the two discourses will converge and make the area rent question even harder to solve for the government and the industry.

The Norwegian aquaculture area access problem certainly falls between the existing 'stove-piped' problem-solving systems, and this clash between a wicked problem and the traditional problem-solving systems calls for either institutional innovations or new ways of working across all stakeholders. Recognizing the wickedness of this problem can provide a more realistic framework for further policy making, as the problem is not necessarily solvable within the current framework. Nevertheless, if wicked problems generate new and "outside the box" solutions, it is likely that these new solutions will contribute to the creation of even more complexity and generate more wicked situations for governments and other stakeholders.

REFERENCES

1. Act of 21. April 1989 no 17 on changes in the Plan and Building Act of 14. June 1985 no 77. Norsk Lovdata, Oslo, Norway.
2. Berkes, F. 2012. Implementing ecosystem-based management: Evolution or revolution? Fish and Fisheries 13: 465–476.
3. Borge, L.E., P. Parmer and R. Torvik (2013) Local Natural Resource Curse? Norwegian Business School. CAMP Working Paper Series No 5/2013.
4. Calabresi, G, and P Bobbit. 1978. Tragic choices: The conflicts societies confront in the allocation of tragically scarce resources. New York: W.W. Norton and Co.
5. Conklin, J. 2006. Dialogue mapping: Building shared understanding of wicked problems. Chichester, England: Wiley.
6. FKD [Ministry of Fisheries and Coastal Affairs] 2013. 'Inntil 4 millioner til kommunene for hver oppdrettstillatelse». Government press release, Ministry of Fisheries and Coastal Affairs, 14 October.
7. Gullestad, P., S. Bjørgo, I. Eithun, A. Ervik, R. Gudding, H. Hansen, R. Johansen, A. B. Osland, M. Rødseth, I. O. Røsvik, H. T. Sandersen, and H. Skarra. 2011. Effektiv og bærekraftig arealbruk i havbruksnæringen - areal til begjær Rapport fra et ekspertutvalg oppnevnt av fiskeri og kystdepartementet. Oslo 4. februar 2011: Fiskeri og kystdepartementet.
8. Henriksen, K, MG Sandberg, T Olafsen, H Bull-Berg, U Johansen, and A Stokka. 2012. Verdiskaping og sysselsetting i norsk sjømatnæring i 2010 - en ringvirkningsanalyse. Sintef rapport A23089. Trondheim: Sintef.
9. Hersoug, B. 2013. The battle for space: The position of Norwegian aquaculture in integrated coastal zone planning. In Global challenges in integrated coastal zone management, ed. E Moksness, E Dahl, and J Støttrup, 159–168. West Sussex, UK: Wiley.
10. Hersoug, B, and JP Johnsen (eds.). 2012. Kampen om plass på kysten. Interesser og utviklingstrekk i kystsoneplanleggingen. Oslo: Universitetsforlaget.
11. Hersoug, B, O Andreassen, JP Johnsen, and R Robertsen. 2014. Hva begrenser tilgangen på sjøareal til havbruksnæringen? Nofima-rapport 37/2014. Tromsø: Nofima.
12. Isaksen, JR, and EI Mikkelsen. 2012. Økonomer i kystsonen: Kan kunnskap om verdiskaping gi bedre arealforvaltning? In Kampen om plass på kysten. Planlegging i kystsonen under nye betingelser, ed. B Hersoug and JP Johnsen, 159–178. Oslo: Universitetsforlaget.
13. Isaksen, JR, O Andreassen, and R Robertsen. 2012. Kommunenes holdning til økt oppdrettsvirksomhet. (The municipalities' attitudes towards increased aquaculture). Nofima Rapport 18/2012. Tromsø: Nofima.
14. Jentoft, S, and A Buanes. 2005. Challenges and myths in Norwegian coastal zone management. Coastal Management 33(2): 151–165.
15. Jentoft, S, and R Chuenpagdee. 2009. Fisheries and coastal governance as a wicked problem. Marine Policy 33(4): 553–560.
16. Khan, AS, and B Neis. 2010. The rebuilding imperative in fisheries: Clumsy solutions for a wicked problem? 87: 347–356.
17. Kontali analyse as. 2011. Arealavgift for oppdrettlokaliteter. Utredning av grunnlaget for innføring av arealavgift eller ressursskatt i akvakulturnæringen (Area rent for aquaculture sites). Kontali analyse as for Nettverket for fjord-og kystkommuner.
18. Ludwig, D. 2001. The era of management is over. Ecosystems 4: 758–764.
19. Lund and Co. 2013. Betenkning: Forslag om endring i akvakulturloven - innføring av areal/naturavgift for oppdrettsvirksomhet (oppdrettsavgift). Oslo: Lund and co på oppdrag av Nettverk for fjord- og kystkommuner.
20. NFD (Ministry of industry and fisheries). 2014a. Høringsnotatet kapasitetsauke i lakse- og aureoppdrett i 2015 (Consultation paper - draft regulations on capacity increases in salmon and trout farming in 2015).
21. NFD (Ministry of industry and fisheries). 2014b. Utkast til forskrift om auke av maksimalt tillaten biomasse for løyve til akvakultur med laks, aure og regnbogeaure. Draft regulations on increase of the maximum rate of biomass for permission for aquaculture of salmon, trout and rainbow trout.
22. NFKK. 2013. http://kystnettverk.no/nyhetsarkiv/oppdrettskommunene_mener_ny_ordning_har_klare_mangler/
23. NFKK. 2014. http://kystnettverk.no/nyhetsarkiv/horingssvar_forskrift_om_auke_av_maksimalt_tillaten_biomasse/
24. Norsk Fiskeoppdrett. 2013. Norge kan slakte 2,5 millioner tonn (Norway can slaughter 2.5 mil tonnes). Norsk Fiskeoppdrett, 52:5.
25. NOU 2004:19 (Official Norwegian Report). 2004. Livskraftige distrikter og regioner. Rammer for en helhetlig og tilpasset politikk. Kommunal og regional departementet, 12 October 2004.
26. Parliamentary bill (Storingsproposisjon) no. 1 (2007–2008). Statsbudsjettet. For budsjettåret 2008. Ministry of Fisheries and Coastal Affairs, Oslo.

27. Parliamentary bill (Stortingsproposisjon) no. 1 (2008–2009). Skatte-. avgifts- og tollvedtak for budsjettåret 2009. Ministry of Finance, Oslo.

28. Parliamentary bill (Stortingsproposisjon) no. 1 (2013–2014). Statsbudsjettet. For budsjettåret 2014. Ministry of Fisheries and Coastal Affairs, Oslo.

29. Parliamentary bill (Stortingsproposisjon), no 1 (2014–2015). Statsbudsjettet. For budsjettåret 2015. Ministry of Fisheries and Coastal Affairs, Oslo.

30. Report to the Storting (Stortingsmelding) no. 19 (2004–2005). Marin næringsutvikling. Den blå åker. Ministry of Fisheries and Coastal Affairs, Oslo.

31. Report to the Storting (Stortingsmelding) no. 22 (2012–2013), Verdens fremste sjømatnasjon. Ministry of Fisheries and Coastal Affairs, Oslo.

32. Rittel, HWJ, and MM Webber. 1973. Dilemmas in a general theory of planning. Policy Sciences 4: 155–169.

33. Robertsen, R, O Andreassen, and A Iversen. 2012. Havbruksnæringens ringvirkninger i Troms. Nofima Rapport 28/2012. Tromsø: Nofima.

34. Robertsen, R, I Kvalvik, O Andreassen, B Hersoug, and JP Johnsen. 2014. Interkommunale planleggingsprosesser - et bedre verktøy for lokalitetstilgang? Nofima Report 27/2014. Tromsø: Nofima.

35. Sandersen, HT, and I Kvalvik. 2014. Sustainable governance of Norwegian aquaculture and the administrative reform: Dilemmas and challenges. Coastal Management 42(5): 447–463.

36. Sandersen, HT, and O Nikolaisen. 2007. Fra kystallmenning til kolonihage - havbruk og arealbruk i tre nordlandskommuner. In Havbruk: Akvakultur på norsk, ed. B Aarset and G Rusten, 173–193. Bergen: Fagbokforlaget.

37. FHL [Norwegian Seafood Federation] 2012. Sjømat 2025—hvordan skape verdens fremste havbruksnæring (Seafood 2025 - how to create the world's best aquaculture industry). Oslo 21 May 2012.

38. Norwegian Seafood Council 2015. Montly statistics 2014. downloaded athttp://www.seafood.no/Markedsinformasjon/Statistikk/(stats)/8809

39. Weber, EP, and AM Khademian. 2008. Wicked problems, knowledge challenges, and collaborative capacity builders in network settings. Public Administrative Review 68(2): 334–349.

40. White Paper (Odelstingsproposisjon) no. 1 (2008–2009). Skatte og avgiftsopplegget 2009 - lovendringer. Ministry of Finance, Oslo.

41. White Paper (Odelstingsproposisjon) no. 65 (2000–2001). Om lov om endring i lov 14. juni 1985 nr. 68 om oppdrett av fisk, skalldyr m.v. (vederlag ved tildeling av konsesjoner for matfiskoppdrett av laks og ørret). Ministry of Fisheries, Oslo.

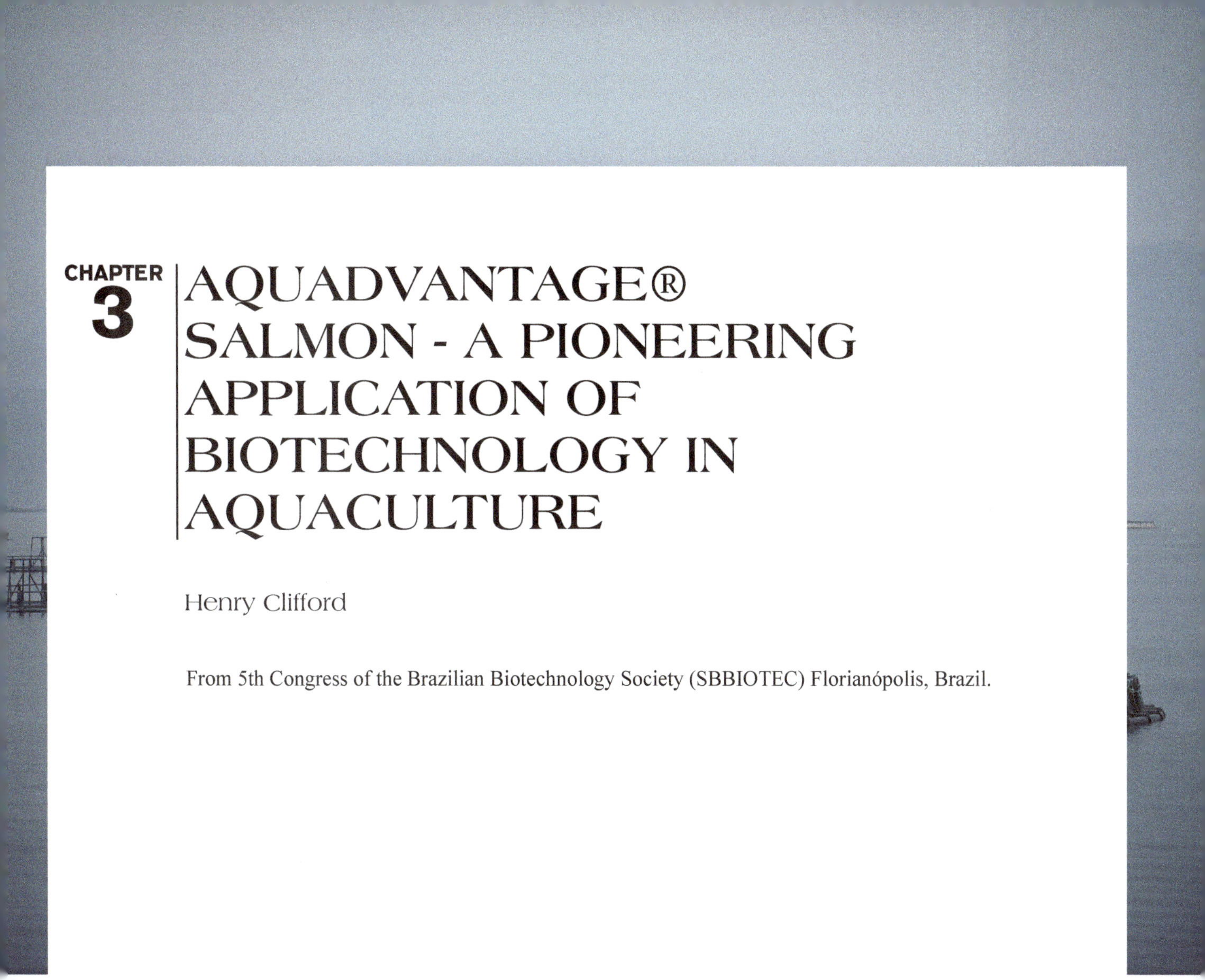

CHAPTER 3

AQUADVANTAGE® SALMON - A PIONEERING APPLICATION OF BIOTECHNOLOGY IN AQUACULTURE

Henry Clifford

From 5th Congress of the Brazilian Biotechnology Society (SBBIOTEC) Florianópolis, Brazil.

The U.S. FDA is currently reviewing AquaBounty Technologies' New Animal Drug Application for a fast growing transgenic Atlantic salmon, known as AquAdvantage® Salmon, which would be the first genetically modified food animal approved for human consumption. AquAdvantage® Salmon was developed by inserting into a fertilized Atlantic salmon egg, a transgene constructed from the growth hormone gene of a related species (Chinook salmon). The AquAdvantage® genetic construct confers upon the Atlantic salmon a rapid growth phenotype and improved metabolic efficiencies associated with low but sustained levels of growth hormone. This allows the fast growing AquAdvantage® Salmon to reach market size in half the time of a conventional Atlantic salmon. AquAdvantage® Salmon will be available only as all female, sterile, eyed eggs, and the resulting fish must be maintained in a freshwater, land based, physically contained facility [1]. The founding line of AquAdvantage® Salmon was created in 1989, and an Investigational New Animal Drug (INAD) dossier was opened with the FDA in 1995. In 2009, AquaBounty submitted to the FDA the last of more than 25 scientific studies in support of our application. In 2010 the FDA presented their findings to the Veterinary Medical Advisory Committee (VMAC), and recommended an approval. The VMAC committee, comprised of independent experts selected from academia, industry, and NGO's, concurred with the FDA's findings, which were that AquAdvantage® Salmon is equivalent to conventional farmed Atlantic salmon, and is safe for human consumption as well as safe for the environment. The regulatory analysis included an environmental assessment of a proposed AquAdvantage® Salmon egg production site in Canada, and an AquAdvantage® Salmon growout site in Panama, both of which were inspected and approved by the FDA. Important biosecurity elements for containing AquAdvantage® Salmon and minimizing potential risk to the environment at the two production sites include physical, biological, and environmental containment. For example, at the proposed growout site in Panama which is more than 120 km from the nearest ocean, sterile, all female populations of AquAdvantage® Salmon are confined to a highly biosecure culture facility possessing 21 individual containment elements, with a minimum of eleven containment barriers in sequence. In addition to the biological and physical containment elements, a natural thermal barrier of lethally high water temperatures (for Atlantic salmon) downstream of the facility would effectively prevent any live, escaped AquAdvantage® Salmon from reaching the Pacific Ocean. In 2012 the FDA published the environmental assessment for public commentary, and a Finding of No Significant Impact (FONSI), indicating that rearing AquAdvantage® Salmon in those sites did not present a risk to the environment when reared under the conditions of use stipulated by the FDA. The public comment period concluded in April of 2013, and a final resolution of the application is expected this year from

Citation: Clifford, H. AquAdvantage® Salmon - a pioneering application of biotechnology in aquaculture. BMC Proceedings 2014 8 (Suppl 4):O31. doi:10.1186/1753-6561-8-S4-O31

the FDA. If approved by the FDA, the product offered by AquaBounty would be eyed eggs of AquAdvantage® Salmon, which would carry an FDA approved label indicating that they are female, triploid, genetically engineered Atlantic salmon eggs carrying a single copy of the AquAdvantage®gene construct. The label would also stipulate that the fish from these eggs could only be grown in pre-approved, land based, physically contained, freshwater aquaculture facilities. As a pioneering application for a genetically modified animal destined for human consumption, AquAdvantage® Salmon has attracted considerable attention and opposition by anti-GMO, anti-biotech activist organizations, as well as the organic food industry which is threatened by the more productive, lower cost, food

REFERENCE

1. Clifford Henry: Pioneering Application for a Genetically Modified Food Animal. International Society for Biosafety Research 2013 [http://isbr.info/ news?news_id = 25].

CHAPTER 4

JAPANESE CHUM SALMON STOCK ENHANCEMENT: CURRENT PERSPECTIVE AND FUTURE CHALLENGES

Shuichi Kitada

Graduate School of Marine BiosciencesTokyo University of Marine Science and Technology

ABSTRACT

This study reviews the present status of the Japanese chum salmon *Oncorhynchus keta* stock enhancement program and considers the ecological sustainability of wild populations while providing fishery production, exemplified by the hatchery-based Kitami region set net fishery. The return rate and the number of returns have been historically high in the Sea of Okhotsk, but have decreased in other regions since 2005. Natural spawning of chum salmon occurred in at least 160 rivers in Hokkaido. The genetic diversity of Japanese chum salmon was similar to or higher than that of other Pacific Rim populations. Numbers of alleles were high at microsatellite loci, but the loss of rare haplotypes was observed in all populations. The estimated N_e/N ratio for the Kitami region was >0.15 % including hatchery and wild fish under the present high fishing pressure. Four regional populations were inferred in Hokkaido, however, genetic differentiation was weak and some river-populations were nested. Substantial changes in run timing were observed, but it has recovered gradually owing to the recent practice of escapement. Our analyses highlight the importance of juvenile quality and the vital roles of escapements in enhanced and non-enhanced rivers. New research is needed to minimize the genetic risks associated with hatchery programs.

INTRODUCTION

The hatchery stock enhancement program in Japan was initiated in the 1880s to increase Japanese chum salmon *Oncorhynchus keta* fishery production and has continued for over 120 years. The numbers of chum salmon returning to Japan have increased remarkably since the late 1970s. Hatchery technology innovation is considered the major cause for the increase [1, 2], and the favorable ocean and climate conditions have supported this increase [3, 4]. The closure of high seas fisheries and improved river water quality may have boosted the recent abundance of chum salmon in Hokkaido [5], the principal salmon production area in Japan. Although past management of Japanese chum salmon focused on producing and releasing hatchery-reared fish (hereafter hatchery fish), Japanese salmon scientists and hatchery managers have become aware of the importance of conserving wild fish. Counting surveys of wild fish recently conducted in Hokkaido have found that natural spawning occurs in many rivers [6]. However, the majority of chum salmon returning to Hokkaido are hatchery-released fish; therefore, the ecological sustainability of the chum salmon stock is controversial [7–9]. While traditional hatcheries have concentrated on fish production, the goals of modern hatcheries

Citation: Shuichi Kitada, Japanese chum salmon stock enhancement: current perspective and future challenges, doi:10.1007/s12562-013-0692-8

are to conserve wild populations and/or provide harvest with minimal risk to the wild population [10, 11]. There are strong ecological and economic incentives in Hokkaido to reform salmon fishery management by focusing on the coexistence of hatchery and wild fish [2,12–14]. Further discussion on the use of hatcheries from the viewpoint of conservation may be useful.

This study reviews the present status of the Japanese chum salmon stock enhancement program and considers the role of hatcheries in the ecological sustainability of wild populations, focusing on the hatchery-based set net fishery in the Kitami region, Hokkaido, the largest chum salmon fishery in Japan. Onsite surveys of the chum salmon stock enhancement program were conducted, and statistics and information were collected from the Salmon and Freshwater Fisheries Research Institute (SFFRI) at Eniwa, the Hokkaido Federation of Fisheries Cooperative Association (HFFCA) at Sapporo, and the Kitami Salmon Enhancement Program Association (KSEPA) at Abashiri. After conducting a brief review of the chum salmon stock enhancement program and the results of wild fish counting surveys in Hokkaido, the ratio of effective population size to the number of returning adult fish was estimated for the Kitami region by combining hatchery broodstock and hatchery/wild escapements. Genetic diversity indices of Japanese chum salmon populations were compared with those of other populations in the Pacific Rim. Fine-scale population structures of Japanese chum salmon populations were estimated based on publicly available mitochondrial (mt) DNA and microsatellite (ms) DNA data. Our analyses highlight the importance of juvenile fish quality and the vital role of hatchery-wild integrated reproduction through escapements to natural streams.

HATCHERY RELEASES AND RETURN RATES TO HOKKAIDO

The chum salmon hatchery program in Hokkaido is one of the largest salmon hatchery programs in the world [2]. The number of juveniles released from Hokkaido has been stable since the 1990s, averaging 1,030 ± 22 (standard deviation) million per annum during the last decade. There are currently 150 private and 10 national hatcheries in Hokkaido, and the majority of the chum salmon released (87 %) are produced by private hatcheries, with the residual portion (13 %) being produced and released by national hatcheries [2]. Hatchery fish are released in 140 rivers, of which 60 rivers use juveniles produced from parent fish collected in adjacent rivers (transplants). In addition, transplants are reared in 80 net pens and released from the rearing sites [2]. Such transplants are allowed within the geographic region in order to minimize genetic disturbance. All juveniles released from national hatcheries have otolith thermal marks. Fish from private hatcheries are marked only when scientific research is conducted. The otolith-marked chum salmon are currently <15 % of the total number released [2].

The number of chum salmon returning to Hokkaido has increased remarkably since the late 1970s, and the average number was 49 ± 10 million (range 38–61 million) during 2008–2012. This number has been decreasing since 2005, and as of 2012 (39 million) was ~64 % of the historical maximum of 61 million in 2004 (Figure. 1a). The average return rate was ~5 % for the brood years after 1996 [2]. In this article, the return rate was simply calculated by the number of returns (i.e., coastal commercial catches plus the number of adults caught for broodstock in rivers) divided by the number of hatchery fish released in the previous 4 years with the assumption that the number of naturally spawned chum salmon was negligible if any [2]. Returning salmon are caught by set net fisheries in the coastal waters off Hokkaido. Approximately 1,000 salmon set nets are operated from late August to late December, with landings peaking in late September and early October, although the fishing season differs among regions [14]. Commercial fishers pay for ~7 % of the landings every year to support the hatchery program. Commercial and recreational fishing for adult chum salmon are prohibited in all rivers and in coastal waters near the river mouths where hatchery broodstock are collected. Therefore, chum salmon that have escaped capture by commercial set nets are only captured at weirs at enhanced rivers [2]. Surplus fish caught in weirs are sold to support hatchery operations.

The Hokkaido Prefectural Government established the salmon enhancement plan in 1999, which prescribes the number of adults to be captured for broodstock, number of eggs to be taken in each river, number of fry to be reared in each hatchery, and the number of fry to be released in each river or net pen location [2]. In the enhancement plan, Hokkaido is divided into five geographical regions (Sea of Okhotsk coast, Nemuro Strait, eastern Pacific coast, western Pacific coast, and Sea of Japan coast) [14], each of which is further subdivided into 2–4 sub-regions, resulting in 14 management regions [2]. The hatcheries are operated by the nine regional salmon enhancement program associations, including KSEPA, which are supported by the Hokkaido Salmon Propagation Association, Sapporo. The salmon enhancement plan is temporally divided into three periods to manage the coastal set net fishery and broodstock collection in rivers [2]. These periods for the coastal fishery include early- (September), middle- (October), and late-periods (November and later). However, this is a period for fishery management, and not for identifying biological differences in returning fish. Here, the term "early run" is tentatively defined for fish that return during September–November, and "late run" for fish that return during December–January. The term "wild fish" is defined as naturally-produced individuals in the wild, even though they could include hatchery descendants [15–17].

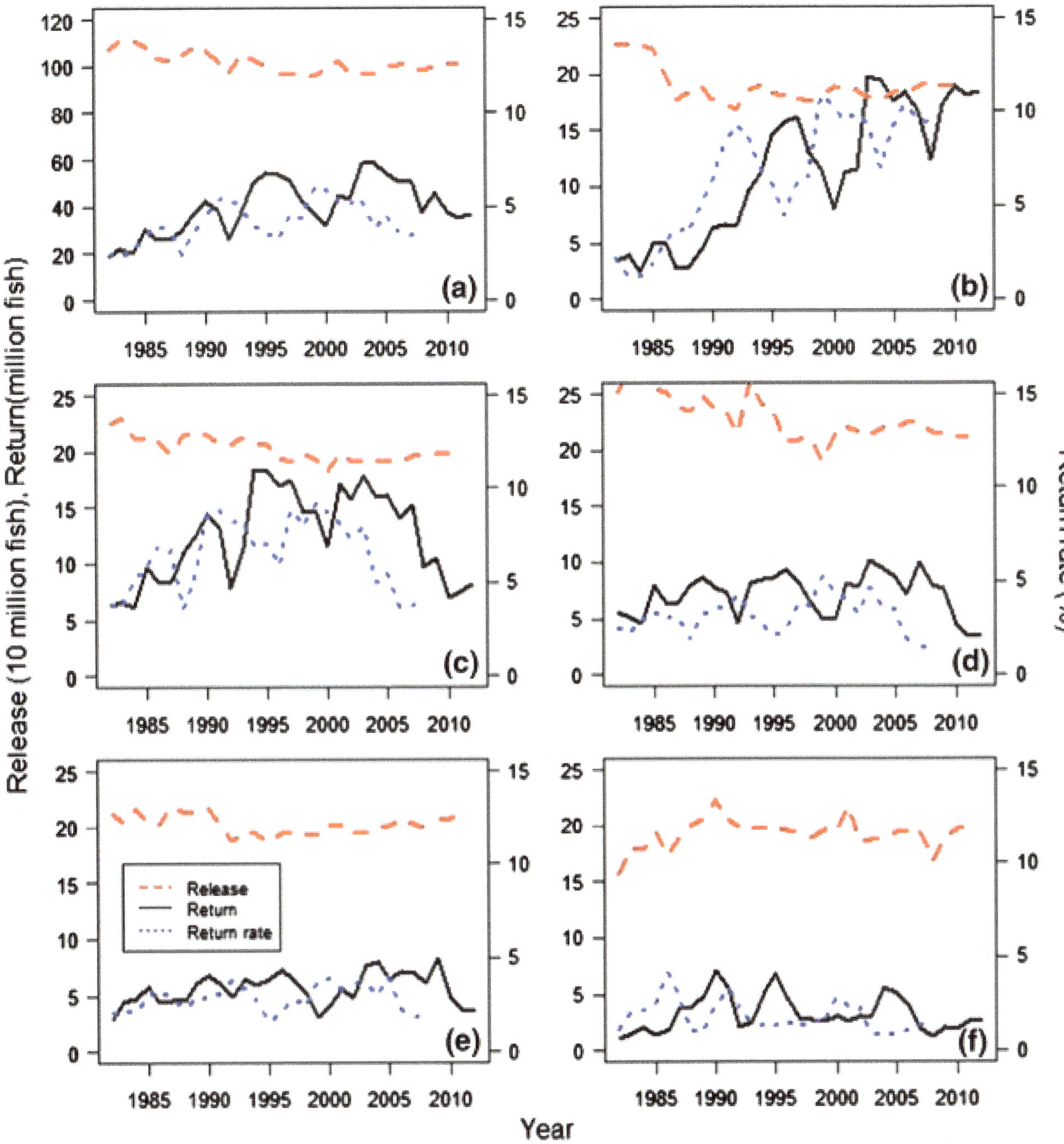

Figure. 1 Numbers of releases and returns with return rates of chum salmon in **a** total, and in the regions of the **b** Sea of Okhotsk coast (Kitami region), **c** Nemuro Strait, **d** eastern Pacific,**e** western Pacific, and **f** Sea of Japan, Hokkaido. The return rate was estimated as (number returned 4 years after release/number released)

A decreasing trend in returns has been observed in all regions except the Sea of Okhotsk. The recent return rates and the number of fish returning have been historically high in the Sea of Okhotsk and relatively low in the Sea of Japan region (Figure. 1). Release sizes have been increasing in all regions, and at most hatcheries they exceed the target size to be reared, i.e., 5 cm in fork length and 1 g in weight [2]. Despite the large release sizes, declining return rates have been observed in the Nemuro Strait, eastern Pacific, and western Pacific regions (Figure. 1). A plausible cause for the regional difference in return rates is the recent changes in the coastal environment around release time, but there remains an insufficient understanding of the causes [2].

The average number of chum salmon returning to the Sea of Okhotsk coast was 1,369,000 during 1959–1970. The average number of returning fish for the past 5 years (2008–2012) was 16,998,000 (12.4 times larger), most of which were hatchery fish. Hatchery innovation techniques are considered the main reason for the remarkable increase, which mimics the wild recruitment process and allows release timing to match favorable conditions [13] in river and coastal areas. In addition, KSEPA, an agency that enhances chum salmon populations in the Sea of Okhotsk region, has improved seed production techniques during the past decade to produce healthy juveniles. The quality of released juveniles may have contributed to the increased number of returning fish and return rates after 2000 (Figure. 1b). It may be useful to evaluate the present management system in the Kitami region and consider chum salmon sustainability, which depends on both hatchery and wild fish.

NATURAL SPAWNING IN NON-ENHANCED AND ENHANCED RIVERS

About 1,500 river systems occur in Hokkaido, of which 140 (9.3 %) rivers currently receive hatchery-releases of chum salmon [6]. The rivers are categorized into four types in relation to hatchery operations: (1) no enhancement to date, (2) currently no enhancement, (3) enhanced without broodstock collection, and (4) enhanced with broodstock collection for seed production [6, 14]. The rivers are relatively short, so hatchery fish are released in most large rivers with lengths >50 km, and naturally spawning chum salmon populations remain in many small (short) rivers. A total of 322 rivers (length > 8 km) were selected for counting surveys throughout Hokkaido, because water levels in smaller rivers may be inadequate to allow the passage of chum salmon [6]. Therefore, the number of rivers possible for chum salmon runs is estimated to be at least 322 in Hokkaido. Among the rivers, the number of enhanced rivers (categories 3 and 4) was 140 (43.4 %), of which 76 (2008) (23.8 %) and 84 (2009) (26.0 %) had weirs to collect hatchery broodstock (category 4).

Miyakoshi et al. [6] conducted foot surveys of 238 rivers during 2008 and 2009 and found that natural spawning occurred in 59 (31.4 %) and 50 (37.6 %) of the non-enhanced rivers, respectively. Furthermore, chum salmon were seen to ascend upstream in 191 and 175 rivers, including rivers from which broodstock are collected (76 and 84 rivers) and rivers <8 km (11 and eight rivers), respectively. These observations show that natural spawning of chum salmon occurs in 30–40 % of non-enhanced rivers (Figure. 2a) and in 70–80 % of 238 rivers surveyed, including categories 1–4.

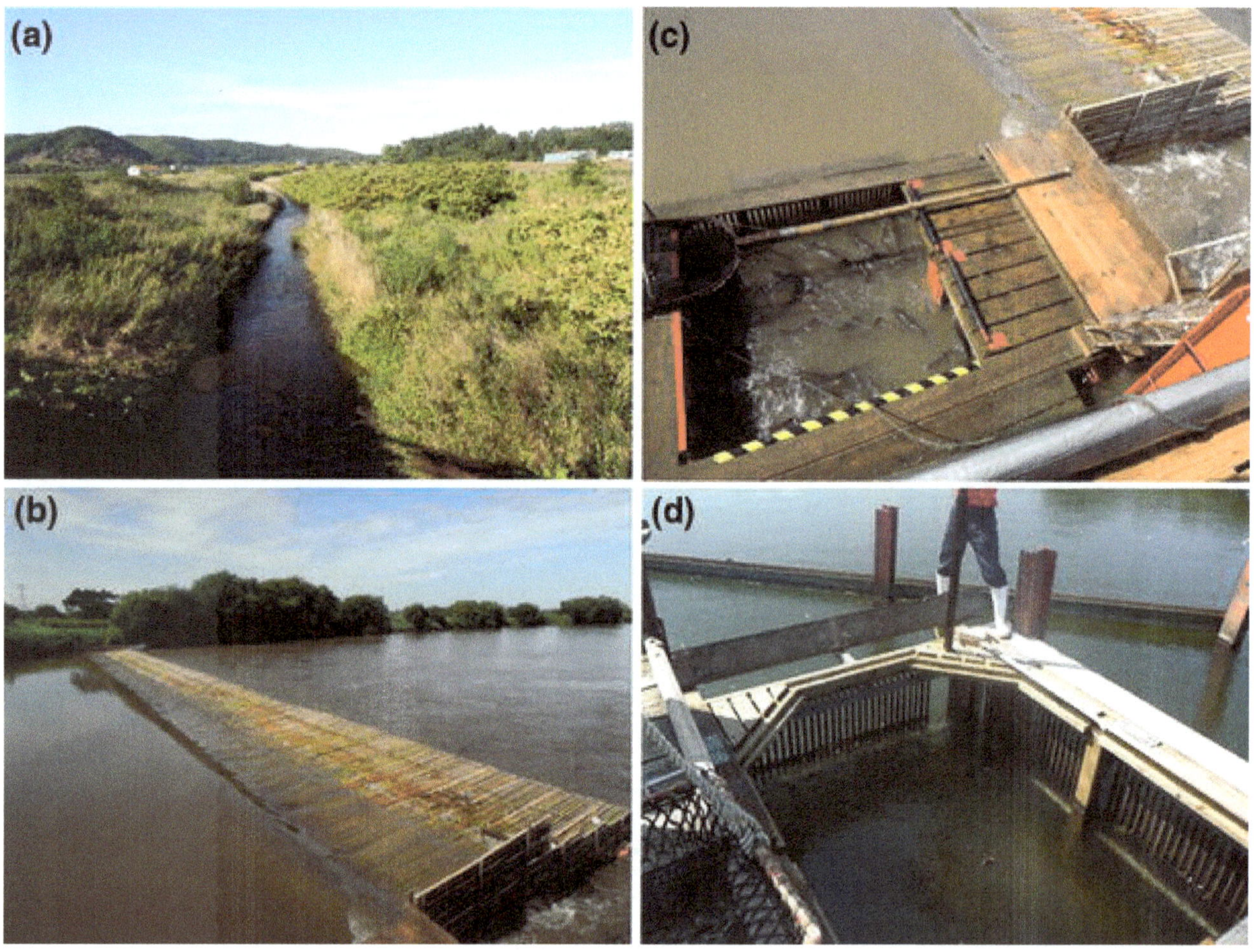

Figure. 2 a A non-enhanced river where natural spawning occurs (Nikura River, a tributary of the Saromabetsu River, where counting surveys of natural spawning fish are conducted), **b**, **c**a weir to collect hatchery broodstock (Tokoro River) and **d** a remodeled fish pool to adjust escapement upstream (Shari River)

The escapement of hatchery fish into enhanced rivers with hatcheries creates wild fish. Escapement during hatchery operations supplements early-run wild fish, and escapement creates late-run wild fish after hatchery operations. All weirs in the enhanced rivers of the Kitami region (Figure. 2b, c) have been remodeled recently to provide passage of fish for spawning (KSEPA, pers. comm., Figure. 2d). KSEPA recorded the number of escapements after 2005 in some enhanced rivers with hatcheries, although the number was low (Figure. 3). Weirs are removed around the end of November, when hatchery operations are finished. Returning hatchery and wild fish freely ascend the stream and spawn in the wild on enhanced rivers, and supplement early and late-run wild fish, respectively. The natural run in enhanced rivers without hatcheries and non-enhanced rivers creates wild fish throughout the run season. Thus, regardless of enhancement, wild fish can be created in all river systems. Therefore, it is important to evaluate how the sustainability of the chum salmon populations is impacted by integrated reproduction in hatcheries and natural streams.

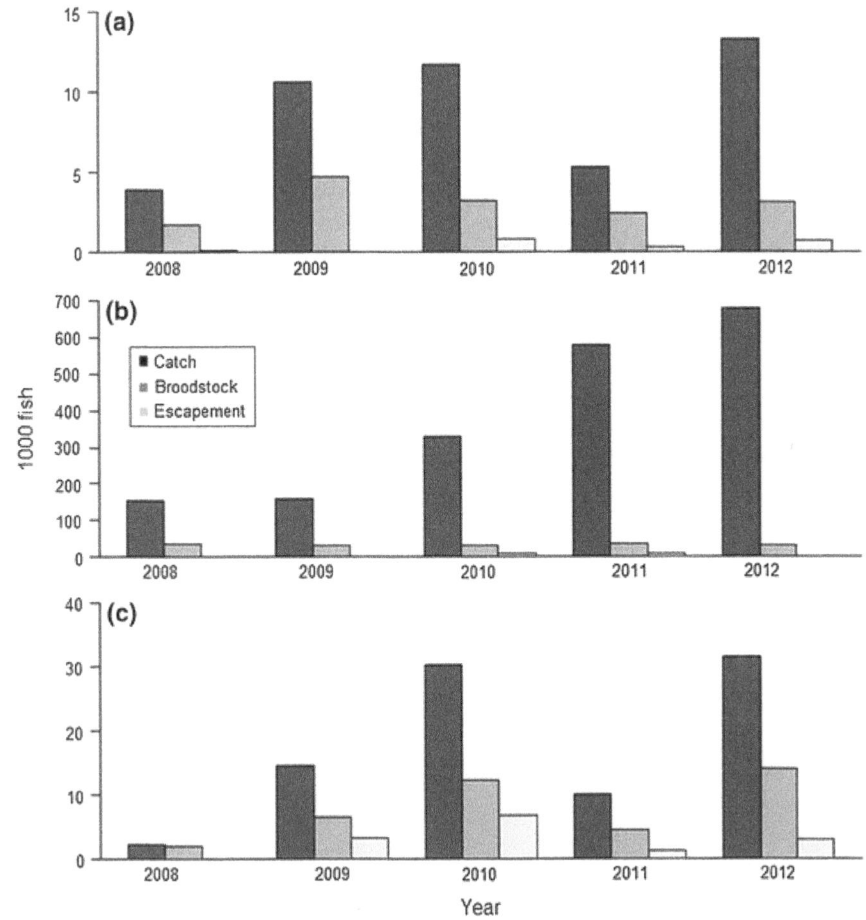

Figure. 3 Numbers of fish caught at the weir, hatchery broodstock, and escapement of chum salmon in the **a** Iwaobetsu, **b** Abashiri, and **c** Horonai Rivers of the Kitami region

INTEGRATED HATCHERY AND WILD REPRODUCTION

Huge numbers of hatchery fish return to hatchery-enhanced rivers, mainly in September–November. In 2012, 18,460,000 fish returned to the Kitami region. Among them, ~16 million fish were caught by set nets in coastal areas, and 2 million fish were caught at weirs on enhanced rivers, of which 114,000 fish were sent to hatcheries as parent fish, which provided an estimate that 4,140 hatchery fish escaped upstream for wild stock conservation. The surplus hatchery fish caught in weirs (1,886,000) were sold to support hatchery operations. The exploitation rates were very high at 86.7 % (set nets) and 94.3 % (weirs). This calculation indicates that 0.62 % of returning hatchery fish were used for hatchery broodstock and 0.02 % escaped from weirs to upstream rivers.

Substantial numbers of wild fish return to some Kitami-region rivers [6, 18]. The largest portion of this natural return occurs during December and early January, after returns of hatchery fish and the hatchery-focused fishery are completed in late November. The management system believes that fish returning during this time frame represent the historical natural return and are not subject to significant harvest. Escapements of these fish have been relatively consistent for the last several years, and current wild salmon escapement was estimated at 20,191 in the Kitami region (SFFRI, unpubl. data, 2013).

The percentage of hatchery breeders among all returned fish is very small (0.62 %). However, this is not surprising because the effective population size (N_e) is very small compared with the census population size (*N*) [19]. It is generally accepted that the effective population size of highly fecund marine teleosts is between two and five orders of magnitude smaller than the census spawning size [20–22]. For example, the N_e/N ratio was 10^{-5} for North Sea cod *Gadus morhua* [22] and New Zealand snapper *Pagrus auratus* [23], 10^{-3} for red drum *Sciaenops ocellatus* [24], and 10^{-4} for Japanese Spanish mackerel *Scomberomorus niphonius* in the Seto Inland Sea [17]. This may be explained by a biased sex ratio caused by dominant males monopolizing females, a disproportionate contribution of genotypes through egg production, an increase in larval viability with increasing female size, overlapping generations, or high variance in individual reproductive success [22]. In addition, the chance spatial and temporal match or mismatch of spawning with suitable oceanic conditions [20, 25] could result in the majority of surviving recruits being spawned by a minority of the population [17, 22].

The N_e/N ratio for salmonids should be larger than that of marine fish because of lower fecundity with larger eggs. The N_e/N ratio was estimated to be 4.3 % for a natural population of Chinook salmon *O. tshawytscha* [26]. The N_e/N ratio is on the order of 0.2 for most natural stocks of salmonid fish, where *N* is the number of spawning adults [27]. No N_e/N ratio estimate for a natural and a hatchery population of chum salmon was found in the literature, so 20 % (~10^{-1}) was used as a tentative estimate of N_e/N for both hatchery and wild spawning chum salmon. The number of spawning adults in hatcheries was 114,000, which was assumed to be hatchery-released fish.

The rough estimate of the N_e/N ratio for the Kitami hatchery program was ~0.12 % (=114,000 × 0.2/18,460,000 × 100), and ~0.03 % (=24,331 × 0.2/18,460,000 × 100) for the estimated escapement (24,331 = 4,140 hatchery + 20,191 wild fish). The total N_e estimate was 27,666 (=(114,000 + 24,331) × 0.2) and the N_e/N ratio was ~0.15 % (10^{-3}) after combining the hatchery broodstock and escapement. Current wild salmon escapement (20,191) was ~0.1 % of the 18,460,000 fish returned to the Kitami region in September–November, which was an underestimate because it did not account for naturally spawning salmon in major enhanced rivers (SFFRI, pers. comm.). Therefore, the N_e/N ratio estimate (~0.15 %) should be an underestimate. In addition, the N_e/N ratio might be much higher in hatcheries, as reported for rainbow trout *O. mykiss* (~90 %) [26]. A large number of adults (114,000) were used for artificial fertilization and the sex ratio was adjusted to ~0.5. However, even if this was the case, the N_e/N ratio of the chum salmon in the Kitami region was very small (~10^{-2}). It may be much smaller than those of Chinook salmon, coho salmon *O. kisutch*, steelhead trout *O. mykiss*, and Atlantic salmon *Salmo salar*, reflecting differences in early life history and the large census population size of chum salmon. To estimate the N_e/N ratio correctly, the effective population size of returned fish and the effective number of breeders in hatcheries needs to be estimated based on genetic monitoring [28].

Genetic Diversity

Preceding studies on chum salmon using different genetic markers have provided valuable information on population sustainability. Sato et al. [29] examined the genetic population structure of chum salmon in the Pacific Rim using the mtDNA control region for 2,154 individuals from 47 rivers in 15 populations from Japan, Korea, Russia, and North America, and found that haplotype diversity was highest in the Japanese populations. Beacham et al. [30] surveyed genetic variation at 14 microsatellite loci in 3,602 individuals from 26 Japanese chum salmon populations, one population from Kamchatka, and three populations from North America, and found that the Japanese populations are more genetically diverse at each of the 14 microsatellites surveyed than that of North American populations in the Yukon River and British Columbia. Seeb et al. [31] surveyed the allelic variation of 10,458 chum salmon at 60 loci of single-nucleotide polymorphisms (SNPs) in 114 locations from Japan (16 sites), Korea, Russia, and North America, and showed that neither marker type suggested that the Japanese populations are inherently less diverse than those from North America, as suggested by the current SNP panel. Genetic diversity indices of Japanese chum salmon for the three markers were generally higher or similar in comparison with Korean, Russian, and North American chum salmon populations (Figure. 4).

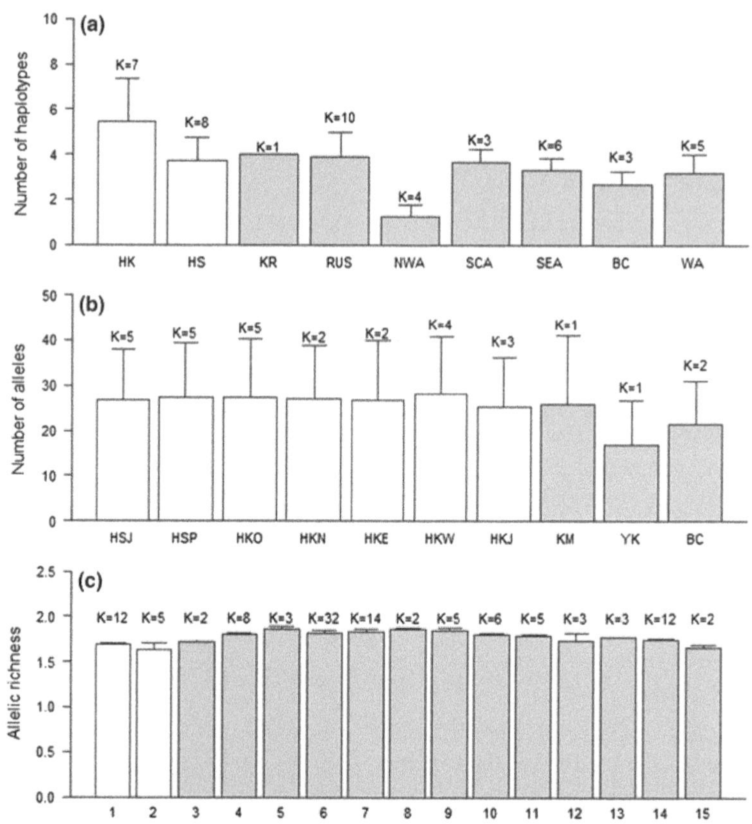

Figure. 4 Genetic diversity of chum salmon in the Pacific Rim. *K* refers to the number of rivers where samples were taken (*white*: Japan, *grey*: other countries). **a** Number of haplotypes for the mitochondrial DNA control region (data from [29], $n = 2,154$). *Error bars* show standard deviations. *HK* Hokkaido, *HS* Honshu, *KR* Korea, *RUS* Russia, *NWA* northwest Alaska and Alaskan peninsula, *SCA* south-central Alaska, *SEA* southeast Alaska, *BC*British Columbia, *WA* Washington. **b** Number of alleles for 14 microsatellite DNA loci (data from [30], $n = 3,602$). *HSJ* Honshu, Sea of Japan coast), *HSP* Honshu, Pacific coast, *HKO* Hokkaido, Sea of Okhotsk coast, *HKN* Hokkaido, Nemuro Strait, *HKE*Hokkaido, eastern Pacific coast, *HKW* Hokkaido, western Pacific coast, *HKJ* Hokkaido, Sea of Japan coast, *KM* Kamchatka, *YK* Yukon, *BC* British Columbia. **c** Number of alleles for 60 single-nucleotide polymorphic loci (data from [31], $n = 10,458$). *1*, Hokkaido; *2*, Honshu; *3*, Amur; *4*, Kamchatka; *5*, Alaska; *6*, western Alaska; *7*, Yukon; *8*, Yukon2; *9*, Bristol Bay; *10*, north Alaskan peninsula; *11*, southern Alaskan peninsula; *12*, Kodiak Island; *13*, Susitna River; *14*, Alaska Bay; *15*, Washington

Population Structure

Analyses of the mtDNA control region and SNPs coincidently show that the Japanese and Korean populations form a single, distinct cluster [29, 31]. Analyses of msDNA [30] found five clusters of populations in Hokkaido (Sea of Okhotsk coast, Nemuro Strait, eastern Pacific coast, western Pacific coast, and Sea of Japan coast) and two in Honshu (Pacific coast and Sea of Japan coast). Honshu Pacific coast had the most distinct cluster. KSEPA and Japanese salmon scientists suggest that the December–January return (late run) represents the native return and that the hatchery-dominated September–November return represents transplants. An analysis of the mtDNA control region from four sites in Hokkaido ($n = 323$) [32] showed that the Yurappu River chum salmon has two populations: the late native-population spawning in December and the early mixed-population introduced from other river populations in October and November.

To infer the fine population structure including early and late-run populations, genetic differentiation was estimated using a common scale in terms of F_{ST} based on mtDNA haplotype frequencies of 15 samples ($n = 273$) [29, 32, 33] and msDNA allele frequencies of 26 samples ($n = 2{,}670$) at 14 loci [30]. The samples [29, 30, 33] were collected in hatcheries (hatchery-released fish), and the Yurappu sample of Yokotani et al. [32] consisted of naturally-spawning fish (Kaeriyama M. pers. comm.). The msDNA allele frequencies were cited from the web site of the Molecular Genetics Lab at the Pacific Biological Station, Fisheries and Ocean Canada (http://www.pac.dfo-mpo.gc.ca/science/facilities-installations/pbs-sbp/mgl-lgm/data-donnees/index-eng.html). We used an empirical Bayesian method [34] with the software POPDIF 1.0 (http://popgene.co-site.jp/software_eng.html) to avoid a biased estimation of pairwise F_{ST} (e.g., negative F_{ST} values), which is a concern for species with high gene flow such as marine fish. The maximum likelihood global F_{ST} over populations, the rate of gene flow ($=1/F_{ST} - 1$), and the means of the posterior distributions of pairwise F_{ST} were estimated based on haplotype and allele frequencies.

Global F_{ST} was estimated as 0.0433 ± 0.0130 for the mtDNA haplotype frequencies. The rate of gene flow was 22.1 ± 6.9, suggesting that 22 females in terms of the effective population size migrate between rivers per generation. The unweighted pair group method with arithmetic mean (UPGMA) dendrogram of pairwise F_{ST} showed a distinct cluster in the samples taken in September and October. Honshu Pacific coast samples also formed a cluster. In contrast, samples taken in November and December were included in a different cluster for except Tokushibetsu, collected in September (Figure. 5). The November samples were distinct from the October samples at Tokoro and Yurappu, suggesting that it was influenced by transplantations from the Chitose hatchery. The pairwise F_{ST} value [95 % credible interval] between October and November was small at 0.0160 [0.0013, 0.0482] for Tokoro and 0.0125 [0.0017, 0.0352] for Yurappu, respectively. F_{ST} between November and December for Yurappu was 0.0290 [0.0050, 0.0708] and between October and December was 0.0529 [0.0158, 0.1072]. The genetic difference was greater for the later run. However, the posterior F_{ST} distributions overlapped, clearly showing that early- and late-run populations were not genetically distinct with high gene flow (Figure. 6). The difference between Yurappu (December) and Tokushibetsu (September) was also similar at 0.0349 [0.0084, 0.0750]. It was unclear why the early sample from Tokushibetsu was close to the late one from Yurappu.

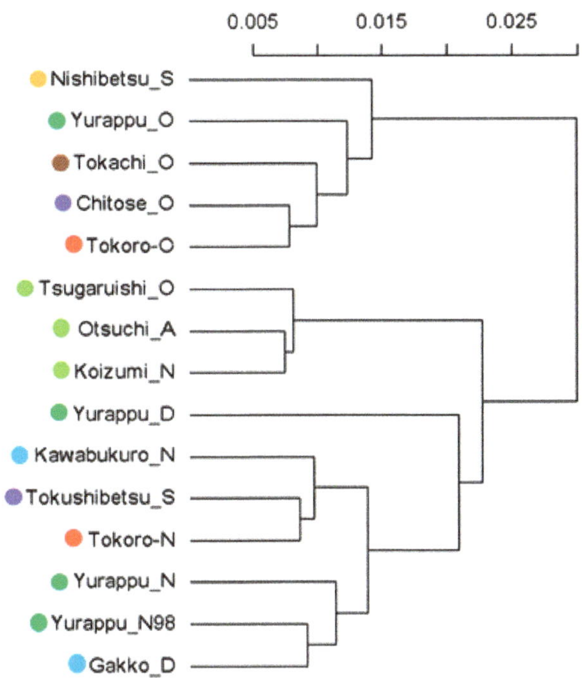

Figure. 5 Genetic differentiation of chum salmon based on the posterior mean of pairwise F_{ST}, inferred from mitochondrial DNA control region haplotype frequencies of 15 geographical samples ($n = 273$) (data from Sato et al. [29, 33], Yokotani et al. [32]). Hokkaido, Sea of Okhotsk Coast (*red circle*); Hokkaido, Nemuro Strait (*yellow circle*); Hokkaido, eastern Pacific coast (*brown circle*); Hokkaido, western Pacific coast (*dark green circle*); Hokkaido, Sea of Japan coast (*purple circle*); Honshu, Pacific coast (*light green circle*); Honshu, Sea of Japan coast (*blue circle*). *S*, September; *O*, October; *N*, November; *D*, December; *A*, April

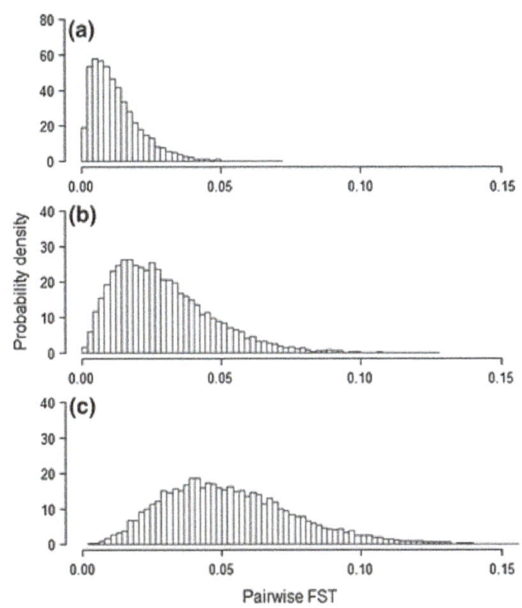

Figure. 6 Posterior distributions of pairwise F_{ST} between early- and late-run samples of chum salmon in the Yurappu River inferred from the mitochondrial DNA control region haplotype frequencies; **a** October vs. November, **b** November vs. December, **c** October vs. December

The global F_{ST} was much smaller (0.0069 ± 0.0002) than that from the mtDNA haplotype frequencies based on the msDNA allele frequencies. The rate of gene flow was 155.3 ± 4.7. The UPGMA dendrogram of pairwise F_{ST} showed three clusters. Honshu Pacific coast samples formed a distinct cluster, coinciding with the results from the mtDNA analysis. Five geographical samples (except from the eastern Pacific coast of Hokkaido) were located closely in the large middle cluster, but some were nested. The Honshu Sea of Japan coast formed one cluster with Teshio and Chitose, but Kawabukuro was close to the western Pacific coast of Hokkaido, and Uono was separated from the cluster. Hokkaido samples from the Sea of Okhotsk, Nemuro Strait, the western Pacific coast, and the Sea of Japan coast were generally located nearby. In contrast, samples from the Pacific coast of Hokkaido (Kushiro and Tokachi) were not close and Horonai was different from other Sea of Okhotsk samples (Figure. 7), though the F_{ST} value between Horonai and Tokoro was very small at 0.0036 [0.0011, 0.0082]. Our results generally coincided with the results of Beacham et al. [30], who found seven regional populations (five on Hokkaido and two on Honshu) within Japan, and learned that the Honshu Pacific coast populations were the most distinct based on Cavalli–Sforza and Edwards chord distance. However, the genetic differentiation was very small and population structure was not rigid in terms of the posterior mean of F_{ST}. The samples were all hatchery fish, as mentioned above. The nested population structure in Hokkaido may have been influenced by the transplantation of hatchery fish. No sample set was available from the same site including both early and late runs, so it was not possible to detect the genetic difference between the two runs. Beacham et al. [30] analyzed the original temporal samples from three populations within 1 year: (1) Gakko River, October and December samples; (2) Yurappu River, September and November samples; and (3) Chitose River, October and December samples. Their F_{ST} values were 0.015 ± 0.002 (Gakko), 0.008 ± 0.002 (Yurappu), and 0.010 ± 0.001 (Chitose). Beacham et al. [30] suggested that transplantation history could influence the observed temporal variation within these populations.

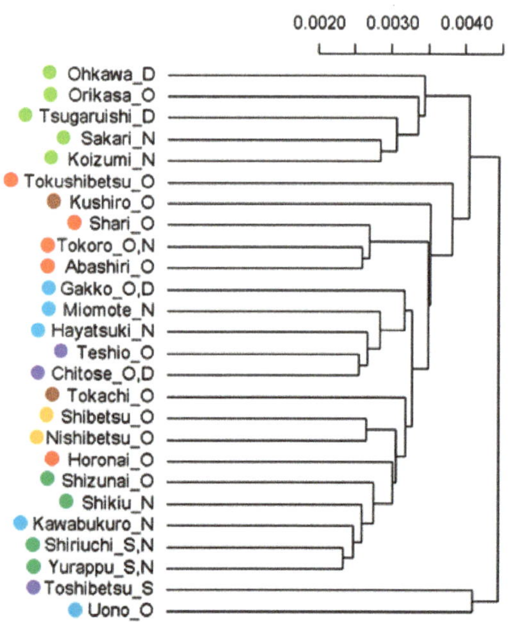

Figure. 7 Genetic differentiation of chum salmon based on the posterior mean of pairwise F_{ST} inferred from microsatellite DNA allele frequencies of 26 geographical samples (n = 2,670) at 14 loci [30] (data source: http://www.pac.dfo-mpo.gc.ca/science/facilities-installations/pbs-sbp/mgl-lgm/data-donnees/index-eng.html). Hokkaido, Sea of Okhotsk coast (*red circle*); Hokkaido, Nemuro Strait (*yellow circle*); Hokkaido, eastern Pacific coast (*brown circle*); Hokkaido, western Pacific coast (*dark green circle*); Hokkaido, Sea of Japan coast (*purple circle*); Honshu, Pacific coast (*light green circle*); Honshu, Sea of Japan Coast (*blue circle*). *S*, September; *O*, October; *N*, November; *D*, December

In the Yurap River, our F_{ST} value based on mtDNA between October and December (0.0529 [0.0158, 0.1072]) was 6.6 times larger than that between September and November based on msDNA (0.008 ± 0.002) [30]. Our estimate of the rate of global gene flow for msDNA (155.3 ± 4.7) was seven times larger than that of mtDNA (22.1 ± 6.9), suggesting that females return to natal rivers and spawning sites, but males stray into non-natal rivers to extend the chance of reproduction. Straying of males should serve to enhance genetic diversity in populations and adaptability to environmental changes. It is unclear if straying reflects a failure to return home or a decision to spawn elsewhere, but in either case, straying has been essential for the persistence and distribution of salmon [35]. Chinook salmon stray more (mean population range 0.11–34.6 %) than coho salmon (0.08–0.94 %), ocean-type Chinook (5.2–18.6 %) stray more than stream-type Chinook (0.11–10 %), and Chinook salmon (0.90–54.9 %) stray more than steelhead (0.30–2.3 %) [36].These patterns are largely the result of species-specific behavioral and endocrine factors during juvenile life stages [36]. The level of straying in hatchery chum salmon has not been assessed in Japan [2]. Further research is needed on chum salmon and pink salmon *O. gorbuscha*, which exhibit similar juvenile life histories (individuals typically migrate soon after emergence) [36]. The distributions of the mean pairwise F_{ST}(0.0037 ± 0.0007) and the mean pairwise rate of gene flow (278 ± 49), estimated from msDNA allele frequencies of 26 samples (n = 2,670), suggested that straying is typical in chum salmon (Figure. 8). The global F_{ST} values for masu salmon *O. masou* were 0.016–0.032 within the Uryu River system in Hokkaido [37], which were estimated from genotypes of 10 msDNA loci using the same software POPDIF. These point estimates were 2.3–4.6 times larger than our estimate of chum salmon (0.0069), which reflect the smaller rate of straying of masu salmon.

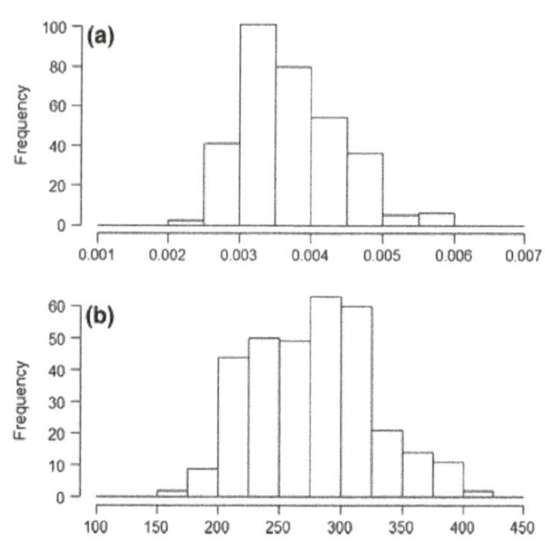

Figure. 8 Distributions of **a** pairwise F_{ST} and **b** the rate of gene flow (= the effective population size per generation) of chum salmon between rivers in Japan, inferred from microsatellite DNA allele frequencies in 26 geographical samples (n = 2,670) at 14 loci [19]

GENETIC EFFECTS OF HATCHERY RELEASE AND ROLES OF INTEGRATED HATCHERY MANAGEMENT

Japanese populations had a similar number of haplotypes as in the Pacific Rim (Figure. 4a). However, the number of haplotypes ranged from two to seven in a population, and the loss of rare haplotypes was observed in all populations. Similar reductions in the numbers of haplotypes were observed in other foreign populations. In contrast, the number of microsatellite alleles was very high at 26.8 ± 11.3 (range 4.8–73.9) [30]. A large number of parents have been used since the mid-1970s when the stock enhancement program became effective. Currently, an annual total of 1.27 million fish (3.2 % of returned fish in 2012) are collected for hatchery broodstock from about 75 rivers in Hokkaido [2]. The large number of parent fish used for artificial fertilization may have contributed to maintaining the high number of microsatellite alleles (26.8 ± 11.3) [30]. However, it is unclear if long-term hatchery release is responsible for the loss of rare haplotypes, as the reproductive ecology of females might also be a factor.

Fish generally remain in a chum salmon hatchery (from eggs to release) for ~6 months. Survival rates in hatcheries are very high at >90 %. In contrast, survival rates should be very low in the wild, particularly during the early life stages. The rearing process in hatcheries should relax natural selection until release. In addition, eggs and sperm are taken artificially, so parent fish do not experience natural spawning. Carrying out this artificial fertilization process for more than 10 generations might reduce the natural spawning behavior performance of hatchery and wild fish. Such artificial selection could cause a reduction in the fitness of hatchery fish in the wild. Evidence for a reduction in reproductive success (RS) of hatchery fish has accumulated for Atlantic salmon [38, 39], steelhead trout [40–42], coho salmon [43], and Chinook salmon [44]. A Bayesian meta-analysis for the steelhead data showed that the average level of relative RS (RRS) implied the reproductive decline of hatchery fish, but RRS varies greatly depending on genetic cross, sex, year of release, and environmental conditions [45]. Reflecting the large variation in RRS in a super population, no RS difference between hatchery and wild fish has been found for steelhead trout [46], brown trout *Salmo trutta* [47], or coho salmon [48]. If the ratio of hatchery fish with lower RS is substantial in the stocked population, the total number of progeny recruited in the stocked population could be reduced in the next generation [49]. The carryover effect of reduced RS in the next generation, as estimated in a steelhead population [50], should amplify the reduction in recruitment.

The mechanisms responsible for reduced RS of hatchery fish are unknown [40]. The most likely cause is thought to be selection pressures associated with domestication in captivity [51]. Björnsson et al. [52] suggested that it is important not to lose sight of the fact that when handling expression data in endocrinology, it is the circulating hormones that stimulate target tissues. Gene expression mirrors the physiological activity of an organism. Highly significant differences in transcription levels have been found between farmed and wild Atlantic salmon [53]. A significant correlation between the domesticated/wild-type gene expression ratio and the transgenic/wild-type gene expression ratio for coho salmon was found [54]. Transgenesis of the growth hormone gene causes inferior reproductive behavior and reduces fertilization success of Atlantic salmon [55]. These findings provide an important suggestion that enhanced growth in captivity causes a reduction in reproductive performance of hatchery fish. Actually, hatchery-reared steelhead are almost ubiquitously released as yearlings, whereas wild steelhead typically spend two or more years in freshwater before migrating to sea [56]. It is crucial to discover the essential genetic cascades related to survival and reproductive traits, and to identify the main trigger genes, which will determine the mechanisms of the fitness reduction caused by captive rearing [57]. The results will contribute to the development of juvenile production techniques that help to conserve natural populations.

The duration of hatchery rearing for chum salmon (~6 months) is about half that for steelhead. The growth of hatchery-reared chum salmon is also similar to wild fish and is not enhanced in the hatchery, as in steelhead and farmed Atlantic salmon. The effect of domestication might be smaller in chum salmon than in other salmon, except in pink salmon. No significant difference in RS between hatchery and wild chum salmon has been found [58]. The RS of wild and hatchery chum salmon has not been evaluated in Japan, so it is unknown if hatchery fish have lower RS in the wild and if releases reduce the RS of the population. It should be noted that DNA parentage assignment for the precise estimation of RS requires a complete sampling of the parent and progenies for hatchery and wild fish [59]. The number of returns and return rates in the Kitami region have continued to increase in the last two decades (Figure. 1b), showing that reductions in the survival rate of hatchery-released fish are unlikely. However, a consistent decreasing trend in the number of returns and the return rate since about 2005 in other regions in Hokkaido clearly indicates a survival decline of hatchery-released fish. Mizuno et al. [60] suggested the need for improving juvenile quality in the areas where recent returns have been decreasing in order to recover the chum salmon stocks.

Efforts to enhance the early-run population have been encouraged since the early 1980s, and resulted in a change in run timing [2], as found in coho salmon [48]. Miyakoshi et al. [2] demonstrated the long-term change (1970–2010) in the distribution of adult chum salmon captured in enhanced rivers in Hokkaido. There were both early- and late-run populations of chum salmon in Hokkaido until the early 1980s, and the numbers of fish in the two runs were almost the same (see Figure. 8 in Miyakoshi et al. [2]). The late-run population had almost disappeared by the late 1990s. The late-run population has continued to decrease because adults captured late in the season for several years in the 1980s were not used for hatchery fertilization, but instead were simply sold. Hatchery managers have recently become aware that such extreme intentional selection is undesirable and have made an effort to escape the late-run fish, as mentioned above. The late-run population has been recovering gradually, suggesting the effectiveness of such escapement in enhanced rivers (Figure. 8 in Miyakoshi et al. [2]). The run-timing distribution for 1970–1975 should reflect that of the natural (original) populations of chum salmon in Hokkaido, because that period was the beginning of the increase in returned fish due to success of the hatchery program. The very small population differentiation of chum salmon was evaluated by neutral genetic markers. The F_{ST} analysis evaluates differences in allele or haplotype frequencies and might not reflect the genetic characteristics of the indigenous demes, but the present run-timing distribution suggests that indigenous individuals remain in many rivers even with hatcheries. Therefore, escapement in enhanced and non-enhanced rivers throughout the run season should enable the recovery of the natural (original) genetic component. Further efforts for the escapement of both early- and late-run fish are key to the recovery and maintenance of natural populations of chum salmon.

The escapement of hatchery [H] fish into the wild [W] population allows for the reproduction between genetic crosses of [HH], [HW], and [WW], which creates three types of wild fish, denoted as W [HH], W [HW], and W [WW]. There are also three types of hatchery fish, such as H [HH], H [HW] and H [WW]. Substantial introgression of hatchery-released fish into a wild population has been observed, even after several generations [17]. Considering long-term intensive hatchery release, almost all chum salmon returning to the Kitami region may be hatchery descendants. The increasing trend in return rate may suggest that hatchery-wild integrated reproduction acts to compensate for domestication selection caused by captive rearing and artificial fertilization in hatcheries. The Sea of Okhotsk region is considered as a genetic management unit, and the genetic difference between hatchery (early-run) and wild fish (late-run) is very small (F_{ST} = 0.008–0.010) [30]. This coincides with the concept of the genetically-integrated hatchery program [10]. In the successful salmon stock enhancement in southeast Alaska, hatcheries are located away from significant wild stocks [61], regarded as a genetically-segregated hatchery program [10]. However, the contribution of hatchery-origin chum salmon is 78 % in southeast Alaska [61], showing that any hatchery program is an integrated one within enhanced regions. Under integrated management, the idea of zoning management for enhanced and non-enhanced rivers in Hokkaido [13, 14] is useful both for conservation and for the production of chum salmon. Spawning habitats suitable for the reproduction of chum salmon are estimated to comprise 891 km in the Kitami region, including the Shiretoko area (Urabe H, SFFRI, unpublished data). It is crucial to allow escapements throughout the run season in order to sustain the habitat capacity of enhanced and non-enhanced rivers. Establishing river-specific escapement goals in enhanced rivers is needed. Marking and genetic monitoring will contribute to an evaluation of the effectiveness of the integrated hatchery program. Furthermore, the restoration of rivers, including dams [13, 14], increases spawning habitat capacity.

CONCLUDING REMARKS

The number of returns and return rates of hatchery fish have continued to increase in the past two decades in the Sea of Okhotsk region, supported by the favorable ocean and climate conditions. The increasing trend in return rates suggests that a survival reduction of hatchery-released fish in the Kitami chum salmon population does not occur. On the other hand, a consistent decreasing trend in the number of returns and the return rate since about 2005 in other regions in Hokkaido clearly indicates a survival decline of hatchery-released fish. However, it is unknown if hatchery fish have lower RS in the wild and whether releases reduce the RS of the stocked population. It is crucial to determine the mechanisms of the fitness reduction caused by captive rearing. New approaches are needed to discover the essential genetic cascades related to the survival and reproductive traits and to identify the main trigger genes, which will contribute to developing juvenile production techniques aimed at conserving natural populations.

The genetic diversity of Japanese chum salmon is higher or similar compared with other populations in the Pacific Rim. However, the number of haplotypes ranged from two to seven in a population, and a loss of rare haplotypes was observed in all populations. In contrast, the number of microsatellite alleles was very high at 26.8 ± 11.3 (range 4.8–73.9). A tentative estimate of N_e/N ratio for the Kitami hatchery program was >0.15 % under the present high fishing pressure. The integrated hatchery-wild reproduction system, realized by a large number of broodstock in hatcheries and escapement into enhanced and non-enhanced rivers, maintains the present genetic diversity. However, it is unclear if long-term hatchery release is responsible for the loss of rare haplotypes, which might be caused by the reproductive ecology of females. Six regional populations (four on Hokkaido and two on Honshu) are assumed to occur within Japan, which suggests the management units of Japanese chum salmon. However, genetic differentiation is very low, and the population structure is not rigid. The nested population structure in Hokkaido might reflect the transplantation history of hatchery fish. The large gene flow suggests that straying, particularly by males, is a chum salmon strategy to maintain genetic diversity. Early and late-run populations are not genetically distinct in terms of F_{ST}.

Intensive hatchery enhancement for the early run populations has changed the distribution of run timing for the chum salmon populations in Hokkaido. Recent escapement practices of late-run fish have gradually recovered the late-run populations. Hatchery releases serve to maintain the genetic composition of the early run, whereas escapement practices recover the late run, which contributes to maintaining the genetic characteristics of the natural (original) populations. Further efforts for escapement of both early- and late-run fish are the key to recovering and maintaining natural populations of chum salmon. Surplus spawning habitat capacity exists in the Kitami region including the Shiretoko area. The challenge is to extend present escapement practices to other enhanced rivers with allowable escapements for habitat capacity. The restoration of rivers and dams increases spawning habitat capacity. A region-wide integrated hatchery management plan with river-specific escapement goals in enhanced rivers will recover the natural populations and contribute to sustainable chum salmon fisheries in the future. New research is needed to evaluate the effectiveness of the integrated hatchery program, which will contribute to the minimization of ecological and genetic risks for hatchery fish.

ACKNOWLEDGMENTS

I am indebted to Y. Miyakoshi (SFFRI), who provided data for Figs. 1 and 3 with valuable information. I thank M. Kaeriyama (Hokkaido University) and M. Nagata (SFFRI) for constructive comments on the earlier version of the manuscript. I also thank H. Urabe and Y. Sasaki (SFFRI), K. Fukuda (HFFCA), N. Masukawa, and O. Ishizuka (KSEPA) for providing useful materials. This study was supported by a JSPS Grant-in-Aid for Scientific Research (B) 25280006.

REFERENCES

1. Kaeriyama M (1999) Hatchery programmes and stock management of salmonid populations in Japan. In: Howell BR, Moksness E, Svåsand T (eds) Stock enhancement and sea ranching. Blackwell, Oxford, pp 153–167
2. Miyakoshi Y, Nagata M, Kitada S, Kaeriyama M (2013) Historical and current hatchery programs and management of chum salmon in Hokkaido, northern Japan. Rev Fish Sci 21:469–479
3. Beamish RJ, Bouillon DR (1993) Pacific salmon production trends in relation to climate. Can J Fish Aquat Sci 50:1002–1016
4. Kaeriyama M, Seo H, Kudo H (2009) Trends in run size and carrying capacity of Pacific salmon in the North Pacific Ocean. N Pac Anadr Fish Comm Bull 5:293–302
5. Morita K, Saito T, Miyakoshi Y, Fukuwaka M, Nagasawa T, Kaeriyama M (2006) A review of Pacific salmon hatchery programmes on Hokkaido Island, Japan. ICES J Mar Sci 63:1353–1363
6. Miyakoshi Y, Urave H, Saneyoshi H, Aoyama T, Sakamoto H, Ando D, Kasugai K, Mishima Y, Takada M, Nagata M (2012) The occurrence and run timing of naturally spawning chum in northern Japan. Environ Biol Fish 94:197–206
7. Hilborn R (1992) Hatcheries and the future of salmon in the northwest. Fisheries 17:5–8
8. Waples RS (1999) Dispelling some myths about hatcheries. Fisheries 24:12–21
9. Brannon EL, Amend DF, Cronin MA, Lannan JE, LaPatra S, McNeil WJ, Noble RE, Smith CE, Talbot AJ, Wedemeyer GA, Westers H (2004) The controversy about salmon hatcheries. Fisheries 29:12–31
10. Mobrand L, Barr J, Blankenship HL, Campton DE, Evelyn TTP, Flagg TA, Mahnken CVW, Seeb LW, Seidel RR, Smoker WW (2005) Hatchery reform in Washington state: principles and emerging issues. Fisheries 30:11–23
11. Waples RS, Drake J (2004) Risk/benefit considerations for marine stock enhancement: a Pacific salmon perspective. In: Leber KM, Kitada S, Svåsand T, Blankenship HL (eds) Stock enhancement and sea ranching, 2nd edn. Blackwell, Oxford, pp 260–306
12. Kaeriyama M, Edpalina RR (2004) Evaluation of the biological interaction between wild and hatchery population for sustainable fisheries management of Pacific salmon. In: Leber KM, Kitada S, Svåsand T, Blankenship HL (eds) Stock enhancement and sea ranching, 2nd edn. Blackwell, Oxford, pp 247–259
13. Kaeriyama M, Seo H, Kudo H, Nagata M (2012) Perspectives on wild and hatchery salmon interactions at sea, potential climate effects of Japanese chum salmon at the need for sustainable salmon fisheries management reform in Japan. Environ Biol Fish 94:165–177
14. Nagata M, Miyakoshi Y, Urabe H, Fujiwara M, Sasaki Y, Kasugai K, Torao M, Ando D, Kaeriyama M (2012) An overview of salmon enhancement and the need to manage and monitor natural spawning in Hokkaido, Japan. Environ Biol Fish 94:311–323
15. Kitada S, Kishino H (2006) Lessons learned from Japanese marine finfish stock enhancement programmes. Fish Res 80:101–112
16. Kitada S, Shishidou H, Sugaya T, Kitakado T, Hamasaki K, Kishino H (2009) Genetic effects of the long-term stock enhancement programs. Aquaculture 290:69–79
17. Nakajima K, Kitada S, Habara Y, Sano S, Yokoyama E, Sugaya T, Iwamoto A, Kishino H, Hamasaki K (2014) Genetic effects of marine stock enhancement: a case study based on the highly piscivorous Japanese Spanish mackerel. Can J Fish Aquat Sci. doi:10.1139/cjfas-2013-0418.21
18. Morita K, Takahashi S, Ohkuma K, Nagasawa T (2013) Estimation of the proportion of wild chum salmon Oncorhynchus keta in Japanese hatchery rivers (in Japanese with English abstract). Nippon Suisan Gakkaishi 79:206–213
19. Frankham R (1995) Effective population size/adult population size ratios in wildlife: a review. Genet Res 66:95–107
20. Hedgecock D (1994) Does variance in reproductive success limit effective population sizes of marine organisms? In: Beaumont AR (ed) Genetics and evolution of aquatic organisms. Chapman and Hall, London, pp 122–134
21. Grant WS, Waples RS (2000) Spatial and temporal scales of genetic variability in marine and anadromous species: implications for fisheries oceanography. In: Harrison PJ, Parsons TR (eds) Fisheries oceanography. Blackwell Science, Cambridge, pp 61–93
22. Hutchinson WF, van Oosterhout C, Rogers SI, Carvalho GR (2003) Temporal analysis of archived samples indicates marked genetic changes in declining North Sea cod (Gadus morhua). Proc Royal Soc B-Biol Sci 270:2125–2132
23. Hauser L, Adcock GJ, Smith PJ, Bernal Ramirez JH, Carvalho GR (2002) Loss of microsatellite diversity and low effective population size in an overexploited population of New Zealand snapper (Pagrus auratus). Proc Natl Acad Sci USA 99:11742–11747
24. Turner TF, Wares JP, Gold JR (2002) Genetic effective size is three orders of magnitude smaller than adult census size in an abundant, estuarine-dependant marine fish (Sciaenops ocellatus). Genetics 162:1329–1339
25. Cushing DH (1995) Population production and regulation in the sea: a fisheries perspective. Cambridge University Press, Cambridge
26. Bartley D, Bagley M, Gall G, Bently B (1992) Use of linkage disequilibrium data to estimate effective size of hatchery and natural fish populations. Cons Biol 6:365–375
27. Allendorf FW, Bayles D, Bottom D, Currens KP, Frissell CA, Hankin D, Lichatowich JA, Nehlsen W, Trotter PC, Williams TH (1997) Prioritizing Pacific salmon stocks for conservation. Cons Biol 11:140–152
28. Schwartz MK, Luikart G, Waples RS (2006) Genetic monitoring as a promising tool for conservation and management. Trends Ecol Evol 22:25–33

29. Sato S, Kojima H, Ando J, Ando H, Wilmot RL, Seeb LW, Efremov V, LeClair L, Buchholz W, Jin DH, Urawa S, Kaeriyama M, Urano A, Abe S (2004) Genetic population structure of chum salmon in the Pacific Rim inferred from mitochondrial DNA sequence variation. Environ Biol Fish 69:37–50

30. Beacham TD, Sato S, Urawa S, Lei KD, Wetklo M (2008) Population structure and stock identification of chum salmon Oncorhynchus keta from Japan determined by microsatellite DNA variation. Fish Sci 74:983–994

31. Seeb LW, Templin WD, Sato S, Abe S, Warheit K, Park JY, Seeb JE (2011) Single nucleotide polymorphisms across a species' range: implications for conservation studies of Pacific salmon. Molec Ecol Res 11(Suppl 1):195–217

32. Yokotani R, Azuma N, Kudo H, Abe S, Kaeriyama M (2009) Genetic differentiation between early- and late-run populations of chum salmon (Oncorhynchus keta) naturally spawned in the Yurappu River inferred from mitochondrial DNA analysis. Fish Genet Breed Sci 39:1–8

33. Sato S, Ando J, Ando H, Urawa S, Urano A, Abe S (2001) Genetic variation among Japanese populations of chum salmon inferred from the nucleotide sequences of the mitochondrial DNA control region. Zool Sci 18:99–106

34. Kitada S, Kitakado T, Kishino H (2007) Empirical Bayes inference of F ST and its distribution in the genome. Genetics 177:861–873

35. Quinn TP (2005) The behavior and ecology of Pacific salmon and trout. University of Washington Press, Seattle

36. Westley PAH, Quinn TP, Dittman AH (2013) Rates of straying by hatchery-produced Pacific salmon (Oncorhynchus spp.) and steelhead (Oncorhynchus mykiss) differ among species, life history types, and populations. Can J Fish Aquat Sci 70:735–746

37. Kitanishi S, Yamamoto Y, Edo K, Higashi S (2012) Influences of habitat fragmentation by damming on the genetic structure of masu salmon populations in Hokkaido, Japan. Conserv Genet 13:1017–1026

38. Fleming I, Hindar K, Mjølnerød IB, Jonsson B, Balstad T, Lamberg A (2000) Lifetime success and interactions of farm salmon invading a native population. Proc R Soc Lond B 267:1517–1523

39. McGinnity P, Prodöhl P, Ferguson A, Hynes R, Ó Maoiléidigh N, Baker N, Cotter D, O'Hea B, Cooke D, Rogan G, Taggart J, Cross T (2003) Fitness reduction and potential extinction of wild populations of Atlantic salmon Salmo salar as a result of interactions with escaped farm salmon. Proc R Soc Lond B 270:2443–2450

40. Araki H, Cooper B, Blouin MS (2007) Genetic effects of captive breeding cause a rapid, cumulative fitness decline in the wild. Science 318:100–103

41. Berntson EA, Carmichael RW, Flesher MW, Ward EJ, Moran P (2011) Diminished reproductive success of steelhead from a hatchery supplementation program (Little Sheep Creek, Imnaha Basin, Oregon). Trans Am Fish Soc 140:685–698

42. Christie MR, Marine ML, French RA, Blouin MS (2012) Genetic adaptation to captivity can occur in a single generation. PNAS 109:238–242

43. Thérault V, Moyer G, Jackson LS, Blouin MS, Banks MA (2011) Reduced reproductive success of hatchery coho salmon in the wild: insights into most likely mechanisms. Molec Ecol 20:1860–1869

44. Anderson JH, Faulds PL, Atlas WI, Quinn TP (2013) Reproductive success of captively bred and naturally spawned Chinook salmon colonizing newly accessible habitat. Evol Appl 6:165–179

45. Kitada S, Kishino H, Hamasaki K (2011) Bias and significance of relative reproductive success estimates based on steelhead trout (Oncorhynchus mykiss) data: a Bayesian metaanalysis. Can J Fish Aquat Sci 68:1827–1835

46. Araki H, Ardren WR, Olsen E, Cooper B, Blouin MS (2007) Reproductive success of captive-bred steelhead trout in the wild: evaluation of three hatchery programs in the Hood River. Conserv Biol 21:181–190

47. Dannewitz J, Petersson E, Dahl J, Prestegaard T, Löf AC, Järvi T (2004) Reproductive success of hatchery-produced and wild-born brown trout in an experimental stream. J Appl Ecol 41(2):355–364

48. Ford MJ, Fuss H, Boelts B, LaHood E, Hard J, Miller J (2006) Changes in run timing and natural smolt production in a naturally spawning coho salmon (Oncorhynchus kisutch) population after 60 years of intensive hatchery supplementation. Can J Fish Aquat Sci 63:2343–2355

49. Chilcote MW, Goodson KW, Falcy MR (2011) Reduced recruitment performance in natural populations of anadromous salmonids associated with hatchery-reared fish. Can J Fish Aquat Sci 68:511–522

50. Araki H, Cooper B, Blouin MS (2009) Carry-over effect of captive breeding reduces reproductive fitness of wild-born descendants in the wild. Biol Lett 5:621–624

51. Araki H, Berejikian BA, Ford MJ, Blouin MS (2008) Fitness of hatchery-reared salmonids in the wild. Evol Appl 1:342–355

52. Björnsson BT, Stefansson SO, McCormick SD (2011) Environmental endocrinology of salmon smoltification. Gen Comp Endocrinol 170:290–298

53. Roberge C, Normandeau E, Einum S, Guderley H, Bernatchez L (2008) Genetic consequences of interbreeding between farmed and wild Atlantic salmon: insights from the transcriptome. Mol Ecol 17:314–324

54. Devlin RH, Sakhrani D, Tymchuk WE, Rise ML, Goh B (2009) Domestication and growth hormone transgenesis cause similar changes in gene expression in coho salmon (Oncorhynchus kisutch). Proc Natl Acad Sci USA 106(9):3047–3052

55. Moreau DTR, Conway C, Fleming IA (2011) Reproductive performance of alternative male phenotypes of growth hormone transgenic Atlantic salmon (Salmo salar). Evol Appl 4:736–748

56. Berejikian BA, Larsen DA, Swanson P, Moore ME, Tatara CP, Gale WL, Pasley CR, Beckman BR (2012)

Development of natural growth regimes for hatchery-reared steelhead to reduce residualism, fitness loss, and negative ecological interactions. Environ Biol Fish 94:29–44

57. Nakamichi R, Kishino H, Kitada S (2013) A novel method to identify key factors of the gene regulatory network behind salmonids reproductive behavior using directed graphical modeling. N Pac Anadr Fish Comm Tec Rep 9:50–

58. Berejikian BA, Van Doornik DM, Scheurer JA, Bush R (2009) Reproductive behavior and relative reproductive success of natural- and hatchery-origin Hood Canal summer chum salmon (Oncorhynchus keta). Can J Fish Aquat Sci 66:781–789

59. Marshall TC, Slate J, Kruuk LEB, Pemberton JM (1998) Statistical confidence for likelihood-based paternity inference in natural populations. Molec Ecol 7:639–655

60. Mizuno S, Nakajima M, Naito K, Koyama T, Saneyoshi H, Kobayashi M, Koide N, Ueda H (2010) Physiological impacts of high rearing density on chum salmon Oncorhynchus keta fry. Aquaculture Sci 58:387–399

61. Heard WR (2012) Overview of salmon stock enhancement in southeast Alaska and compatibility with maintenance of hatchery and wild stocks. Environ Biol Fish 94:273–283

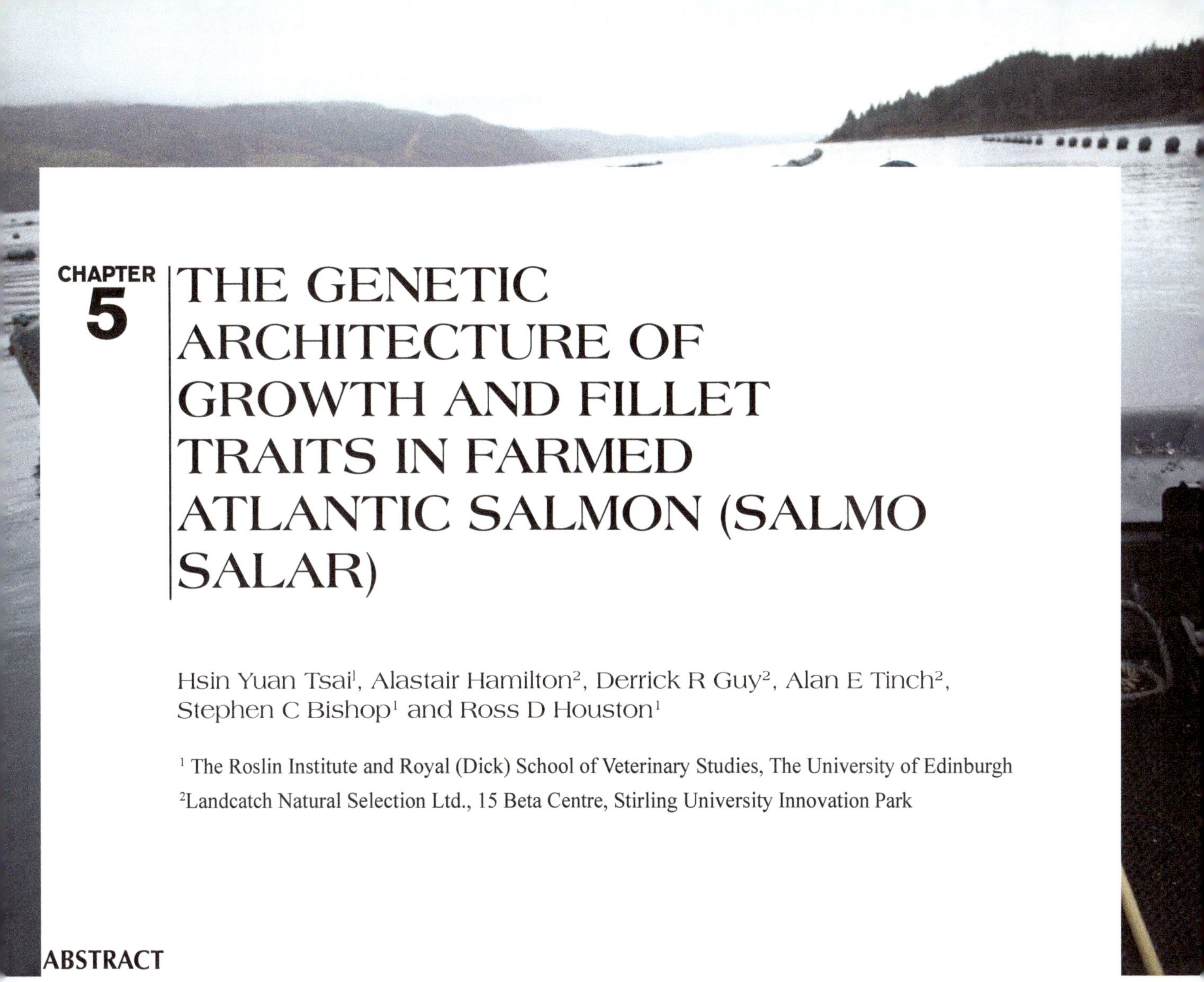

CHAPTER 5

THE GENETIC ARCHITECTURE OF GROWTH AND FILLET TRAITS IN FARMED ATLANTIC SALMON (SALMO SALAR)

Hsin Yuan Tsai[1], Alastair Hamilton[2], Derrick R Guy[2], Alan E Tinch[2], Stephen C Bishop[1] and Ross D Houston[1]

[1] The Roslin Institute and Royal (Dick) School of Veterinary Studies, The University of Edinburgh

[2]Landcatch Natural Selection Ltd., 15 Beta Centre, Stirling University Innovation Park

ABSTRACT

Background

Performance and quality traits such as harvest weight, fillet weight and flesh color are of economic importance to the Atlantic salmon aquaculture industry. The genetic factors underlying these traits are of scientific and commercial interest. However, such traits are typically polygenic in nature, with the number and size of QTL likely to vary between studies and populations. The aim of this study was to investigate the genetic basis of several growth and fillet traits measured at harvest in a large farmed salmon population by using SNP markers. Due to the marked heterochiasmy in salmonids, an efficient two-stage mapping approach was applied whereby QTL were detected using a sire-based linkage analysis, a sparse SNP marker map and exploiting low rates of recombination, while a subsequent dam-based analysis focused on the significant chromosomes with a denser map to confirm QTL and estimate their position.

Results

The harvest traits all showed significant heritability, ranging from 0.05 for fillet yield up to 0.53 for the weight traits. In the sire-based analysis, 1695 offspring with trait records and their 20 sires were successfully genotyped for the SNPs on the sparse map. Chromosomes 13, 18, 19 and 20 were shown to harbor genome-wide significant QTL affecting several growth-related traits. The QTL on chr. 13, 18 and 20 were detected in the dam-based analysis using 512 offspring from 10 dams and explained approximately 6–7 % of the within-family variation in these traits.

Conclusions

We have detected several QTL affecting economically important complex traits in a commercial salmon population. Overall, the results suggest that the traits are relatively polygenic and that QTL tend to be pleiotropic (affecting the weight of several components of the harvested fish). Comparison of QTL regions across studies suggests that harvest trait QTL tend to be relatively population-specific. Therefore, the application of marker or genomic selection for improvement in these traits is likely to be most effective when the discovery population is closely related to the selection candidates (e.g. within-family genomic selection).

Citation: Hsin Yuan Tsai, Alastair Hamilton, Derrick R Guy, Alan E Tinch, Stephen C Bishop and Ross D Houston, The genetic architecture of growth and fillet traits in farmed Atlantic salmon (Salmo salar), DOI: 10.1186/s12863-015-0215-y

BACKGROUND

Traditional selective breeding has rapidly improved economically important traits in aquaculture species, such as growth and disease resistance in aquaculture species [1]. Atlantic salmon have been more extensively studied than most other aquaculture species due to its high economic value and the significant scientific interest in salmonid species [2]. However, the genetic factors affecting some complex traits of economic importance, such as size, morphology and composition, are not yet well known. The limitations to detecting and defining these genetic factors may include a previous lack of genomic resources, the polygenic nature of the traits in question, and the relatively recent whole genome duplication (e.g. [3, 4]) in the salmonid lineage.

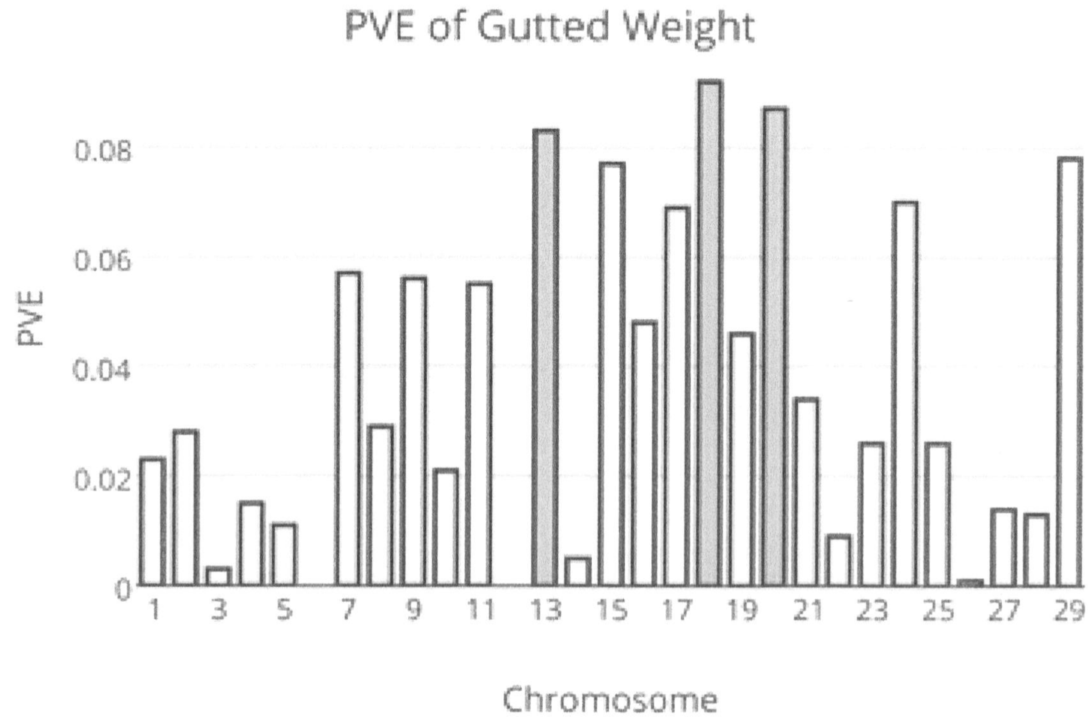

Figure. 1 The distribution of PVE according to chromosome in the sire-based analysis for the representative weight trait of gutted weight. Gray represents the chromosome showing genome-wide significance ($p<0.05$) in sire-based analysis. Chromosome 20 also showed chr-wide significance in dam-based analysis ($p<0.05$)

Genomic resources for salmonids are rich in comparison to most aquacultural species [5]. Benefitting from the development of next generation sequencing (e.g. [6]), abundant genetic markers have been discovered in most salmonid species (e.g. [7, 8, 9, 10]). Many other genomic resources and salmonid-specific databases are available, e.g. the Genomics Research on All Salmon (GRASP, http://web.uvic.ca/grasp/) and SalmonDB (http://salmondb.cmm.uchile.cl/). Furthermore, the genomes of rainbow trout [3] and Atlantic salmon [2] have been sequenced and assembled, which provide reference sequences for genomic studies of these and other salmonid species [11].

Understanding the genetic basis of phenotypic variation is a fundamental goal of biological research. Quantitative genetic analysis has been widely used to apportion variation in the traits of interest into genetic and environmental factors [12]. A further goal is to ascertain the genetic architecture of these traits, and quantitative trait loci (QTL) mapping is useful for this purpose. This approach has been widely applied in most farmed animal and plant species to improve genetic breeding programs [13, 14, 15, 16]. To date, QTL mapping relating to the growth performance of farmed salmonid species have been undertaken in Atlantic salmon [17, 18, 19, 20, 21], Coho salmon [22, 23], Arctic char [24], Chinook salmon [25] and Rainbow trout [26, 27]. The loci associated with these apparently polygenic growth traits tend to vary between studies, which may reflect population differences or gene by environment interaction.

Traits of economic interest in aquaculture species include those pertaining to the efficient production of high quality fillets. As such, overall growth rate is important, alongside the relative proportion of particular components of the fish (fillet, guts, and head, etc.). Ultimately, fillet weight is a key economic trait, and variation in this characteristic significantly depends on the proliferation and composition of white and red muscle. Muscle cell development and proliferation are part of a complex regulatory process and intricately linked with the development of the skeleton. These processes are typically controlled by networks involving many genes and biological pathways [28]. As such, a polygenic architecture of variation in this trait may be expected. Previous studies have shown that the less desirable parts of Atlantic salmon (e.g. head weight and vertebral weight) have a significant positive correlation with desirable traits such as harvest and fillet yields [29]. By detecting and selecting haplotypes at specific QTL, it may be possible to improve the proportion of fillet within the fish for any given growth rate (albeit caution should also be applied to ensure overall wellbeing and robustness of the fish).

The objective of this study was to detect and characterize QTL affecting growth and fillet characteristics in farmed salmon, using SNP markers genotyped in several large families reared under commercial aquaculture conditions. Due to the lower recombination rate observed throughout much of the genome in male salmon, compared to female salmon [30], the efficiency of QTL detection is increased by using a two stage analysis. In this strategy, QTL are first detected in a sire analysis using few markers per chromosome, and the chromosomes harbouring significant QTL are then genotyped for additional markers and analysed using dam mapping parents. Here, we use this approach with the overall target of improving understanding of the genetic regulation of growth and fillet characteristics in Atlantic salmon, and providing candidate regions for potential application in marker-based selection to capture within-family variation in these traits.

METHODS

Animals and Phenotype Measurement

A commercial salmon population comprising 198 full-sibling families derived from 136 sires and 198 dams (Landcatch Natural Selection, Ormsary, UK) was utilized in this experiment. Details of this population have been previously published [31, 32, 33]. Briefly, approximately 5000 fish were harvested at ~3 years of age and measured for overall and component weight traits: harvest weight (kg), gutted weight (kg), deheaded weight (kg), fillet weight (kg), gutted yield (%), fillet yield (%), head weight (kg), gut weight (kg), body waste weight (kg) and total waste weight (kg), fat percentage [% as estimated using a Torry Fatmeter (Distell Ltd, Aberdeen, Scotland)]; and fillet color [assessed visually using the Roche SalmoFan scale (Hoffmann-La Roche, U.K.), ranging from 20 (Yellow) to 34 (Red)]. Details of trait measurements at harvest are given in Powell *et al.* [29]. A fin clip sample of each fish was retained for DNA extraction. All animals were reared and harvested in accordance with all relevant national and EU legislation concerning health and welfare. Landcatch are accredited participants in the RSPCA Freedom Foods standard, the Scottish Salmon Producers Organization Code of Good Practice, and the EU Code-EFABAR Code of Good Practice for Farm Animal Breeding and Reproduction Organizations. The traits of fat percentage and gut weight were $\log_{10}$ transformed to approximate a normal distribution. Two generation pedigree records were available for all fish and the sex of the offspring was not observable at harvest and processing. Heritability estimates for some of the traits have been estimated previously in the larger population from which the QTL families were sampled [31, 32]. For gut, head, waste and total waste weight, the polygenic heritability was estimated in this larger population using a simple animal model, $Y_{ij} = \mu + A_i + e_{ij}$, where Y_{ij} is the trait value measured in the individual i, μ is the overall mean value of the trait, A_i is the additive genetic effect of the individual based on the pedigree information and e_{ij} is the residual error. The heritability for each of the traits was estimated using the above model, and the procedure was described in Tsai *et al.* [32].

Snp Marker Selection and Genotyping

To account for the large differences in recombination rate between male and female salmon, a two-stage QTL detection and mapping strategy was employed [30, 34]. Stage 1 used sire mapping parents (low recombination), with few markers per chromosome to detect chromosomes containing putative QTL. Stage 2 used dam mapping parents, with a denser marker coverage, to confirm QTL on significant chromosomes and estimate QTL position. For stage 1, the 20 sires in the population with the most progeny were chosen for analysis (total n=1695). The sparse panel of SNP markers described in Gonen *et al.* [35], largely taken from Moen *et al.* [36], were provided to LGC Genomics (Herts, U.K.) for the design of Kompetitive Allele Specific PCR (KASP) assays (see details athttp://www.lgcgroup.com/products/kasp-genotyping-chemistry/kasp-technical-resources/#.VVUKo_1waM8) for genotyping. From these, a total of 51 informative SNPs, with one to three SNPs per chromosome, were genotyped in all 1695 offspring (Additional file 1: Table S1). Stage 2 aimed to confirm the QTL detected in stage 1 and to estimate their position on the chromosome. Therefore, stage 2 focused on three putative QTL-containing chromosomes (chr. 13, 18, and 20) detected in stage 1. Thirty additional segregating SNP markers (Additional file1: Table S1) [9] were chosen to be positioned at regular (~10 cM) intervals across the candidate chromosome according to published linkage maps. As such, it was anticipated that this marker density would be sufficient to estimate approximate position of QTL on chromosomes. These SNPs were selected for genotyping in the ten dams with the largest number of offspring. A total of ten, eight and eight informative SNPs from chr. 13, 18 and 20 respectively were genotyped in the 512 offspring of these dams (which were a subset of the offspring genotyped offspring in stage 1).

Linkage and QTL Mapping

Sex-specific genetic maps were constructed using Crimap version 2.4 [37]. The 'prepare' option was used to create the input files (markers had previously been assigned to linkage groups based on a LOD score of >4.0), followed by the 'build' option to estimate marker order, and 'fixed' option to estimate the map distance between the markers. Where relevant, the 'flipsn' option was used to test different order permutations and determine the most likely marker order.

For both sire and dam based QTL detection, a two stage linear regression-based linkage analysis was performed using the GridQTL software [38]. The conditional probability of inheriting a particular haplotype from the sire or dam was inferred from the marker genotypes in all offspring, at 1 cM intervals. Subsequently, the trait value was regressed on the probability that a particular haplotype allele was inherited from the sire (stage 1) or the dam (stage 2). At each genomic location, the model containing a single QTL is compared to a model with no QTL resulting in an F Ratio statistic. The chromosome-wide significance thresholds for each trait were computed by permutation using 10,000 iterations per chromosome. With 29 chromosomes, the expected number of false positive was 1.45 at the 5 % significance threshold, and 0.29 at the 1 % significance threshold per genome scan respectively. The genome-wide thresholds were determined by applying the Bonferroni correction [39] to 29 independent chromosomes. In addition, in the stage 2 dam-based analysis, the confidence intervals for the QTL were estimated using bootstrapping with 10,000 permutations. In order to estimate the size of the effect of the significant QTL on the traits, the within-family variation explained by the QTL (PVE) was calculated using the following equation: $h^2_{QTL} = 4[1-(MSE_{full} / MSE_{reduced})]$ for sire-based analysis. For the dam-based analysis, because the dams were nested within sires (full-sibling families), the estimated equation was revised to $h^2_{QTL}=2[1-(MSE_{full} / MSE_{reduced})]$, where the MSE_{full} is the mean square error of the performed model including the QTL and $MSE_{reduced}$ is the model including the family mean only.

For traits related to the component weights of the fish, the QTL analyses were repeated including harvest weight as a covariate in the analysis. This was done to assess and distinguish QTL associated with an overall growth effect on the fish, versus QTL associated with proportional growth of specific components (e.g. fillet and waste, etc.)

RESULTS

Trait records of 1695 offspring derived from 20 sire families were obtained from a larger dataset of ~5000 salmon measured at harvest (~3 years old). The heritability of the weight traits was significant and consistent with previous estimated (h^2=0.52 to 0.53). For the traits not previously analysed in this population (i.e. gut, head, waste and total waste weight) the heritabilities ranged from 0.15 to 0.32. Summary statistics from the QTL-mapping offspring and population-wide estimates of heritability for these traits are given in Table 1. The weight traits showed a high phenotypic and genetic correlation (Table 2) and fitting overall harvest weight as a covariate in the animal model reduced the estimated h^2 for the component traits to 0.02 – 0.05 (although these were still significantly different from zero).

Table 1 Summary statistics and heritabilies for the phenotypes used in this study

Trait	Sample Size[†]	Mean (SD)	Heritability (SE) [29]
Harvest Weight	1524	2.57 (0.63)	0.52 (0.05)
Gutted Weight	1616	2.35 (0.58)	0.53 (0.05)
Gutted Yield	1447	0.92 (0.02)	0.04 (0.01)
Deheaded Weight	1604	2.06 (0.52)	0.52 (0.05)
Fillet Weight	1516	1.70 (0.42)	0.53 (0.05)
Fillet Yield	1363	0.66 (0.04)	0.05 (0.02)
Fat Percentage	1679	12.2 (5.58)	0.18 (0.03)
Fillet Colour	1322	29.0 (0.73)	0.14 (0.03)
Head Weight	1475	0.32 (0.08)	0.21 (0.03)[a]
Gut Weight	1447	0.42 (0.08)	0.30 (0.04)[a]
Body Waste Weight	1426	0.33 (0.12)	0.15 (0.02)[a]
Total Waste Weight	1422	0.65 (0.17)	0.32 (0.04)[a]

Gut weight (kg)=harvest weight - gutted weight; Head weight (kg)=gutted weight - deheaded weight

Waste weight (kg)=deheaded weight - fillet weight (weight of vertebrae and caudal fin); Total waste weight (kg)=head weight+body waste weight

[a]The heritability was estimated in this study and the used population was the same as Tsai *et al.* [32]

[†]: Only the number of individuals used in the calculation is shown, after removal of missing data

Table 2 Genetic and phenotypic correlation of traits using in this study

Genetic/Phenotypic	Harvest weight	Gutted weight	Gutted yield	Deheaded weight	Fillet weight	Fillet yield	Fat per-centage	Fillet colour	Gut weight	Head weight	Body waste weight	Total waste weight
Harvest Weight	-	1.00	0.16	1.00	1.00	0.35	0.84	−0.17	−0.96	0.97	1.00	0.98
Gutted Weight	1.00	-	0.19	1.00	1.00	0.33	0.83	−0.20	−0.95	0.98	0.99	0.99
Gutted Yield	−0.02	0.06	-	0.19	0.20	0.53	0.05	−0.27	0.13	0.08	0.06	0.09
Deheaded Weight	0.98	0.98	0.06	-	0.99	0.37	0.83	−0.19	−0.95	0.97	1.00	0.98
Fillet Weight	0.97	0.97	0.05	0.97	-	0.41	0.82	−0.20	−0.95	0.95	1.00	0.98
Fillet Yield	0.02	0.06	0.31	0.08	0.27	-	0.21	−0.15	−0.21	0.09	0.23	0.19
Fat Percentage	0.41	0.41	0.04	0.41	0.42	0.07	-	−0.19	−0.82	0.76	0.84	0.80
Fillet Colour	−0.08	−0.08	−0.02	−0.07	−0.08	0.03	−0.06	-	0.10	−0.24	−0.13	−0.12
Gut Weight	−0.77	−0.72	0.56	−0.71	−0.72	0.12	−0.30	0.05	-	−0.94	−0.99	−0.96
Head Weight	0.61	0.62	0.04	0.47	0.92	−0.09	0.21	−0.11	−0.45	-	0.99	1.00
Body Waste Weight	0.62	0.61	0.09	0.63	0.41	−0.65	0.25	−0.04	−0.42	0.63	-	1.00
Total Waste Weight	0.83	0.83	0.08	0.83	0.67	−0.48	0.31	−0.07	−0.59	0.88	0.93	-

In total, 51 SNP markers dispersed over all 29 chromosomes were successfully genotyped in the parents and offspring. In the sire-based QTL mapping analysis, a total of 13 chromosomes showed suggestive evidence for a QTL (chromosome-wide $p<0.05$), while four chromosomes showed a significant effect on growth-related traits at the genome-wide level (chr. 13, 18, 19, and 20; Table 3, Figure 1). The QTL typically affected several of the weight measurements and, given the high phenotypic correlations between these traits ($r \sim 0.97$-1.00), it is plausible that these results reflect single pleiotropic QTL on each chromosome, rather than distinct linked QTL. Interestingly, when harvest weight was fitted as a covariate (as a proxy for an overall measure of growth), the QTL affecting the component traits on chr. 18, 19 and 20 were no longer significant, suggesting these QTL affect overall growth of the fish. In contrast, on chr. 13, most of the QTL effects for the component traits remained after fitting the covariate, suggesting putative proportional differences in the growth of components of the fish. In addition, four new QTLs (chr. 12, 22, 23, and 25) reached chromosome-wide significance in the sire-based analysis with the inclusion of harvest weight as a covariate in the analysis (Table 3). The proportion of within-family phenotypic variance explained (PVE) varied between 8 and 10 % for the genome-wide significant QTL in the sire-based analysis, suggesting QTL of moderate but not large effect in this population.

Table 3 Results of sire-based QTL mapping analysis and proportion of phenotypic variance explained by each chromosome

	Harvest Weight		Fillet Weight				Gutted Weight				Deheaded Weight				Fillet Yield			
					covariate$				covariate$				covariate$				covariate$	
Chr	F ratio	PVE	F ratio	PVE	F ratio	PVE	F ratio	PVE	F ratio	PVE	F ratio	PVE	F ratio	PVE	F ratio	PVE	F ratio	PVE
1	-	-	1.88*	0.047	-	-	-	-	-	-	-	-	-	-	-	-	-	-
6	-	-	-	-	-	-	1.96*	0	-	-	-	-	-	-	1.90*	n.a.†	1.89*	n.a.†
7	-	-	-	-	-	-	2.03*	0.057	-	-	2.07*	0.06	1.79*	0.037	-	-	-	-
9	-	-	2.04*	0.055	-	-	2.12*	0.056	-	-	2.26*	0.062	-	-	-	-	-	-
10	-	-	2.08*	0.053	-	-	-	-	-	-	-	-	-	-	-	-	-	-
11	2.10*	0.065	2.27*	0.077	2.26**	0.071	1.96*	0.055	-	-	1.91*	0.052	-	-	2.32*	n.a.†	2.31**	n.a.†
12	-	-	-	-	-	-	-	-	-	-	-	-	2.05*	0.078	-	-	-	-
13	2.49**	0.074	2.53**	0.075	1.90*	0.07	2.78**	0.083	-	-	2.67**	0.08	1.81*	0.034	2.37*	n.a.†	2.37**	n.a.†
15	2.12*	0.076	-	-	-	-	2.24*	0.077	-	-	2.43*	0.091	-	-	-	-	-	-
16	2.23*	0.06	-	-	-	-	2.06*	0.048	-	-	2.19*	0.055	-	-	-	-	-	-
17	2.22*	0.082	2.47*	0.096	-	-	2.10*	0.069	-	-	2.37*	0.085	-	-	-	-	-	-
18	2.59**	0.083	2.76**	0.092	-	-	2.89**	0.092	-	-	2.82**	0.089	-	-	-	-	-	-
19	2.00*	0.05	2.62**	0.078	-	-	2.00*	0.046	-	-	2.00*	0.049	-	-	-	-	-	-
20	2.66**	0.09	2.42*	0.077	-	-	2.76**	0.087	-	-	2.76**	0.09	-	-	-	-	-	-
21	-	-	-	-	-	-	1.72*	0.034	1.89*	0.114	-	-	-	-	-	-	-	-
22	-	-	-	-	-	-	-	-	-	-	-	-	1.95*	0.063	-	-	-	-
23	-	-	-	-	1.86*	0.033	-	-	-	-	-	-	-	-	-	-	-	-
24	2.52*	0.068	2.58*	0.07	-	-	2.66*	0.07	2.45*	n.a.†	2.84*	0.079	2.56*	0.066	-	-	-	-
27	1.77*	0.039	-	-	-	-	-	-	-	-	-	-	-	-	-	-	-	-

28	1.88*	0.043	-	-	-	-	-	-	-	-	-	-	-	-	-	-	-	-
29	-	-	1.97*	0.06	-	-	2.31*	0.078	-	-	2.32*	0.079	-	-	-	-	-	-

	Head Weight				Gut Weight				Body Waste Weight		SalmoFan Scale		Total Waste Weight			
			covariate$				covariate$								covariate$	
Chr	F ratio	PVE	F ratio	PVE	F ratio	PVE	F ratio	PVE	F ratio	PVE	F ratio	PVE	F ratio	PVE	F ratio	PVE
1	1.76*	0.07	-	-	-	-	-	-	-	-	-	-	1.80*	0.068	-	-
3	-	-	-	-	-	-	-	-	-	-	2.30*	0.088	-	-	3.20**	0.154
6	2.12*	0.074	-	-	-	-	-	-	-	-	-	-	2.05*	0.078	-	-
7	-	-	-	-	1.73*	0.044	1.72*	0.048	2.01*	0.063	-	-	2.49*	0.075	-	-
9	2.05*	0.07	-	-	-	-	-	-	-	-	-	-	-	-	-	-
11	1.96*	n.a.†	-	-	-	-	-	-	-	-	-	-	-	-	2.00*	0.099
13	2.72**	0.072	-	-	-	-	-	-	-	-	-	-	2.44*	0.046	1.91*	0.052
16	-	-	-	-	1.98*	0.044	-	-	-	-	-	-	-	-	-	-
17	-	-	-	-	2.16*	0.093	-	-	-	-	2.11*	0.085	-	-	-	-
18	2.11*	0.069	-	-	-	-	-	-	-	-	-	-	1.88*	0.047	-	-
19	2.49**	0.068	-	-	-	-	-	-	-	-	-	-	2.22*	0.033	-	-
20	2.02*	0.067	-	-	2.19**	0.063	-	-	-	-	-	-	1.99*	0.048	-	-
21	-	-	1.72*	n.a.†	-	-	-	-	1.85*	0.03	-	-	1.76*	0.046	-	-
22	-	-	1.90*	0.4	-	-	-	-	-	-	-	-	-	-	-	-
24	2.24*	0.068	-	-	3.03**	0.094	3.13**	0.126	-	-	-	-	-	-	-	-
25	-	-	1.97*	0.4	-	-	-	-	-	-	2.04*	0.067	-	-	-	-

*: chromosome-wide significance at $p<0.05$; **: genome-wide significance at $p<0.05$; PVE: proportion of phenotypic variance for half-sib analysis

†: Due to the MSE_{full} value being equal to $MSE_{reduced}$

$: Harvest weight was fitted as covariate

Three of the genome-wide significant QTL in the sire-based analysis (chr. 13, 18, and 20) were tested in a dam-based analysis using 512 offspring from ten dams, and a denser SNP marker map of the significant chromosomes (Additional file 1: Table S1). The genome-wide significant QTL affecting gutted, deheaded and total waste weight on chr. 20 was confirmed in the dam-based analysis and mapped to a best estimated position of 21, 19 and 14 cM respectively, although the 95 % confidence intervals encompassed the entire linkage map for this chromosome (Table 4). The evidence for QTL on chr. 13 and 18 was not as strong in the dam-based analysis, with only gut weight (chr. 13) and gutted weight (chr. 18) showing chromosome-wide significance (in the analysis with harvest weight included as a covariate). For the chr. 20 QTL, there were three sires and three dams segregating for a QTL affecting at least one weight trait, and the average size of the allelic substitution effect for deheaded weight of the salmon in segregating parents was consistent across all segregating parents, with an average effect of 620 grams (Table 5).

Table 4 Results of dam-based QTL mapping analysis and proportion of phenotypic variance explained for significant trait/chromosome combinations

Chr	Trait	Dam F-ratio	PVE	Average QTL position (cM)	95 % C.I. for QTL Position (cM)
20	Gutted Weight	2.48*	0.06	20.8	0.0-43.0
	Deheaded Weight	2.71*	0.07	19.4	
	Total Waste Weight	2.35*	0.06	14.0	
	Body Waste Weight	2.18*	0.06	12.4	0.0 - 40.0
13[†]	Gut Weight	2.49*	0.07	42.0	0.0-64.0
18[†]	Gutted Weight	2.62*	0.07	20.2	0.0 - 39.0

*: chromosome-wide significance at $p<0.05$; †: QTLs found in the analysis fitting harvest weight as covariate

PVE: proportion of phenotypic variance for full-sib analysis

Table 5 The QTL effect on growth traits and associated absolute T values in segregating individual parents for the significant QTL at chr. 20

Sire-based analysis	Traits	QTL effect estimate (SE)[a] (g)	Absolute T value
J9L2M0088	Harvest Weight	−580 (170)	3.43
	Fillet Weight	−430 (110)	4.01
	Gutted Weight	−650 (150)	4.41
	Deheaded Weight	−580 (130)	4.41
	Head Weight	−80 (20)	3.61
	Total Waste Weight	−140 (50)	2.98
J9L2M0091	Harvest Weight	−360 (160)	2.19
J9L3M3080	Total Waste Weight	170 (70)	2.29
Dam-based analysis			
J9L2F0144	Gutted Weight	570 (170)	3.36
	Deheaded Weight	480 (150)	3.28
	Total Waste Weight	140 (50)	2.53
J9L2F1295	Gutted Weight	590 (250)	2.35
	Deheaded Weight	480 (200)	2.35
	Body Waste Weight	200 (60)	3.56
	Total Waste Weight	270 (80)	3.52
J9L2F0695	Deheaded Weight	−940 (400)	2.32

[a] The sign + or – is arbitrary when compared across families but indicates the direction of the allelic effect within families (e.g. an allele decreasing harvest weight in sire J9L2M0088 also decreased fillet, gutted, deheaded, head and total waste weight)

DISCUSSION

In this study, the genetic basis and architecture of growth-related traits was investigated in a large commercial population of Atlantic salmon using a two-stage QTL mapping approach. All traits measured showed significant evidence for heritability and significant weight-related QTLs were observed on chr. 13, 18, 19 and 20 in the sire-based analysis. These QTL typically affected several of the weight measurements taken at harvest, which reflects the high positive correlation between these traits and suggests that their effect is related to overall size of the fish. However, the QTL on chr. 13 may have effects on the weight of components of the fish independent of an overall growth effect, as indicated by an analysis including harvest weight as a covariate.

A QTL affecting several of the growth-related traits on chr. 20 was confirmed in the dam-based analysis. This chromosome has previously been shown to harbor QTL affecting body weight of Atlantic salmon at younger age (10 months; [16]). However, a comparison of the QTL detected in the current study with those observed in previous studies (Table 6) shows that, even amongst populations of salmon measured at similar age, QTL tend to be rather population-specific. This may reflect differing underlying quantitative trait nucleotide affecting growth of the populations, genotype by environment interaction, or simply that a proportion of QTL identified in most studies are likely to be 'false positives'. The weight traits measured at harvest had high positive genetic and phenotypic correlations (r~0.97-1.00 in phenotypic and ~0.99-1.00 in genetic correlation), and this is generally reflected in the QTL results, because individual QTL tended to affect the weight of several components of the fish. This is a phenomenon observed in several other studies (e.g. [18]) and suggests that improvement of the growth of all components of the fish in breeding programs can be made by simply measuring overall harvest weight. This will improve harvest weight, the most important trait, but is likely to also improve potentially undesirable traits such as gut weight. Achieving different rates of gain in individual components of the fish using QTL or conventional family-based selection is likely to be challenging and may require more detailed or accurate measures of these component traits. However, the existence of QTL affecting fillet weight seemingly independent of overall harvest weight (e.g. chr. 11) suggests that there are potentially some genes affecting component traits partially independently of harvest weight that could be targets for further study.

Table 6 Comparison of harvest weight QTL chromosomes in Atlantic salmon from this and previous studies

	Gutierrez *et al.* [17]		**Baranski *et al.* [18]**	**Houston *et al.* [19]**	**This study**	
				Sire	**Dam**	
Chr/Age	~27 months	~38 months	~36 months	~30 months	~36 months	
1		C	C			
2	G	C		C/G		
3				C		
4			G			
5	C	C	G			
6				C		
7			C			
8	C			C		
9	C					
10	C		G			
11			C		C	
12						
13			C	C/G	G	
14						

15	C	C			C	
16		C	C		C	
17					C	
18			C		G	C
19	C				C	
20					G	C
21	C			C		
22			C			
23			C			
24					C	
25		C	C			
26		G				
27		C			C	
28					C	
29	C		C			

C: chromosome-wide significance; G: genome-wide significance

Atlantic salmon are closely related to rainbow trout and previous studies in trout have reported several QTLs affecting body mass [25, 40-41]. There was some overlap between these QTL and the genome-wide significant QTL identified in the current study, in particular for body mass QTL mapped to trout chromosomes 1q and 16q/12p [26], chr. 9p and 21p [42] and chr. 16q [40], which correspond to chr. 13 and 18 in salmon. In addition, corresponding QTL regions showing chromosome-wide significance with body weight were also discovered between Chinook salmon (chr. 25) [25] and Atlantic salmon (chr. 28) (this study). These results raise the possibility that some of the QTL affecting complex growth traits may be conserved across salmonid species. However, clearly some overlap between studies will occur by chance and the likelihood of the underlying QTL being common in both species will become more apparent with further studies and a finer mapping resolution. The confidence intervals associated with the QTL in the current study were large which precluded the meaningful identification of potential underlying candidate genes. However, known candidate genes explaining a small percentage of variation in growth in this population (*myostatin* [31] and *IGF1* [32]) do not coincide with the QTL identified here.

The size of the QTL effects in the current study was typically around 5-9 % and 6-7 % of the within-family phenotypic variance in the sire and dam-based analysis respectively. While this may be an overestimate due to the Beavis effect [41], it is certainly plausible that markers linked to these QTL may be of use in selective breeding programs. However, the confidence intervals were large and this indicates that while the two-stage mapping approach employed appears to be effective at detecting QTL, the fine mapping to smaller chromosome regions in the dam analysis may benefit from additional markers. The results of this and other studies support the hypothesis that complex traits such as weight are polygenic, which may reflect the involvement of diverse regulation pathways related to energy balance, muscle cell proliferation and skeletal growth. The fact that the proportion of variation explained by the QTL is smaller than in previous studies (e.g. [19]) is probably due to the large sample size of the current study (i.e. ~1700 offspring for the sire-based analysis), and hence potentially more reliable estimates of QTL effect size [41]. Further, the two-step approach provided some degree of within-study validation for the detected QTL on chr. 18. The traits of most commercial interest in salmon production, such as fillet weight were affected by the QTL on chr. 13, 18, 19, and 20 (genome-wide significance) in the sire-based analysis. Notably, except chr. 19 in sire-based analysis - for which further study may be merited - all of these QTL regions showed a significant effect on gutted weight and deheaded weight.

No QTL affecting fat content were detected in our study. Interestingly, components of fat content of salmon, such as n-3 long chain polyunsaturated acid, are highly heritable [43]. Therefore, perhaps more consideration could be given to the investigation of the genetic architecture of the specific components of the fat content of the fillet, as opposed to a more crude overall measure of fat levels. Naturally, this refinement of phenotype would incur a greater cost. In addition, only three QTL (chr. 3, 7 and 26) were shown to affect fillet colour at the chromosome-wide significance level. Chromosomes 3 and 26 have previously been suggested to harbor QTL associated with fillet colour traits [18]. The heritability of this trait is relatively low in this study ($h^2 \sim 0.1$ - 0.2) when compared with weight related traits in Atlantic salmon [44], although recently published studies have given higher heritabilities [45] and fillet color has been suggested to show a significant association with a single locus SCAR marker [46]. It has also been suggested that fillet colour is positively correlated with overall body weight in farmed Coho salmon ($r \sim 0.4 \pm 0.5$) [46] and Atlantic salmon ($r \sim 0.49$) [47]. This may be related to the inclusion of dietary additives such as astaxanthin, canthaxanthin and carotenoid, which are included in feed to enhanced fillet pigmentation [48]. As such, protein/muscle gains may be accompanied by an associated increase in colour additives. However, we did not observe a correlation between harvest weight and fillet colour in our study. In part, this may be due to a lack of fillet colour variation observed in the population (coefficient of variation ~ 0.025). Of the putative colour QTL in the current study, only chr. 17 showed some evidence for an effect on growth-related traits, while chr. 3 and 26 were associated with fillet colour independent of the other traits measured. Given the economic importance of this trait, further study of these putative QTL and other aspects of the genetic regulation of colour are merited.

Marker-assisted selection has been applied in the salmon aquaculture industry for several traits, the foremost example being resistance to the Infectious Pancreatic Necrosis virus [34, 49, 50]. However, the genetic architecture of resistance to this disease was unusually monogenic, with a single QTL explaining most of the genetic variation. For more typical complex traits such as growth or fillet component traits, the optimal use of markers in selective breeding programs has yet to be established. Clearly, the advantages of using markers in selection for aquaculture are maximal where the traits are difficult or impossible to measure on the selection candidates themselves, and some of the harvest traits fall into this category. However, due to the lack of large-effect QTL and the putative population-specificity of those QTL, it is unlikely that QTL-targeted, across population marker-assisted selection will be a highly effective tool for breeding. With the recent development of high density SNP arrays (e.g. [7]), genomic selection may be a more effective (albeit expensive) means of capturing variation at QTL of small effect, but is likely to be the most effective when the training and selection population are closely related. Within family genomic selection using lower marker density may be a more cost-effective method of capturing the within-family genetic variation associated with QTLs that are relatively population-specific [51]. The large full-sibling family sizes and routine sib-testing in salmon breeding schemes makes such approaches feasible and powerful.

CONCLUSIONS

This study investigated the genetic basis of traits measured at harvest in a large commercial population of Atlantic salmon. The traits showed significant heritability and four genome-wide significant QTL were identified on chr. 13, 18, 19 and 20. The QTL on chr. 20 had relatively large effects on several weight-related traits that were consistent in the sire and dam analysis. The abundant putative QTLs provide a broad view of the genetic architecture of body weight and component traits in salmon. It is likely that weight traits in salmon are controlled by a finite number of partially population-specific loci of moderate-effect, in addition to a large polygenic component. These factors should be accounted for when considering the optimal methods of applying genomic markers in selective breeding programs.

ACKNOWLEDGEMENTS

The authors acknowledge funding from the Biotechnology and Biological Sciences Research Council (BBSRC) via grant number BB/H022007/1 and an Institute Strategic Funding Grant to The Roslin Institute, and the Studying Funding from Ministry of Education, Taiwan. We also thank Oswald Matika for assistance with the data analysis.

AUTHORS' CONTRIBUTIONS

RDH and SCB proposed and designed the experiments. Tissue samples and phenotype measurements were supported by AET, DRG and AH. Data analyses were performed by HYT and RDH. Manuscript was written by HYT and RDH. All authors read and approved the final manuscript.

REFERENCES

1. Gjedrem T, Robinson N, Rye M. The importance of selective breeding in aquaculture to meet future demands for animal protein: A review. Aquaculture. 2012;350–353:117–29.
2. Davidson WS, Koop BF, Jones SJM, Iturra P, Vidal R, Maass A, et al. Sequencing the genome of the Atlantic salmon (Salmo salar). Genome Biol. 2010;11:403.
3. Berthelot C, Brunet F, Chalopin D, Juanchich A, Bernard M, Noël B, et al. The rainbow trout genome provides novel insights into evolution after whole-genome duplication in vertebrates. Nat Commun. 2014;5:3657.
4. Macqueen DJ, Johnston IA. A well-constrained estimate for the timing of the salmonid whole genome duplication reveals major decoupling from species diversification A well-constrained estimate for the timing of the salmonid whole genome duplication reveals major decoupling from spe. Proc R Soc B. 2014;281:20132881.
5. Yáñez JM, Houston RD, Newman S. Genetics and genomics of disease resistance in salmonid species. Front Genet. 2014;5(November):415.
6. Baird NA, Etter PD, Atwood TS, Currey MC, Shiver AL, Lewis ZA, et al. Rapid SNP discovery and genetic mapping using sequenced RAD markers. PLoS One. 2008;3:e3376.
7. Houston RD, Taggart JB, Cézard T, Bekaert M, Lowe NR, Downing A, et al. Development and validation of a high density SNP genotyping array for Atlantic salmon (Salmo salar). BMC Genomics. 2014;15:90.
8. Palti Y, Gao G, Liu S, Kent MP, Lien S, Miller MR, *et al.* The Development and Characterization of a 57 K

SNP Array for Rainbow Trout. Mol Ecol Resour. 2015;15(3):662–72.

9. Lien S, Gidskehaug L, Moen T, Hayes BJ, Berg PR, Davidson WS, et al. A dense SNP-based linkage map for Atlantic salmon (Salmo salar) reveals extended chromosome homeologies and striking differences in sex-specific recombination patterns. BMC Genomics. 2011;12:615.
10. Gonen S, Lowe NR, Cezard T, Gharbi K, Bishop SC, Houston RD. Linkage maps of the Atlantic salmon (Salmo salar) genome derived from RAD sequencing. BMC Genomics. 2014;15:166.
11. Danzmann RG, Cairney M, Davidson WS, Ferguson MM, Gharbi K, Guyomard R, et al. A comparative analysis of the rainbow trout genome with 2 other species of fish (Arctic charr and Atlantic salmon) within the tetraploid derivative Salmonidae family (subfamily: Salmoninae). Genome. 2005;1051:1037–51.
12. Weller JI. Quantitative Trait Loci Analysis in Animals. 2nd ed. London: CABI Publishing; 2009.
13. Bishop SC, Woolliams JA. Genomics and disease resistance studies in livestock. Livest Sci. 2014;166:190–8.
14. Mittler R, Blumwald E. Genetic engineering for modern agriculture: challenges and perspectives. Annu Rev Plant Biol. 2010;61:443–62.
15. Khatkar MS, Thomson PC, Tammen I, Raadsma HW. Quantitative trait loci mapping in dairy cattle: review and meta-analysis. Genet Sel Evol. 2004;36:163–90.
16. Yue GH. Recent advances of genome mapping and marker-assisted selection in aquaculture. Fish Fish. 2014;15(3):376–96.
17. Gutierrez AP, Lubieniecki KP, Davidson EA, Lien S, Kent MP, Fukui S, et al. Genetic mapping of quantitative trait loci (QTL) for body-weight in Atlantic salmon (Salmo salar) using a 6.5K SNP array. Aquaculture. 2012;358–359:61–70.
18. Baranski M, Moen T, Våge DI. Mapping of quantitative trait loci for flesh colour and growth traits in Atlantic salmon (Salmo salar). Genet Sel Evol. 2010;42:17.
19. Houston RD, Bishop SC, Hamilton A, Guy DR, Tinch AE, Taggart JB, et al. Detection of QTL affecting harvest traits in a commercial Atlantic salmon population. Anim Genet. 2009;40:753–5.
20. Boulding EG, Culling M, Glebe B, Berg PR, Lien S, Moen T. Conservation genomics of Atlantic salmon: SNPs associated with QTLs for adaptive traits in parr from four trans-Atlantic backcrosses. Heredity (Edinb) 2008;101:381–91.
21. Reid DP, Szanto A, Glebe B, Danzmann RG, Ferguson MM. QTL for body weight and condition factor in Atlantic salmon (Salmo salar): comparative analysis with rainbow trout (Oncorhynchus mykiss) and Arctic charr (Salvelinus alpinus). Heredity (Edinb) 2005;94:166–72.
22. McClelland EK, Naish KA. Quantitative trait locus analysis of hatch timing, weight, length and growth rate in coho salmon, Oncorhynchus kisutch. Heredity (Edinb). 2010;105:562–73.
23. O'Malley KG, McClelland EK, Naish KA. Clock genes localize to quantitative trait loci for stage-specific growth in juvenile coho salmon, Oncorhynchus kisutch. J Hered. 2010;101:628–32.
24. Küttner E, Moghadam HK, Skúlason S, Danzmann RG, Ferguson MM. Genetic architecture of body weight, condition factor and age of sexual maturation in Icelandic Arctic charr (Salvelinus alpinus). Mol Genet Genomics. 2011;286:67–79.
25. Everett MV, Seeb JE. Detection and mapping of QTL for temperature tolerance and body size in Chinook salmon (Oncorhynchus tshawytscha) using genotyping by sequencing. Evol Appl. 2014;7:480–92.
26. Wringe BF, Devlin RH, Ferguson MM, Moghadam HK, Sakhrani D, Danzmann RG. Growth-related quantitative trait loci in domestic and wild rainbow trout (Oncorhynchus mykiss). BMC Genet. 2010;11:63.
27. Haidle L, Janssen JE, Gharbi K, Moghadam HK, Ferguson MM, Danzmann RG. Determination of quantitative trait loci (QTL) for early maturation in rainbow trout (Oncorhynchus mykiss). Mar Biotechnol (NY). 2008;10:579–92.
28. Shi J, Luo L, Eash J, Ibebunjo C, Glass DJ. The SCF-Fbxo40 complex induces IRS1 ubiquitination in skeletal muscle, limiting IGF1 signaling. Dev Cell. 2011;21:835–47.
29. Powell J, White I, Guy D, Brotherstone S. Genetic parameters of production traits in Atlantic salmon (Salmo salar). Aquaculture. 2008;274:225–31.
30. Hayes BJ, Gjuvsland A, Omholt S. Power of QTL mapping experiments in commercial Atlantic salmon populations, exploiting linkage and linkage disequilibrium and effect of limited recombination in males. Heredity (Edinb). 2006;97:19–26.
31. Peñaloza C, Hamilton A, Guy DR, Bishop SC, Houston RD. A SNP in the 5′ flanking region of the myostatin-1b gene is associated with harvest traits in Atlantic salmon (Salmo salar). BMC Genet. 2013;14:112.
32. Tsai HY, Hamilton A, Guy DR, Houston RD. Single nucleotide polymorphisms in the insulin-like growth factor 1 (IGF1) gene are associated with growth-related traits in farmed Atlantic salmon. Anim Genet. 2014;45:709–15.
33. Guy DR. Genetic Resistance to Infectious Pancreatic Necrosis virus in Pedigreed Atlantic salmon (Salmo salar), PhD thesis: University of Edinburgh. 2011. p. 184.
34. Houston RD, Haley CS, Hamilton A, Guy DR, Tinch AE, Taggart JB, et al. Major quantitative trait loci affect resistance to infectious pancreatic necrosis in Atlantic salmon (Salmo salar). Genetics. 2008;178:1109–15.
35. Gonen S, Baranski M, Thorland I, Norris A, Grove H, Arnesen P, *et al.* Mapping and validation of a major QTL affecting the resistance to pancreas disease (salmonid alphavirus) in Atlantic salmon (Salmo salar). Heredity 2015, In press.
36. Moen T, Hayes B, Baranski M, Berg PR, Kjøglum S, Koop BF, et al. A linkage map of the Atlantic salmon

(Salmo salar) based on EST-derived SNP markers. BMC Genomics. 2008;9:223.

37. Green P, Falls K, Crooks S. Documentation for CRI-MAP, Version 2.4. USA: Washington School of MedicineSt Louis, MO; 1990. p. 63110.

38. Seaton G, Hernandez J, Grunchec J, White I, Allen J, Koning D, Wei W, Berry D, Haley C, Knott S. 2006. (http://www.ngs.ac.uk)

39. Knott SA, Marklund L, Haley CS, Andersson K, Davies W, Ellegren H, et al. Multiple marker mapping of quantitative trait loci in a cross between outbred wild boar and large white pigs. Genetics. 1998;149:1069–80.

40. O'Malley KG. Quantitative trait Loci for spawning date and body weight in rainbow trout: testing for conserved effects across ancestrally duplicated chromosomes. J Hered. 2003;94:273–84.

41. Beavis W. The Power and Deceit of QTL Experiments: Lessons from Comparative QTL Studies. In: Proceedings of the Forty-Ninth Annual Corn & Sorghum Industry Re- search Conference. Washington, DC: American Seed Trade Association; 1994. p. 250–66.

42. Martyniuk CJ, Perry GML, Mogahadam HK, Ferguson MM, Danzmann RG. The genetic architecture of correlations among growth-related traits and male age at maturation in rainbow trout. J Fish Biol. 2003;63:746–64.

43. Leaver MJ, Taggart JB, Villeneuve L, Bron JE, Guy DR, Bishop SC, et al. Heritability and mechanisms of n-3 long chain polyunsaturated fatty acid deposition in the flesh of Atlantic salmon. Comp Biochem Physiol Part D Genomics Proteomics. 2011;6:62–9.

44. Norris AT, Cunningham EP. Estimates of phenotypic and genetic parameters for flesh colour traits in farmed Atlantic salmon based on multiple trait animal model. Livest Prod Sci. 2004;89:209–22.

45. Odegård J, Moen T, Santi N, Korsvoll S a, Kjøglum S, Meuwissen THE. Genomic prediction in an admixed population of Atlantic salmon (Salmo salar). Front Genet. 2014;5(November):402.

46. Araneda C, Neira R, Iturra P. Identification of a dominant SCAR marker associated with colour traits in Coho salmon (Oncorhynchus kisutch). Aquaculture. 2005;247:67–73.

47. Quinton CD, McMillan I, Glebe BD. Development of an Atlantic salmon (Salmo salar) genetic improvement program: Genetic parameters of harvest body weight and carcass quality traits estimated with animal models. Aquaculture. 2005;247:211–7.

48. Albert G.J. Tacon. The Nutrition and Feeding of Farmed Fish and Shrimp - a Training Manual. Rome: FAO 1987.

49. Moen T, Baranski M, Sonesson AK, Kjøglum S. Confirmation and fine-mapping of a major QTL for resistance to infectious pancreatic necrosis in Atlantic salmon (Salmo salar): population-level associations between markers and trait. BMC Genomics. 2009;10:368.

50. Houston RD, Haley CS, Hamilton A, Guy DR, Mota-Velasco JC, Gheyas AA, et al. The susceptibility of Atlantic salmon fry to freshwater infectious pancreatic necrosis is largely explained by a major QTL. Heredity (Edinb). 2010;105:318–27.

51. Lillehammer M, Meuwissen THE, Sonesson AK. A low-marker density implementation of genomic selection in aquaculture using within-family genomic breeding values. Genet Sel Evol. 2013;45:39.

CHAPTER 6

A HIGHLY REDUNDANT BAC LIBRARY OF ATLANTIC SALMON (SALMO SALAR): AN IMPORTANT TOOL FOR SALMON PROJECTS

Jim Thorsen[1,2], Baoli Zhu[2], Eirik Frengen[3,4], Kazutoyo Osoegawa[2], Pieter J de Jong[2], Ben F Koop[5], William S Davidson[6] and Bjørn Høyheim[1,7]

[1]Department of Basic Sciences and Aquatic Medicine, Section of Genetics, Norwegian School of Veterinary Science, Dep, N-0033 Oslo, Norway,

[2]BACPAC Resources, Children's Hospital Oakland Research Institute, Oakland, CA, USA,

[3]Biotechnology Centre of Oslo, University of Oslo, Oslo, Norway,

[4]Institute of Medical Genetics, University of Oslo, Blindern, N-0315 Oslo, Norway,

[5]Centre for Biomedical Research, Department of Biology, University of Victoria, Victoria, BC, V8P 5C2, Canada, [6]Department of Molecular Biology and Biochemistry, Simon Fraser University, Burnaby, BC, V5A 1S6, Canada and [7]Department of Dermatology, National Hospital, Rikshospitalet, N- 0027 Oslo, Norway

ABSTRACT

Background

As farming of Atlantic salmon is growing as an aquaculture enterprise, the need to identify the genomic mechanisms for specific traits is becoming more important in breeding and management of the animal. Traits of importance might be related to growth, disease resistance, food conversion efficiency, color or taste. To identify genomic regions responsible for specific traits, genomic large insert libraries have previously proven to be of crucial importance. These large insert libraries can be screened using gene or genetic markers in order to identify and map regions of interest. Furthermore, large-scale mapping can utilize highly redundant libraries in genome projects, and hence provide valuable data on the genome structure.

Results

Here we report the construction and characterization of a highly redundant bacterial artificial chromosome (BAC) library constructed from a Norwegian aquaculture strain male of Atlantic salmon (*Salmo salar*). The library consists of a total number of 305 557 clones, in which approximately 299 000 are recombinants. The average insert size of the library is 188 kbp, representing 18-fold genome coverage. High-density filters each consisting of 18 432 clones spotted in duplicates have been produced for hybridization screening, and are publicly available [1]. To characterize the library, 15 expressed sequence tags (ESTs) derived overgos and 12 oligo sequences derived from microsatellite markers were used in hybridization screening of the complete BAC library. Secondary hybridizations with individual probes were performed for the clones detected. The BACs positive for the EST probes were fingerprinted and mapped into contigs, yielding an average of 3 contigs for each probe. Clones identified using genomic probes were PCR verified using microsatellite specific primers.

Citation: Thorsen, J et al. (2005). A highly redundant BAC library of Atlantic salmon (Salmo salar): an important tool for salmon projects. BMC Genomics, v. 6:50. DOI: 10.1186/1471-2164-6-50.

CONCLUSION

Identification of genes and genomic regions of interest is greatly aided by the availability of the CHORI-214 Atlantic salmon BAC library. We have demonstrated the library's ability to identify specific genes and genetic markers using hybridization, PCR and fingerprinting experiments. In addition, multiple fingerprinting contigs indicated a pseudo-tetraploidity of the Atlantic salmon genome. The highly redundant CHORI-214 BAC library is expected to be an important resource for mapping and sequencing of the Atlantic salmon genome.

BACKGROUND

Atlantic salmon (*Salmo salar*) is a fish native to the basin of the North Atlantic Ocean. As farming of Atlantic salmon is growing as an agricultural enterprise, identification of genetic regions and genes responsible for economically important traits might be of importance for future agriculture of salmon.

Genome projects in farm animals aim at identifying regions of the genome responsible for traits of economic importance in order to implement this into breeding and management programs. This is for instance achieved by performing QTL studies to identify genomic regions that are linked to traits of economic importance. Genetic and physical maps, publicly available databases and cDNA libraries from various tissues are a few of the tools needed to identify genomic regions responsible for traits of interest. Currently there are two published linkage maps for Atlantic salmon [2, 3] and several EST databases available to the public [4, 5,6, 7, 8]. After a genomic region has been shown to be linked to a trait of economic importance, a high quality BAC library resource is crucial in the identification and functional characterization of the genetic variation.

Atlantic salmon along with all fish of the family *Salmonidae* shows residual tetraploidity after a duplication event that occurred 25–100 Myr ago [9]. However, rapid chromosome divergence has been observed between different salmonid species, involving Robertsonian changes as well as other structural rearrangements [9]. Hence, the pseudo-tetraploid state is challenging for researchers working with the Atlantic salmon genome.

Here we describe the construction and characterization of a highly redundant BAC library of Atlantic salmon. To our knowledge, this is the first reported Atlantic salmon BAC library, which will be an important tool for constructing physical maps and in the identification and sequencing of regions of the Salmon genome. The large insert BAC clones will be useful as to provide better understanding of the pseudo-tetraploid state, identify regions of interest to the aquaculture industry as well as to the basic science community.

RESULTS AND DISCUSSION

The current BAC library, CHORI 214, consists of three segments. The average insert size was estimated based upon analysis of NotI digested DNA isolated from 249, 218, and 220 clones from segment 1, 2 and 3 respectively. The distribution of insert sizes is presented in Figure 1. Based on these results, the average insert size was estimated to be 188 kbp. With a total of approximately 299 000 clones the BAC library provides 18.8 haploid genome equivalents (see Table 3) given a genome size of 3.27 pg (6.55 pg/N) [10] or similar 3.10 pg at the Genome Size Database [11].

Table 1 Details of the three segments of the BAC library

Segment	1	2	3	All segments
Plate numbers	1–288	289–576	577–816	1–816
Plate count	288	288	240	816
Empty wells	1982 (1.79%)	2830 (2.56%)	2975 (3.23%)	7787 (2.49%)
Non-recombinant clones	0	5	4	9
Non-insert clones [1]	1.2% [2]	2.8% [3]	2.3% [4]	2.0%
Recombinant clones [5]	~107 000	~105 000	~87 000	~299 000
Average insert size	189 Kbp	190 Kbp	186 Kbp	188 Kbp
Genomic coverage	6.8X	6.6X	5.4X	18.8X

Genomic DNA obtained from Salmon sperm was partially digested with EcoRI restriction enzyme and cloned in pTARBAC2.1 vector.

[1] Estimated numbers for each segment is based upon NotI digestion.

[2] 3 out of 249 clones were identified as non-insert clones

[3] 6 out of 218 clones were identified as non-insert clones

[4] 5 out of 220 clones were identified as non-insert clones

[5] Recombinant clones have been estimated from the total well number subtracting non-insert clones, non-recombinant clones and empty wells.

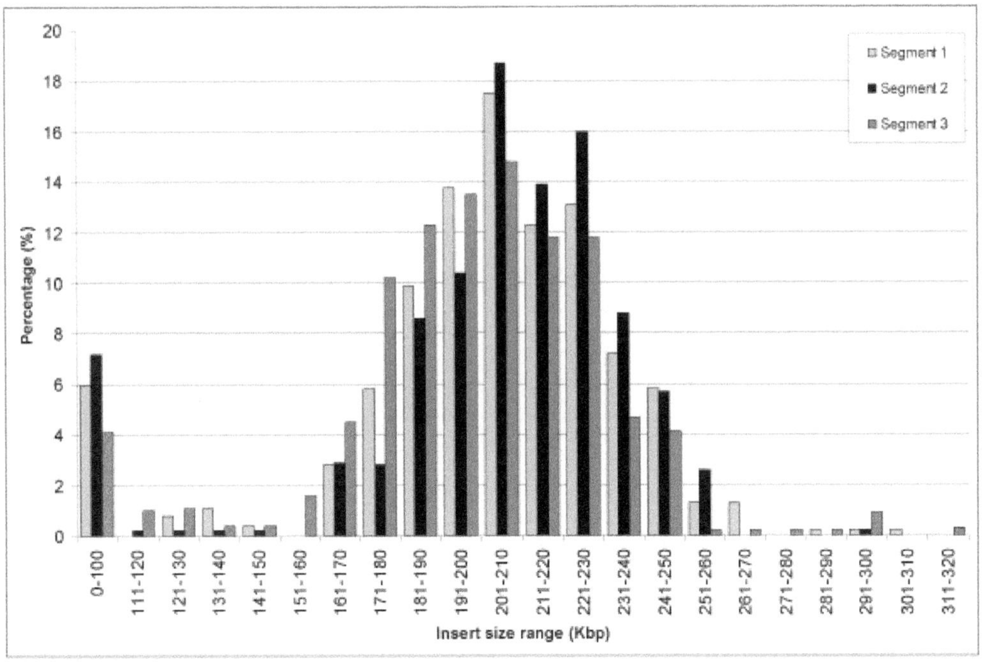

Figure 1 Size distribution of BAC clone insert sizes in the CHORI-214 library. A total of 249, 218, and 220 clones from each of the segment 1–3 was digested with NotI and analyzed by PFGE. The horizontal axis refers to the range of insert sizes, and the vertical axis indicates percentage of clones corresponding to each size range in each of the three segments. The grey tones represent the segments as indicated

In the construction of the library, great care was made to estimate the amount of enzymes to be used for the digestion and dephosphorylation of the vector. Excessive amounts of either enzyme showed an increase in ligation products containing non-insert clones with a smaller vector fragment. These products are possibly due to star activity and deletion in the SacB gene or its promoter region, causing the clones to survive on media containing sucrose. The number of non-insert clones were estimated to represent about 2% of the library based upon the results from the NotI digestion analysis (see Table 3). Nine non-recombinant clones, which contain the complete BAC vector only, were detected in the entire library by hybridization using pUC19 DNA onto the filter sets. Reducing the frequency of non-recombinant and non-insert clones is important as these clones can easily dominate the clone collection of the library due to reduced transformation efficiency of large DNA fragments in *E. coli.* High frequency of non-insert and non-recombinant clones will increase the cost for storage and utilization of the library. In addition, large scale random fingerprinting and sequencing will have a significant reduced cost using libraries with low frequency of non-insert and non-recombinant clones.

To further characterize the library, a set of 34 oligo probes generated from sequence flanking known non-linked microsatellite markers was selected to be used in screening of the 17 BAC library colony filters. These filters contain the complete BAC library gridded in duplicate. In an attempt to reduce the probability of these probes to contain repeat sequences, each probe was hybridized to one BAC filter in order to check the quality of the probe as a unique marker. From these test hybridizations, in combination with Southern blots of genomic DNA digested with TaqI, PstI and MspI, 12 probes were characterized as unique with 0 to 4 signals onto a single BAC filter and single bands onto the genomic filters (results not shown). The three enzymes TaqI, PstI and MspI were chosen because of a high frequency of restriction sites for these enzymes in the salmon genome. As many as 20 of the 32 probes yielded several hundreds to thousands of positive double signals onto one BAC filter and multiple bands or smear onto the genomic southern blots indicating that the probes contained repetitive sequence elements.

The 12 probes, which were assumed not to contain repeat elements based on the initial hybridization results, were hybridized as one pool to the filters and resulted in the identification of 396 positive BAC clones. To identify the relationships between the probes and the clones, a second hybridization experiment was performed for each probe individually onto new small nylon filters containing all 396 clones. Out of the 396 BAC clones identified in the primary screening, 233 were verified in these secondary screening experiments (Table 1). In addition, all the clones that were not assigned to a probe in the second hybridization experiments, possible false positives, were analyzed individually in PCR experiments with all the microsatellite PCR primer pairs, resulting in the assignment of an additional 91 BAC clones to the microsatellite markers (Table 1). See additional file 1 for a complete data set of BAC clones linked to each microsatellite marker.

Table 2 Number of BAC clones detected after hybridization and PCR verification of probes flanking microsatellites

GenBank accession no.	Mic.sat. clone name	No. of clones observed in secondary hybridisation	Verified by PCR	Positive PCR product on unassigned clones	Total no. of clones verified by PCR
AF256671	BHMS-175	28	28	7	35
AF256676	BHMS-189	24	21	5	26
AF256678	BHMS-201	9	9	4	13
AF256741	BHMS-289	19	18	5	23
AF256698	BHMS-304	19	18	6	24
AF256748	BHMS-330	12z	12	1	13
AF256746	BHMS-349	23	23	4	27
AF256719	BHMS-429	42	42	29	71
AF256693	BHMS-278	8	*	*	*
AF256848	BHMS-255	24	#	#	#
AF256750	BHMS-337	21	5	29	34
AF256714	BHMS-396	4	3	1	4
Total number of clones:		233	179	91	270
Average		19.4	19.9	9.1	27.0

* PCR products was not observed from BAC template, and a weak smear was observed using genomic DNA

Weak smear was observed from BAC template and genomic DNA

Furthermore, we also screened the CHORI 214 BAC library using probes specific to known expressed sequences. Hence, 15 overgos were constructed from Atlantic salmon EST sequences (Table 2). These probes were hybridized to the BAC filters as one pool, and 1203 BAC clones were verified as positive for the 15 probes after a second hybridization experiment. All the BAC clones were fingerprinted in order to provide additional information about their organization. The fingerprinting results generated multiple contigs for 14 of the probes while the clones for the last probe resulted in six single clones. Out of the 1203 clones, 160 clones were not mapped into contigs and are here referred to as singletons. The observation of singletons is either caused by clones containing small insert sizes, possible false positives or lower quality DNA loaded onto the agarose gel. The FPC software will not be able to map small insert clones into contigs using the selected cut off value. For a complete data set of BAC clones linked to each gene marker and the respective fingerprinting contigs see additional file 1.

Table 3 Contigs assembled from the hybridization of 15 EST derived overgo probes onto the BAC filters

					No. of BAC clones in each contig						
GenBank accession no.	Putative gene	No. of positive BACs identified	Singletons	No. of contigs	1	2	3	4	5	6	7
BG 934178	Elongation factor 2	79	19	3	28	18	14				
BG934439	Eukaryotic translation initiation factor 5	91	3	2	78	10					
BG935804	Ubiquitin	69	8	2	54	3					
BG934353	Phosphate transfer protein	108	14	4	47	31	8	4			
BG935917	cyclin L ania-6a	6	6	0							
AF201470	Retinal rod opsin	100	16	3	64	17	3				
AJ344158	Myostatin, isoform II	104	10	4	42	22	17	10			

BG936489	Actin related protein (P16arc)	122	10	3	51	39	20				
BG933799	AMP deaminase	76	12	5	38	7	7	4	3		
BG933794	TAR DNA-binding protein	69	11	2	35	22					
BG934675	Helicase	142	25	7	35	25	22	16	9	5	4
BE518514	Translation initiation factor 3	17	0	2	13	4					
X14305	Growth hormone	74	8	3	42	20	3				
BG935839	Collagen type I	97	7	6	37	28	12	4	4	3	
BG935084	Transducin alpha subunit	49	11	3	15	12	9				
Average:		80.2	10.7	3.3							

Contigs with 3 or more clones are shown

An average of 80 BAC clones was observed in the screening of the EST derived overgo probes, ranging from 6 to 142 BAC clones and 3 contigs on average for each of these probes. However, the results obtained using the genomic oligos resulted in an average of 19.4 clones after secondary hybridization screening, and an average of 27 clones from PCR verification (Table 1). The average observed with the genomic probes corresponds to the estimated genome coverage of 18.8 fold. Because of selection pressure one would expect duplicated expressed DNA sequences to be more preserved than non-coding sequences. Consequently, probes containing expressed sequences might more often identify the pseudo-duplicated regions (in addition to possible pseudo genes) than non-coding probes. However, using fingerprinting, these pseudo-duplicated regions containing restriction enzyme differences might still be mapped to different contigs. Observation of multiple fingerprinting contigs has previously been reported in a BAC library for Rainbow trout (*Onchorhynchus mykiss*), a member of the *Salmonidae* family, using gene and EST probes in screening of the library [12]. The authors present their results as evidence of locus duplication in Rainbow trout. Our results are in agreement with these findings [12]. Our approach, using both expressed sequences as well as non-coding sequences, might be more suited in characterization of BAC libraries of pseudo-ploidity state genomes than expressed sequences exclusively.

CONCLUSION

We have constructed a BAC library from Atlantic salmon (*Salmo salar*). Based on the number of clones and the observed average insert size of 188 kb, we estimate the library to have in excess of 18-fold genome representation.

Providing a publicly available highly redundant genomic large insert library for Atlantic salmon is important for several reasons. First, genomic large insert clones can be used to construct contigs for regions of interest, fingerprinted in large scale and thus create large physical maps of the salmon genome, or be used in shotgun sequencing approaches. Secondly, utilizing one library would be important in exchange of data between researchers. The characterization reported here illustrates the usefulness of the library in identifying genomic clones and the possibility of utilizing clones in fingerprinting analysis. The very large average insert size of the clones combined with the high redundancy will provide researchers with a possibility of obtaining complete gene sequences within a single BAC clone. This might be useful for expression studies of genes and their regulatory elements. The BAC library of Atlantic salmon will be an important tool for future salmon projects and to the salmon industry.

METHODS

Library construction

The BAC library was constructed following the protocols from [13, 14, 15] using the pTARBAC2.1 vector [16]. Transformation of the ligation products was performed using electrocompetent *E. coli* DH10B T1 phage resistant cells (ElectroMAX DH10B T1 resistant, Invitrogen). High-density replica filters were prepared as previously described [14].

High molecular weight DNA preparation

Sperm cells were isolated from one single Atlantic salmon which was provided by Aqua Gen AS [17], and embedded in agarose plugs at a final concentration of 5×10^7 cells/ml followed by a sequence of treatments [15]. DNA used in the library construction was partially digested with EcoRI in the presence of EcoRI Methylase, and size fractionated by pulsed-field gel electrophoresis (PFGE) using a CHEF apparatus (BioRad). The size fractioned agarose gels were stored in 0.5 M EDTA until use. Electroelution procedures were used in obtaining the partially digested DNA from the gel slices [15].

BAC vector preparation

The pTARBAC2.1 plasmid DNA was isolated using cesium chloride gradient purification [15], digested with ApaLI and EcoRI and treated with calf intestine phosphatase (CIP, from New England Biolabs) [15], and separated on 1.0% agarose CHEF gel. The vector fragment was purified from the gel as previously described [15].

Insert size analysis

One clone from each 384 plate in the library was inoculated in LB medium containing 20 µg/ml chloramphenicol. The clones were grown for 18 hours and the BAC DNA was purified using the automated plasmid isolation machine AutoGen 960 (AutoGen). BAC DNA was digested with NotI and analyzed by PFGE [13]. Low Range PFG Marker (New England Biolabs) was used as DNA size marker. The molecular weight determination was achieved using Alpha Innotech MultiImage digital imager and AlphaEase computer software (Alpha Innotech).

Hybridization screening

A set of twelve 36-mer oligo probes from unique genomic sequence flanking twelve different microsatellite markers were end labeled using PNK4 and hybridized as one pool onto the filters. In addition, 15 probes from Atlantic salmon EST sequences were used to design overlapping oligonucleotide probes (overgo; [18]). All probes used are presented in Table 4. The microsatellite oligos and the EST overgos were used in hybridization as two separate pools. Hybridization was carried out in tubes at 65°C overnight in Church buffer without BSA [19]. The filters were washed 4 times at 65°C each for 15 minutes using 1.5 X SSC and 0.1% SDS. Identification of positive signals was achieved by exposing the filters to either X-ray film (Agfa) or to Phosphor Image cassettes (Amersham Biosciences). All the clones identified in the screening were arrayed into new 384 plates, spotted onto nylon filters using a 384 pin tool, grown overnight on LB agar containing 20 µg/ml chloramphenicol, and processed following established procedures [19]. Each filter was hybridized with the individual probes, and the clones identified were re-arrayed in 96 well plates.

Genomic DNA was isolated using established phenol/chloroform extraction procedures [19], digested with TaqI, PstI and MspI and separated on a 1% TBE agarose gel. DNA was transferred to nylon filters using established Southern blot techniques [19].

Table 4 Primer sequences for EST and genomic probes

	EST probe	
GenBank accession no.	**Overgo A primer**	**Overgo B primer**
15844006	5'-AAGCCTGTGCTGATGATGAACAAG-3'	5'-GCAGGGCACGGTCCATCTTGTTCA-3'
15844267	5'-GTCGTTTATGTCATCCCCTCTTCT-3'	5'-GGTATCTTCTGTCTGGAGAAGAGG-3'
15845632	5'-GAAGGCATCCCTCCTGATCAGCAG-3'	5'-CAGCGAAGATCAACCTCTGCTGAT-3'
15844181	5'-TACTCCATGCAGGGACTCTGCAAG-3'	5'-CCTCATAGAAGCCAAACTTGCAGA-3'
15845745	5'-GCGACCAGCTACATTTACCAAAGC-3'	5'-ATTCCACATCACCCAGGCTTTGGT-3'
7271780	5'-GCTTCCCCATCAACTTCCTCACGC-3'	5'-TCGATGGTGACGTAGAGCGTGAGG-3'
16604728	5'-ACTGGATTATTGCCCCTAAGCGCT-3'	5'-CAGTAGTTGGCCTTGTAGCGCTTA-3'
15846317	5'-TGCCACAAGTTCATGCGCTTCATG-3'	5'-TCTCGGCTCTCATCATCATGAAGC-3'
15843627	5'-ATGTCTCCGCTCAGCAACAACAGC-3'	5'-GGTAGCTGAGGAAGAGGCTGTTGT-3'
15843622	5'-GAGCCTAAGCACAATAATAGTAGG-3'	5'-CACGATCCATCATTTGCCTACTAT-3'
15844503	5'-CCATCAAGAAGGACGAGGACGTGC-3'	5'-GGGCAGTTCTTCTTCAGCACGTCC-3'
15967287	5'-GAGCTTCCAGCTGGTGGACACTGC-3'	5'-AGTCTTCTGCGTCTTGGCAGTGTC-3'
15845667	5'-CACTTGCTTAAGCTGGGCTCTATC-3'	5'-TCCATTGGTCCTCTCCGATAGAGC-3'
15843919	5'-AAGATCCCAGGTGGGCGAGGGAAT-3'	5'-TGTGATCCCGCTGACCATTCCCTC-3'
15844912	5'-AACATCCTGCAGTCTGCTCTGGCC-3'	5'-CCATGCCTCTGATGATGGCCAGAG-3'
	PCR primers for microsatellites	
	Forward primer	Reverse primer
AF256671	5'-TCACATCCCTTAGCTCCC-3'	5'-CCTTTTTTGTGTCTTCAGC-3'

AF256676	5'-AAACACCCTTCCCTTCAC-3'	5'-CAATTCAGGTCAAACCAAC-3'
AF256678	5'-CCCCATGATGTGTTCTTC-3'	5'-CACAATGAGGCTTGACAC-3'
AF256741	5'-TTGAGCCATCCTCACCTC-3'	5'-CACTGGTTTGTTGTTGTTG-3'
AF256698	5'-CAGAACCGTGATCTGAAG-3'	5'-TGGACATTCTCTGGCGTC-3'
AF256748	5'-CTAGATCACTCACCCAGG-3'	5'-GTGCTTTTGGCTTATGTTAG-3'
AF256746	5'-GCTGTGATTTCTCTCTGC-3'	5'-AAAGGTGGGTCCAAGGAC-3'
AF256719	5'-CCCCTGTCAAACGTCTTC-3'	5'-AGCACACTGGATTCAAGG-3'
AF256693	5'-AGGCACAAACATGCAAGC-3'	5'-TCACCCCTGTGTCATCAC-3'
AF256848	5'-TCCAAACCTGAATCCAGG-3'	5'-TTGTAGTGAAAGCCGCTG-3'
AF256750	5'-TCCCACTGCCAACTACAG-3'	5'-GTTTAATCAAAGCATTCGCC-3'
AF256714	5'-CCTGCCATCATCCAACTC-3'	5'-TCCACACCCAACATACTC-3'
	Genomic probe sequence	
AF256671	5'-GCAGCTCAGTGACTATGACTTCTCCG-GTTTCCTGTTCTCT-3'	
AF256676	5'-GCCCTAGAGATTGAAATAG-CATCCTCTTTCACGCCATGCA-3'	
AF256678	5'-CTGCAAGACAGAGAACACCATGACA-CACAGACCTCTGGAT-3'	
AF256741	5'-AGTGAAGACCTCAACCCACAAAGGC-GCTATAATCGGCAAC-3'	
AF256698	5'-TCCTGTGTATCTGCAGTCAGTTCCAG-GAAATGGAGGAGCA-3'	
AF256748	5'-TGAGGGGGCTTACAAGAGGTCTTC-GCTTTGCCCCAGAAAA-3'	
AF256746	5'-CACAGTTGCCAGTTGAGAGAAGAGA-AAGACGTTAGGGACA-3'	
AF256719	5'-TGGCAAAGCCTAGAGAGGTTTATCT-CAGCACCACATTGCA-3'	
AF256693	5'-TCACCCCTCATTCACACAATCTC-CAGCTGTCACATCAAGC-3'	
AF256848	5'-TTCTCGGCTAGATCACTTGCTCT-GTCTCTCTTCCCCACTC-3'	
AF256750	5'-TCCTGTAGCATGCTGACATTCTG-GCAGTCAGACACACAAG-3'	
AF256714	5'-CGCTGACTTGATTTGCCTTAATGCAG-TATGTGTCAACCCC-3'	

PCR verification of positive clones

PCR primers were constructed based on the genomic sequence of the microsatellites used in the screening of the library. Cultures containing 20 µg/ml chloramphenicol and 7% glycerol of the re-arrayed clones identified by the microsatellite oligos were transferred to new 96 well plates containing PCR reaction mix. PCR reactions were carried out in 96 well plates in a total volume of 25 µl, containing 0.2 mM dNTPs (Amersham Biosciences), 1X PCR buffer (Qiagen), 7 pmol of each primer and 0.25 units of Taq DNA polymerase (Qiagen). The PCR reactions were incubated at 95°C for 3 minutes, followed by 35 cycles at 95°C for 30 seconds, annealing at 54°C for 30 seconds, and extension at 72°C for 30 seconds with a final incubation at 72°C for 10 minutes. After thermocycling, 2 µl of each PCR reaction was analyzed on a 1.5% agarose gel (ReadyGel, Amersham Biosciences).

Fingerprinting and contig assembly

DNA from all the clones identified in the screening using the EST overgos was isolated using the AutoGen 960 (AutoGen), digested with HindIII and DNA fragments were separated in a 1.2% agarose gel using established protocols [20]. The gels were stained using SYBRGreen and scanned using a FluorImager 595 (Amersham Biosciences). Gel images were analyzed using Image 3.10 [6] and the data obtained were used with FPC v4.7 [6]. A fixed tolerance of seven, cut-off value of e-14 and a bury value of 10% was used to build contigs. To merge contigs, the cut-off value was raised to e-12 to contigs that corresponded to the individual EST sequences

ACKNOWLEDGEMENTS

We gratefully acknowledge Barbara Swiatkiewicz and Dorota Kostecka for filter processing assistance. We thank Gery Vessere for assistance with informatics. This work was supported by grants from the Research Council of Norway, NFR 139617/140 Salmon Genome Project, NFR 145847/140 The construction of an Atlantic salmon (*Salmo salar*) BAC library. This work was also supported by grant from Genome Canada (Genome Research on Atlantic Salmon Project). We would like to thank Aqua Gen AS [17] for providing the sperm used in the library construction.

AUTHORS' CONTRIBUTIONS

JT performed the experiments and drafted the manuscript. BZ, EF and BH provided supervision. KO, PJD, BFK, WSD and BH coordinated the project.

REFERENCES

1. BACPAC Resources Center [http://bacpac.chori.org]
2. Gilbey J, Verspoor E, Mclay A, Houlihan D: A microsatellite linkage map for Atlantic salmon (Salmo salar). Animal Genetics 2004, 35:98-105.
3. Moen T, Hoyheim B, Munck H, Gomez-Raya L: A linkage map of Atlantic salmon (Salmo salar) reveals an uncommonly large difference in recombination rate between the sexes. Animal Genetics 2004, 35:81-92.
4. Norwegian Salmon Genome Project [http://www. salmonge nome.no]
5. NCBI - National Center for Biotechnology Information [http://www.ncbi.nlm.nih.gov]
6. The Sanger Institute [http://www.sanger.ac.uk]
7. GRASP (The Genomic Research on Atlantic Salmon Project) [http://www.uvic.ca/cbr/grasp]
8. TIGR A. salmon Gene Index [http://www.tigr.org/tigr-scripts/tgi/ T_index.cgi?species=salmon]
9. Allendorf FW, Thorgaard GH: Tetraploidy and the evolution of Salmonid fishes. In Evolutionary genetics of fishes. Edited by: Turner BJ. New York, Plenum Publishing Corporation; 1984:1-53.
10. Hardie DC, Hebert PDN: The nucleotypic effects of cellular DNA content in cartilaginous and ray-finned fishes. Genome 2003, 46:683-706.
11. Genome Size Database [http://www.genomesize.com]
12. Palti Y, Gahr SA, Hansen JD, Rexroad CEIII: Characterization of a new BAC library for rainbow trout: evidence for multi-locus duplication. Animal Genetics 2004, 35:130-133.
13. Osoegawa K, Woon PY, Zhao B, Frengen E, Tateno M, Catanese JJ, de Jong PJ: An improved approach for construction of bacterial artificial chromosome libraries. Genomics 1998, 52:1-8.
14. Osoegawa K, Tateno M, Woon PY, Frengen E, Mammoser AG, Catanese JJ, Hayashizaki Y, de Jong PJ: Bacterial artificial chromosome libraries for mouse sequencing and functional analysis. Genome Research 2000, 10:116-128.
15. Osoegawa K, de Jong P, Frengen E, Ioannu PA: Construction of bacterial artificial chromosome (BAC/PAC) libraries. In Current Protocols in Molecular Biology Edited by: Ausubel FM, Brent R, Kingston RE, Moore DD, Seidman JG, Smith JA and Struhl K. New York, Wiley; 2001:5.15.1-5.15.33.
16. Osoegawa K, Zhu BL, Shu CL, Ren T, Cao Q, Vessere GM, Lutz MM, Jensen-Seaman MI, Zhao SY, de Jong PJ: BAC resources for the rat genome project. Genome Research 2004, 14:780-785.
17. AquaGen AS [http://www.aquagen.no]
18. Ross MT, Labrie S, McPherson JP, Stanton VPJ: Screening largeinsert libraries by hybridization. In Current protocols in human genetics Edited by: Dracopoli NC, Haines JL, Korf BR, Morton CC, Seidman CE, Seidman JG and Smith DR. New York, Wiley; 1999:5.6.1-5.6.52.

CHAPTER 7

EVALUATION OF A CANOLA PROTEIN CONCENTRATE AS A REPLACEMENT FOR FISHMEAL AND POULTRY BY-PRODUCT MEAL IN A COMMERCIAL PRODUCTION DIET FOR ATLANTIC SALMON (SALMO SALAR)

Gary S Burr[1], William R Wolters[1], Frederic T Barrows[2] and Alan W Donkin[3]

[1]ARS National Cold Water Marine Aquaculture Center, USDA, 25 Salmon Farm Road, Franklin, ME 04634, USA

[2]Agricultural Research Service, Fish Technology Center, USDA, Bozeman, MT 59715, USA.

[3] Northeast Nutrition Inc, 494 Willow St, Truro, Nova Scotia B2N 6X8, Canada.

ABSTRACT

Atlantic salmon (*Salmo salar*) is an important cultured carnivorous species that, in the past, has not tolerated high levels of most plant protein feed ingredients in their diet. In order to increase efficiency and production to meet global demand, new sources of protein must be incorporated into the aquafeeds. A 38-week feeding trial was conducted at the National Cold Water Marine Aquaculture Center (Franklin, ME, USA) with juvenile Atlantic salmon (133 g per fish, initial weight) to determine the effect of feeding graded levels of canola protein concentrate (CPC) in a commercial-type diet. A commercial diet (Signature Salmon 3.5 mm, Northeast Nutrition, Truro, Nova Scotia, Canada) was modified and analyzed by the manufacturer to contain 10% or 20% canola protein concentrate, replacing fishmeal and poultry by-product meal. Fish fed diet containing 20% canola protein concentrate had significantly lower growth compared with those fed 0% canola protein concentrate diet ($p = 0.04$). There was not any significant difference in feed efficiency ($p = 0.22$) or protein efficiency ratio ($p = 0.21$). There was not a significant difference in growth comparing the salmon fed the 0% CPC and the 10% CPC diets ($p > 0.05$). Canola protein concentrate significantly depressed growth when included in the diet at 20%, but not at 10%, indicating that canola may be used as a minor feed ingredient when available.

BACKGROUND

Atlantic salmon (*Salmo salar*) is an important cultured carnivorous species that, in the past, has not tolerated high levels of most plant protein feed ingredients in their diet Gatlin et al. (2007). The inclusion of plant protein sources in aquafeeds is expanding due to the limited amount of fishmeal available for the production of animal feeds Glencross et al. (2005; Gatlin et al. 2007). Juvenile Senegalese sole (*Solea senegalensis*) fed diets that contain 5% fishmeal or less have been shown to have comparable growth to fish fed a 37% fishmeal diet when amino acids were supplemented Silva et al. (2009). For inclusion in the diets of rainbow trout (*Oncorhynchus mykiss*), many different plant protein sources have been examined, including plant protein meals and plant protein concentrates Lim and Webster (Lim et al. 2008). Canola protein concentrate (CPC), soybean meal and soy protein concentrates, pea protein concentrate, barley protein concentrate, rice protein concentrate, and wheat gluten meal have all been tested as fishmeal replacements with

Citation: Burr et al: Evaluation of a canola protein concentrate as a replacement for fishmeal and poultry by-product meal in a commercial production diet for Atlantic salmon (Salmo salar). International Aquatic Research 2013 5:5. doi:10.1186/2008-6970-5-5

varying degrees of success Forster et al. (1999; Thiessen et al.2003; Barrows et al. 2007; Lim et al. 2008; Gaylord and Barrows 2009). Many of these same protein sources have been tested in Atlantic salmon (Lim et al. 2008). Atlantic salmon was shown to have slightly reduced, but not significantly lower, performance when fed diets that were fishmeal-free and contained wheat gluten, corn gluten, fish soluble, and crystalline amino acids (Espe et al. 2006). Atlantic salmon fed a diet that contained 12% fishmeal had slightly decreased growth but produced 2 kg of fish for every 1 kg of fishmeal consumed (Torstensen et al. 2008). Stickwater, the water-soluble fraction that is produced in making fishmeal, has been shown to stimulate the growth of Atlantic salmon when fed low fishmeal diets (10% of the diet) and high plant protein diets (62% of the diet) (Kousoulaki et al. 2009). Amino acid supplementation is often needed to maintain the growth performance of fish fed plant protein-based feeds (Espe et al. 2007; Lim et al. 2008; Gaylord and Barrows 2009). A major challenge is to increase the amount of plant or other non-animal protein in the diet of carnivorous fishes while maintaining acceptable growth, feed conversion, and production costs compared to traditional diets.

Recently, multiple studies have been reported on the digestibility of various plant feed ingredients by Atlantic salmon () (Storebakken et al. *Salmo salar*2000; Glencross et al. 2004a; Refstie et al. 2005; Aas et al. 2006; Aslaksen et al. 2007; Refstie et al. 2008). The protein digestibility was not significantly lower for the plant feed ingredients, except for extracted soybean meal, oat, canola, and sunflower. Canola meal contains a high level of carbohydrates and other antinutritional factors, such as phytic acid, tannins, sinapine, and goitrogenic compounds, lowering the nutrient availability of this feed ingredient to aquatic species (Kousoulaki et al. 2009). Further processing into a protein concentrate should remove more of the antinutritional factors and carbohydrates; however, significant amounts of phytic acid could remain (Burel and Kaushik 2008). In rainbow trout (*O. mykiss*), experimental diets that contained 20% fishmeal and 19.4% canola protein concentrate significantly lowered growth (Drew et al. 2007).

Both existing and new aquaculture operations are trying to increase the sustainability and reduce the environmental impacts by optimizing water use and reducing waste. Recirculating systems for cold water fish have undergone significant improvements within the last decade and can provide culture systems with reduced environmental impact. Such systems have greater control of the rearing environment, minimize water use, produce a concentrated and relatively small volume effluent, and can have improved biosecurity measures to control disease (Summerfelt et al. 2001).

The purpose of this study was to determine the effect of feeding multiple levels of canola protein concentrate in a commercial-type diet on the growth and nutrient efficiency of Atlantic salmon in a recirculating aquaculture system.

METHODS

The experimental protocols and methods used in the research at the NCWMAC were in compliance with Animal Welfare Act (9CFR) requirements and were approved by the location's Institutional Animal Care and Use Committee. A 38-week feeding trial was conducted at the National Cold Water Marine Aquaculture Center (Franklin, ME, USA) with healthy juvenile Atlantic salmon. A commercial diet (Signature Salmon 3.5 mm, Northeast Nutrition, Truro, Nova Scotia, Canada) was used and modified by the manufacturer to contain 10% or 20% canola protein concentrate (insoluble canola protein concentrate, CanPro, Arborfield Saskatchewan, Canada). The three diets were formulated to contain 45% to 46% protein and 21% to 23% lipid. Diets were formulated and prepared by the manufacturer to ensure practical relevance, and the proportional composition is shown in Table 1. The 10% CPC diet contained only 75% of the fish protein ingredients, and the 20% CPC diet contained only 57% of the fish protein ingredients compared to the commercial diet. Amino acids and lipids were balanced by the manufacturer. The feed was prepared before the trial and was stored in a cool dry room (temperature, 4°C to 12°C).

Table 1 Nutrient composition of canola protein concentrate and proportional ingredient composition of reference and experimental diets

	Canola protein concentrate	0%	10%	20%
		Canola protein	Canola protein	Canola protein
Ingredient[a] (g/kg)				
Canola protein concentrate		0	100	200
Plant protein ingredients (other than canola)		270.2	253.0	241.4
Animal protein ingredients		398.4	336.8	260.3
Fish protein ingredients		140.0	105.0	80.0
Fish oil and lecithin		165.2	177.9	189.7
Amino acids		4.4	5.5	6.8
Additives (nutritional)		18.3	18.3	18.3

Vitamin + trace mineral package		3.5	3.5	3.5
Analyzed levels (as fed basis)				
Dry matter (g/kg)	950.6	950.5	947.5	961.7
Protein (g/kg)	604.1	459.6	449.6	455.7
Fat (g/kg)	24.2	228.6	209.4	211.3
Ash (g/kg)	54.1	66.1	61.9	57.1
Calcium (g/kg)	6.0	15.3	13.8	10.6
Phosphorus (g/kg)	8.7	10.9	10.1	8.5
Sodium (g/kg)	6.2	3.0	3.1	3.1
Potassium (g/kg)	3.6	6.1	5.5	5.2
Magnesium (g/kg)	4.5	1.2	1.5	1.7
Manganese (ppm)	57.25	143.46	194.83	128.95
Copper (ppm)	11.52	16.03	11.97	14.36
Zinc (ppm)	79.07	113.08	104.99	131.19

[a]The basal diet contained poultry meal, wheat, fishmeal, fish oil, blood meal, crab meal, feather meal, CPSP special G fishmeal (fish hydrolysate from fresh whole fish or fresh offals; Sopropêche, Cedex, France), lecithin, betaine, XPC yeast culture (Diamond V, Cedar Rapids, Iowa, USA), choline chloride, amino acids, vitamin, and mineral premixes. The CPSP special G fishmeal was omitted from the CPC-containing diets.

Each experimental diet was randomly assigned to triplicate tanks (0.26 m^3 each) containing 20 Atlantic salmon (133.0 ± 7.8 g SD per fish, initial weight) at an initial stocking density of 10 kg/m^3. All tanks were connected to a common particulate (rotating drum filter) and a biological filter (fluidized sand) to maintain optimal water quality. The water in the system was brackish water (3%) and obtained from a ground source. Temperature (°C) and dissolved oxygen (mg/l) were continuously monitored with stationary probes. The temperature averaged to 10°C using well water and varied seasonally, ranging from 8°C in January to 13°C in July. Fish were subjected to a natural photoperiod (9 h light (L):15 h dark (D) initial to 15L:9D final) maintained by artificial lighting controlled using a computer. Dissolved oxygen was maintained at or above saturation using an oxygen cone. The water quality of the system was monitored weekly and maintained within acceptable levels for Atlantic salmon.

The ArvoPro computer program (Arvo-tec, Huutokoski, Finland) was used to control the feed being dispensed using automatic feeders four times per day to the tanks, and the feed was adjusted daily with fish fed to apparent satiation. The computer program was developed from experimental growth models validated from commercial data for conditions and different genetic stocks (Ursin 1967; From and Rasmussen 1984; Ruohonen and Makinen 1992). The feed was slowly sinking. Fish were group weighed approximately every 4 weeks to confirm the projected weight (g) using the computer software. Throughout the trial, weight gain (g) was monitored. At the end of the trial, feed efficiency (weight gain/feed intake), protein efficiency ratio (g of weight gain/g of protein fed), and specific growth rate values were computed, and three fish were collected from each tank for subsequent measurement of whole-body composition according to established procedures (Webb et al. 2003).

RESULTS

Atlantic salmon fed the diet containing 20% canola protein concentrate showed the lowest weight gain compared to fish fed the diet containing 0% canola protein concentrate ($p = 0.04$; Figure 1; Table 2). These fish fed the 0% canola protein concentrate diet had the highest weight gain but were not statistically different from the fish fed the diet containing 10% canola protein concentrate (Figure 1; Table 2). Fish fed the fishmeal reference diet had the highest average weight throughout the study (Figure 1). Feed efficiency and protein efficiency ratio did not differ among the treatments (Table 2), indicating that the fish fed the reference diet consumed the most feed. Fish fed the higher fish protein diet had a significantly greater specific growth rate compared to fish fed the 20% canola concentrate diet (Table 2).

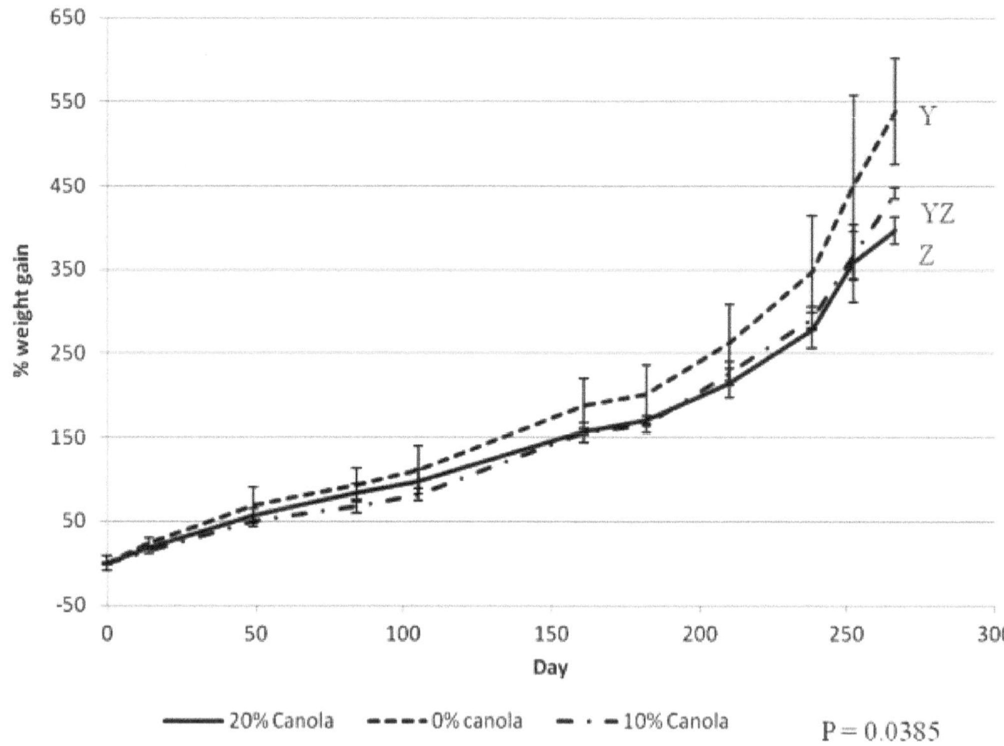

Figure 1 Percent weight gain over 38 weeks of Atlantic salmon fed the three diets. Values with common superscript letter do not differ significantly ($p > 0.05$).

Table 2 Feed efficiency (FE), protein efficiency ratio (PER), specific growth rate(SGR), and weight increase (%)

Diet	FE	PER	SGR	Weight increase (%)
0	0.80 ± 0.07	1.74 ± 0.29	0.73 ± 0.05 Y	539.6 ± 62.9 Y
10	0.78 ± 0.01	1.74 ± 0.05	0.67 ± 0.01 YZ	441.2 ± 6.8 YZ
20	0.68 ± 0.01	1.50 ± 0.04	0.64 ± 0.02 Z	397.3 ± 16.1 Z
ANOVA *p*	0.2170	0.2081	0.0268	0.0385

Values within columns that have a common uppercase letter do not differ significantly ($p > 0.05$). ANOVA, analysis of variance.

Proximate composition of the whole fish lipid, moisture, or ash concentrations did not vary among the treatments ($p = 0.05$; Table 3). Fish fed with the control diet had significantly lower crude protein concentrations compared to fish fed with the canola protein concentrate diets ($p = 0.04$; Table 3).

Table 3 Proximate composition of whole fish (g/kg of wet weight)

Diet	Protein	Lipid	Ash	Moisture
0	140.1 ± 15.9 b	107.5 ± 21.2	61.4 ± 36.3	716.0 ± 7.3
10	175.2 ±12.3 a	108.6 ± 15.0	57.9 ± 12.6	689.2 ± 2.8
20	167.7 ± 11.4 a	95.0 ± 3.6	67.8 ± 13.9	694.6 ± 3.0
ANOVA *p*	0.0401	0.5697	0.3961	0.1102

Values within columns that have a common lowercase letter do not differ significantly ($p > 0.05$).

DISCUSSION

Atlantic salmon fed the diet containing 20% canola protein concentrate had significantly lower weight gain and specific growth rate compared to fish fed diets containing higher levels of fish protein ingredients. However, feed efficiencies were similar, indicating that the increased weight gain of the fish on the fishmeal diet was the result of consuming more feed. There are several antinutritional factors associated with canola that could have influenced the consumption and resulting weight gain in this study, such as tannins, phytic acid, and glucosinolates (Kousoulaki et al. 2009). If antinutritional factors are indicated in future research, then each batch used in aquafeeds should be analyzed for antinutritional factors.

Several plant proteins have been evaluated in the diets of Atlantic salmon (Refstie et al. 2001; Refstie and Tiekstra 2003; Mundhein et al. 2004; Denstadli et al. 2007; Torstensen et al. 2008; Øverland et al. 2009). Potato protein concentrate (included up to 21% of the diet) and soy protein concentrate (included at 19.7% of the diet) did not inhibit the growth of Atlantic salmon compared to fishmeal-based diets (Refstie et al. 2001; Refstie and Tiekstra 2003). The growth of Atlantic salmon was inhibited by increasing plant proteins (full-fat soybean meal and corn gluten meal) up to 42% of the diet compared to a fishmeal-based diet (Kraugerud et al. 2007). Kraugerud et al. (2007) reported that a diet containing 46% soybean meal stunted the growth of Atlantic salmon (Torstensen et al. 2008). When included at 20% of the diet, canola protein concentrate also inhibited the growth of Atlantic salmon. These results indicate that a certain level of plant protein concentrates is tolerated in the diets of Atlantic salmon, but when the threshold is exceeded, either antinutritional factors or digestibility of the plant proteins hinders the intake or uptake of nutrients and limit growth. This would indicate that the digestibility of plant proteins need to be improved or the formulae require further modification to compensate. The antinutritional factors need to be mitigated or removed from the plant protein product prior to feeding (Denstadli et al. 2007).

Canola meal and canola protein concentrate have been investigated in a variety of finfish (Burel and Kaushik 2008). Protein apparent digestibility coefficients for rainbow trout fed rapeseed meal are similar to fishmeal (Mwachireya et al. 1999; Burel et al. 2000). However, apparent digestibility coefficients for energy and dry matter are significantly lower compared to a fishmeal reference diet (Mwachireya et al. 1999). There have been contradictory results obtained for rainbow trout fed a canola protein concentrate. Rainbow trout grew only slightly slower when fed a diet containing up to 42% canola protein concentrate compared to fish fed a fishmeal-based diet (Forster et al. 1999). Drew et al. (2007) also found that the growth of rainbow trout fed a canola protein concentrate was not affected when the levels were below 29%. However, Teskeredžić et al. (Teskeredzić et al. 1995) reported that rapeseed protein concentrate included in the diet above 26% did significantly lower the specific growth rate of rainbow trout. Hajen et al. (1993) reported that Chinook salmon (*Oncorhynchus tshawytscha*) fed a diet containing 30% rapeseed protein concentrate had the same growth rate as fish fed a fishmeal diet. The growth of the juvenile Chinook salmon fed diets that contained up to 23% canola meal was similar to a fishmeal reference diet (Higgs et al. 1983). Canola meal inclusion rates could be increased up to 30% when the diet was supplemented with 5 ppm 3,5,3'-triiodo-L-thyronine (Webster et al. 1997). In channel catfish, fish fed a diet containing 48% canola meal had significantly lower growth compared to fish fed a fishmeal diet or diets containing 12% or 36% canola meal (Webster et al. 1997). Sunshine bass (*Morone chrysops* × *Morone saxatilis*) fed a diet containing 20% canola meal did not have a significant difference in growth compared to fish fed a fishmeal diet (Webster et al. 2000). The processing method affected the digestibility of canola meals for red seabream, and protein digestibility was similar to soybean meal (Glencross et al. 2004a). The growth of red seabream fed diets that contained up to 60% canola meal was not adversely affected compared to a fishmeal reference diet (Glencross et al. 2004b). These results along with the results in the present study indicate that the source and processing of the canola meal/protein concentrate and species of fish affect growth performance. However, due to limited published studies, more research is needed to determine the upper limit of canola meal/protein concentrate that can be included in aquafeeds.

CONCLUSIONS

Overall, Atlantic salmon does not appear to be able to tolerate canola protein concentrate above 10% when included in their diet. Canola protein concentrate could be a candidate for a diet if included in the diet at low levels. This study did not address the antinutritional factors associated with canola meal/protein concentrate, but the viability of canola as a feed ingredient could be enhanced if the antinutritional factors could be removed or the effects could be mitigated.

ACKNOWLEDGEMENTS

We would also like to thank the staff of the National Cold Water Marine Aquaculture Center; Melissa Albert, Sharon Baron, Ryan Hastey, Davin O'Connell, and Amber Rankin for their technical assistance.

DISCLAIMER

The mention of trade names or commercial products in this publication is solely for the purpose of providing specific information and does not imply recommendation or endorsement by the US Department of Agriculture.

REFERENCES

1. Aas TS, Grisdale-Helland B, Terjesen BF, Helland SJ (2006) Improved growth and nutrient utilization in Atlantic salmon (Salmo salar) fed diets containing a bacterial protein meal. Aquaculture 259:365–376
2. Aslaksen MA, Kraugerud OF, Penn M, Svihus B, Denstadli V, Jøgensen HY, Hillestad M, Krogdahl Å, Storebakken T (2007) Screening of nutrient digestibilities and intestinal pathologies in Atlantic salmon, Salmo salar, fed diets with legumes, oilseeds, or cereals. Aquaculture 272:541–555
3. Barrows FT, Gaylord TG, Stone DAJ, Smith CE (2007) Effect of protein source and nutrient density on growth efficiency, histology, and plasma amino acid concentration of rainbow trout (Oncorhynchus mykiss). Aquacult Res 38:1747–1758
4. Burel C, Kaushik SJ (2008) Use of rapeseed/canola in diets of aquaculture species. In: Lim CE, Webster CD, Lee C-S (ed) Alternative protein sources in aquaculture diets. Haworth, New York, p 343
5. Burel C, Boujard T, Tulli F, Kaushik SJ (2000) Digestibility of extruded peas, extruded lupin and rapeseed meal in rainbow trout (Oncorhynchus mykiss) and turbot (Psetta maxima). Aquaculture 188:285–298

6. Denstadli V, Storebakken T, Svihus B, Skrede A (2007) A comparison of online phytase pretreatment of vegetable feed ingredients and phytase coating in diets for Atlantic salmon (Salmo salar L.) reared in cold water. Aquaculture 269:414–426
7. Drew MD, Ogunkoya AE, Janz DM, Van Kessel AG (2007) Dietary influence of replacing fish meal and oil with canola protein concentrate and vegetable oils on growth performance, fatty acid composition and organochlorine residues in rainbow trout (Oncorhynchus mykiss). Aquaculture 267:260–268
8. Espe M, Lemme A, Petri A, El-Mowafi A (2006) Can Atlantic salmon (Salmo salar) grow on diets devoid of fish meal? Aquaculture 255:255–262
9. Espe M, Lemme A, Petri A, El-Mowafi A (2007) Assessment of lysine requirement for maximal protein accretion in Atlantic salmon using plant protein diets. Aquaculture 263:168–178
10. Forster I, Higgs DA, Dosanjh BS, Rowshandeli M, Parr J (1999) Potential for dietary phytase to improve the nutritive value of canola protein concentrate and decrease phosphorus output in rainbow trout (Oncorhynchus mykiss) held in 11°C fresh water. Aquaculture 179:109–125
11. From J, Rasmussen G (1984) A growth model, gastric evacuation, and body composition in rainbow trout, Salmo gairdneri Richardson, 1836. Dana 3:61–139
12. Gatlin DM, III, Barrows FT, Brown P, Dabrowski K, Gaylord TG, Hardy RW, Herman E, Hu G, Krogdahl Å, Nelson R, Overturf K, Rust M, Sealey W, Skonberg D, Souza EJ, Stone D, Wilson R, Wurtele E (2007) Expanding the utilization of sustainable plant products in aquafeeds: a review. Aquacult Res 38:551–579
13. Gaylord TG, Barrows FT (2009) Multiple amino acid supplementation to reduce dietary protein in plant-based rainbow trout, Oncorhynchus mykiss, feeds. Aquaculture 287:180–184
14. Glencross B, Hawkins W, Curnow J (2004a) Nutritional assessment of Australian canola meals. I. Evaluation of canola oil extraction method and meal processing conditions on the digestible value of canola meals fed to the red seabream (Pagrus auratus, Paulin). Aquacult Res 35:15–24
15. Glencross B, Hawkins W, Curnow J (2004b) Nutritional assessment of Australian canola meals. II. Evaluation of canola oil extraction method and meal processing conditions on the digestible value of canola meals fed to the red seabream (Pagrus auratus, Paulin). Aquacult Res 35:25–34
16. Glencross B, Evans D, Dods K, Mcafferty P, Hawkins W, Maas R, Sipsas S (2005) Evaluation of the digestible value of lupin and soybean protein concentrates and isolates when fed to rainbow trout, Oncorhynchus mykiss, using either stripping or settlement faecal collection methods. Aquaculture 245:211–220
17. Hajen WE, Beames RM, Higgs DA, Dosanjh BS (1993) Digestibility of various feedstuffs by post-juvenile Chinook salmon (Oncorhynchus tshawytscha) in sea water. Aquaculture 113:321–332
18. Higgs DA, Fagerlund UHM, Mcbride JR, Plotnikoff MD, Dosanjh BS, Markert JR, Davidson J (1983) Protein quality of
19. Altex canola meal for juvenile Chinook salmon (Oncorhynchus tshawytscha) considering dietary protein and 3,5, 3'-triiodo-L-thyronine content. Aquaculture 34:213–238
20. Kousoulaki K, Albrektsen S, Langmyhr E, Olsen HJ, Campbell P, Aksnes A (2009) The water soluble fraction in fish meal (stickwater) stimulates growth in Atlantic salmon (Salmo salar L.) given high plant protein diets. Aquaculture 289:74–83
21. Kraugerud OF, Penn M, Storebakken T, Refstie S, Krogdahl Å, Svihus B (2007) Nutrient digestibilities and gut function in Atlantic salmon (Salmo salar) fed diets with cellulose or non-starch polysaccharides from soy. Aquaculture 273:96–107
22. Lim C, Webster CD, Lee C-S (ed) (2008) Alternative protein sources in aquaculture diets. Haworth, New York. 571 pp
23. Mundhein H, Aksnes A, Hope B (2004) Growth, feed efficiency and digestibility in salmon (Salmo salar L.) fed different dietary proportions of vegetable protein sources in combination with two fish meals qualities. Aquaculture 237:315–331
24. Mwachireya SA, Beames RM, Higgs DA, Dosanjh BS (1999) Digestibility of canola protein products derived from physical, enzymatic and chemical processing of commercial canola meal in rainbow trout Oncorhynchus mykiss (Walbum) held in fresh water. Aquacult Nutr 5:73–82
25. Øverland M, Sørensen M, Storebakken T, Penn M, Krogdahl Å, Skrede A (2009) Pea protein concentrate substituting fish meal or soybean meal in diets for Atlantic salmon (Salmo salar)—effect on growth performance, nutrient digestibility, carcass composition, gut health, and physical feed quality. Aquaculture 288:305–311
26. Refstie S, Tiekstra HAJ (2003) Potato protein concentrate with low content of solanidine glycoalkaloids in diets for Atlantic salmon (Salmo salar). Aquaculture 216:283–298
27. Refstie S, Storebakken T, Baeverfjord G, Roem AJ (2001) Long-term protein and lipid growth of Atlantic salmon (Salmo salar) fed diets with partial replacement of fish meal by soy protein products at medium or high lipid level. Aquaculture 193:91–106
28. Refstie S, Sahlström S, Bråthen E, Baeverfjord G, Krogedal P (2005) Lactic acid fermentation eliminates indigestible carbohydrates and antinutritional factors in soybean meal for Atlantic salmon (Salmo salar). Aquaculture 246:331–345
29. Refstie S, Glencross B, Landsverk T, Sørensen M, Lilleeng E, Hawkins W, Krogdahl Å (2008) Digestive function and intestinal integrity in Atlantic salmon (Salmo salar) fed kernel meals and protein concentrates made from yellow or narrow-leafed lupins. Aquaculture 261:1382–1395
30. Ruohonen K, Makinen T (1992) Validation of rainbow trout (Oncorhynchus mykiss) growth model under Finnish circumstances. Aquaculture 105:353–362

31. Silva JMG, Espe M, Conceição LEC, Dias J, Valente LMP (2009) Senegalese sole juveniles (Solea senegalensis Kaup, 1858) grow equally well on diets devoid of fish meal provided the dietary amino acids are balanced. Aquaculture 296:309–317
32. Storebakken T, Shearer KD, Baeverfjord G, Nielsen BG, Åsgård T, Scott T, De Laporte A (2000) Digestibility of macronutrients, energy and amino acids, absorption of elements and absence of intestinal enteritis in Atlantic salmon, Salmo salar, fed diets with wheat gluten. Aquaculture 184:115–132
33. Summerfelt ST, Bebak-Williams J, Tsukuda S (2001) Controlled Systems (2001) Water reuse and recirculation. In: Wedermeyer G (ed) Fish hatchery management, 2nd edition. American Fisheries Society, Bethesda, pp 285–395
34. Teskeredzić Z, Higgs DA, Dosanjh BS, McBride JR, Hardy RW, Beames RM, Jones JD, Sirnell M, Vaarae T, Bridges RB (1995) Assessment of undephytinized and dephytinized rapeseed protein concentrate as sources of dietary protein for juvenile rainbow trout (Oncorhynchus mykiss). Aquaculture 131:261–277
35. Thiessen DL, Campbell GL, Adelizi PD (2003) Digestibility and growth performance of juvenile rainbow trout (Oncorhynchus mykiss) fed with pea and canola products. Aquacult Nutr 9:67–75
36. Torstensen BE, Espe M, Sanden M, Stubhaug I, Waagbø R, Hemre G-I, Fontanillas R, Nordgarden U, Harvøy EM, Olsvik P,
37. Berntssen MHG (2008) Novel production of Atlantic salmon (Salmo salar) protein based on combined replacement of fish meal and fish oil with plant meal and vegetable oil blends. Aquaculture 285:193–200
38. Ursin E (1967) A mathematical model of some aspects of fish growth, respiration, and mortality. J Fisheries Res Board Canada 24:2355–2453. 14
39. Webb KA, Jr, Gatlin DM, III (2003) Effects of dietary protein level and form on production characteristics and ammonia excretion of red drum Sciaenops ocellatus. Aquaculture 225:17–26
40. Webster CD, Tiu LG, Tidwell JH, Grizzle JM (1997) Growth and body composition of channel catfish (Ictalurus punctatus) fed diets containing percentages of canola meal. Aquaculture 150:103–112
41. Webster CD, Thompson KR, Morgan AM, Grisby EJ, Gannam AL (2000) Use of hempseed meal, poultry by-product meal and canola meal in practical diets without fish meal for sunshine bass (Morone chrysops × M. saxatilis). Aquaculture 188:299–309

CHAPTER 8

DIFFERENTIAL RESPONSES OF THE GUT TRANSCRIPTOME TO PLANT PROTEIN DIETS IN FARMED ATLANTIC SALMON

Elżbieta Król[1], Alex Douglas[1], Douglas R. Tocher[2], Viv O. Crampton[3], John R. Speakman[1], Christopher J. Secombes[1] and Samuel A. M. Martin[1]

[1] Institute of Biological and Environmental Sciences, University of Aberdeen, Aberdeen AB24 2TZ, UK.

[2] Institute of Aquaculture, University of Stirling, Stirling FK9 4LA, UK.

[3] EWOS Innovation AS, Dirdal, Norway.

ABSTRACT

Background

The potential for alternative plant protein sources to replace limited marine ingredients in fish feeds is important for the future of the fish farming industry. However, plant ingredients in fish feeds contain antinutritional factors (ANFs) that can promote gut inflammation (enteritis) and compromise fish health. It is unknown whether enteritis induced by plant materials with notable differences in secondary metabolism is characterised by common or distinct gene expression patterns, and how using feeds with single vs mixed plant proteins may affect the gut transcriptome and fish performance. We used Atlantic salmon parr to investigate the transcriptome responses of distal gut to varying dietary levels (0–45 %) of soy protein concentrate (SPC) and faba bean (Vicia faba) protein concentrate (BPC) following an 8-week feeding trial. Soybean meal (SBM) and fish meal (FM) were used as positive and negative controls for enteritis, respectively. Gene expression profiling was performed using a microarray platform developed and validated for Atlantic salmon.

Results

Different plant protein materials (SPC, BPC and SBM) generated substantially different gut gene expression profiles, with relatively few transcriptomic alterations (genes, pathways and GO terms) common for all plant proteins used. When SPC and BPC were simultaneously included in the diet, they induced less extensive alterations of gut transcriptome than diets with either SPC or BPC singly, probably due to reduced levels of individual ANFs. The mixed plant protein diets were also associated with improved body composition of fish relative to the single plant protein diets, which may provide evidence for a link between the magnitude of changes in gut transcriptome and whole-animal performance.

Citation: Elżbieta Król , Alex Douglas , Douglas R. Tocher , Viv O. Crampton, John R. Speakman , Christopher J. Secombes and Samuel A. M. Martin, Differential responses of the gut transcriptome to plant protein diets in farmed Atlantic salmon, DOI 10.1186/s12864-016-2473-0

Conclusions

Our results indicate that gut transcriptomic profiling provides a useful tool for testing the applicability of alternative protein sources for aquaculture feeds and designing diets with reduced impact of ANFs on fish health. Ultimately, understanding diet-gut interactions and intestinal homeostasis in farmed fish is important to maximize performance and to ensure that aquaculture continues to be a sustainable source of food for a growing world population.

BACKGROUND

The Atlantic salmon (*Salmo salar* L.) is a carnivorous species and its natural diet consists primarily of aquatic invertebrates in early life and then crustaceans and small fish during the marine phase of growth in later life. Farmed salmon were historically fed diets containing fish meal (FM) as the primary protein source to mimic their natural feeding habits [1]. With the rapid expansion of the salmon farming industry and the limited availability of wild-caught fish, the global contribution of FM to salmon feeds has decreased from ~45 % in 1995 to ~20 % in 2012 and is predicted to reduce to ~12 % by 2020 [2]. Thus, the future of salmon farming will depend to a large extent on sustainable dietary protein alternatives to FM.

In recent years, significant progress has been made towards replacing FM with plant protein meals and plant protein concentrates [3]. The large-scale evaluation of plant protein sources for aquaculture feeds has focused on legumes such as beans, peas and lupins, as these plants are rich in digestible proteins and have favourable amino acid profiles [4]. While the soybean (*Glycine max*) from North and South America continues to dominate as a plant protein source for commercial and research aquaculture feeds, attempts are being made to diversify plant protein sources, with the emphasis on locally grown crops that would reduce the environmental impact of the fish farming industry [5].

Plants produce a wide range of secondary metabolites to defend themselves from predation by herbivores [6]. When consumed, these antinutritional factors (ANFs) typically interfere with digestion, absorption and/or utilisation of nutrients and have numerous adverse effects on animal health and performance [7]. Studies of animal transcriptome responses to ANFs have revealed that these effects are more pronounced when radical changes in the diet are made [8, 9] and in animals with no prior evolutionary experience to the plant materials and associated ANFs [10]. Indeed, feeding carnivorous fish like salmon with plant-based diets and exposing them to the evolutionary novel set of dietary toxins is one of the major challenges facing sustainable aquaculture, coupled with the need for each new plant ingredient to be evaluated for its potential health risks in a dose-dependent manner and in the context of potential interactions with other feed ingredients [11, 12].

Legumes are generally high in ANFs, but their specific profile varies from plant to plant and depends on the method used to extract proteins [13]. When salmon are fed diets containing full-fat or solvent extracted soybean meal (SBM), they develop an inflammatory condition in distal intestine, termed gut inflammation (enteritis), commonly characterized by shortening of mucosal folds, infiltration of the lamina propria by various inflammatory cells and decreased numbers of absorptive vacuoles in the enterocytes [14]. Further processing of SBM to produce soy protein concentrate (SPC) typically uses an aqueous alcohol wash, which removes alcohol-soluble ANFs, such as saponins. Since feeding salmon diets containing SPC does not induce major changes in gut histology or inflammatory responses that resemble the SBM-induced enteritis, development of this condition has been linked to the presence of saponins [15]. Recent studies with purified soy saponins supplemented to salmon diets containing FM, lupin kernel meal or pea protein concentrate (PPC) demonstrated that saponins do not necessarily cause enteritis on their own or without being potentiated by other ANFs [16–19]. Instead, they may act to increase gut permeability and therefore expose the local immune system to antigens that would not normally cross the intestinal epithelial barrier or would cross it at lower rates. Despite intensive research, the antigens responsible for triggering the SBM-induced inflammatory reaction in salmon gut have not yet been identified [20].

Further complexity is added by the fact that enteritis in salmon may also be induced by high dietary inclusions of plant materials that are relatively low in saponins. These include PPC made from the field pea (*Pisum sativum*) and bean protein concentrate (BPC) made from the faba bean (*Vicia faba*), both produced by fine grinding of dehulled seeds into flour, followed by air classification [5, 17]. It remains unknown whether enteritis induced by plant materials with notable differences in secondary metabolism has a common gene expression profile that generates altered gut histology or whether the mechanisms underlying the enteropathy are different (i.e., plant- or ANF-specific), but they subsequently converge at the level of altered gut histology as the inflammatory condition progresses.

Numerous approaches are available to investigate the relationship between dietary toxin exposure, toxicity and disease states [21]. Recent advances in molecular approaches enable the characterization of these effects at the cellular level, by studying the modulation of gene expression by bioactive substances in vivo. Given the number of saponins, proteinase inhibitors, lectins, oligosaccharides and other ANFs present in legumes, their potential interactions and dose-dependent effects on gut function in carnivorous fish are largely unknown. Furthermore, aquaculture feeds that utilise plant proteins are likely to affect multiple rather than individual genes and signalling pathways [22–24]. Global gene expression profiling studies are therefore required to provide insights into diet-gut interactions to ultimately aid protection of fish health in aquaculture.

In this study, we used Atlantic salmon parr to investigate transcriptome responses of distal gut to plant protein diets relative to a reference FM diet, following an 8-week feeding trial described in detail previously [5]. Fish were exposed to plant proteins with distinct sets of ANFs to establish whether similar or distinct alterations of the gut transcriptome were generated, and to characterise the gut transcriptome responses once the plant proteins were mixed. The single plant protein diets contained either SPC or BPC. SPC is typically low in the ANFs contributing to enteritis due to extensive processing [25], while BPC is characterised by high levels of condensed tannins and the presence of faba bean-specific glucosides such as vicine and convicine [26]. The mixed plant protein diets contained both SPC and BPC in three different proportions to generate graded inclusion levels of the plant materials under investigation. We also included an SBM diet to determine whether a distinction could be made between the gene expression profiles of gut inflammation induced by SBM and high levels of BPC. Finally, we investigated potential associations between the magnitude of changes in the gut transcriptome and aspects of fish performance, to provide insights into how diet-induced alterations of the gut transcriptome may affect the whole-animal physiology.

METHODS

Diets

Ingredients and composition of the five experimental and two reference diets are described in detail elsewhere [5] and summarised in Table 1 and Additional file 1. The experimental diets were isonitrogenous, isolipidic and isocaloric and had similar contents of fish meal (~22 %) and similar total plant protein contents (~45 %), but differed in the proportions of soy protein concentrate (SPC) and bean protein concentrate (BPC). These two ingredients were used to replace each other (either partially or fully) at levels of incorporation ranging from 0 to 45 %. As a result, the experimental diets contained either 45 % SPC and 0 % BPC (S_{45}), 34 % SPC and 11 % BPC ($S_{34}B_{11}$), 22 % SPC and 22 % BPC ($S_{22}B_{22}$), 11 % SPC and 34 % BPC ($S_{11}B_{34}$) or 0 % SPC and 45 % BPC (B_{45}). The S_{45} and B_{45} diets were single plant protein diets, while the others ($S_{34}B_{11}$, $S_{22}B_{22}$ and $S_{11}B_{34}$) were mixed plant protein diets. The reference diets were enriched with fish meal (FM diet) or soybean meal (SBM diet) to provide negative and positive controls for gut inflammation, respectively. Diets S_{45}, $S_{34}B_{11}$, $S_{22}B_{22}$, $S_{11}B_{34}$, B_{45}, FM and SBM were the same as the previously described diets 20:80:00 (FM:SPC:BPC), 20:60:20, 20:40:40, 20:20:60, 20:00:80, FMref and HiSBM, respectively [5].

Table 1 Diet formulation, fish performance and histological evaluation of gut health

Parameter	Experimental plant protein diets					Reference diets	
	S_{45}	$S_{34}B_{11}$	$S_{22}B_{22}$	$S_{11}B_{34}$	B_{45}	FM	SBM
Key ingredient differences (g/100 g) between diets[1]							
Soy protein concentrate (SPC)	45	34	22	11	0	16	0
Bean protein concentrate (BPC)	0	11	22	34	45	0	0
Fish meal (FM)	22	22	22	22	22	56	44
Soybean meal (SBM)	0	0	0	0	0	0	36
Fish performance (growth rate and body composition)[2]							
Growth rate (g over 56 days)[3]	9.8^{a} (0.5)	9.7^{a} (0.3)	9.9^{a} (0.3)	10.1^{a} (0.2)	7.0^{b} (0.7)	13.1^{c} (0.2)	11.5^{d} (0.4)
Protein content (%)	16.2^{ab} (0.1)	16.5^{c} (0.1)	16.5^{ac} (0.2)	16.5^{c} (0.3)	16.1^{b} (0.2)	16.4^{ac} (0.1)	16.4^{abc} (0.1)
Lipid content (%)	9.6^{a} (0.2)	10.3^{b} (0.3)	10.0^{ab} (0.4)	10.2^{b} (0.4)	9.9^{ab} (0.1)	11.3^{c} (0.3)	11.0^{c} (0.4)
Ash content (%)	2.29^{a} (0.002)	2.25^{bc} (0.022)	2.26^{b} (0.023)	2.26^{b} (0.016)	2.27^{ab} (0.024)	2.22^{d} (0.017)	2.23^{cd} (0.013)
Histological scores from 1 (no enteritis) to 5 (severe enteritis) in distal gut[4]							
Sub-epithelial mucosa	1 (1–1)	1 (1–1)	1 (1–2)	1 (1–1)	1 (1–2)	1 (1–1)	1 (1–2)
Mucosal folds	2 (1–4)	2 (1–3)	3 (1–4)	2 (1–3)	3 (2–4)	3 (1–4)	3 (2–5)
Lamina propria	2 (1–2)	2 (1–3)	2 (1–3)	2 (1–3)	2 (1–2)	2 (1–3)	2 (2–3)
Eosinophilic granulocytes	2 (2–3)	2 (1–3)	2 (1–3)	2 (1–3)	2 (2–4)	2 (2–3)	2 (2–3)
Goblet cells[5]	3^{a} (1–4)	3^{a} (2–5)	3^{a} (2–4)	3^{a} (1–4)	3^{b} (2–4)	3^{a} (2–4)	3^{b} (2–4)
Supranuclear vacuoles[5]	2^{a} (1–5)	2^{b} (1–4)	2^{a} (1–4)	2^{b} (1–3)	4^{c} (2–5)	3^{a} (1–5)	5^{c} (3–5)
Gut health (based on histological scores)							
Enteritis presence and severity	-	-	-	-	mild	-	moderate

[1]See Additional file 1 for complete diet composition; [2]values are means and standard deviations (in PARENTHESES), see [5] for details; [3]different letters indicate significant differences between the diets (P-value < 0.05); [4]values are medians and ranges (in PARENTHESES), see [5] for details; [5]different letters indicate significant differences between the diets (P-value < 0.05), as assessed by multilevel ordinal regression [5]

Animals and Feeding Trial

The study was conducted at the EWOS Innovation Research Facility (Dirdal, Norway), as described in detail elsewhere [5]. Briefly, Atlantic salmon of the SalmoBreed strain were supplied as fertilized eggs and hatched on site. Mixed-sex juvenile salmon in groups of ~150 fish were transferred to 26 indoor tanks (0.6 × 0.6 × 0.6 m), with ~60 L of freshwater flowing at a rate of 3.9 L/min and continuous aeration. The temperature of water was regulated at 13 °C and all animals were exposed to a constant light regime (24 h light/day). Fish were fed a commercial diet and acclimated to the experimental conditions for 2 weeks, after which (at body weight ~1.5 g) they were subjected to an 8-week feeding trial, consistent with other nutritional studies involving farmed salmon [22, 24, 27]. Tanks were randomly assigned to the dietary treatments, with four replicate tanks per each experimental diet (S_{45}, $S_{34}B_{11}$, $S_{22}B_{22}$, $S_{11}B_{34}$ and B_{45}) and three replicate tanks per each reference diet (FM and SBM diets). Fish were fed to satiation prior to and during the feeding trial using automatic band feeders (Holland Technology, Norway). During the feeding trial, all fish in each tank had their biomass recorded on days 0 (start of experiment), 14, 28, 42 and 56 (end of experiment) for evaluation of growth performance.

Sampling

At the end of the feeding trial, fish from each tank were sampled for body composition and distal gut histology (as described in detail elsewhere [5]). For gene expression profiling of distal gut, five fish per tank for each experimental diet and seven fish per tank for each reference diet were killed and sampled. Prior to sampling, fish were starved for 24 h. For sampling, the peritoneal cavity was opened, alimentary tract excised and gut contents removed. The distal gut (defined as the region from the increase in intestinal diameter and presence of visible folds to the rectum) was subsequently cleaned of mesenteric and adipose tissue. Distal gut samples were immediately transferred to RNA*later*® (Sigma-Aldrich, St. Louis, MO, USA), kept at 4 °C overnight and then stored at −80 °C prior to RNA extraction.

Statistical Analysis of Fish Performance

The effects of diets on growth rate and protein, lipid and ash contents were determined using one-way ANOVA. Fisher post hoc pairwise comparisons were used to compare individual treatment contrasts where required. Modelling assumptions were assessed using standard plots of residuals. All analysis was performed using Minitab 16 (Minitab Inc., State College, PA, USA) and statistical significance was determined at P-value<0.05.

RNA Isolation

Total RNA from distal gut was isolated by homogenization of ~100 mg of tissue in TRIzol® Reagent (Ambion by Life Technologies, Carlsbad, CA, USA), using 3 mm tungsten carbide beads and a TissueLyser II Disruption System (Qiagen GmbH, Hilden, Germany). Following isolation, the RNA was quantified by spectrophotometry (NanoDrop Technologies, Wilmington, DE, USA), had integrity confirmed by electrophoresis (Agilent Technologies, Santa Clara, CA, USA) and was then stored at − 80 °C for later processing.

Design of Microarray Experiment, RNA Amplification and Labelling

The individual RNA samples were pooled to generate five biological replicates per each dietary treatment, with an equimolar contribution of RNA from four fish to each pool (~10 µg of RNA per fish per pool). Fish from the same pool were either from four replicate tanks (experimental diets; one fish per tank) or from three replicate tanks (reference diets; 1-2 fish per tank, with one randomly selected tank contributing twice to the same pool). As a result, each dietary treatment was represented by RNA samples from 20 fish (5 RNA pools × 4 fish per pool). All 35 experimental RNA pools (7 diets × 5 biological replicates) were subsampled to generate a common control, with contribution of RNA samples from 140 fish (7 diets × 20 fish per diet). Antisense amplified RNA (aRNA) with amino allyl UTP incorporation was generated from ~2 µg total RNA per experimental or common control pool, using the Amino Allyl MessageAmp™ II aRNA Amplification Kit (Ambion by Life Technologies, Carlsbad, CA, USA), as described previously [28].

Pools of aRNA were subjected to a coupling reaction between the amino allyl-modified UTP residues and amine reactive Cy fluorescent dyes (Amersham™ Cy™3 and Cy™5 Mono-reactive Dye Packs; GE Healthcare UK Limited, Little Chalfont, UK), as described previously [29]. All experimental samples were labelled with Cy3, and Cy5 was used to label the common control. The reaction products were purified using a DyeEx® 2.0 Spin Kit (Qiagen GmbH, Hilden, Germany). Dye incorporation and post-labelling aRNA yield were quantified by spectrophotometry (NanoDrop Technologies, Wilmington, DE, USA).

Microarray Hybridisation, Scanning and Feature Extraction

The current experiment used a custom designed Agilent oligonucleotide microarray platform Salar_2 (Agilent design ID: 025520), developed for Atlantic salmon and validated by two independent research groups using RT-qPCR [24, 28–30]. The experiment consisted of 35 hybridisations (7 diets × 5 biological replicates), which were performed on nine microarray slides in a semi-randomised order (each slide with four different dietary treatments).

Each hybridisation was performed using 825 ng of Cy3-labelled experimental sample and 825 ng of Cy5-labelled common control. The aRNA was first fragmented and then hybridised at 65 °C for 17 h in an Aligent hybridization oven, as described previously [29]. Following hybridisation, slides were subjected to washing steps, after which they were air dried in the dark and scanned within 2 h.

Scanning was carried out at 5 µm resolution on a GenePix Personal 4100A scanner (Axon Instruments, Molecular Devices Corp., Sunnyvale, CA, USA), with the PMT values adjusted manually to ensure the mean intensity ratio of Cy3:Cy5 signal was close to one. Agilent Feature Extraction Software (version 9.5.3) was used to identify features and extract raw intensity values associated with these features for subsequent statistical analysis.

Microarray Data Analysis

Feature intensities were pre-processed and differential gene expression between dietary treatments identified using R [31] and the Bioconductor package limma [32]. Briefly, loess normalisation was performed within each array to account for intensity-dependent variation in dye bias and quantile normalisation was used to stabilize experimental variances across arrays. Normalised data were then filtered to remove control features and, additionally, non-responsive RNA targets with equal expression across treatments were removed by non-specific filtering [33]. Differential expression of RNA targets between diets was assessed using linear modelling and empirical Bayes methods [34]. Comparisons between diet groups were made by extracting specific contrasts from the linear model. These contrasts were set up to compare each plant protein diet with the reference FM diet. Multiple testing was accounted for by controlling the false discovery rate (FDR) at 5 % using the Benjamini-Hochberg procedure. The RNA targets with adjusted P-value<0.2 and absolute fold change≥1.4 were considered as differentially expressed. The relatively low stringency of the adopted cut-off criteria is consistent with other nutrigenomic studies [35]. All RNA targets meeting these criteria (n=1558) are listed in Additional file 2.

Gene Annotation and Functional Analysis

The salmon RNA targets were mapped to human orthologs to generate HGNC (HUGO Gene Nomenclature Committee) gene symbols for downstream functional analysis. This approach has been demonstrated to improve functional analysis of fish genes by providing access to well-annotated databases and tools for mammalian model organisms, despite limitations of the mapping due to the extra genome duplication events in teleost fish and species differences in gene function and pathways [36]. Each salmon sequence was annotated with the best hit in the human proteome from the Ensembl database (BLASTX), but only hits with$E$-value<0.001 were accepted. As a result, 1341 of 1558 (86 %) differentially expressed RNA targets were annotated, with a number of redundant RNA targets that were mapped to the same gene. The fold changes of the redundant features were then averaged to provide a single fold change for each gene. A total number of 1058 differentially expressed genes (Additional file 2) were identified for subsequent functional analysis.

Sets of differentially up-regulated and down-regulated genes (140, 80, 33, 71, 254 and 750 genes for S_{45}, $S_{34}B_{11}$,

$S_{22}B_{22}$, $S_{11}B_{34}$, B_{45} and SBM dietary treatments, respectively) were analysed using IPA (Ingenuity Pathway Analysis, QIAGEN Redwood City, www.qiagen.com/ingenuity) to identify pathways that were significantly altered by dietary manipulation at *P*-value (calculated by Fisher exact test right-tailed)<0.01. These pathways are reported in Additional file 3, along with their ratios (number of differentially regulated genes divided by the total number of genes within a specific pathway in the IPA reference gene set, called the Ingenuity® Knowledge Base) and IPA activation z-scores (an algorithm to predict the direction of change for a given pathway, with z-score≥2 suggesting an overall increase in the pathway activity and z-score≤−2 predicting an overall decrease in the pathway activity). Furthermore, the IPA downstream effects analysis was performed to identify biological functions that were expected to increase or decrease, taken into account genes' direction of change, i.e., up-regulated or down-regulated status of the differentially expressed genes. This analysis is based on expected causal effects between genes and functions derived from the literature compiled in the Ingenuity® Knowledge Base, and uses the z-score algorithm to make predictions whether the function is significantly increased (z-score≥2), significantly decreased (z-score≤2), non-significantly modified (−2<z-score<2) or with no clear pattern of activation related to the literature (no z-score calculated).

The same sets of differentially expressed genes were also submitted to DAVID Bioinformatics Resources 6.7 [37, 38] to examine the enrichment of Gene Ontology (GO) terms for biological processes against the human reference gene set. The analysis was performed on category GOTERM_BP_FAT using the functional annotation tool with a default threshold of 2 counts. The GO terms with a modified Fisher exact *P*-value (EASE score)<0.01 were considered significantly enriched (Additional file 4).

RESULTS

Phenotypic Characteristics of Fish

Changes in the phenotype of the fish subjected to the dietary manipulation are described in detail elsewhere [5] and summarised in Table 1. Briefly, salmon with higher content of fish meal in the diet grew faster, while salmon fed the same level of fish meal but different levels of soy and bean protein concentrates had similar growth rates, apart from the group fed the diet containing 45 % BPC (B_{45}), which grew significantly slower (Figure. 1). High levels of fish meal in the reference FM and SBM diets (56 and 44 %, respectively) had no effect on salmon protein content, but fish fed these diets tended to have higher lipid and lower ash contents than fish fed the plant protein diets with 22 % of fish meal in the diet. Furthermore, fish fed the mixed plant protein diets ($S_{34}B_{11}$, $S_{22}B_{22}$ and $S_{11}B_{34}$) had significantly higher protein (16.5 % *vs* 16.1–16.2 %) and lipid (10.0–10.3 % *vs* 9.6–9.9 %) contents and lower ash contents (2.25–2.26 % *vs* 2.27–2.29 %) compared with fish fed the single plant protein diets (S_{45} and B_{45}).

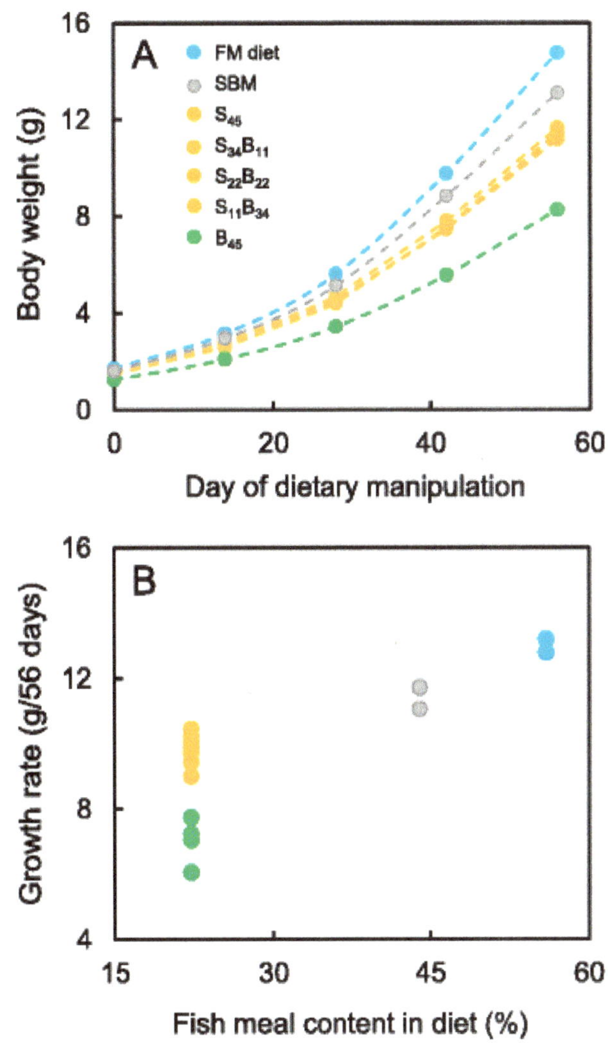

Figure. 1 Body weight and growth of fish during an 8-week feeding trial. Different colours indicate significant differences in growth rate of fish between diets (*P*-value<0.05, see Table 1 for details). Dietary manipulations were .

Histological examination of distal gut in fish fed experimental and reference diets involved scoring six histological parameters (sub-epithelial mucosa, mucosal folds, lamina propria, eosinophilic granulocytes, goblet cells and supranuclear vacuoles) from 1 (no enteritis) to 5 (severe enteritis). Neither sub-epithelial mucosa, mucosal folds, lamina propria nor eosinophilic granulocytes were affected by dietary manipulation. In contrast, both goblet cells and supranuclear vacuoles had elevated scores for enteritis in fish fed B_{45} and SBM diets. Overall, fish fed the FM diet (a negative control for enteritis) as well as fish fed S_{45}, $S_{34}B_{11}$, $S_{22}B_{22}$ and $S_{11}B_{34}$ diets had no signs of gut inflammation, fish fed B_{45} diet showed signs of mild enteritis, while fish on the SBM diet (a positive control for enteritis) developed moderate enteritis.

Magnitude of Gut Transcriptome Responses to Single *vs.* Mixed Plant Protein Diets

Among all seven diets used in this study, the reference FM diet most closely resembled the natural diet of salmon. We therefore considered the distal gut of fish fed the FM diet as being characterised by gene expression typical for a healthy digestive system. All responses of gut transcriptome to the plant protein diets were expressed relative to the FM diet, with up-regulation and down-regulation referring to the higher and lower levels of gene expression in fish fed plant protein diets than in fish fed the FM diet, respectively.

Single plant protein diets with either 45 % SPC (S_{45}) or 45 % BPC (B_{45}) induced more extensive changes in gut transcriptome than mixed plant protein diets ($S_{34}B_{11}$, $S_{22}B_{22}$ and $S_{11}B_{34}$), based on the number of differentially expressed genes (Figure. 2a). Specifically, the S_{45} and B_{45} diets were associated with changes in 140 and 254 genes, respectively, whereas only 80, 33 and 71 genes were altered by $S_{34}B_{11}$, $S_{22}B_{22}$ and $S_{11}B_{34}$ diets, respectively (Additional file 2). Similar trends with more extensive changes in the gut transcriptome induced by single plant protein diets compared with mixed plant protein diets were also observed for the numbers of enriched pathways (Figure. 2b, Additional file 3) and the numbers of enriched GO terms for biological processes (Figure. 2c, Additional file 4). Importantly, fish fed the mixed plant protein diets (with smaller alterations of gut transcriptome) had higher protein and lipid contents and lower ash content than fish fed either S_{45} or B_{45} diets (with larger alterations of gut transcriptome) (Table 1, Figure. 2). Overall, the most extensive changes in gut transcriptome (alterations of 750 genes, 68 pathways and 78 GO terms) were found in the fish fed the SBM diet that induced moderate enteritis (details below).

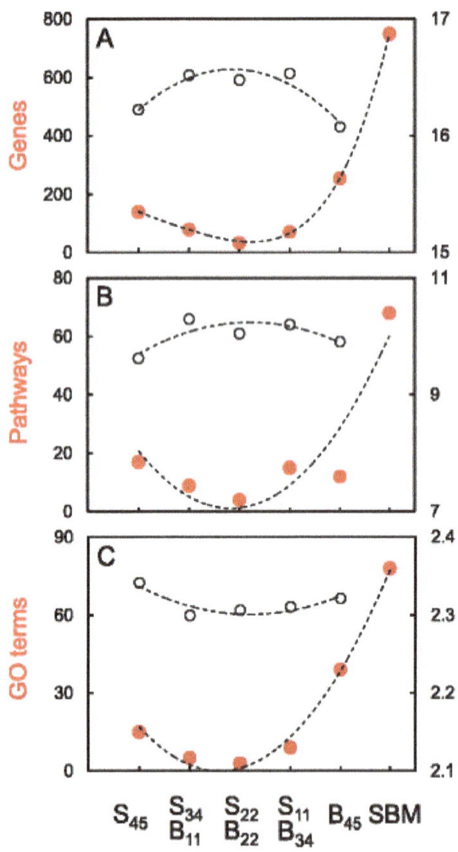

Figure. 2 Magnitude of gut transcriptome responses to dietary plant proteins. **a**Number of differentially expressed genes (adjusted *P*-value<0.2, absolute fold change≥1.4) for single (S_{45} and B_{45}) *vs* mixed (S_{34} **...**

Analysis of the common and unique features altered in the gut transcriptome by single and mixed plant protein diets revealed no obvious synergistic interactions between soy and bean protein concentrates (Additional file 5). Instead, many features (at gene, pathway and GO term levels) altered in the gut of fish fed mixed plant protein diets ($S_{34}B_{11}$, $S_{22}B_{22}$ and $S_{11}B_{34}$) were also altered by single plant protein (S_{45} and B_{45}) or SBM diets. The focus of subsequent analysis was therefore to characterise the gut transcriptome responses to the diets with single plant protein ingredients, either highly processed (SPC), moderately processed (SBM) or air-classified (BPC), and determine whether the observed effects persisted, disappeared or existed at the dose-dependent intermediate levels after the plant proteins were mixed.

Specificity of gut transcriptome Responses to Soy and Bean Proteins

Analysis of common and unique features altered in the distal gut of fish fed diets containing 45 % SPC (S_{45}), 45 % BPC (B_{45}) or 36 % SBM revealed that all these ingredients generated highly specific gut transcriptome responses, reflecting the differences in the origin of the plant protein (soy *vs* bean) as well as differences in the level of processing protein from the same plant (highly processed SPC *vs* moderately processed SBM) (Figure. 3, Additional files 6, 7 and 8). The S_{45} diet was associated with significant alterations of 140 genes, 17 pathways and 15 GO terms, of which changes in 85 genes, six pathways and six GO terms were S_{45}-specific. The majority of non-specific responses to this diet were shared with the SBM treatment rather than with the B_{45} diet, indicating the presence of the common signature of the soy protein that was not affected by the extraction method used to produce SPC. Among 254 genes, 12 pathways and 39 GO terms altered by the B_{45} diet, 160 genes, six pathways and 30 GO terms were B_{45}-specific. The B_{45} diet shared more alterations with the SBM diet rather than with the S_{45} diet, suggesting the common signature of gut inflammation induced by both these diets. Finally, the gut transcriptome responses to the SBM diet was also highly specific, with 643 of 750 genes, 55 of 68 pathways and 61 of 78 GO terms being unique for that treatment.

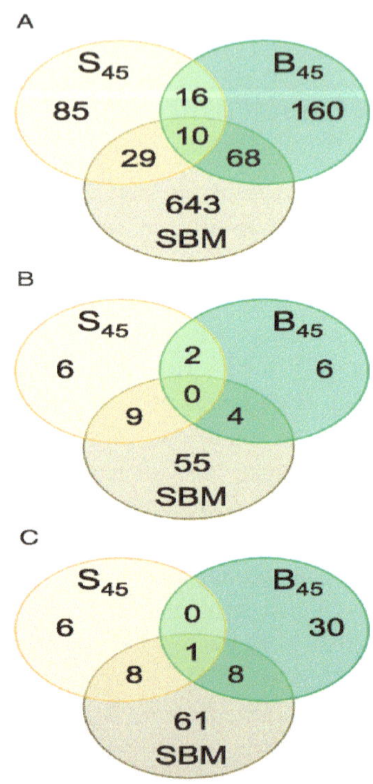

Figure. 3 Characterisation of gut transcriptome responses to dietary plant proteins. Venn diagrams show the numbers of common and unique features altered in gut transcriptome by single plant protein diets, enriched with either soy protein concentrate (S_{45}), bean **...**

Effects of diet with 45 % of Soy Protein Concentrate (S_{45})

Among 17 pathways significantly altered in the distal gut of fish fed S_{45} diet, cholesterol biosynthesis pathway (details below) was the most enriched (Additional file 9). Consistent with these changes were also changes mapped to biosynthesis of zymosterol, an intermediate in the biosynthesis of cholesterol. Furthermore, the long-chain polyunsaturated fatty acid (LC-PUFA) biosynthesis pathway and the oleate biosynthesis pathway were also highly enriched.

The S_{45} diet had relatively little impact on gut inflammatory and immune responses, as only one pathway (complement system) related to these functions was affected. The alteration of this pathway was unique for the S_{45} diet and associated with the altered expression of CFD, SERPING1 and C7 genes, all of which were down-regulated. Other pathways specific for the S_{45} diet included mismatch repair in eukaryotes, hypoxia signalling in the cardiovascular system and PCP pathways.

Neither differential expression of genes encoding the cytochrome P450 superfamily of enzymes (CYP51A1, CYP2F1 and CYP2A7) nor pathways associated with these genes were specific to the S_{45} diet. These pathways included bupropion degradation, acetone degradation, estrogen biosynthesis, nicotine degradation and melatonin degradation, and their significant enrichments were shared either with the B_{45} or SBM diets.

The IPA downstream effects analysis predicted significant changes in the gut of fish fed the S_{45} diet, related to two functional categories (Table 2, Additional file 10). The first was an increase in synthesis of steroid, based on the expression profile of 8 genes (ACAT2, CFTR, CREB1, CYP51A1, DHCR7, FDFT1, G6PC and SLC27A2). The second predicted change was a decrease in concentration of triacylglycerol, with nine genes involved (ALOX5, CACNA1B, FABP6, FADS2, G6PC, JUN, MGAT3, PRF1 and TXNIP).

Table 2 IPA downstream effects analysis performed on genes differentially expressed in distal gut of Atlantic salmon

Diet/Function	Predicted activation state	Activation z-score	*P*-value	Number of genes[a] consistent[b]	inconsistent[c]	unknown[d]
S_{45} (diet with 45 % of soy protein concentrate)						
synthesis of steroid	Increased	2.0	2.2E-03	4	0	4
concentration of triacylglycerol	Decreased	−2.4	1.1E-03	8	1	0
B_{45} (diet with 45 % of bean protein concentrate)						
cell-cell adhesion	Decreased	−2.6	8.0E-07	7	0	7
growth of epithelial tissue	Decreased	−2.4	3.2E-05	17	7	2
development of blood vessel	Decreased	−2.1	3.5E-05	16	6	4
cell movement of epithelial cells	Decreased	−2.0	2.7E-05	8	2	0
cell death of smooth muscle cells	Decreased	−2.1	1.3E-03	6	1	0
apoptosis of connective tissue cells	Decreased	−2.1	3.3E-03	8	2	1
Soybean meal (SBM) diet						
proliferation of cells	Increased	2.4	4.7E-15	125	90	52
binding of DNA	Increased	2.3	1.5E-05	28	14	6
synthesis of protein	Increased	2.1	6.6E-07	22	10	12
synthesis of lipid	Increased	2.4	1.6E-06	30	15	20
metabolism of carbohydrate	Increased	2.6	1.5E-04	16	4	30
organization of cytoplasm	Increased	2.3	2.4E-04	38	21	38
cellular homeostasis	Increased	2.4	2.0E-05	43	27	37
cell movement	Increased	3.4	1.1E-07	87	47	15
cell death	Decreased	−2.1	6.5E-14	124	88	36
autophagy	Increased	2.5	1.6E-04	18	6	6

[a]Genes are listed in Additional file 10; [b]Genes with the direction of expression consistent with the predicted activation state of the function (e.g., gene encoding CD28 molecule is known to decrease cell proliferation and is down-regulated in the SBM dataset, therefore predicted to increase the function); [c]Genes with the direction of expression inconsistent with the predicted activation state of the function (e.g., gene encoding DIABLO is known to decrease cell proliferation and is up-regulated in the SBM dataset, therefore predicted to decrease the function); [d]Genes involved in the function with currently unknown effects

Effects of diet with 45 % of bean protein concentrate (B_{45})

Among 12 pathways altered in the gut transcriptome of fish fed high inclusion of BPC, seven of these pathways were associated with differential expression of three genes encoding integrins, the cell adhesion molecules (Additional file 11). Two of these genes (ITGA6 and ITGB1) were B_{45}-specific, while ITGB3 was also differentially expressed in the distal gut of fish fed the SBM diet. Among other genes that substantially contributed to the gut signature of the high-bean diet were TNF, a multifunctional proinflammatory cytokine, and its receptor TNFRSF1A.

Consistent with the altered gut histology (mild enteritis), the B_{45} diet triggered in the gut transcriptome a number of inflammatory and immune responses that were either B_{45}-specific or common for the B_{45} and SBM diets. In total, six pathways were specific to the high-bean diet, including granulocyte adhesion and diapedesis (Figure. 4a) and agranulocyte adhesion and diapedesis (no graphic representation). Both these pathways are involved in the migration of granulocytes (neutrophils, basophils and eosinophils) and agranulocytes (lymphocytes and monocytes) from the vascular system to sites of pathogenic exposure and are the key events in the process of inflammation. Highly enriched and specific to the high-bean diet was also caveolar-mediated endocytosis signalling pathway, important for endocytic internalization of a variety of particles (including molecules, viruses and bacteria) and acting through caveolae rather than clathrin lattices. Finally, the significant alterations of the tight junction signalling pathway (Figure. 4b) suggested the potential of B_{45} diet to modify the gut epithelial permeability, especially via a paracellular route, for which tight junctions are the main regulators.

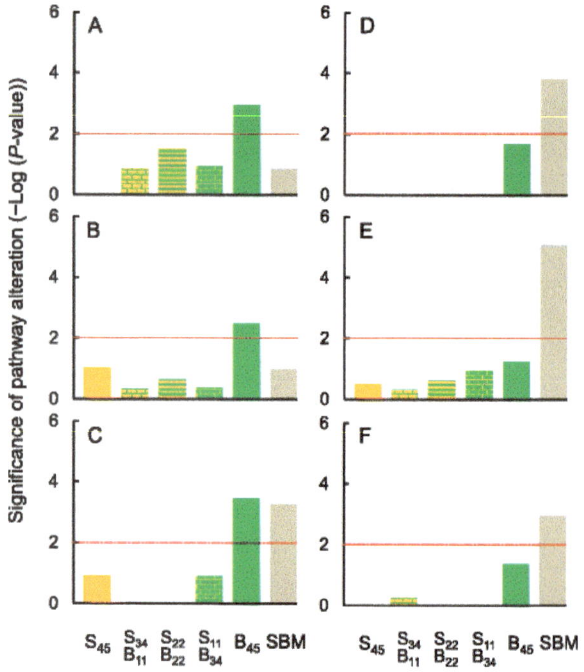

Figure. 4 Top pathways altered in distal gut of Atlantic salmon. **a**, **b** and **c**Pathways altered by the diet with 45 % of bean protein concentrate (B_{45}) included **a** granulocyte adhesion and diapedesis, **b** tight junction signalling and **c** ILK signalling. **d**, **e** and **...**

The gut transcriptome responses common for the B_{45} and SBM diets included alteration of ILK (integrin-linked kinase) signalling pathway (Figure. 4c), implicated in cellular responses to a wide variety of extracellular stimuli. Other aspects of the common signature of gut inflammation induced by these diets resembled the alterations of signalling between the junction of germ cells and Sertoli cells for timely germ cell movement within the seminiferous epithelium (germ cell-Sertoli cell junction signalling pathway). Furthermore, both diets induced gut transcriptome changes related to clathrin-mediated endocytosis signalling pathway, with clathrin-mediated endocytosis being the major pathway for the internalization of nutrients, hormones and other signalling molecules from the plasma membrane into intracellular compartments.

According to the IPA downstream effects analysis, the high-bean diet affected many aspects of intestinal epithelial development and function (Table 2, Additional file 10). Among the most affected were decreased cell-cell adhesion and impaired growth of epithelial tissue, coupled with reduced development of blood vessel and decreased cell movement of epithelial cells. The prediction was also made with regard to cell death of smooth muscle cells and apoptosis of connective tissue cells, with both functions decreased.

Effects of Soybean meal (SBM) Diet

As expected, the SBM diet altered gut histology (moderate enteritis) and induced extensive changes in gut transcriptome, represented in this study by alterations of 750 genes, 68 pathways and 78 GO terms (Figure. 3). These changes are important to understand the potential mechanisms underlying SBM-induced enteropathy and provide insights into a coordinated series of molecular, cellular, tissue and systemic responses that aim to control inflammation and restore gut homeostasis and function.

In total, 20 of 68 pathways were associated with various aspects of inflammatory and immune responses (both innate and acquired), 18 of which were unique to the SBM treatment (Additional file 12). Despite a relatively long exposure to the SBM diet (8 weeks), some aspects of gut transcriptome responses were typical for the early defence systems, such as acute phase response, NF-κB and IL-8 signalling pathways, known to stimulate the synthesis of various proinflammatory cytokines and mediators. These alterations may suggest the excessive translocation of luminal bacteria, viruses and antigens across the intestinal epithelium, which was further supported by the enrichment of the pathways related to endocytosis, such as clathrin-mediated endocytosis signalling (shared with B_{45} diet), macropinocytosis signalling and virus entry via endocytic pathways. The potential contribution of microbial translocation to SBM-induced enteritis was reinforced by the substantial up-regulation of the IFNG transcript and the enrichment of associated pathways (e.g., role of PKR in interferon induction and antiviral response) as well as modification of pathways directly linked to the intestinal barrier function and permeability (e.g., remodeling of epithelial adherens junctions and epithelial adherens junction signalling). Furthermore, our transcriptomic data were consistent with the increased phagocytic activity of macrophages (Fcγ receptor-mediated phagocytosis in macrophages and monocytes pathway) and neutrophils (fMLP signalling in neutrophils pathway), coupled with the highly significant alteration of the pathway related to production of nitric oxide and reactive oxygen species in macrophages (Figure. 4d). Finally, the SBM-induced enteritis was associated with the transcriptomic activation of T cells by alteration of pathways related to CD28 signalling and CTLA4 signalling, as well as modification of the T cell receptor signalling pathway. According to our gene expression data, some of these T cells may have potentially responded to activation by killing other cells via the induction of a programmed cell death known as apoptosis (enrichment of the pathway related to cytotoxic T lymphocyte-mediated apoptosis of target cells).

The distal gut of fish fed the SBM diet showed at the transcriptomic level clear signs of increased cellular damage and injury. Part of this damage was likely to result from the increased production of nitric oxide and reactive oxygen species, known for their microbicidal and tumoricidal properties and generated by phagocytic cells to control infection, but inevitably affecting also the host cells. Indeed, the pathways associated with the detoxification of reactive intermediates and alleviation of oxidative stress (e.g., NRF2-mediated oxidative stress response and eNOS signalling) were highly enriched in the gut transcriptome of fish fed the SBM diet (Figure. 4e). The effects of the SBM diet on gut gene expression were also consistent with the accumulation of unfolded and misfolded proteins in endoplasmic reticulum (ER) lumen, as suggested by the alteration of ER stress pathway. The presence of the ER stress was further supported by the transcriptomic activation of unfolded protein response pathway, indicating the attempts to resolve the protein folding defect and restore ER homeostasis or promoting apoptosis (under unresolvable ER stress conditions). In fact, some aspects of the SBM-induced ER stress were probably unresolvable, because the inflamed gut tissue generated (at the transcriptomic level) multiple signals known to initiate apoptosis, with the up-regulation of DIABLO and CASP8 genes among the most important. The pathways associated with the apoptosis signalling (e.g., Myc-mediated apoptosis signalling, apoptosis signalling and death receptor signalling) were also highly enriched.

The changes induced by the SBM diet in the gut transcriptome were consistent with increased cellular proliferation and migration. These changes included the up-regulation of PCNA (a universal marker of proliferating cells) as well as alterations of the pathways associated with cell cycle control, such as cell cycle: G2/M DNA damage checkpoint regulation. An increased migration of cells within the inflamed gut tissue was supported by the enrichment of pathways related to actin cytoskeleton signalling and integrin signalling, essential for cytoskeletal remodelling during cell adhesion, loss of attachment and successive re-adhesion of migrating cells (Figure. 4f).

Increased cell proliferation of the intestinal tissue with the SBM-induced enteritis was also predicted by the IPA downstream effects analysis, along with concomitant increases in binding of DNA, synthesis of protein, synthesis of lipid, metabolism of carbohydrate, organization of cytoplasm, cellular homeostasis and cell movement (Table 2, Additional file 10). The prediction was also made with regard to increased autophagy. Despite the activation of apoptosis signalling pathways, cell death in the inflamed gut tissue was predicted to decrease.

Dose-dependent Alterations of Cholesterol and LC-PUFA Biosynthesis Pathways

Dose effects of dietary plant proteins on distal gut gene expression were established by examining the relationship between SPC/BPC content and the significance of pathway alteration ($-\mathrm{Log}_{10}$ (*P*-value)). The effects of SPC could not be separated from the effects of BPC, because both ingredients were changing simultaneously to replace each other.

The dose effects of SPC/BPC were most clearly visible in the cholesterol biosynthesis pathway, with the highest levels of alteration in the gut of fish fed S_{45} and $S_{34}B_{11}$ diets, the lowest levels when fish were fed $S_{11}B_{34}$ and B_{45} diets, and intermediate levels for $S_{22}B_{22}$ and SBM diets (Figure. 5a). Correlations between the significance of the pathway alteration and the highly or moderately processed soy protein contents ($r=0.94$, *P*-value$=0.006$, $n=6$ diets, Figure. 5c) as well as the BPC content ($r=-0.88$, *P*-value$=0.020$, $n=6$ diets, no graphic representation) were both highly significant. In total, 7 of 39 genes mapped to the cholesterol biosynthesis pathway were significantly altered by our dietary manipulations (Table 3). All these genes were up-regulated relative to the FM diet, which may suggest an overall increase in the pathway activity.

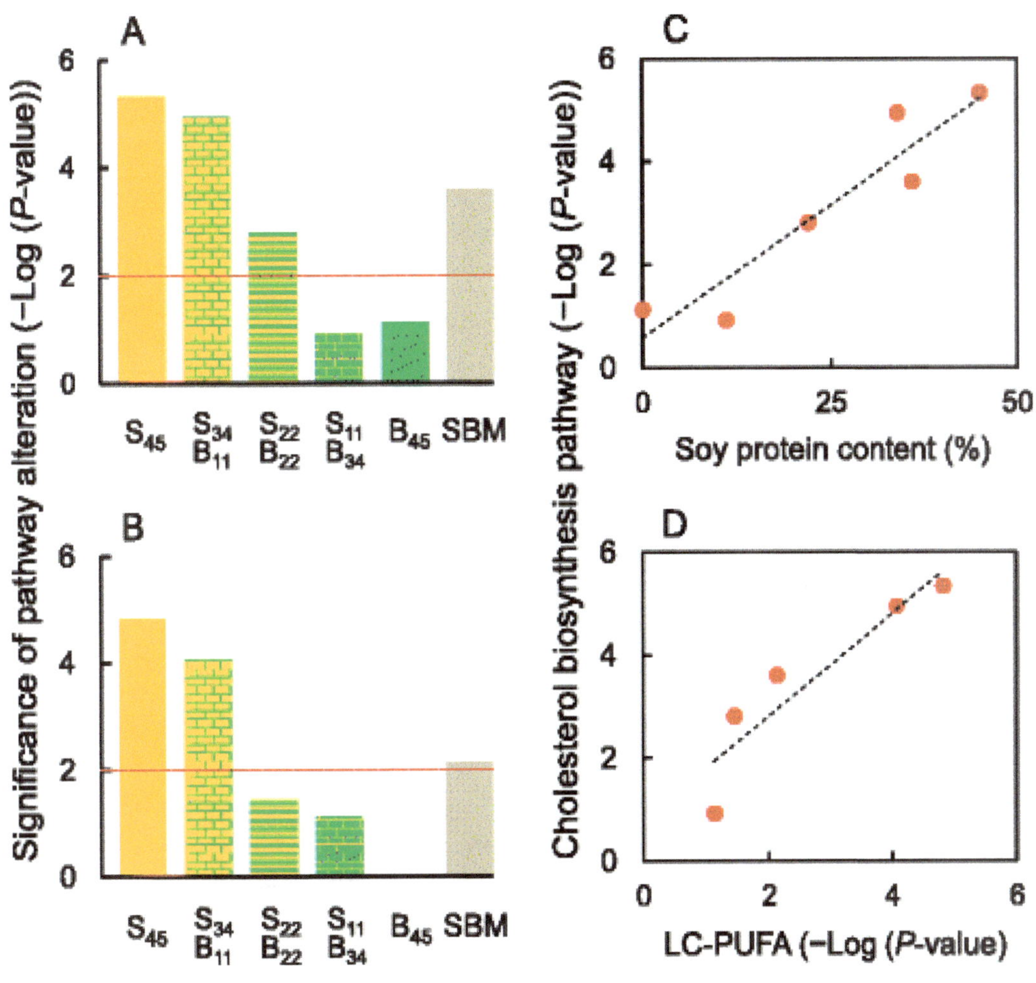

Figure. 5 Transcriptomic alterations of cholesterol and LC-PUFA biosynthesis pathways in distal gut of Atlantic salmon. **a** and **b** Significance of pathway alteration ($-\mathrm{Log}_{10}$ (*P*-value)) for **a** cholesterol biosynthesis and **b** LC-PUFA biosynthesis pathway in relation ...

Table 3 Transcriptomic alterations of cholesterol and LC-PUFA biosynthesis pathways in distal gut of Atlantic salmon

Pathway/Gene	Gene fold change relative to fish meal (FM) diet[a]					
	S_{45}	$S_{34}B_{11}$	$S_{22}B_{22}$	$S_{11}B_{34}$	B_{45}	SBM
Cholesterol biosynthesis						
FDFT1, farnesyl-diphosphate farnesyltransferase 1	1.8					2.9
CYP51A1, cytochrome P450, family 51, subfamily A, polypeptide 1	3.1	2.3				2.6
TM7SF2, transmembrane 7 superfamily member 2	2.1	1.8	1.7			1.5
MSMO1, methylsterol monooxygenase 1					2.5	2.4
NSDHL, NAD(P) dependent steroid dehydrogenase-like						2.4
SC5D, sterol-C5-desaturase	2.5	1.9				1.9
DHCR7, 7-dehydrocholesterol reductase	1.9	1.9	2.2	1.9	1.6	1.6
LC-PUFA biosynthesis						
ACSBG2, acyl-CoA synthetase bubblegum family member 2				15.3		
ACSL5, acyl-CoA synthetase long-chain family member 5						−4.2
ACSL6, acyl-CoA synthetase long-chain family member 6						−1.6
CYB5A, cytochrome b5 type A (microsomal)						1.7
CYB5R3, cytochrome b5 reductase 3	1.6					1.5
FADS2, fatty acid desaturase 2	2.1	2.0	2.4			
SLC27A2, solute carrier family 27 (fatty acid transporter), member 2	1.7	1.6				

S_{45}, $S_{34}B_{11}$, $S_{22}B_{22}$, $S_{11}B_{34}$, B_{45} and SBM are diets with 45 % of soy protein concentrate (SPC), 34 % SPC and 11 % of bean protein concentrate (BPC), 22 % SPC and 22 % BPC, 11 % SPC and 34 % BPC, 45 % BPC, and enriched with soybean meal (SBM), respectively

[a]Only genes considered as differentially expressed (adjusted *P*-value < 0.2, absolute fold change ≥ 1.4) are reported

The dose-dependent alteration profile of the LC-PUFA biosynthesis pathway (annotated as γ-linolenate biosynthesis pathway) (Figure. 5b) closely resembled the profile of cholesterol biosynthesis pathway (Figure. 5a), despite different sets of genes involved in these pathways (Table 3). In total, seven of 24 genes mapped to the LC-PUFA biosynthesis pathway were altered, including up-regulation of the FADS2 gene, known to catalyse biosynthesis of 18:4n-3 from 18:3n-3 and 18:3n-6 from 18:2n-6. The alteration levels of LC-PUFA and cholesterol biosynthesis pathways were highly correlated ($r=0.92$, P-value$=0.025$, $n=5$ diets, Figure. 5d).

Discussion

The present study is the first to demonstrate that mixing plant proteins with distinct sets of ANFs induces less extensive changes in the gut transcriptome of carnivorous fish than using diets with a single plant protein source. In our investigation, we assumed that the gut of fish fed the reference FM diet was characterised by levels of gene expression typical of a healthy digestive system. At the level of the transcriptome, the guts from fish on diets containing both SPC and BPC were closer to that healthy baseline than the guts from fish on diets with either SPC or BPC (Figure. 2). Further support for this assumption comes from the changes in gut transcriptome induced by plant proteins relative to the FM diet, where there was a good agreement between the transcriptomic and histological data. Specifically, the transcriptomic changes were the greatest in the distal gut showing histological signs of moderate enteritis (SBM diet), intermediate in the gut showing mild enteritis (B_{45} diet), and the smallest in the guts having no histological signs of gut inflammation (S_{45}, $S_{34}B_{11}$, $S_{22}B_{22}$ and $S_{11}B_{34}$ diets). Despite the gene expression profile and histological changes consistent with gut inflammation, fish fed the SBM diet grew significantly faster than fish with healthy guts on the experimental diets (Table 1, Figure. 1). This result may be explained by the high levels of FM in the SBM diet (44 % *vs* 22 % in the experimental diets), which are known to promote growth of carnivorous fish and are commonly added in excess to mitigate the inflammatory condition induced by SBM during nutritional experiments [39]. In contrast, mild enteritis was likely implicated in the reduced growth rate of fish fed the B_{45} diet, which contained the standard levels of FM (22 %).

The primary aim of the present study was to determine whether plant proteins with distinct sets of ANFs produced similar or different changes in the gut transcriptome of carnivorous fish. We generated gene expression profiles corresponding to three diets with either 45 % SPC (S_{45}), 45 % BPC (B_{45}) or SBM, and demonstrated that these diets evoked profoundly different alterations of the gut transcriptome, with a relatively small number of features (genes, pathways and GO terms) that were altered commonly by any two or all three treatments (Figure. 3). Importantly, the two soy-based diets shared more pathways with each other ($n=9$) than with the bean-based diet (two pathways common for S_{45} and B_{45} diets, and four pathways common for SBM and B_{45} diets). Furthermore, the pathways shared by the soy-based diets had typically similar sets of genes altered, while the pathways common for the soy- and bean-based diets had usually different sets of genes affected by the dietary manipulations (Additional files 9, 11 and 12), supporting our findings about the ANF-specific alterations of gut transcriptome in Atlantic salmon. Overall, our results are consistent with the use of microarray technology to distinguish between gene expression patterns generated by structurally unrelated toxicants in the livers of laboratory rats [21], natural and novel plant secondary compounds in the livers of desert woodrats [10] and acute and chronic mycobacterial infections in zebrafish [40].

Since the gut transcriptome alterations induced by SPC and BPC were either dose-dependent (e.g., cholesterol and LC-PUFA biosynthesis pathways), associated with the high dose of SPC (e.g., complement system pathway) or associated with the high dose of BPC (e.g., granulocyte adhesion and diapedesis pathway), mixing those two plant protein ingredients reduced the absolute amount of each of them in the diet, which subsequently resulted in reduced transcriptomic changes in the gut (Figure. 2). It is important to note that even the plant proteins considered as most suitable for use in aquaculture and showing no associated histological abnormalities in the gut, such as SPC, altered the gut transcriptome when used at high doses (45 % in this study). Hence, elimination of those high dose effects by diversification of plant protein ingredients used for aquaculture feeds may be important for controlling individual and cumulative levels of ANFs to which fish are exposed, leading to improvements in fish health and performance. Indeed, fish fed the mixed plant protein diets with relatively small alterations of gut transcriptome were characterised by higher protein and lipid contents and lower ash content than fish fed diets with either 45 % SPC or 45 % BPC (Figure. 2). The causality of this relationship was not addressed in our study and requires further investigation.

Both SBM and B_{45} diets significantly altered the histology of the distal gut and induced moderate and mild gut inflammation, respectively (Table 1). These diets had distinct sets of ANFs, which in turn generated profoundly different alterations of gut transcriptome, with some aspects of transcriptomic modifications common for both diets. The nature of these common responses (especially alterations of ILK signalling and germ cell-Sertoli cell junction signalling pathways) suggests that although the mechanisms by which different ANFs affect gut health may be different, they are all likely to contribute to the overall loss of intestinal integrity, thus facilitating antigenic translocation and promoting the persistence of enteritis. We believe that the overlapping gut transcriptomic responses to SBM and B_{45} diets are of special interest for aquaculture as they may harbour biomarkers that characterise all types of gut inflammatory diseases, independent of their origin and aetiology. Identification of these potential biomarkers and their further validation is beyond the scope of this study.

CONCLUSIONS

In summary, we have demonstrated that different plant protein materials, such as SPC and SBM made from soybean and BPC made from faba bean, generated substantially different gene expression profiles in the distal gut of Atlantic salmon relative to guts from fish fed a FM-based diet, with relatively few transcriptomic alterations common for all plant proteins used. These different expression patterns likely reflect different sets of ANFs present in the highly (SPC) and moderately (SBM) processed soybean and air-classified faba bean (BPC). When different plant proteins (SPC and BPC) were mixed, they generated less extensive alterations of the gut transcriptome relative to single plant protein diets with either SPC or BPC, probably due to reduced levels of individual ANFs. The mixed plant protein diets were also associated with an improved body composition of the fish relative to the effects seen with single plant protein diets, which may provide evidence for a link between the magnitude of changes in the gut transcriptome and whole-animal performance. Furthermore, fish with more advanced gut inflammation (moderate enteritis induced by the SBM diet, assessed histologically) had more extensive alterations of gut transcriptome than fish with mild enteritis, induced by the high levels of BPC in the diet. The common aspects of these predominantly different gene expression profiles of SBM- and BPC-induced enteritis may have additional diagnostic value when fish are screened for gut health and potential enteropathies. Our results clearly indicate that gut transcriptomic profiling provides a useful tool for testing the applicability of alternative protein sources for aquaculture feeds and for designing diets that have a reduced impact of ANFs on fish health.

Perspectives and Significance

Our paper addresses the complex diet-gut interactions and intestinal homeostasis that are of special interest for the fish farming industry for a number of reasons [41]. Firstly, farmed fish kept at high stocking densities are susceptible to intestinal infections, with the gut being an important entry point for pathogens. Secondly, farmed fish are typically fed commercial pelleted feeds, which opens up avenues for manipulating fish health through the incorporation of various drugs and vaccines into the feed. Thirdly, the gut immune system of teleost fish allows microbial colonization by symbionts, and this microbial community is a potential platform to modulate fish pathogens. Finally, gut microbiota in fish are likely to respond to dietary manipulations [42]. Hence, a comprehensive understanding of the diet-gut interactions and immunoregulatory properties of intestinal epithelium in fish could aid in the development of new strategies to prevent and treat their multiple infectious and inflammatory diseases. Ultimately, understanding the diet-gut interactions and intestinal homeostasis in farmed fish is important to maximise performance and to ensure that aquaculture continues to be a sustainable source of food for the growing world population.

Attempts to understand the diet-gut interactions are important because of their widespread ramifications across many areas of inquiry beyond aquaculture. The gastrointestinal tract of vertebrates along with its single layer of epithelial cells constitutes the largest and most important barrier against the external environment [43]. The intestinal epithelium acts as a selectively permeable barrier for dietary nutrients, electrolytes and water, while maintaining an effective defence against luminal toxins, antigens and microbial communities. The epithelial cells are also crucial mediators of mucosal innate and adaptive immunity, and they actively contribute to maintaining intestinal homeostasis [44]. The latter may be at risk, when humans and other animals consume evolutionary 'mismatched' diets, i.e., diets they did not evolve to digest, absorb and utilise, and which typically come with novel sets of dietary toxins, antigens and microbial challenges. These factors are likely to contribute to the increased intestinal epithelial permeability, which has been implicated in the predisposition to intestinal inflammation and a number of gastrointestinal diseases in humans [43], zoo animals [45] and pets [46], laboratory rodents [47], domesticated livestock [48] and poultry [49]. Evidence is also growing that many wild animals are facing new dietary challenges resulting from global climate change [50, 51]. Our results clearly indicate that farmed Atlantic salmon may provide an attractive new animal model for investigating the complex interactions between the digestive system and evolutionary mismatched diets in vertebrates, at both whole-animal and molecular levels.

Availability of Supporting Data

The data sets supporting the results of this article are available in the ArrayExpress repository (http://www.ebi.ac.uk/arrayexpress/) under accession numbers A-MEXP-2065 (Salar_2 array design) and E-MTAB-3613 (microarray raw data).

Ethical Statement

The feeding trial and sampling procedures were approved by the Ethical Committee of the Institute of Aquaculture (University of Stirling, UK) and performed in compliance with laws regulating experimentation with live animals in Norway, as overseen by the Norwegian Animal Research Authority (Forsøksdyrutvalget). All diets used in this trial met the nutritional requirements of the Atlantic salmon.

ACKNOWLEDGEMENTS

The study was supported by the UK Technology Strategy Board (TSB Grant 11974-81166). We are grateful to the three anonymous referees who made valuable comments on the earlier version of the manuscript.

Abbreviations

General

ANF: antinutritional factor; aRNA: amplified RNA; B_{45}: diet with 0 % SPC and 45 % BPC; BPC: bean protein concentrate; Cy: cyanine; ER: endoplasmic reticulum; FM: fish meal; GO: gene ontology; HGNC: HUGO Gene Nomenclature Committee; IPA: ingenuity pathway analysis; PMT: photomultiplier tube; PPC: pea protein concentrate; $S_{11}B_{34}$: diet with 11 % SPC and 34 % BPC; $S_{22}B_{22}$: diet with 22 % SPC and 22 % BPC; $S_{34}B_{11}$: diet with 34 % SPC and 11 % BPC; S_{45}: diet with 45 % SPC and 0 % BPC; SBM: soybean meal; SPC: soy protein concentrate.

GENE NAMES

ACAT2: acetyl-CoA acetyltransferase 2; ALOX5: arachidonate 5-lipoxygenase; C7: complement component 7; CACNA1B: calcium channel, voltage-dependent, N type, alpha 1B subunit; CASP8: caspase 8, apoptosis-related cysteine peptidase; CD28: CD28 molecule; CFD: complement factor D (adipsin); CFTR: cystic fibrosis transmembrane conductance regulator (ATP-binding cassette sub-family C, member 7); CREB1: cAMP responsive element binding protein 1; CYP2A7: cytochrome P450, family 2, subfamily A, polypeptide 7; CYP2F1: cytochrome P450, family 2, subfamily F, polypeptide 1; CYP51A1: cytochrome P450, family 51, subfamily A, polypeptide 1; DHCR7: 7-dehydrocholesterol reductase; DIABLO: diablo, IAP-binding mitochondrial protein; FABP6: fatty acid binding protein 6, ileal; FADS2: fatty acid desaturase 2; FDFT1: farnesyl-diphosphate farnesyltransferase 1; G6PC: glucose-6-phosphatase, catalytic subunit; IFNG: interferon, gamma; ITGA6: integrin, alpha 6; ITGB1: integrin, beta 1 (fibronectin receptor, beta polypeptide, antigen CD29 includes MDF2, MSK12); ITGB3: integrin, beta 3 (platelet glycoprotein IIIa, antigen CD61); JUN: jun proto-oncogene; MGAT3: mannosyl (beta-1,4-)-glycoprotein beta-1,4-N-acetylglucosaminyltransferase; PCNA: proliferating cell nuclear antigen; PRF1: perforin 1 (pore forming protein); SERPING1: serpin peptidase inhibitor, clade G (C1 inhibitor), member 1; SLC27A2: solute carrier family 27 (fatty acid transporter), member 2; TNF: tumor necrosis factor; TNFRSF1A: tumor necrosis factor receptor superfamily, member 1A; TXNIP: thioredoxin interacting protein.

PATHWAY-RELATED

CTLA4: cytotoxic T-lymphocyte associated protein 4; eNOS: endothelial nitric oxide synthase; fMLP: N-formylmethionyl-leucyl-phenylalanine; IL-8: interleukin 8; ILK: integrin-linked kinase; LC-PUFA: long-chain polyunsaturated fatty acid; NF-κB: nuclear factor kappa-light-chain-enhancer of activated B cells; NRF2: nuclear factor-erythroid 2-related factor; PCP: planar cell polarity; PKR: protein kinase R

COMPETING INTERESTS

The authors declare that they have no competing interests.

AUTHOR' CONTRIBUTIONS

EK performed the microarray experiment, interpreted results and wrote the manuscript with SAM and AD, who performed statistical analysis of microarray data. JRS performed IPA analysis and interpreted results. SAM, VOC, DRT and CJS conceived and designed the study, analysed and interpreted results and gave final approval of the manuscript. All authors discussed and commented on the manuscript.

REFERENCES

1. Lovell T. Nutrition and feeding of fish. New York: Springer Science+Business Media; 1998.
2. Bostock J, McAndrew B, Richards R, Jauncey K, Telfer T, Lorenzen K, Little D, Ross L, Handisyde N, Gatward I, Corner R. Aquaculture: global status and trends. Philos Trans R Soc Lond B Biol Sci. 2010;365:2897–912.
3. Klinger D, Naylor R. Searching for solutions in aquaculture: charting a sustainable course. Annu Rev Environ Resour. 2012;37:247–76. doi: 10.1146/annurev-environ-021111-161531.
4. Hardy RW. Utilization of plant proteins in fish diets: effects of global demand and supplies of fishmeal. Aquacult Res. 2010;41:770–6. doi: 10.1111/j.1365-2109.2009.02349.x.
5. De Santis C, Ruohonen K, Tocher DR, Martin SAM, Król E, Secombes CJ, Bell JG, El-Mowafi A, Crampton VO. Atlantic salmon (*Salmo salar*) parr as a model to predict the optimum inclusion of air classified faba bean protein concentrate in feeds for seawater salmon. Aquaculture. 2015;444:70–8.
6. Bennett RN, Wallsgrove RM. Secondary metabolites in plant defence mechanisms. New Phytol. 1994;127:617–33. doi: 10.1111/j.1469-8137.1994.tb02968.x.
7. Liener EI. Toxic constituents of plant foodstuffs. New York: Academic; 1980.
8. Carsten LD, Watts T, Markow TA. Gene expression patterns accompanying a dietary shift in *Drosophila melanogaster*. Mol Ecol. 2005;14:3203–8. doi: 10.1111/j.1365-294X.2005.02654.x.
9. Matzkin LM, Watts TD, Bitler BG, Machado CA, Markow TA. Functional genomics of cactus host shifts in *Drosophila mojavensis*. Mol Ecol. 2006;15:4635–43. doi: 10.1111/j.1365-294X.2006.03102.x.
10. Magnanou E, Malenke JR, Dearing MD. Hepatic gene expression in herbivores on diets with natural

and novel plant secondary compounds. Physiol Genomics. 2013;45:774–85. doi: 10.1152/physiolgenomics.00033.2013.

11. Francis G, Makkar HPS, Becker K. Antinutritional factors present in plant-derived alternate fish feed ingredients and their effects in fish. Aquaculture. 2001;199:197–227. doi: 10.1016/S0044-8486(01)00526-9.

12. Krogdahl Å, Penn M, Thorsen J, Refstie S, Bakke AM. Important antinutrients in plant feedstuffs for aquaculture: an update on recent findings regarding responses in salmonids. Aquacult Res. 2010;41:333–44. doi: 10.1111/j.1365-2109.2009.02426.x.

13. Champ MM. Non-nutrient bioactive substances of pulses. Br J Nutr. 2002;88:S307–19. doi: 10.1079/BJN2002721.

14. Baeverfjord G, Krogdahl Å. Development and regression of soybean meal induced enteritis in Atlantic salmon, *Salmo salar* L, distal intestine: a comparison with the intestines of fasted fish. J Fish Dis.1996;19:375–87. doi: 10.1111/j.1365-2761.1996.tb00376.x.

15. Knudsen D, Urán P, Arnous A, Koppe W, Frøkiaer H. Saponin-containing subfractions of soybean molasses induce enteritis in the distal intestine of Atlantic salmon. J Agric Food Chem. 2007;55:2261–7. doi: 10.1021/jf0626967.

16. Knudsen D, Jutfelt F, Sundh H, Sundell K, Koppe W, Frøkiær H. Dietary soya saponins increase gut permeability and play a key role in the onset of soyabean-induced enteritis in Atlantic salmon (*Salmo salar*L.) Br J Nutr. 2008;100:120–9. doi: 10.1017/S0007114507886338.

17. Penn MH, Bendiksen EÅ, Campbell P, Krogdahl Å. High level of dietary pea protein concentrate induces enteropathy in Atlantic salmon (*Salmo salar* L.) Aquaculture. 2011;310:267–73. doi: 10.1016/j.aquaculture.2010.10.040.

18. Chikwati EM, Venold FF, Penn MH, Rohloff J, Refstie S, Guttvik A, Hillestad M, Krogdahl Å. Interaction of soya saponins with plant ingredients in diets for Atlantic salmon, *Salmo salar* L. Br J Nutr. 2012;107:1570–90.

19. Krogdahl Å, Gajardo K, Kortner TM, Penn M, Gu M, Berge GM, Bakke AM. Soya saponins induce enteritis in Atlantic salmon (*Salmo salar* L.). J Agric Food Chem. 2015;63:3887–902.

20. Couto A, Kortner TM, Penn M, Bakke AM, Krogdahl Å, Oliva-Teles A. Effects of dietary phytosterols and soy saponins on growth, feed utilization efficiency and intestinal integrity of gilthead sea bream (*Sparus aurata*) juveniles. Aquaculture. 2014;432:295–303. doi: 10.1016/j.aquaculture.2014.05.009.

21. Hamadeh HK, Bushel PR, Jayadev S, Martin K, DiSorbo O, Sieber S, Bennett L, Tennant R, Stoll R, Barrett JC, Blanchard K, Paules RS, Afshari CA. Gene expression analysis reveals chemical-specific profiles. Toxicol Sci. 2002;67:219–31.

22. Kortner TM, Skugor S, Penn MH, Mydland LT, Djordjevic B, Hillestad M, Krasnov A, Krogdahl Å. Dietary soyasaponin supplementation to pea protein concentrate reveals nutrigenomic interactions underlying enteropathy in Atlantic salmon (*Salmo salar*). BMC Vet Res. 2012;8:101.

23. Tacchi L, Bickerdike R, Douglas A, Secombes CJ, Martin SA. Transcriptomic responses to functional feeds in Atlantic salmon (*Salmo salar*) Fish Shellfish Immunol. 2011;31:704–15. doi: 10.1016/j.fsi.2011.02.023.

24. Tacchi L, Secombes CJ, Bickerdike R, Adler MA, Venegas C, Takle H, Martin SA. Transcriptomic and physiological responses to fishmeal substitution with plant proteins in formulated feed in farmed Atlantic salmon (*Salmo salar*). BMC Genomics. 2012;13:363.

25. Peisker M. Manufacturing of soy protein concentrate for animal nutrition. In: Brufau J, editor. Feed manufacturing in the Mediterranean region. Improving safety: From feed to food. Zaragoza: CIHEAM; 2001. pp. 103–7.

26. Helsper JPFG, Hoogendijk JM, Van Norel A, Burger-Meyer K. Antinutritional factors in faba beans*(Vicia faba* L.) as affected by breeding toward the absence of condensed tannins. J Agric Food Chem.1993;41:1058–61. doi: 10.1021/jf00031a008.

27. Betancor MB, Sprague M, Sayanova O, Usher S, Campbell PJ, Napier JA, Caballero MJ, Tocher DR. Evaluation of a high-EPA oil from transgenic *Camelina sativa* in feeds for Atlantic salmon (*Salmo salar* L.): effects on tissue fatty acid composition, histology and gene expression. Aquaculture. 2015;444:1–12.

28. Pooley NJ, Tacchi L, Secombes CJ, Martin SA. Inflammatory responses in primary muscle cell cultures in Atlantic salmon (*Salmo salar*) BMC Genomics. 2013;14:747. doi: 10.1186/1471-2164-14-747.

29. Tacchi L, Bron JE, Taggart JB, Secombes CJ, Bickerdike R, Adler MA, Takle H, Martin SAM. Multiple tissue transcriptomic responses to *Piscirickettsia salmonis* in Atlantic salmon (*Salmo salar*). Physiol Genomics. 2011;43:1241–54.

30. De Santis C, Taylor JF, Martinez-Rubio L, Boltana S, Tocher DR. Influence of development and dietary phospholipid content and composition on intestinal transcriptome of Atlantic salmon (*Salmo salar*) PLoS One. 2015;10:e0140964. doi: 10.1371/journal.pone.0140964.

31. R Core Team (2014). R: A language and environment for statistical computing. R Foundation for Statistical Computing, Vienna, Austria. URL http://www.R-project.org/.

32. Ritchie ME, Phipson B, Wu D, Hu Y, Law CW, Shi W, et al. Limma powers differential expression analyses for RNA-sequencing and microarray studies. Nucleic Acids Res. 2015;43:e47. doi: 10.1093/nar/gkv007.

33. Hackstadt AJ, Hess AM. Filtering for increased power for microarray data analysis. BMC Bioinformatics. 2009;10:11. doi: 10.1186/1471-2105-10-11.

34. Smyth GK. Linear models and empirical Bayes methods for assessing differential expression in microarray experiments. Stat Appl Genet Mol Biol. 2004;3:1–26.

35. Betancor MB, Sprague M, Usher S, Sayanova O, Campbell PJ, Napier JA, Tocher DR. A nutritionally-enhanced oil from transgenic *Camelina sativa* effectively replaces fish oil as a source of eicosapentaenoic acid for fish. Sci Rep. 2015;5:8104.

36. Yadetie F, Karlsen OA, Lanzén A, Berg K, Olsvik P, Hogstrand C, Goksøyr A. Global transcriptome analysis of Atlantic cod (*Gadus morhua*) liver after in vivo methylmercury exposure suggests effects on energy metabolism pathways. Aquat Toxicol. 2013;126:314–25.

37. Huang DW, Sherman BT, Lempicki RA. Systematic and integrative analysis of large gene lists using DAVID Bioinformatics Resources. Nat Protoc. 2009;4:44–57. doi: 10.1038/nprot.2008.211.

38. Huang DW, Sherman BT, Lempicki RA. Bioinformatics enrichment tools: paths toward the comprehensive functional analysis of large gene lists. Nucleic Acids Res. 2009;37:1–13. doi: 10.1093/nar/gkn923.

39. Kortner TM, Gu J, Krogdahl Å, Bakke AM. Transcriptional regulation of cholesterol and bile acid metabolism after dietary soyabean meal treatment in Atlantic salmon (*Salmo salar* L.) Br J Nutr.2013;109:593–604. doi: 10.1017/S0007114512002024.

40. van der Sar AM, Spaink HP, Zakrzewska A, Bitter W, Meijer AH. Specificity of the zebrafish host transcriptome response to acute and chronic mycobacterial infection and the role of innate and adaptive immune components. Mol Immunol. 2009;46:2317–32. doi: 10.1016/j.molimm.2009.03.024.

41. Salinas I, Parra D. Fish mucosal immunity: intestine. In: Beck BH, Peatman E, editors. Mucosal health in aquaculture. London: Academic; 2015. pp. 136–70.

42. Ingerslev H-C, von Gersdorff Jørgensen L, Lenz Strube M, Larsen N, Dalsgaard I, Boye M, Madsen L. The development of the gut microbiota in rainbow trout (*Oncorhynchus mykiss*) is affected by first feeding and diet type. Aquaculture. 2014;424-425:24–34.

43. Groschwitz KR, Hogan SP. Intestinal barrier function: molecular regulation and disease pathogenesis. J Allergy Clin Immunol. 2009;124:3–20. doi: 10.1016/j.jaci.2009.05.038.

44. Peterson LW, Artis D. Intestinal epithelial cells: regulators of barrier function and immune homeostasis. Nat Rev Immunol. 2014;14:141–53. doi: 10.1038/nri3608.

45. Seeley KE, Baitchman E, Bartlett S, DebRoy C, Garner MM. Investigation and control of an attaching and effacing *Escherichia coli* outbreak in a colony of captive budgerigars (*Melopsittacus undulatus*) J Zoo Wildl Med. 2014;45:875–82. doi: 10.1638/2012-0281.1.

46. Walker D, Knuchel-Takano A, McCutchan A, Chang YM, Downes C, Miller S, Miller S, Stevens K, Verheyen K, Phillips AD, Miah S, Turmaine M, Hibbert A, Steiner JM, Suchodolski JS, Mohan K, Eastwood J, Allenspach K, Smith K, Garden OA. A comprehensive pathological survey of duodenal biopsies from dogs with diet-responsive chronic enteropathy. J Vet Intern Med. 2013;27:862–74.

47. de La Serre CB, Ellis CL, Lee J, Hartman AL, Rutledge JC, Raybould HE. Propensity to high-fat diet-induced obesity in rats is associated with changes in the gut microbiota and gut inflammation. Am J Physiol Gastrointest Liver Physiol. 2010;299:G440–8. doi: 10.1152/ajpgi.00098.2010.

48. Petschow BW, Blikslager AT, Weaver EM, Campbell JM, Polo J, Shaw AL, Burnett BP, Klein GL, Rhoads JM. Bovine immunoglobulin protein isolates for the nutritional management of enteropathy. World J Gastroenterol. 2014;20:11713–26.

49. Wu SB, Stanley D, Rodgers N, Swick RA, Moore RJ. Two necrotic enteritis predisposing factors, dietary fishmeal and *Eimeria* infection, induce large changes in the caecal microbiota of broiler chickens. Vet Microbiol. 2014;169:188–97. doi: 10.1016/j.vetmic.2014.01.007.

50. Saino N, Ambrosini R, Rubolini D, von Hardenberg J, Provenzale A, Hüppop K, Hüppop O, Lehikoinen A, Lehikoinen E, Rainio K, Romano M, Sokolov L. Climate warming, ecological mismatch at arrival and population decline in migratory birds. Proc Biol Sci. 2011;278:835–42.

51. Kerby JT, Post E. Advancing plant phenology and reduced herbivore production in a terrestrial system associated with sea ice decline. Nat Commun. 2013;4:2514. doi: 10.1038/ncomms3514.

CHAPTER 9

TRANSCRIPTOMIC AND PHYSIOLOGICAL RESPONSES TO FISHMEAL SUBSTITUTION WITH PLANT PROTEINS IN FORMULATED FEED IN FARMED ATLANTIC SALMON (SALMO SALAR)

Luca Tacchi[1], Christopher J Secombes[1] , Ralph Bickerdike[2] , Michael A Adler[2] , Claudia Venegas[3] , Harald Takle[3,4] and Samuel AM Martin[1]

[1] Institute of Biological and Environmental Sciences, University of Aberdeen, Tillydrone Avenue, Aberdeen AB24 2TZ, UK.

[2] BioMar Ltd, Grangemouth Docks, Grangemouth FK3 8UL, UK.

[3] AVS Chile S.A., Casilla 300, Puerto Varas, Chile.

[4] Nofima, Aas, Norway.

ABSTRACT

Background

Aquaculture of piscivorous fish is in continual expansion resulting in a global requirement to reduce the dependence on wild caught fish for generation of fishmeal and fish oil. Plant proteins represent a suitable protein alternative to fish meal and are increasingly being used in fish feed. In this study, we examined the transcriptional response of Atlantic salmon (Salmo salar) to a high marine protein (MP) or low fishmeal, higher plant protein replacement diet (PP), formulated to the same nutritional specification within previously determined acceptable maximum levels of individual plant feed materials.

Results

After 77 days of feeding the fish in both groups doubled in weight, however neither growth performance, feed efficiency, condition factor nor organ indices were significantly different. Assessment of histopathological changes in the heart, intestine or liver did not reveal any negative effects of the PP diet. Transcriptomic analysis was performed in mid intestine, liver and skeletal muscle, using an Atlantic salmon oligonucleotide microarray (Salar_2, Agilent 4x44K). The dietary comparison revealed large alteration in gene expression in all the tissues studied between fish on the two diets. Gene ontology analysis showed, in the mid intestine of fish fed PP, higher expression of genes involved in enteritis, protein and energy metabolism, mitochondrial activity/kinases and transport, and a lower expression of genes involved in cell proliferation and apoptosis compared to fish fed MP. The liver of fish fed PP showed a lower expression of immune response genes but a higher expression of cell proliferation and apoptosis processes that may lead to cell reorganization in this tissue. The skeletal muscle of fish fed PP vs MP was characterized by a suppression of processes including immune response, energy and protein metabolism, cell proliferation and apoptosis which may reflect a more energy efficient tissue.

Citation: Luca Tacchi, Christopher J Secombes, Ralph Bickerdike, Michael A Adler, Claudia Venegas, Harald Takle and Samuel AM Martin. Transcriptomic and physiological responses to fishmeal substitution with plant proteins in formulated feed in farmed Atlantic salmon (Salmo salar), DOI: 10.1186/1471-2164-13-363

Conclusions

The PP diet resulted in significant effects on transcription in all the 3 tissues studied. Despite of these alterations, we demonstrated that high level of plant derived proteins in a salmon diet allowed fish to grow with equal efficiency as those on a high marine protein diet, and with no difference in biometric quality parameters.

BACKGROUND

Aquaculture production has been the fastest growing animal food-producing sector globally for over half a century, with production growing at an average rate of 8.1% per year since 1961[1], representing almost 50% of human consumed fish[1, 2]. Additionally, the capture from wild fisheries has plateaued with approximately three quarters of wild fisheries fished to capacity, overfished or recovering[3]. Atlantic salmon (Salmo salar) production alone now reaches over 1.5 million tonnes per year[4]. Total production in 2008 is estimated at 1.4 million tonnes, representing a 6% increase on 2007[5]. The positive growth trend of the industry is expected to continue, reflecting the rising demand for healthy human food products high in protein and marine oils.

Piscivorous fish, which include salmonids, require high protein diets and until recently this was almost exclusively derived from wild caught non food fish such as anchovies and sardines. It has become apparent that with increasing fish meal demand due to the expanding aquaculture industry the proportion of protein in diets cannot be sustained by use of fish meal alone. During 2003, 20% of total fish meal usage for aquaculture feed production was utilized by the salmon industry[6], but continued development of diets has reduced the inclusion of fish meal in salmon diets from 500 $g kg^{-1}$ to 350 $g kg^{-1}$ by replacement with alternative protein sources[7]. Plant proteins currently represent the only economic and sustainable protein alternative to fish meal and are increasingly being used in commercial fish feed, with the most common being soybean meal (SBM) or soybean protein concentrate (SPC) which have a high protein content and contain the majority of essential amino acids required by salmonids. All plants contain a number of anti nutritional factors (ANFs) as part of their inherent defence mechanisms, effected by lectins, saponins, phytic acid and proteinase inhibitors amongst others[8]. In early attempts to use plant proteins in fish diets ANFs were often co purified with the proteins, leading to metabolic dysfunction in fish liver[9], inflammation in the intestine[10, 11], reduced protein deposition[12], and an impaired immune response[13]. Currently used plant replacement diets use a combination of different plant proteins, with SPC often the major component[14]. Processing of soy protein to SPC with alcohol-extraction removes the majority of ANFs facilitating use of these products in fish feeds at a high inclusion level without causing enteritis or other gross morphological changes[15]. Replacement of up to 50% of the fish meal with a mixture of plant proteins is possible in rainbow trout without affecting fish growth, and even complete replacement of fish meal with SPC has been reported[16, 17]. Other successful plant proteins used regularly in aquaculture include wheat gluten at up to 40% of fish meal replacement in feed for salmon and trout (Hardy 1996), corn gluten and sunflower meal at up to 30% replacement, which show no adverse effects on salmonid growth performance[18, 19, 20]. Extensive studies investigating responses of Atlantic salmon to such diets indicate that a replacement of fish meal close to 100% can be used without negative effects on salmon growth providing the proteins are highly purified and the amino acid profile is well balanced[21]. High levels of pea protein concentrate do still cause enteropathy in salmon though[22].

In this paper we report a whole transcriptome based study on post-smolt Atlantic salmon fed on a low marine protein diet, where marine protein fish meal was partially replaced with a combination of plant derived proteins, compared to fish fed a diet high in fish meal. The high plant protein (PP) diet contained a balanced combination of soy protein concentrate, corn gluten and wheat gluten, and was supplemented with lysine and methionine to ensure the requirement for essential amino acids was met. The fish were fed for 77 days following which growth, feed efficiency and quality biometric parameters were measured and tissues were collected for histopathological changes and gene transcription analysis. The transcriptional response was examined using an Atlantic salmon"salar_2" microarray platform (Agilent)[23], enriched for genes involved in protein metabolism, lipid metabolism and immune function.

RESULTS

Feeding Trial

During the 77 day feeding trial fish were fed either a high marine protein (MP) diet or a plant protein replacement (PP) diet (Table1), and both groups doubled in weight over this period. No significant differences were seen between the groups in growth performance, feeding efficiency, condition factor or organ indexes (Table2). There were also no significant mortalities in any of the tanks.

Table 1 Dietary formulations

	PP	MP
Raw Materials (% of inclusion)		
Fish meal (%)	26.0	51.0
Fish oil (%)	19.2	17.3
Wheat (%)	6.3	11.9
Maize gluten (%)	10.0	3.0
Sunflower meal (%)	15.0	10.0
Rapeseed meal (%)	3.0	3.0
Wheat gluten (%)	1.3	
Soy protein concentrate(%)	18.0	3.0
Premix (%)	0.4	0.4
Monocalcium phosphate (%)	0.5	0.3
Lysine (%)	0.3	
Methionine(%)	0.1	
Proximate Analysis (% in feed)		
Protein (%)	44.8	44.3
Oil(%)	23.5	24
Ash(%)	6.8	8.9
Moisture(%)	7.7	8.5
Marine protein (as % of total Crude Protein)	39%	78%

Dietary formulation of high marine protein diet (MP) and plant protein diet (PP). Values represent percentage inclusion or content. Marine content calculated for total protein. The Fish oil is a commercial Chilean fish oil.

Table 2 Growth performance and somatic indexes

	PP		MP	
	Mean	**+/− SD**	**Mean**	**+/− SD**
Initial Weight (g)	110.6	0.8	110.8	0.4
Final Weight (g)	231.4	1.9	237.0	3.2
SGR (%)	1.09	0.02	1.12	0.02
FCR	0.87	0.01	0.90	0.03
SFR (%)	0.95	0.02	1.00	0.04
Condition factor	1.08	0.00	1.06	0.03
CSI	0.08	0.00	0.08	0.01
HSI	1.09	0.07	1.06	0.02
ISI	2.32	0.20	2.40	0.20

Growth performance and somatic indexes. SGR represents the specific growth rate of fish, FCR is the feed conversion ratio, SFR is the specific feeding rate, CSI is the cardiac somatic indice, HSI is the hepatic somatic indice and ISI is the intestinal somatic indice.

Histology

A number of histological changes were found, however different types of anomalies were observed in tissues of individuals from both treatment groups and such changes were observed in low frequency indicating that the fish were in a good condition (Table3). There was no significant dietary effect for any histological change except the frequency of perivascular and peri-biliar duct infiltrations in the liver, where 21% of fish fed the MP diet exhibited this change compared to 3.8% of fish fed the PP diet (Table3). A number of fish from both diets showed a mild fatty change in the liver, but this was not restricted to a particular dietary group (Table3). Equally, some anomalies in heart were observed at a low level in both dietary treatments. For the intestine, no changes were observed in pyloric caeca or mid intestine, whereas for the distal intestine an inflammatory infiltration in the submucosa and lamina propria was observed but at a low level in both diets, with no significant difference in occurrence between dietary groups (Table3).

Table 3 Histopathological observations/changes

Organ	Changes	PP	MP	*P*-value*
Liver	No significant findings	39.0%	33.3%	0.39
	Mild fatty change	38.1%	31.4%	0.312
	Mild to moderate fatty change	8.6%	7.6%	0.801
	Moderate fatty change	8.6%	8.6%	n.d.
	Moderate to severe fatty change	0.0%	1.9%	0.156
	Severe fatty change	0.0%	0.0%	n.d.
	Perivascular infiltration	2.9%	6.7%	0.196
	Peri biliar ducts infiltration	3.8%[a]	21.0%	<0.001
	Necrotic foci w/infiltration	1.0%	0.0%	0.317
	Mild congestion	0.0%	1.0%	0.317
	Focal necrosis	0.0%	2.9%	0.082
	Mild sinusoidal dilatation	0.0%	1.0%	0.317
	Biliar ducts esclerosis	0.0%	2.9%	0.082
	Biliar ducts proliferation	0.0%	0.0%	n.d.
	Foci of inflammatory cells	0.0%	1.0%	0.317
Heart (ventricle)	No significant findings	99.0%	98.1%	0.562
	Severe infiltration and myodegeneration	1.0%	1.0%	n.d.
	Infiltration, no myodegeneration	0.0%	1.0%	0.317
	Presence of melanin granules	0.0%	0.0%	n.d.
Mid-intestine and Pyloric caeca	No significant findings	100%	100%	n.d.
Distal intestine	No significant findings	97.1%	95.2%	0.472
	Infiltration of submucosa and lamina propria, widened of lamina propria within folds (simple and complex), prominent goblet cells/low number of supranuclear vacuoles	2.9%	4.8%	0.472

Histopathological observations/changes. Values (in percentage) represent occurrence where n=105. Significantly different when $P<0.05$, different superscript letters indicate significant differences within each row-wise comparison.

Transcriptomic Responses

Fish fed the PP diet showed changes in transcriptome in all 3 tissues examined compared to fish fed the marine protein rich diet MP. In total 8,151 oligo features were found to be expressed at different levels as a result of the dietary manipulation. For subsequent analysis only significant ($p<0.05$) features with a fold difference of greater than 2 were analysed further (1961 mRNAs). The intestine exhibited the greatest differential gene expression with 1,236 genes significantly different in expression level between fish fed the PP diet and MP diet. Of these 615 genes were higher in expression and 621 lower in PP fed fish. In skeletal muscle 505 genes had an expression higher than two fold with 132 higher in PP fed fish and 373 lower and in liver 220 genes were found modified with 161 higher and 59 genes lower in PP fed fish vs MP fed fish (Table4).

Table 4 Genes significantly different in mid intestine of fish fed PP diet

Probe name[1]	ACC[2]	FC±SEM[3]	Identity[4]
Immune function and stress related			
Ssa#CK897125	CK897125	7.6±1.1	B-cell linker
Ssa#CL60Contig3	X70167	5.2±1.2	MHC class II antigen beta chain
Ssa#STIR13675	TC71772	5.0±1.3	CD209f
Ssa#NP9934055	NP9934055	4.8±1.4	T cell receptor alpha
Ssa#STIR21272	TC82967	4.3±1.2	Interferon inducible mx protein
Ssa#S35685629	S35685629	4.3±1.0	TCR-gamma
Ssa#S30239635	S30239635	2.8±1.2	IRF1
Ssa#STIR10385	TC67231	2.5±1.1	Vig-2 protein
Ssa#STIR15805	TC74805	2.4±1.0	cd3 epsilon
Ssa#STIR05606	BT056756	2.4±1.1	β2 microglobulin
Ssa#gi156446662	EF579742	2.0±1.2	MyD88
Ssa#CL81Contig1	BT072778	2.0±1.1	Virus induced TRIM protein
Ssa#S37438814	S37438814	2.0±1.0	CD3 gamma/delta
Ssa#STIR00071_2	DW580947	2.0±1.2	Interferon induced protein 35
Ssa#S18833713	S18833713	−7.5±1.2	Serum lectin 2
Ssa#CK882427	CK882427	−3.2±1.3	Serum amyloid A
Ssa#S35474845	S35474845	−3.0±1.1	HSP β-7
Ssa#S35677496	S35677496	−2.9±1.1	CCR3
Ssa#S31986130	S31986130	−2.7±1.2	Macrophage colony stimulating factor receptor
Ssa#S18892409	S18892409	−2.7±1.1	IgM
Ssa#STIR03818	NM_001141099	−2.4±1.3	CXC13
Ssa#STIR19205	TC79827	−2.3±1.1	HSP 70a
Ssa#STIR13083	TC70912	−2.3±1.2	SAMHD1
Ssa#S35663823	S35663823	−2.2±1.2	Complement c1q
Ssa#S35583279	S35583279	−2.1±1.0	IL-17D
Ssa#S35517748	S35517748	−2.0±1.1	HSP 30
Ssa#S35536386	S35536386	−2.0±1.1	HSP β-11
Omy#gi31087931	AY160984	−2.0±1.2	IL-8
Ssa#TC106255	TC106255	−2.0±1.1	Galectin-4
Cell proliferation and apoptosis			
Ssa#S35582566	S35582566	7.1±1.2	Caspase-14
Ssa#CL201Ctg1	NM_001139921	2.4±1.0	Caspase-3
Ssa#S35693335	S35693335	2.2±1.0	Caspase 8

Ssa#S32005165	S32005165	−9.4 ± 1.0	DNA topoisomerase 2-alpha
Omy#S34421775	S34421775	−6.6 ± 1.5	Replication factor C subunit 3
Omy#CA346576	CA346576	−4.5 ± 1.2	Tetraspanin-14
Ssa#CA038824	CA038824	−3.9 ± 1.1	Fgfr3 protein
Omy#S34313679	S34313679	−3.6 ± 1.4	Cyclin A1
Omy#S22901990	S22901990	−3.4 ± 1.0	Chromodomain-helicase-DNA-binding protein 7
Ssa#S47728937	S47728937	−3.1 ± 1.1	Tetraspanin-16
Ssa#S35496360	S35496360	−2.9 ± 1.0	DNA-repair protein complementing XP-A cells
Ssa#S35490761_S	S35490761	−2.7 ± 1.1	Transcription factor HES-1
Ssa#STIR38390	TC108636	−2.7 ± 1.3	RNA helicase
Ssa#S35531441	S35531441	−2.4 ± 1.1	Melanoma-derived growth regulatory protein
Ssa#STIR17200	TC76867	−2.4 ± 1.0	Tumor necrosisalpha-induced protein 2
Protein metabolism			
Ssa#S35499913	S35499913	6.5 ± 1.0	Titin
Ssa#S31974046	S31974046	4.2 ± 1.1	Cathepsin C
Ssa#S30293144	S30293144	3.9 ± 1.5	Proteasome subunit alpha type-5
Ssa#STIR05302	BT046757	3.2 ± 1.1	Proteasome subunit beta type-7
Ssa#CL233Ctg1	BT072668	3.0 ± 1.0	Cathepsin Z
Omy#S15290109	S15290109	3.0 ± 1.2	Keratin e1
Ssa#STIR25562	TC89420	3.0 ± 1.1	Troponinskeletal muscle
Ssa#STIR20536	TC81827	2.9 ± 1.1	Cathepsin A
Ssa#STIR24947	TC88495	2.8 ± 1.2	Keratin 18
Ssa#DY714088	DY714088	2.7 ± 1.0	Ribosomal protein S6 kinase b
Ssa#KSS4531	KSS4531	2.6 ± 1.1	Ubiquitin-conjugating enzyme E2
Ssa#STIR08978	TC65424	2.2 ± 1.0	Eukaryotic translation initiation factor 4e
Ssa#DY712052	DY712052	2.0 ± 1.1	Gamma-tubulin complex component 4
Ssa#S30279979	S30279979	2.0 ± 1.2	Eukaryotic translation initiation factor 2 subunit 1
Ssa#S30240560	S30240560	2.0 ± 1.2	β-actin
Ssa#KSS1565	KSS1565	2.0 ± 1.0	Proteasome subunit alpha type-6
Ssa#KSSb2668	KSSb2668	2.0 ± 1.2	Tubulin beta-1 chain
Ssa#STIR39880	TC110797	2.0 ± 1.1	Proteasome subunit alpha type-7
Ssa#S31996856	S31996856	2.0 ± 1.1	E3 ubiquitin-protein ligase RNF128
Ssa#STIR19643	TC80514	−3.9 ± 1.3	Myosin light chain 1–3
Ssa#TC76471	TC76471	−3.6 ± 1.2	Tropomyosin-1 alpha chain
Ssa#AJ425777	AJ425777	−2.7 ± 1.3	60 S ribosomal protein L34
Ssa#STIR11900	TC69277	−2.5 ± 1.1	Myosin ic
Ssa#S35582711	S35582711	−2.3 ± 1.0	Tropomyosin-1 alpha

Ssa#S35505113	S35505113	−2.2 ± 1.1	60 S ribosomal protein L30
Ssa#FC072705	FC072705	−2.1 ± 1.1	40 S ribosomal protein S10
Omy#CX150249	CX150249	−2.1 ± 1.0	60 S ribosomal protein L36
Lipid metabolism			
Ssa#CB509140	CB509140	25.3 ± 1.2	Fatty acid-binding protein
Ssa#S31963508	S31963508	14.5 ± 1.5	Apolipoprotein A-I
Ssa#STIR00045_2	AY170327	5.2 ± 1.0	PUFA elongase 5A
Ssa#STIR00100_2	CK887422	4.7 ± 1.0	Delta-6 fatty acyl desaturase
Omy#S18154618	S18154618	4.4 ± 1.2	Glycolipid transfer protein
Ssa#STIR00093_4	AF478472	4.3 ± 1.1	Delta-5 fatty acyl desaturase
Energy and mitochondrial activity/kinases			
Ssa#STIR00144_4	TC64612	4.2 ± 1.0	Glyceraldehyde-3-phosphate dehydrogenase
Ssa#STIR39924	TC110855	2.6 ± 1.2	Cytochrome P450
Ssa#S32006874	S32006874	2.3 ± 1.1	Peroxisomal membrane protein 11 C
Ssa#S30263228	S30263228	2.4 ± 1.1	Succinate dehydrogenase
Ssa#S35661441	S35661441	2.0 ± 1.0	Fructose-1.6-bisphosphatase 1
Ssa#S32000347	S32000347	−2.5 ± 1.2	Hemoglobin subunit beta-1
Cellular transport			
Ssa#S30284157	S30284157	23.6 ± 1.2	Solute carrier family 22
Ssa#STIR21606	C83482	9.1 ± 1.2	Solute carrier family 16
Ssa#DW575876	DW575876	7.2 ± 1.3	Solute carrier family 39
Ssa#STIR19539	TC80341	4.6 ± 1.0	Solute carrier family 31
Ssa#STIR19539	S35536215	2.8 ± 1.0	Solute carrier family 25
Ssa#S35454393	S35454393	2.1 ± 1.1	Solute carrier family 27
Omy#TC172518	TC172518	−2.2 ± 1.1	Beta globin
Ssa#CK991045	CK991045	−2.0 ± 1.0	Hemoglobin subunit alpha

There were no genes expressed at different levels in common between all the tissues examined indicating there was not a systematic response to the diets. However there was a co regulation of features between the mid intestine and skeletal muscle with 9 genes higher in both tissues of PP fed fish vs MP fed fish and 10 lower in both. Similarly in the liver and intestine 6 genes were found to be significantly different in common, with 5 genes higher in expression and 1 gene lower in the PP fed fish. To further investigate the biological significance of the differently expressed gene sets we used gene ontology analysis to help identify biological processes that were significantly different in tissues following the feeding trial.

Annotation of the microarray allowed 77% of the features to be allocated to a functional protein and 55% of these proteins were assigned a gene ontology (GO) identifier. This allowed statistical analysis for enrichment for GO biological processes to help interpret the changes in the transcriptome following the feeding trial (Figure1) to gain a holistic view of which biological processes were significantly modified following feeding the PP diet.

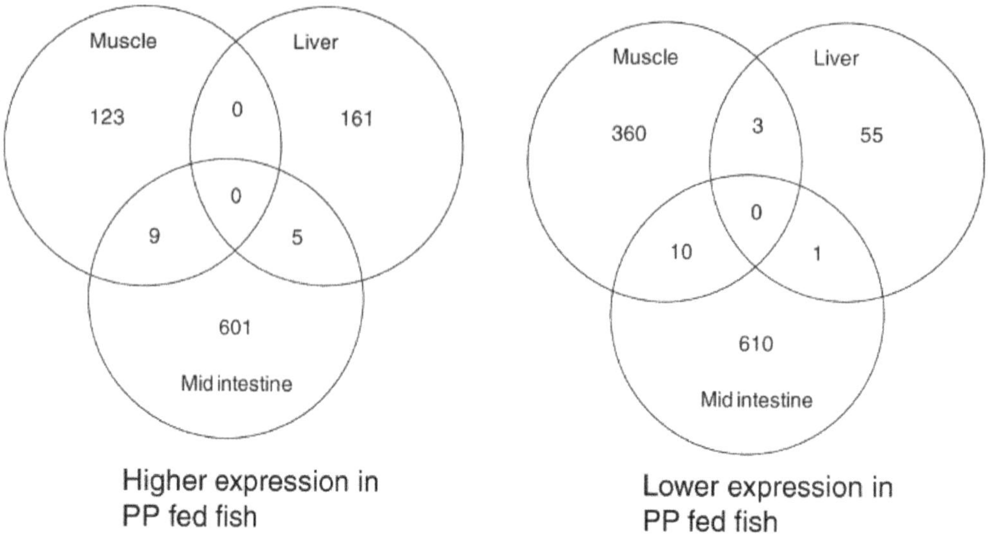

Figure 1 Venn diagram showing numbers of genes identified as expressed at different levels by microarray analysis.Summary of numbers of genes higher and lower expressed. The genes presented are all significantly different in expression ($p < 0.05$) with >2 fold change in expression.

Genes expressed at different Levels in mid Intestine

The intestine is the major tissue to come into direct contact with any feed components and for this reason may be very sensitive to dietary changes. This is shown here by the intestine having the greatest transcriptional response in terms of the number of genes expressed at different levels and the magnitude of different expression of these genes. Several biological processes were significantly different in the PP diet fed fish, related to intestinal functions including immune and stress related processes, protein metabolism, energy and mitochondrial activity, lipid metabolism and transport (Table4).

Immune and Stress Response

Genes encoding proteins related to both innate and acquired immune function were expressed at different levels by the dietary treatments. For fish fed PP innate immune parameters were found to be both higher and lower in expression indicating a dynamic response. Modulators of the immune response were higher in PP fed fish, such as MyD88, a key transcription factor associated with induction of an inflammatory response and P105 subunit of an inhibitory protein that sequesters NF-κ B in the cytoplasm. Another transcription factor, interferon regulatory factor (IRF) 1 was also higher, and potentially increased expression of interferon responsive genes including Mx, Vig-2, interferon induced protein 35 and virus induced TRIM protein. Genes encoding cytokine receptors (IL-17R) and regulators of cytokine function including the suppressor of cytokine signalling (SOCS)-7) were also higher. A number of genes related to innate immune responses were also lower, and included IL-17D, MCSFR and chemokines such as CCL3, CXC13 and IL-8 (CXCL8). Other innate immune serum proteins were also lower in expression including serum amyloid A, and a number of lectins. Genes related to proteins in the acquired immune response were mostly higher in expression in fish fed the PP diet compared to the MP diet. For example, T cell receptor chains (α and γ) and their signalling subunits CD3 epsilon and gamma/delta were higher as were genes involved with antigen presentation such as β2 microglobulin and MHC class II). Related to antibody production by B cells, a B-cell linker protein that regulates B-cell function and development was higher in PP fed fish as was a B cell enhancing factor.

Regarding stress related genes, PP fed fish showed an elevated expression of glutathione-S-transferase and a thioredoxin interacting protein, involved in reducing oxidative stress, while heat shock proteins (HSP) β-7, β-11, 30 and 70a were all suppressed.

Cell proliferation and Apoptosis

The intestinal transcription of genes encoding proteins related to cell proliferation (tumor necrosis alpha-induced protein 2, tetraspanins-14 and −16 and melanoma-derived growth regulatory protein) and the cell cycle (DNA topoisomerase 2-alpha, replication factor C subunit 3, FGFR3 protein, cyclin A1) were found to be higher in PP fed fish; whilst transcription of caspases 3, 8 and 14,which are directly involved in apoptosis, were found to be lower in expression relative to fish fed MP.

Protein Metabolism

Feeding the PP diet the expression of genes involved in protein metabolism was significantly different compared to MP. For example, protein synthesis related genes including elongation factor, translation initiation factors 2 and 4e and ribosomal protein S6 kinase b were higher, while a number of 60 S ribosomal protein encoding genes were lower relative to MP fed fish. Protein degradation encoding genes were more highly expressed as shown by lysosomal proteases, cathepsins A, C, and Z, ubiquitin proteasome route (UbP) of proteolysis, and proteasome subunits α5, α6, α7 and β7. Genes involved in targeting of proteins for degradation in the UbP were also higher including ubiquitin conjugating enzyme E2 and ubiquitin E3 ligases. Genes encoding the structural proteins troponin, tubulin, titin, actin and keratin were all found to have a higher expression level whereas a number of myosin mRNAs were lower in expression.

Lipid metabolism

A higher transcript level of genes involved in lipid transport and metabolism was observed, with higher transcription of fatty acid-binding protein, apolipoprotein A-I, apolipoprotein B and delta-6 fatty acyl desaturase, delta-5 fatty acyl desaturase, PUFA elongase 5A and glycolipid transfer protein, respectively.

Energy and Mitochondrial activity/kinases

There was an induction of many genes related to energy metabolism in pathways such as glycolysis (glyceraldehyde-3 phosphate dehydrogenase), the tricarboxylic acid cycle (succinate dehydrogenase) and gluconeogenesis (fructose 1-6-bisphosphate) which were expressed at a higher level in PP fed fish relative to the MP fed fish.

Cellular Transport

Transport of solutes into and from a cell is an energy demanding process. Genes involved in membrane transport were higher in expression in fish fed the PP diet, and included a number of solute carrier family members such as SCF22 (organic cation transporter), SCF 27 (fatty acid transporter), SCF 25 (mitochondrialphosphate carrier), SCF 39 (zinc transporter), SCF 16 (monocarboxylic acid transporter), and SCF 31 (copper transporter). Two genes encoding haemoglobin subunits (hemoglobin subunit alpha-4 and hemoglobin subunit beta-1) and beta globin, however, were found to have a lower relative expression level.

Genes Expressed at Different Levels in Liver

Several biological processes were significantly different in liver that can be related to immune parameters, stress responses, and protein and lipid metabolism, with key genes shown in Table5.

Table 5 Genes significantly different in liver of fish PP diet

Probe name[1]	ACC[2]	FC±SEM[3]	Identity[4]
Immune function and stress response			
Ssa#S35601811	S35601811	2.7±1.0	MPV17 protein
Ssa#gi156446662	EF579742	2.4±1.1	MyD88
Omy#S18150823	S18150823	2.2±1.1	Amine oxidase
Ssa#STIR13675	TC71772	2.0±1.0	CD209f
Ssa#CK874360	CK874360	2.0±1.1	Heat shock 70
Ssa#STIR36546	TC105929	−2.8±1.1	C-type MBL-2 protein
Ssa#NP797925	NP797925	−2.4±1.1	MHC class I alpha 2
Ssa#STIR04816	NM_001140849	−2.4±1.2	Hepcidin
Ssa#NP9934311	NP9934311	−2.2±1.1	T cell receptor alpha
Ssa#S48440415	S48440415	−2.2±1.3	SMAD3
Ssa#S35663823	S35663823	−2.1±1.1	Complement c1q
Cell proliferation and apoptosis			
Ssa#CA041082	CA041082	2.6±1.2	Transforming growth factor beta receptor
Ssa#S35486979	S35486979	2.5±1.0	Cell death activator CIDE-3
Ssa#S35559076	S35559076	2.3±1.0	Angiopoietin-related protein 4
Ssa#S30276711	S30276711	2.3±1.1	Activin receptor type-1B
Ssa#S35582821	S35582821	2.3±1.0	Serine protease HTRA1
Ssa#STIR31305	TC98147	2.2±1.1	Annexin A3
Omy#S18150823	S18150823	2.2±1.0	Amine oxidase
Protein metabolism			
Ssa#DW574268	DW574268	2.4±1.1	Ribosomal protein S18
Ssa#STIR03071	BT048999	2.3±1.1	Proteasome beta type 3
Ssa#KSSb2684	KSSb2684	2.2±1.2	Peptidyl-prolyl cis-trans isomerase
Ssa#STIR16404	TC75662	2.2±1.1	Calpain
Ssa#STIR04151	NM_001141015	2.1±1.0	Ribosomal protein l39

Immune and Stress Response

Genes encoding proteins involved in innate immunity such as complement c1q, C-type MBL-2 and a dendritic cell specific lectin CD209f which binds mannose carbohydrate molecules, and hepcidin a major liver associated antimicrobial peptide were lower in PP fed fish. A limited number of genes related to adaptive immunity were also found to be expressed at different levels, with MHC class I and SMAD 3 lower in expression level, whereas MyD88 was higher in fish fed the PP diet relative to MP fed fish. Moreover, genes involved in the oxidative response (MPV17 protein, amine oxidase and heat shock cognate 70 kDa protein) were also higher in fish fed the PP diet.

Cell Proliferation and Apoptosis

The PP diet stimulated/modified hepatic cell turnover as indicated by higher expression of serine protease HTRA1 and annexin A3 and the stimulation of apoptotic processes through genes involved in TGF beta pathways (cell death activator CIDE-3 and angiopoietin-related protein 4).

Protein Metabolism

Hepatic protein metabolism was clearly stimulated in PP fed fish since all genes significantly different in expression were higher in these fish, including ribosomal protein S18a, ribosomal protein l39 and peptidyl-prolyl cis-trans isomerise. Similarly genes involved in protein degradation were more highly expressed as seen with the non-lysosomal protein calpain and the proteasome subunit β3.

Genes Expressed at Different Levels in Skeletal Muscle

The GO analysis showed a number of processes to be significantly different in skeletal muscle that can be related to protein metabolism, immune function and energy metabolism (Table6).

Table 6 Genes significantly different in skeletal muscle of fish PP diet

Probe name[1]	ACC[2]	FC±SEM[3]	Identity[4]
Immune function			
Ssa#STIR12634	TC70300	5.0±1.2	Vig-2
Ssa#STIR00067_2	U66477	3.2±1.3	Interferon inducible Mx protein
Ssa#CK894557	CK894557	−11.2±1.4	MHC class Ib antigen
Ssa#STIR00132_2	NM_001140254	−7.4±1.3	Tumor necrosis alpha-induced protein 2
Omy#NP565601	NP565601	−6.4±1.4	T-cell receptor beta
Ssa#S31982089	S31982089	−5.4±1.1	Platelet-activating factor receptor
Ssa#STIR12498	TC70105	−4.6±1.2	γ-ip (CXCL10)
Ssa#S48440415	S48440415	−3.4±1.1	SMAD3
Omy#S15331473	S15331473	−3.2±1.2	Interferon-inducible protein Gig2-like
Ssa#S35544087	S35544087	−3.1±1.3	B-cell CLL/lymphoma 7 protein family member B
Ssa#STIR13083	TC70912	−3.6±1.0	SAMHD1
Omy#S18153399	S18153399	−2.4±1.1	CD80
Ssa#KSS5035	KSS5035	−2.2±1.2	Interferon regulatory factor 2-binding protein
Ssa#STIR00071_3	DW580947	−2.3±1.1	Interferon -induced protein 35
Ssa#STIR08451	TC64790	−2.2±1.2	Heat shock protein 47
Ssa#S35480903	S35480903	−2.0±1.2	Heat shock protein 30
Ssa#STIR29454	TC95297	−2.1±1.1	Beta defensin 1
Cell proliferation and apoptosis			
Ssa#S35602638	S35602638	3.7±1.4	Tripartite motif 39

Ssa#TC110067	TC110067	3.5±1.5	Cell death inducing protein
Ssa#STIR08668	TC65039	3.1±1.1	Bh3 interacting domain death agonist
Ssa#STIR12507	TC70118	3.0±1.3	Syndecan 4
Ssa#STIR02208	BT049868	−38.6±1.2	Caspase 14
Ssa#S35559333	S35559333	−4.1±1.3	Ankyrin repeat domain-containing protein 54
Omy#TC151190	TC151190	−2.9±1.2	Nucleostemin
Ssa#S35550715	S35550715	−2.8±1.2	Growth hormone-inducible transmembrane protein
Ssa#S35585784	S35585784	−2.6±1.1	Caspase-8
Ssa#KSS5035	KSS5035	−2.4±1.2	Interferon regulatory factor 2-binding protein 2-B
Protein metabolism			
Ssa#S35682089	S35682089	4.2±1.3	Ubiquitin-like protein 1
Ssa#STIR17445	TC77227	2.6±1.3	Acta1 protein
Ssa#STIR21465	TC83266	2.5±1.2	Actin-binding
Ssa#STIR03710	BT048358	2.0±1.2	Proteasome subunit beta type-9
Ssa#STIR00115_4	BT045917	2.0±1.2	Tropomyosin-1 alpha chain
Ssa#S32008331	S32008331	−3.6±1.7	40 S ribosomal protein S16
Ssa#CL224Ctg1	NM_001140473	−2.8±1.5	Receptor-interacting serine/threonine-protein kinase 4
Omy#CX150460	CX150460	−2.7±1.1	60 S ribosomal protein L27
Omy#S15320960	S15320960	−2.6±1.1	Cathepsin D
Ssa#TC111443	TC111443	−2.5±1.1	Serine/threonine-protein kinase 35
Ssa#FC072705	FC072705	−2.3±1.0	40 S ribosomal protein S10
Ssa#S35583213	S35583213	−2.1±1.2	Serine/threonine-protein kinase PINK1
Ssa#S35591236	S35591236	−2.1±1.1	Histidine triad nucleotide-binding protein 3
Energy metabolism			
Ssa#STIR39924	TC110855	2.5±1.1	Cytochrome P450
Ssa#S32006874	S32006874	2.2±1.1	Peroxisomal membrane protein 11 C
Ssa#KSS5038	KSS5038	−4.0±1.3	Phosphoglycerate kinase
Ssa#STIR19484	TC80264	−3.1±1.3	Cytochrome c oxidase polypeptide viii
Ssa#CL200Ctg1	BT059338	−2.8±1.3	L-lactate dehydrogenase B chain
Ssa#S30282925	S30282925	−2.7±1.1	Peroxisomal 3.2-trans-enoyl-CoA isomerase
Omy#CX150319	CX150319	−2.4±1.2	Cytochrome c oxidase subunit II
Ssa#STIR13627	TC71700	−2.1±1.2	Cox18 cytochrome c oxidase
Ssa#EL698167	EL698167	−2.0±1.1	Malate dehydrogenase 1

List of selected mRNAs found to be higher or slower in expression following the feeding trial in mid intestine (a), liver (b) or skeletal muscle (c) and grouped according to functional classes (shown in bold). The selection was based on manual assignment of function and genes with greatest fold differences in expression are presented, the genes that are lower in expression are denoted by (−) value. The genes shown were significant at $p<0.05$ following t- tests and greater than 2-fold change. [1]Indicates the unique code for the feature on the microarray, [2]Accession number of the cDNA sequence. [3]Fold-change for genes higher or lower expressed in fish fed PP relative to fish fed MP diet. [4]Identity of the probe target as determined by BLASTX and BLASTN searches.

Immune and Stress Response

Regarding the innate immune status, PP fed fish showed lower expression level of inflammatory related genes such as platelet-activating factor receptor, SAMHD1, tumor necrosis alpha-induced protein 2, and the antimicrobial peptide beta defensin 1. Likewise, fish fed the PP diet showed a suppression of genes involved in the adaptive immune response, i.e. TCR β, MHC class I, SMAD 3, CD80 (a T cell activator). Interferon γ induced proteins γ-ip (CXCL10), Gig2-like and both interferon-induced protein 35 and IRF2-binding protein were also lower in fish fed the PP diet. In contrast, Mx and Vig-2 which are responsive to type I interferon were higher expressed in PP compared to MP. Lastly, the stress induced genes HSP 30 and 47 were lower in expression in PP fed fish relative to MP fed fish.

Cell Proliferation and Apoptosis

Genes involved in cell proliferation were lower expressed in the PP fed fish, including ankyrin repeat domain-containing protein 54, nucleostemin and growth hormone-inducible transmembrane protein. Several genes involved in apoptosis were found to be higher such as cell death inducing protein and bh3 interacting domain death agonist although in contrast whereas both caspase 8 and 14 showed lower transcript levels in the PP fed fish.

Protein Metabolism

Genes involved in protein synthesis such as several 60 S and 40 S ribosomal proteins were lower in expression in skeletal muscle of PP fed fish. A number of enzyme genes that modify essential amino acids (eg serine/threonine-protein kinase PINK1, serine/threonine-protein kinase 35, receptor-interacting serine/threonine-protein kinase 4 amongst others) were also lower in expression. Likewise, a major lysosomal peptidase, cathepsin D was lower in expression in PP fed fish, however, two components of the UbP were higher expressed in PP (ubiquitin-like protein precursor and proteasome subunit beta type-9).

Genes encoding proteins involved in muscle structure and physiology (acta1 protein, actin-binding homolog 1a and tropomyosin-1 alpha chain) were also more highly expressed in skeletal muscle of fish fed the PP diet.

Energy Metabolism

Genes related to energy metabolism were transcriptionally modified with peroxisomal membrane protein 11 C and cytochrome P450 being higher in expression whereas cytochrome c oxidases (COX2, 8 and 18) were found to be lower in PP fed fish compared to MP fed fish. In addition genes involved in glycolysis (phosphoglycerate kinase 1 and L-lactate dehydrogenase B chain) and gluconeogenesis (malate dehydrogenase 1) were found to be expressed lower in PP fed fish.

Co-regulated Genes

Only a small number of genes were co regulated between tissues following the feeding trial. No genes were significantly higher or lower regulated in all the tissues in this experiment. The greatest co regulation was observed between mid-intestine and skeletal muscle, with 9 genes significantly higher and 10 genes significantly lower in PP compared to MP. Genes that were more highly expressed in mid intestine and liver of fish fed the PP diet compared to the MP diet play a role in energy metabolism, such as cytochrome P450, a membrane-associated protein located in the inner membrane of mitochondria and peroxisomal membrane protein 11 C that induces peroxisomal proliferation proteins involved in the pentose phosphate pathway. Genes that showed lower expression in mid intestine and liver have a role in protein synthesis (40 S ribosomal protein S10) and in the immune response, such as SAMHD1 a nuclease involved in innate immune responses by acting as a negative regulator of the cell-intrinsic antiviral response. Five genes were higher expressed in both intestine and liver of fish fed the PP diet, including immune related genes, such as CD209f and MyD88 and only 1 gene (complement c1q) showed lower expression in both tissues. Skeletal muscle and liver did not have any higher expressed gene in common and only 3 genes (membrane-bound O-acyltransferase domain-containing protein 7, SMAD 3 and the non-annotated gene (**NP_998306**) were significantly lower between these two tissues (Figure2; full list of genes expressed at different levels is given in supplementary Table1).

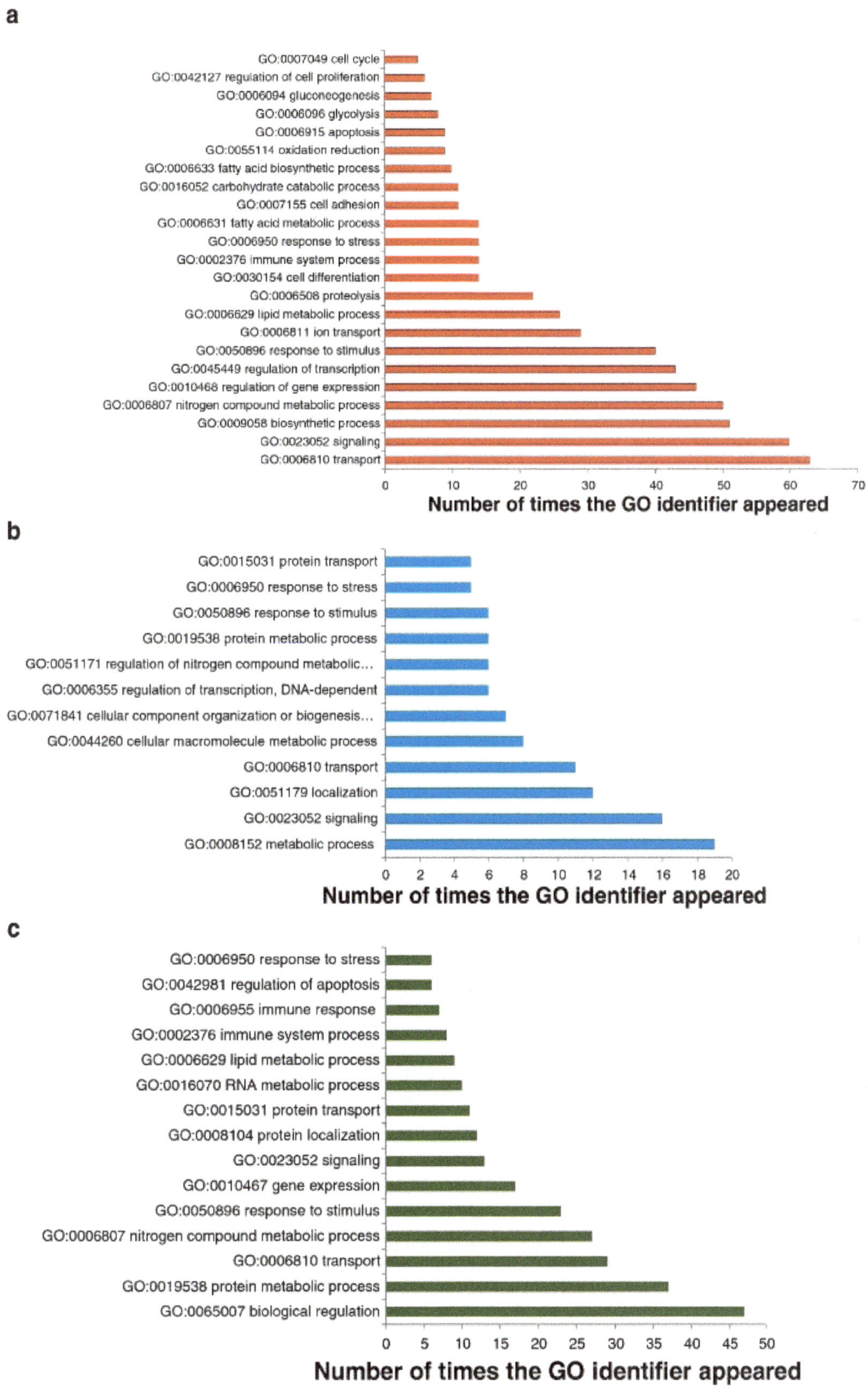

Figure 2 Gene ontology (biological processes) found to be significantly enriched following the feeding experiment in mid intestine (**a**), **liver** (**b**) **and skeletal muscle** (**c**). Only GO categories for which >3 genes were represented are included.

Confirmation of expression by real time PCR

Real time PCR analysis was performed on a number of genes for each tissue to confirm microarray data (Table7). The real time PCR expression data was normalized with HPRT1 as the expression of this housekeeping gene was not found modulated by microarray analysis. For all genes the expression pattern showed the same trend between microarray and real time PCR analysis (Figure3).

Table 7 Primers used for real time PCR for expression analysis, Acc is accession number from NCBI

Gene name	Primer name	Primer sequence(5-prime to 3-prime)	Acc	Product size	Ann T	Tissue
Fabp2	FatF	GCTCTGTACTAGCTCTCCTCCC	CB509140	156 bp	55 °C	Mid intestine
	FatR	GGCGTACAGTTTGACTATGCAC				
Caspase-14	Cas14F	CGATTATACACCCGGACTATGG	S35582566	155p	55 °C	Mid intestine
	Cas14R	CCTATCAAGTGTGAATCCATGC				
TCR alpha	TCRaF	GGAAGACTCTGCTCTGTACCAC	U50991	147 bp	55 °C	Mid intestine
	TCRaR	GCTGTGGTATTTCTTGACTTC				
IgM heavy chain	IgMF	GCTTATAGCCATAGTACTACTG	S18892409	169 bp	55 °C	Mid intestine
	IgMR	GCATAGCTGCCCCATATCGC				
Tpm1	Trop1F	CGAAGATGAGAGAGATAAAGTGC	TC76471	134 bp	55 °C	Mid intestine
	Trop1R	CCTCCTCAACCAGCTGGATACG				
RFC3	ReplF	GCTGACTCACTGCATTCCTCCTG	S34421775	163 bp	55 °C	Mid intestine
	ReplR	GAAGGCCTCTAGGTGGTAAATGG				
HTRA1	HTRA1F	GGTCATCTTCATACAGAGAGG	S35582821	152 bp	55 °C	Liver
	HTRA1R	GCTTAGAGAATACCATCTTGC				
TGF beta receptor	TGBF	CCACAAGAAGCCAGCTGTCAG	gi|209735249	135 bp	55 °C	Liver
	TGFBR	CTAGCCAGGTATCTCTATCATGG				
MRPS18A	28 SF	CCATTGATTCAGTGAAGCCCATC	DW574268	158 bp	55 °C	Liver
	28 S R	CCTGCTGTGAGTTGACATGC				
Timd2	TcellF	CCATGGACAACCACACACACTG	CA368982	141 bp	55 °C	Liver
	TcellR	CCAGTAGAATGGACACCAGGATC				
Hepcidin I	HepF	GCTTCTGCTGCAAATTCTGAGG	gi|209736931	157 bp	55 °C	Liver
	HepR	GTACAAGATTGAGGTTGTGCAG				
TCR alpha	TCRaF	GGAAGACTCTGCTCTGTACCAC	U50991	147 bp	55 °C	Liver
	TCRaR	GCTGTGGTATTTCTTGACTTC				
CDIP	CellF	CCATGTCTGAGACCTACTCTATG	TC110067	243 bp	55 °C	Skeletal muscle
	CellR	GATAGTCACTTGATGTCCAGTG				
acta1 protein	ActaF	CCTGTAAACTGTGAATGCGTC	TC77227	156 bp	55 °C	Skeletal muscle
	ActaR	CCAAAGTTTTATATCAGCTGC				
TGF beta 1	TGF1bF	GCTCGGAGTGTGAGACAGAACTG	S15319964	187 bp	55 °C	Skeletal muscle
	TGF1bR	CACTTGACGCAACAGAAACACTCC				
RT1-CE5	MHC1bF	GGAAAGATCTCCTGAAGACTTGAG	CK894557	101 bp	55 °C	Skeletal muscle
	MHC1bR	CGTTTATGAGAAGTTCAGC				
60 S rib prot	60SF	GCTTCTTACCATGGTTCTTCAG	DR695852	140 bp	55 °C	Skeletal muscle
	60SR	GGTCAAGATCCCATCCACCATC				
Heat shock protein 30	HeatF	CCATCCAACCAGTCTCCTACAAGC	EG804126	303 bp	55 °C	Skeletal muscle
	HeatR	CCTCCTCAGCAGATAATGGATTC				
Hprt	HprtF	CCGCCTCAAGAGCTACTGTAAT	EG866745	255 bp	55 °C	All tissues
	HprtR	GTCTGGAACCTCAAACCCTATG				

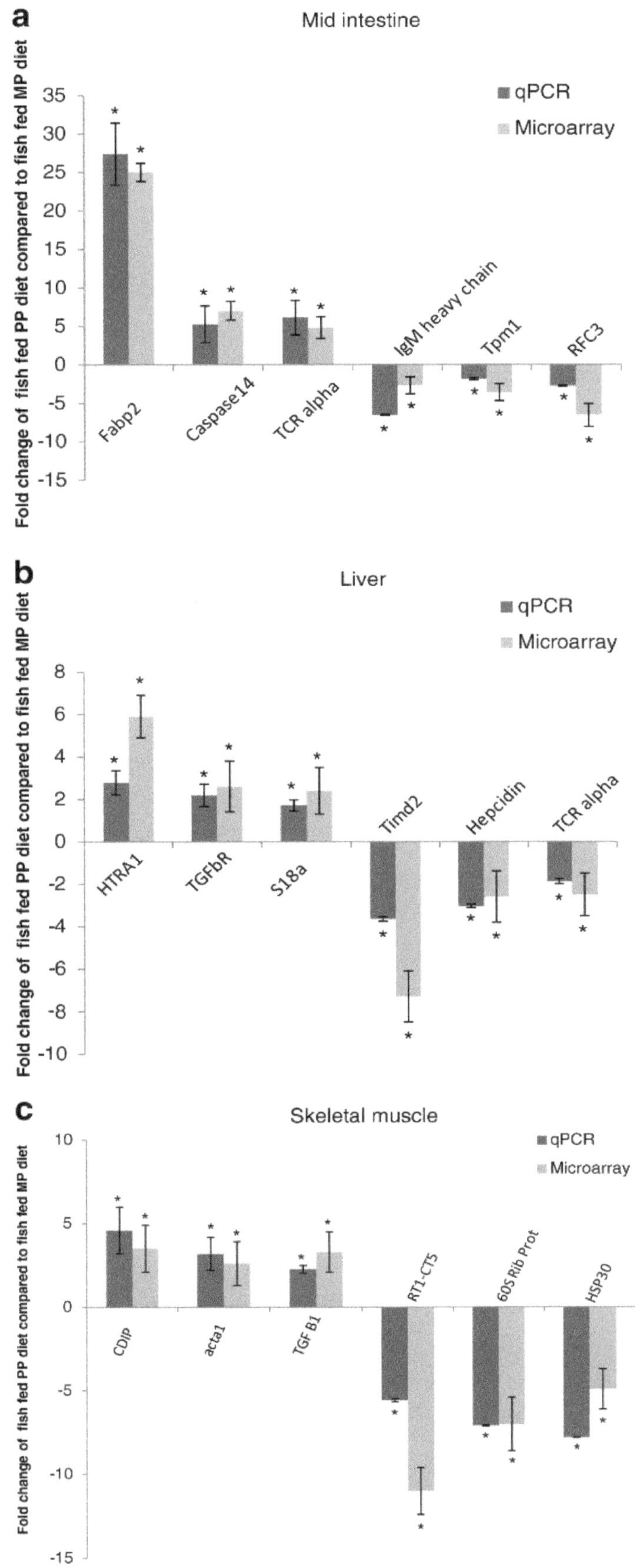

Figure 3 Quantitative real-time PCR confirmation of genes expressed at different levels in mid intestine (a), liver (b) and skeletal muscle (c) of fish fed the PP diet compared to fish fed the MP diet for 6 genes identified by microarray analysis. Bars represent mean±standard error of five fish, asterisks indicate significant ($p<0.05$) differences. The genes chosen for the intestine were: Fatty acid-binding protein (Fabp2), Caspase-14 precursor, TCR alpha, IgM heavy chain, Tropomyosin-1 alpha chain (Tpm1) and Replication factor C subunit 3 (RFC3). The genes selected for the liver were: Serine protease HTRA1, TGF beta receptor, 28 S Ribosomal protein S18a (MRPS18A), T-cell immunoglobulin and mucin domain (Timd2), hepcidin I and TRC alpha. The gene chosen for the skeletal muscle were: cell death inducing protein (CDIP), acta1 protein, Transforming growth factor beta-1 (TGF beta 1), MHC class Ib antigen (RT1-CE5), 60Sribosomal protein L6 (60 S rib prot) and Heat shock protein 30.

DISCUSSION

With the continual expansion of aquaculture of piscivorous fish there is a global requirement to reduce the dependence on wild caught fish for the generation of fishmeal and fish oil. There are a number of studies that have explored the transcriptional response to plant and vegetable oil replacements in fish diets, often when 100% replacement experimental diets are used results in reduced growth[24, 25]. When both proteins and oils were replaced[24] in rainbow trout microarray analysis of liver indicated processes such as protein metabolism and cell cycle being altered, however the reasons for these changes could either have been combined effect of lowered essential fatty acids or changes in amino acid profile. There is also growing body of work demonstrating the genotype diet interaction in rainbow trout[25, 26, 27] and Atlantic salmon[28], with the latter concluding transcription of key metabolic regulators that respond to plant based feeds depend on the fish's genotype. Commercial salmon feed formulations partially replace fish meal with plant derived proteins, with varying effects on fish physiology and performance including fish growth and food conversion[24, 29]. This paper examined the physiological and performance effects with transcriptional responses in mid intestine, liver and skeletal muscle in salmon to a post smolt feed containing high levels of plant derived proteins with low fishmeal content compared to a low plant protein-high fishmeal diet. The tissues examined play different physiological roles in the fish all of which may be affected by dietary changes as reflected in the different response profiles seen in these tissues. In common with a number of other transcriptomic studies on nutritional effects (where extreme dietary manipulations are avoided) the impact of the different diets appears quite subtle suggesting limited changes to physiological and metabolic pathways[30]. Although, in general, the scale of gene response is quite low, distinct changes in all tissues examined do indicate that there are responses occurring as a result of the dietary components.

When interpreting the findings it is important to take a comparative holistic approach whereby apparent down regulation of genes in fish fed one diet may in fact be the result of up regulation in fish fed the other dietary treatment and vice versa. In this paper the expression is analysed relative to the marine profile (MP) diet.

Histological Changes

The occurrence of histological changes was very similar for fish fed either diet, with low frequencies observed for the majority of parameters measured. The main histological observation was a mild fatty change in liver, which is considered normal under intensive rearing conditions. This anomaly appeared to have no clinical effect, and was not prevalent for either dietary group. Although there was a higher incidence of peri-biliar duct infiltration in liver of fish fed the MP diet no other histological changes were observed to indicate an inflammatory response. This reduced infiltration in the liver by the biliary ducts may be related to the altered immune gene expression in the liver. In mammals viral infections can result in biliary atresia[31] and specifically can be related to immune related injury to bile ducts following infiltration of $CD4^+$ T and production of interferon γ[32]. At this stage we are unable to tell if the reduction in the structures is a direct result of the nutritional components or due to the altered hepatic immune gene expression in between the two diets. The distal intestine is often the target organ for anti-nutriental factors, in particular those present in SBM and pea protein concentrate[22], inducing histological changes including shortening of simple and complex mucosal folds with widening of central stroma, inflammatory cell infiltration in the submucosa and lamina propria with a mixed leukocyte population[33, 34]. In this study both dietary groups presented fish with distal intestine anomalies but with low frequencies, and no significant difference between diets. These results together with those reported by Sanden et al. (2005) [35] indicate soybean products, including the alcohol soluble fraction, may be used within formulation constraints without inflicting gross histopathological anomalies in the intestine of Atlantic salmon.

Transcriptome Changes in the Mid Intestine

The fish intestine has multiple functions which will be the first to respond to changes in nutritional intake; particularly digestion and absorption of nutrients and immune responsiveness to ingested pathogens, antigens and new antigens generated by the gut flora via gut associated lymphoid tissue[36]. The intestine contains three distinct regions: the proximal intestine containing pyloric caeca, the mid and distal intestine. Nutrient absorption occurs in all regions via enterocytes, by passive and facilitated diffusion and active transport[37], with the highest rates of uptake in the proximal section[37, 38]. The intestine of piscivorous fish can be particularly sensitive to plant derived ANFs and non-starch polysaccharides (NSP) in feed, resulting in altered permeability in trout feed 44% 44% SBM[7, 39] and enteritis in the distal intestine of salmon given a high dietary inclusion of SBM[33]. The inflammation/enteritis may be similar to a hypersensitivity reaction[33,40]. Often these effects are temporary and quickly disappear when the intestinal tract is no longer exposed to the ANFs[11, 33, 41].

Processing of plant products for fish feed is under continual improvement and some current plant derived protein concentrates have very low contaminating factors or botanical impurities. In addition knowledge of ANF containing plant feed materials has increased to the point where commercial feed formulations permit plant derived proteins at acceptable inclusion levels where no negative health effects or impacts on growth and performance occur. This was confirmed in the current study by the histology assessment where no gross morphological changes associated with plant ANFs were observed. In addition there was no difference in growth or feed utilisation efficiency during the feeding period, where a doubling of weight was achieved. However there were more subtle changes to biological processes that were not apparent during the classical physiological evaluation but were detected by global transcriptomic analysis. In this trial the mid intestine showed the greatest transcriptome response of the tissues studied, reflecting the sensitivity of the intestinal cells to dietary factors. Processes modified in the intestine were related to immune parameters, cell proliferation, apoptosis, protein metabolism, energy metabolism, transport and lipid metabolism.

Fish fed the PP diet showed higher expression of genes involved in inflammation suggesting a possible dysfunction in immune regulation. Our findings support previous studies on gut intraepithelial and systemic T cells in fish which showed rainbow trout IELs are rich in T cells[10, 42]. Additionally the expression of TSC22D3, a regulator of T cell receptor mediated cell death, was found higher in PP, this protein may be induced by glucocorticoids[43, 44] activated by components in the PP diet. Together these results support previous reports that trace levels of substances with allergenic properties may cause expression of genes indicative of a hypersensitivity reaction in the intestine[45].

Interestingly genes involved in the inflammatory response were both higher and lower expressed in PP compared to MP. In particular genes involved in NF-κ B pathway were induced such as the signalling adaptor molecule MyD88 and the inhibitory proteins of the IkB family, NF-kB1 p105 which sequesters NFκ B in an inactive form in the cytoplasm[46]; inhibition of this pathway results in the production of proinflammatory cytokines[47]. MyD88 is also part of the signalling pathway that induces type I IFNs through the interaction of the MyD88–TRAF6–IRF7 complex[48]. IRF1[49] had higher expression in fish fed the PP diet, this transcription factor may have activated interferon responsive genes in PP fed fish including Mx[50], virus induced gene (vig) -2[51] and a virus induced TRIM protein[52]. PP fed fish also had higher expression of (SOCS)-7, which functions to reduce inflammation[53], potentially counteracting the expression of genes related to the inflammatory response in the intestine. Other genes involved in the innate immune response were expressed at lower level in fish fed PP. IL-8 is a chemokine that attracts neutrophils to a site of inflammation[54], whereas IL-17D coordinates the clearance of extracellular bacteria and contributes to the pathology of many autoimmune and allergic conditions in Atlantic salmon[55]. The lower expression of these genes in fish fed the PP diet may indicate there was not a proinflammatory response to the PP diet compared to MP diet.

Several antioxidant genes were expressed higher in PP fed fish mid intestine indicating protection against oxidative damage. The free radicals could be either endogenously produced by immune cells or present in the diet. Alternatively a potential lower concentration of antioxidants in the PP feed, due to the lower fishmeal content, may require the antioxidant system within the fish to be increased accordingly. The overall low expression of heat shock protein mRNAs, suggests a limited stress response in the intestine. Additionally, two genes directly involved in apoptosis process (caspase-3 and14) were expressed at higher level in PP potentially indicating increased apoptotic activity of mid intestinal cells in these fish compared to MP fed fish.

The intestine has an extremely high rate of cellular turnover and hence generally high levels of protein synthesis and protein degradation. Protein metabolism genes relating to both synthesis and degradation were found generally to be higher in PP fed fish suggesting an increase in intestinal protein turnover. Genes related to both transcription and translations were at a higher level such as translation initiation factors, elongation factors and the ribosomal protein S6 kinase. Interestingly, a number of mRNAs encoding ribosomal proteins were at a lower expression in PP. This may relate to the stability of the mRNAs or multiple use of the ribosomal subunits during translation. In parallel to general increase in synthesis genes related to protein degradation were also at a higher level in PP such as cathepsins and components of the ubiqutin proteasome pathway[56, 57]. The higher protein turnover is likely to be energy demanding and this is related to an increase in genes encoding proteins involved in oxidative energy metabolism. Together these changes in expression suggest modulation in control of protein turnover in the intestine with components of both synthesis and degradation being altered which may reflect an increased activity of the intestine in fish fed the PP compared to the MP diet.

Cellular membrane transport related genes were higher expressed in PP, which could suggest that salmonids are able to adaptively modulate the densities of transporters to match changes in diet composition. Lipid metabolism and transport were also affected even though the PP diet contained the same fish oil content as the MP diet. mRNAs encoding two apolipoproteins were higher in expression in PP fed fish reflecting a greater mobilization and transport of cholesterol and fatty acids in the intestine, possibly an adaptive response to the lower dietary cholesterol content in PP compared to MP diet. Fatty acid metabolism genes were higher expressed in the PP diet including both FAD5 and 6, a PUFA elongase and other genes related to cholesterol metabolism. FADs are critical enzymes in the biosynthesis of long-chain highly unsaturated fatty acids (HUFA) from shorter chain PUFA[58, 59]. These lipid metabolism differences between PP and MP are of interest as the intestine is often over looked regarding these processes and cholesterol, even if the essential fatty acids are present at required levels, other factors including cholesterol may change, revealing the complex nature of the early digestion and modifications of nutrients in the mid intestine.

Transcriptome Changes in the Liver

The liver receives nutrients and compounds from the intestine and needs to respond to any substances that may have detrimental effects on the fish.

Transcriptome and proteomic studies on salmonids show nutritionally related modifications in both liver mRNAs and proteins due to feeding status[60, 61] and diet composition[9, 62, 63, 64].

The PP fed fish showed a significant difference to fish fed MP for genes related to immune function with a lower expression of innate factors such as lectins and hepcidin. Acquired immune system components were also found at lower level including T cell receptors, MHC I and II and components of the TGF-β pathway. TGF-β has an important role in the maintenance of T-cell[65] and B-cell homeostasis[66] by regulating cell proliferation process and apoptosis in these cells. This result, with the higher expression of genes involved in cell death such as CIDE-3 and angiopoietin-related protein 4, indicates that apoptosis may be a mechanism induced by the PP diet salmon liver. This is not surprising as apoptosis plays a central role in the differentiation and maintenance of the liver[67]. A balancing effect on the apoptotic TGF-β pathway is seen in the induction of several genes encoding proteins related to cell proliferation (such as HTRA1 serine protease and annexin A3). In particular, HTRA1 serine protease inhibits signalling mediated by TGF-β family proteins[68], playing an important role in contrast to cell death, whereas annexin A3 has a role in the signalling cascade during liver regeneration[69]. Other researchers have found the immune system to be altered following vegetable oil replacement in salmon[28] and in sea bass[70].

Genes involved in oxidative stress response (MPV17 protein, amine oxidase and HSP 70 kDa protein) were higher in liver of fish fed PP compared to MP. This is interesting to note as increases in antioxidant genes were also noted in a salmon diets that had marine oil replaced with vegetable oils[28] in the liver. In rainbow trout HSP expression in liver was increased following SBM rich replacement diets[9, 62, 63, 71], the induction of these genes may indicate a diet-induced stress response in fish fed the PP diet. Moreover, during general high protein turnover to deal with misfolded proteins[72] as may be the case of fish fed PP diet.

It is interesting to note that few lipid related metabolic genes were found significantly different in the liver. Vigilin, a protein implicated in both biosynthesis and metabolism of lipids and steroids, facilitates removal of excess cholesterol from cells[73] and secondly apolipoprotein A IV which facilitates transport of cholesterol to the liver were both expressed at higher level in fish fed PP compared to MP. The high expression of apolipoproteins in fish fed diets containing high levels of plant proteins has previously been observed[9, 74] and is most likely indicates reduced cellular cholesterol in fish fed the PP diet due to decreased dietary cholesterol, associated with low fishmeal content, and/or in response to trace levels of phytoestrogens[75] and phorbol esters[76] co purified with the plant proteins.

Transcriptome Changes in the Skeletal Muscle

Genes related to processes such as protein metabolism, energy metabolism, cell proliferation, apoptosis and immune function were all significantly different in PP fed fish compared to MP. Transcripts related to protein metabolism such as ribosomal protein mRNAs, transcription and translation initiation factors were generally lower in PP fed fish compared to MP fed fish indicating lower protein synthesis. In parallel, protein degradation related genes were also less, for example the ubiquitin proteasome pathway and lysosomal peptidase proteins, together these would suggest a lower protein turnover activity in the muscle tissue in PP group. Both protein synthesis and degradation are highly energy demanding processes[77] and the indication of lower protein turnover, may suggest reduced energy wastage[61, 78, 79]. This idea is strengthened by lower expression of COX2 and COX8 and other genes encoding proteins involved in glycolysis and gluconeogenesis in fish fed PP. Genes involved in cell proliferation were also expressed at lower levels in PP fed fish indicating further energy saving. Together these changes in biological processes may indicate efficient metabolic activity following feeding of the plant protein enriched diet.

Biological tissues with high metabolic rate require mechanisms to deal with free oxygen radicals, on the other hand those tissues where the metabolic rate is reduced, for example when protein turnover is decreased a reduction in oxidative stress response may be observed as was observed in this study. Additionally a number of HSPs 30, 47 and a heat shock transcription factor 1a were all at lower levels in PP reflecting the reduced protein turnover and requirement of stabilising many newly translated proteins. Genes involved in cell proliferation and related to cell death including two caspases (caspase-8 and 14) were also lower expressed in fish fed PP compared to the MP diet. The induction of bh3 interacting domain death agonist, a pro-apoptotic member of the Bcl-2 protein family[80] and the suppression of caspases is in accordance with apoptosis of skeletal muscle in mammals where the Bcl2/bax system was found crucial for muscle apoptosis, whilst the caspase activity appeared inhibited[81].

Differential expression of a number of immune related genes, particularly a decrease in interferon responsive genes including γ-ip[82], Gig2-like[83] and interferon-induced protein 35[84] were also expressed at lower level in PP. Other pro-inflammatory agents including platelet-activating factor receptor was also at a lower level PP, which regulates several pro-inflammatory functions such as chemokine and eicosanoid receptors[85]. An antimicrobial peptide beta defensin 1which is a central component of the non-specific defences[86] was also found lower in PP. Relating to the adaptive immune factors, TCRβ, MHCI and CD80 were all at lower levels in PP. Although we have observed differences in genes related to immune function in skeletal muscle of fish fed PP, the low level of inflammation and the subsequent immune response observed in the intestine did not cause a large immune shift in the muscle tissue. Instead, the lower expression of such genes, may allow fish fed PP to spend less energy resources on immune function for use in growth [61].

CONCLUSIONS

In conclusion, the present study confirm that a moderate level of plant protein derived proteins in a salmon diet allowed fish to grow with equal efficiency as those on a high marine protein diet, and with no difference in biometric quality parameters. The PP diet formulated with higher levels of soy protein concentrate, corn gluten, sunflower meal and wheat gluten resulted in significant effects on transcription in the mid intestine, liver, and skeletal muscle. The PP diet induced tissue specific changes in gene expression, with the mid intestine showing activation of the adaptive immune response indicating potential for hypersensitivity and an increase in protein turnover, although no difference in histopathological changes were observed in the proximate, mid or distal intestine. In liver cell proliferation and apoptosis indicate cellular reorganization and the general suppression of processes such as immune response. In contrast skeletal muscle tissue showed reduced protein metabolism and decrease in immune gene expression suggesting less energy expenditure in this tissue. The presence of only few genes in common between tissues may be due to the relatively mild changes that occurred and the complex nature of studying gene expression between tissue types on conservative dietary changes. These results improve the understanding of mechanisms and pathways activated by fishmeal replacement; in particular substitution by plant derived proteins and suggests that such diets can function well in Atlantic salmon aquaculture, hence to some extent reducing the burden on wild caught fish for fish meal. Additionally these results can assist in selection of molecular biomarkers useful for the development of new alternative feeds in salmonid aquaculture.

METHODS

Fish husbandry and Sampling

One hundred and five juvenile mixed sex Atlantic salmon of approximately 100 g were maintained in 6 replicate 1

m^3 tanks at SGS Chile Ltd., Puerto Mont, Chile, at 10.1°C and 27.6% salinity. Fish were fed a plant protein diet (PP) or a high marine protein diet (MP), both of which were formulated to the same digestible protein and energy content within formulation constraints of a commercial feed specification (CPK 100, 3 mm, 24/44 lipid: protein, BioMar S.A., Chile) (Table1). The feeding trial was conducted in triplicate tanks per dietary treatment and lasted for 77 days. At the end of the feeding period all fish in each tank were bulk weighed and 35 fish from each tank were killed by percussive stunning for individual biometric measurements of round weight, fork length, gutted weight, and weight of liver, heart or intestine for calculation of specific growth rate (SGR), feed conversion ratio (FCR), specific feeding rate (SFR), condition factor, and somatic indices of hepatic (HSI), cardiac (CSI) or intestinal (ISI) organs. Liver, heart and intestine was excised from distinct anatomical regions of the same individuals and fixed in 10% buffered formalin for histological assessment. All fish were sampled 30 minutes following the final meal to ensure there were no differences in postprandial gene expression. An additional 12 fish from each tank were sampled for mid intestine, skeletal muscle and liver and immediately stabilised in RNAlater (Ambion) at 4°C overnight then stored at −20°C until RNA extraction for gene expression analysis.

Growth/feed Parameters were Calculated as Follows

FCR=feed consumed (g)/biomass increase (g) SGR (% body wt d^{-1})=[($\ln W_2$ − $\ln W_1$)/days]×100;W_1=start weight (g), W_2=final weight (g), days=days in the growth periodSFR (%)=SGR×FCRHSI, CSI, ISI (%)=[organ weight (g)/round weight (g)]×100 Condition factor=[round weight (g)×100]/[fork length (cm)]^{3}Biometric parameters were analyzed for significant differences by Anova using InfoStat v 2009 software (University of Córdoba, Argentina), with $p < 0.05$ considered significant. Data is presented as means±standard deviation for each dietary treatment.

Histology Assessment

Fixed tissues were submitted to dehydration process following paraffin embedding (Aquagestión, Puerto Montt, Chile). Sections (5 µm) were stained with hematoxylin and eosin (H & E) and examined under a light microscope. Micrographs were examined "blind" by the same experienced pathologist. Liver sections were evaluated for lipid degeneration level (fatty change) and the integrity of the whole organ. Heart tissue was evaluated for the presence of inflammatory infiltrates, myodegeneration and other possible abnormalities. Intestinal morphology was evaluated according to the following criteria: a) widening and shortening of intestinal folds, 2) loss of supranuclear vacuolization in enterocytes in the intestinal epithelium, 3) widening of central lamina propria within the intestinal folds, and 4) infiltration of (mixed) leukocytes in the lamina propria and submucosa[33]. The occurrence of histological changes for each dietary treatment were analysed by Kruskal-Wallis non-parametric Anova with InfoStat v 2009 (University of Córdoba, Argentina), with $p<0.05$ considered significant.

RNA Isolation

RNA was extracted from 100 mg of tissue by homogenisation in 1 ml TRIZol (Invitrogen) using tungsten carbide beads (3 mm, Qiagen) and shaking (300 times per min) following the manufacturer's instructions. The RNA pellet was washed in 500 µl 80% ethanol, air dried and resuspended in RNase free H_2O. The concentration was determined by spectrophotometry (Nanodrop ND1000, LabTech) and the integrity of the RNA was determined by electrophoresis (Agilent Bioanalyser 2100). The RNA was then stored at −80° until required.

Microarray analysis

Microarray Platform:

Microarray experiments were performed using a custom-designed, Agilent-based microarray platform with 4×44 K probes per slide (Salar_2; Agilent Design ID:025520). The array contained primarily an Atlantic salmon resource with 34,441 features from Atlantic salmon coding sequences but additionally a further 9,111 features from rainbow trout (Oncorhynchus mykiss) sequences - the latter being selected when no homologues appeared to be available within Salmo salar datasets. Full details of the microarray platform and design are shown in Tacchi et al. 2011[23].

Hybridization and Analysis

For microarray analysis, 4 pools of RNA were produced for each tissue from fish fed PP and MP diets. Each RNA pool was an equimolar RNA mix from four different fish chosen randomly from each group. The microarray hybridisation was performed using a reference design, using a reference RNA sample, which comprised an equimolar mix of RNA extracted from all individual fish and tissue samples. Each experimental sample (labelled with Cy3™) was hybridised against this reference sample (labelled with Cy5™) in a 2-colour experiment. mRNA amplification, labelling and hybridization was performed as follows: mRNA was amplified using a MessageAMP™ aRNA Amplification Kit (Ambion). Briefly, 2 µg total RNA was reverse transcribed and the cDNA was used as a template for in vitro transcription in the presence of amino allyl modified dUTP, which allowed the generation of amplified antisense RNA (aRNA). For labelling, aRNA (3 µg) was denatured at 70°C for 2 min in a volume of 10 µl to which 3 µl of 0.5 M $NaHCO_3$ and 2 µl Cy dye (dye Cy3™ or Cy5™ mono-reactive dye pack, Amersham) was added. Incorporation of dyes was performed for 1 h in the dark, and after excess label was removed using a DyeEx tm 2.0 spin kit (QIAGEN) the amount incorporated was checked with a Nanodrop ND1000 (LabTech) spectrophotometer. Prior to hybridisation, 825 ng of each labelled template was fragmented in the presence of 11 µl of 10X blocking agent, 2.2 µl of 25X Fragmentation buffer (Agilent), and made up to a final volume of 20 µl with nuclease-free dH_2O. The solution was then incubated in the dark at 60°C for 30 min, after which 57 µl of 2X GEx Hybridisation buffer (Agilent) was added to each sample and 103 µl of each hybridisation solution was dispensed on the Agilent 4x44K Atlantic salmon "Salmo salar2" oligo array (Agilent array design, 025520, Array express platform A-MEXP-1940). The hybridisations were performed in a Microarray Hybridisation Oven (Agilent) overnight (18 h) at 65°C. Following hybridisation, the slides were washed in Gene Expression Wash buffers 1 and 2 (Agilent) following the manufacturer's instructions. The slides were then scanned using a GenePix personal 4100A Scanner (Axon Instruments) at a resolution of 5 μm and saved as *. TIF files. Images were extracted and initial analysis was performed by Feature extraction v9.5.3 (Agilent) performing background correction of feature intensities (within the software). A Lowess normalisation of background corrected data was next conducted and all intensity values <1.0 were set to 1.0. Statistical analysis of the arrays was performed using Genespring GX analysis platform (version 9.5; Agilent Technologies). Quality control of the data was performed within Genespring and included removal of saturated probe features, non-uniform features, population outliers and those features showing intensities not significantly different from background in the Cy3 or Cy5 channels. After these relatively stringent procedures, 20,095 of the original 43,730 array features were maintained for subsequent analyses. The experimental hybridisations are at European Bioinformatics Institute archived under accession number E-MEXP: E-TABM-1207.

Significant differential expression between fish fed the PP diet and fish fed the control diet was established byt-test analysis ($p<0.05$). Further filtering on fold change was conducted, and only transcripts showing more than two-fold change in expression were further characterised.

Analysis of Gene Ontology

Enrichment for gene ontology (GO) biological processes was performed on all cDNA features that had GO identifiers associated using the GOEAST program[87]. Fisher's exact test was performed within the GOEAST program to determine if GO identifiers occurred more often in a group than would appear by chance. For GO analysis only biological process GO identifiers were considered that occurred more than 3 times.

Real Time PCR

The cDNA was synthesized using 2 μg of total RNA as previously described[88]. Real time PCR was performed on a number of genes to confirm the microarray analysis results using the same RNA samples. For cDNA synthesis, 2 μg of total RNA was denatured (70°C, 3 min) in the presence of 1 μl of oligo-dT$_{17}$ (500 ng μl^{-1}), left at room temperature for 5 min to allow annealing, then stored on ice. The resulting cDNA was diluted to a final volume of 50μl in RNA/DNA free water (Sigma). For real time PCR, 3μl of cDNA was used as template with gene specific primers (Table7). A 2x iQ SYBR Green supermix (Bio-Rad) was used for qPCR, which was performed in a 96-well plate using the DNA Engine OpticonTM system (MJ Research, Inc.) with the following program: 95°C for 5 min, then 35 cycles of 94°C for 30° s, 55°C for 30 s and 72°C for 30 s, with a final extension of 72°C for 5 min. A negative control (no template) reaction was also performed for each primer pair. A sample from the serial dilution was run on a 2% agarose gel and stained with ethidium bromide and viewed under UV light to confirm a band of the correct size was amplified. A melting curve for each PCR was determined by reading fluorescence every degree between 72°C and 95°C to ensure only a single product had been amplified. Atlantic salmon hypoxanthine phosphoribosyl transferase 1 (HPRT1) [89] was used as control for normalization of expression since this gene was found not to be modulated by the diet treatments from the microarray analysis. The relative expression level of the genes was determined using the Pfaffl method[90]. Efficiency of the amplification was determined for each primer pair using serial 10 fold dilutions of pooled cDNA, performed on the same plate as the experimental samples. The efficiency was calculated as $E = 10^{(-1/s)}$ where s is the slope generated from the serial dilutions, when Log dilution is plotted against ΔCT (threshold cycle number). Primers were design to have a Tm of 55°C, and where possible, to cross an exon-exon junction sites to avoid amplification of genomic DNA. Exon-intorn junction sites were determined comparing Salmo salar cDNA with genomic sequences for orthologous genes from Danio rerio, Gasterosteus aculeatus, Oryzias latipes, Takifugu rubripes and Tetraodon nigroviridis obtained from Ensembl (http://ensembl.org/).

The results obtained by real time PCR were analyzed using the Pfaffl method[90]. The qPCR measurements were analyzed by T-test, performed using R software, with $p<0.05$ considered significant. The expression data is presented as means ± standard error.

AUTHOR'S CONTRIBUTIONS

LT performed microarray experiments, analyzed the data, carried out the real time PCR and wrote the manuscript. MAC, CV and HT were involved in fish maintenance, feeding trial and sampling. SAMM, CJS, RB and MAC were involved in the experimental design and drafting of the manuscript. All the authors read and approved the final manuscript.

ACKNOWLEDGEMENTS

The research was supported by an industrial studentship between the University of Aberdeen and BioMar Ltd (for L. Tacchi). The authors are grateful to AVS Chile S.A. and the staff at SGS Chile for conducting the feeding experiment and assistance during sampling.

REFERENCES

1. Tacon AGJ: Aquaculture: A catch for all?. OECD Obs. 2010, 278: 32-
2. Tacon AGJ: FAO Fisheries Circular (FAO), no. 881 / FAO. Feed ingredients for carnivorous fish species: alternatives to fishmeal and other fishery resources. 1994, Fishery Resources and Environment Div, Rome (Italy), 39-
3. F.A.O: State of the World Fisheries and Aquaculture. 2007, Rome: FAO
4. ICES: Report of the Working Group on North Atlantic Salmon (WGNAS),13 April 2006,ICES Headquarters. ICES Headquarters 2006: ICES CM 2006/ACFM. 2023
5. ICES:Report of the Working Group on North Atlantic Salmon (WGNAS),30 March-8 April 2009,Copenahagen, Denmark. ICES CM 2009/ACOM:06. 2009,
6. F.A.O: FAO fisheries technical paper. 2005, FAO, Rome, 235-Review of the state of world marine fishery resources, 457

7. Gatlin DM, Barrows FT, Brown P, Dabrowski K, Gaylord TG, Hardy RW, Herman E, Hu G, Krogdahl A, Nelson R, et al: Expanding the utilization of sustainable plant products in aquafeeds: A review. Aquac Res. 2007, 38 (6): 551-579.

8. Krogdahl Å, Penn M, Thorsen J, Refstie S, Bakke AM: Important antinutrients in plant feedstuffs for aquaculture: An update on recent findings regarding responses in salmonids. Aquac Res. 2010, 41 (3): 333-344.

9. Martin SAM, Vilhelmsson O, Médale F, Watt P, Kaushik S, Houlihan DF: Proteomic sensitivity to dietary manipulations in rainbow trout. Biochim Biophys Acta Proteins Proteomics. 2003, 1651 (1–2): 17-29.

10. Bakke-McKellep AM, Frøystad MK, Lilleeng E, Dapra F, Refstie S, Krogdahl Å, Landsverk T: Response to soy: T-cell-like reactivity in the intestine of Atlantic salmon, Salmo salar L. J Fish Dis. 2007, 30 (1): 13-25.

11. Urán PA, Aydin R, Schrama JW, Verreth JAJ, Rombout JHWM: Soybean meal-induced uptake block in Atlantic salmonSalmo salar distal enterocytes. J Fish Biol. 2008, 73 (10): 2571-2579.

12. Gómez-Requeni P, Mingarro M, Kirchner S, Calduch-Giner JA, Médale F, Corraze G, Panserat S, Martin SAM, Houlihan DF, Kaushik SJ, et al: Effects of dietary amino acid profile on growth performance, key metabolic enzymes and somatotropic axis responsiveness of gilthead sea bream (Sparus aurata). Aquaculture. 2003, 220 (1–4): 749-767.

13. Francis G, Makkar HPS, Becker K: Antinutritional factors present in plant-derived alternate fish feed ingredients and their effects in fish. Aquaculture. 2001, 199 (3–4): 197-227.

14. Zhao H, Jiang R, Xue M, Xie S, Wu X, Guo L: Fishmeal can be completely replaced by soy protein concentrate by increasing feeding frequency in Nile tilapia (Oreochromis niloticus GIFT strain) less than 2 g. Aquac Nutr. 2010, 16 (6): 648-653.

15. Escaffre AM, Kaushik S, Mambrini M: Morphometric evaluation of changes in the digestive tract of rainbow trout (Oncorhynchus mykiss) due to fish meal replacement with soy protein concentrate. Aquaculture. 2007, 273 (1): 127-138.

16. Kaushik SJ, Cravedi JP, Lalles JP, Sumpter J, Fauconneau B, Laroche M: Partial or total replacement of fish meal by soybean protein on growth, protein utilization, potential estrogenic or antigenic effects, cholesterolemia and flesh quality in rainbow trout, Oncorhynchus mykiss. Aquaculture. 1995, 133 (3–4): 257-274.

17. Gaylord TG, Teague AM, Barrows FT: Taurine supplementation of all-plant protein diets for rainbow trout (Oncorhynchus mykiss). J World Aquaculture Soc. 2006, 37 (4): 509-517.

18. Gill N, Higgs DA, Skura BJ, Rowshandeli M, Dosanjh BS, Mann J, Gannam AL: Nutritive value of partially dehulled and extruded sunflower meal for post-smolt Atlantic salmon (Salmo salar L.) in sea water. Aquac Res. 2006, 37 (13): 1348-1359.

19. Mente E, Deguara S, Begoña Santos M, Houlihan D: White muscle free amino acid concentrations following feeding a maize gluten dietary protein in Atlantic salmon (Salmo salar L.). Aquaculture. 2003, 225 (1–4): 133-147.

20. Thiessen DL, Campbell GL, Adelizi PD: Digestibility and growth performance of juvenile rainbow trout (Oncorhynchus mykiss) fed with pea and canola products. Aquac Nutr. 2003, 9 (2): 67-75.

21. Espe M, Lemme A, Petri A, El-Mowafi A: Assessment of lysine requirement for maximal protein accretion in Atlantic salmon using plant protein diets. Aquaculture. 2007, 263 (1–4): 168-178.

22. Penn MH, Bendiksen EA, Campbell P, Krogdahl AS: High level of dietary pea protein concentrate induces enteropathy in Atlantic salmon (Salmo salar L.). Aquaculture. 2011, 310 (3–4): 267-273.

23. Tacchi L, Bron JE, Taggart JB, Secombes CJ, Bickerdike R, Adler MA, Takle H, Martin SAM: Multiple tissue transcriptomic responses to Piscirickettsia salmonis in Atlantic salmon (Salmo salar). Physiol Genom. 2011, 43 (21): 1241-1254.

24. Panserat S, Hortopan GA, Plagnes-Juan E, Kolditz C, Lansard M, Skiba-Cassy S, Esquerré D, Geurden I, Médale F, Kaushik S, et al: Differential gene expression after total replacement of dietary fish meal and fish oil by plant products in rainbow trout (Oncorhynchus mykiss) liver. Aquaculture. 2009, 294 (1–2): 123-131.

25. Le Boucher R, Quillet E, Vandeputte M, Lecalvez JM, Goardon L, Chatain B, Médale F, Dupont-Nivet M: Plant-based diet in rainbow trout (Oncorhynchus mykiss): Are there genotype-diet interactions for main production traits when fish are fed marine vs. plant-based diets from the first meal?. Aquaculture. 2011, 321 (1–2): 41-48.

26. Pierce LR, Palti Y, Silverstein JT, Barrows FT, Hallerman EM, Parsons JE: Family growth response to fishmeal and plant-based diets shows genotype x diet interaction in rainbow trout (Oncorhynchus mykiss). Aquaculture. 2008, 278 (1–4): 37-42.

27. Kolditz CI, Plagnes-Juan E, Quillet E, Lefèvre F, Médale F: Changes in white muscle transcriptome induced by dietary energy levels in two lines of rainbow trout (Oncorhynchus mykiss) selected for muscle fat content. Br J Nutr. 2010, 103 (5): 629-642.

28. Morais S, Pratoomyot J, Taggart J, Bron J, Guy D, Bell JG, Tocher D: Genotype-specific responses in Atlantic salmon (Salmo salar) subject to dietary fish oil replacement by vegetable oil: a liver transcriptomic analysis. BMC Genom. 2011, 12 (1): 255-

29. Tacon AGJ, Metian M: Global overview on the use of fish meal and fish oil in industrially compounded aquafeeds: Trends and future prospects. Aquaculture. 2008, 285 (1–4): 146-158.

30. Morais S, Pratoomyot J, Taggart J, Bron J, Guy D, Bell J, Tocher D: Genotype-specific responses in Atlantic salmon (Salmo salar) subject to dietary fish oil replacement by vegetable oil: a liver transcriptomic analysis. BMC Genom. 2011, 12 (1): 255-

31. Jafri M, Donnelly B, Bondoc A, Allen S, Tiao G: Cholangiocyte secretion of chemokines in experimental

biliary atresia. J Pediatr Surg. 2009, 44 (3): 500-507.

32. Mack CL, Tucker RM, Sokol RJ, Kotzin BL: Armed CD4+ Th1 effector cells and activated macrophages participate in bile duct injury in murine biliary atresia. Clin Immunol. 2005, 115 (2): 200-209.
33. Baeverfjord G, Krogdahl Å: Development and regression of soybean meal induced enteritis in Atlantic salmon, Salmo salar L., distal intestine: A comparison with the intestines of fasted fish. J Fish Dis. 1996, 19 (5): 375-387.
34. van den Ingh TSGAM, Krogdahl Å, Olli JJ, Hendriks HGCJM, Koninkx JGJF: Effects of soybean-containing diets on the proximal and distal intestine in Atlantic salmon (Salmo salar): a morphological study. Aquaculture. 1991, 94 (4): 297-305.
35. Sanden M, Berntssen MHG, Krogdahl Å, Hemre GI, Bakke-McKellep AM: An examination of the intestinal tract of Atlantic salmon, Salmo salar L., parr fed different varieties of soy and maize. J Fish Dis. 2005, 28 (6): 317-330.
36. Salinas I, Meseguer J, Esteban MÁ: Assessment of different protocols for the isolation and purification of gut associated lymphoid cells from the gilthead seabream (Sparus aurata L.). Biol Proced Online. 2007, 9 (1): 43-55.
37. Collie NL, Ferraris RP: Chapter 9 Nutrient fluxes and regulation in fish intestine. Biochemistry and Molecular Biology of Fishes. 1995, 4: 221-239.
38. Bakke-McKellep AM, Nordrum S, Krogdahl Å, Buddington RK: Absorption of glucose, amino acids, and dipeptides by the intestines of Atlantic salmon (Salmo salar L.). Fish Physiol Biochem. 2000, 22 (1): 33-44.
39. Bureau DP, Harris AM, Young Cho C: The effects of purified alcohol extracts from soy products on feed intake and growth of chinook salmon (Oncorhynchus tshawytscha) and rainbow trout (Oncorhynchus mykiss). Aquaculture. 1998, 161 (1–4): 27-43.
40. Rumsey GL, Siwicki AK, Anderson DP, Bowser PR: Effect of soybean protein on serological response, non-specific defense mechanisms, growth, and protein utilization in rainbow trout. Vet Immunol Immunopathol. 1994, 41 (3–4): 323-339.
41. Krogdahl A, Bakke-McKellep AM, Baeverfjord G: Effects of graded levels of standard soybean meal on intestinal structure, mucosal enzyme activities, and pancreatic response in Atlantic salmon (Salmo solar L.). Aquac Nutr. 2003, 9 (6): 361-371.
42. Bernard D, Six A, Rigottier-Gois L, Messiaen S, Chilmonczyk S, Quillet E, Boudinot P, Benmansour A: Phenotypic and functional similarity of gut intraepithelial and systemic T cells in a teleost fish. J Immunol. 2006, 176 (7): 3942-3949.
43. Riccardi C, Cifone MG, Migliorati G: Glucocorticoid hormone-induced modulation of gene expression and regulation of T-cell death: Role of GITR and GILZ, two dexamethasone-induced genes. Cell Death Differentiation. 1999, 6 (12): 1182-1189.
44. Delfino DV, Agostini M, Spinicelli S, Vacca C, Riccardi C: Inhibited cell death, NF-κB activity and increased IL-10 in TCR-triggered thymocytes of transgenic mice overexpressing the glucocorticoid-induced protein GILZ. Int Immunopharmacol. 2006, 6 (7): 1126-1134.
45. Wilson S, Blaschek K, Gonzalez De Mejia E: Allergenic proteins in soybean: Processing and reduction of P34 allergenicity. Nutr Rev. 2005, 63 (2): 47-58.
46. Jacobs MD, Harrison SC: Structure of an IκBα/NF-κB complex. Cell. 1998, 95 (6): 749-758.
47. Takeda K, Kaisho T, Akira S: Toll-like receptors. Annu Rev Immunol. 2003, 21: 335-376.
48. Honda K, Yanai H, Negishi H, Asagiri M, Sato M, Mizutani T, Shimada N, Ohba Y, Takaoka A, Yoshida N, et al: IRF-7 is the master regulator of type-I interferon-dependent immune responses. Nature. 2005, 434 (7034): 772-777.
49. Sun B, Skjaeveland I, Svingerud T, Zou J, Jorgensen J, Robertsen B: Antiviral Activity of Salmonid Gamma Interferon against Infectious Pancreatic Necrosis Virus and Salmonid Alphavirus and Its Dependency on Type I Interferon. J Virol. 2011, 85 (17): 9188-9198.
50. Robertsen B, Trobridge G, Leongt J-A: Molecular cloning of double-stranded RNA inducible MX genes from atlantic salmon (Salmo salar l.). Dev Comp Immunol. 1998, 21 (5): 397-412.
51. Boudinot P, Salhi S, Blanco M, Benmansour A: Viral haemorrhagic septicaemia virus induces vig −2, a new interferon-responsive gene in rainbow trout. Fish Shellfish Immunol. 2001, 11 (5): 383-397.
52. van der Aa LM, Levraud JP, Yahmi M, Lauret E, Briolat V, Herbomel P, Benmansour A, Boudinot P: A large new subset of TRIM genes highly diversified by duplication and positive selection in teleost fish. BMC Biol. 2009, 7: 7-
53. Martens N, Uzan G, Wery M, Hooghe R, Hooghe-Peters EL, Gertler A: Suppressor of Cytokine Signaling 7 Inhibits Prolactin, Growth Hormone, and Leptin Signaling by Interacting with STAT5 or STAT3 and Attenuating Their Nuclear Translocation. J Biol Chem. 2005, 280 (14): 13817-13823.
54. Jimenez N, Coll J, Salguero FJ, Tafalla C: Co-injection of interleukin 8 with the glycoprotein gene from viral haemorrhagic septicemia virus (VHSV) modulates the cytokine response in rainbow trout (Oncorhynchus mykiss). Vaccine. 2006, 24 (27–28): 5615-5626.
55. Kumari J, Larsen AN, Bogwald J, Dalmo RA: Interleukin-17D in Atlantic salmon (Salmo salar): Molecular characterization, 3D modelling and promoter analysis. Fish Shellfish Immunol. 2009, 27 (5): 647-659.
56. Godiksen H, Morzel M, Hyldig G, Jessen F: Contribution of cathepsins B, L and D to muscle protein profiles correlated with texture in rainbow trout (Oncorhynchus mykiss). Food Chem. 2009, 113 (4): 889-896.
57. Tacchi L, Bickerdike R, Secombes CJ, Martin SAM: Muscle-Specific RING Finger (MuRF) cDNAs in Atlantic Salmon (Salmo salar) and Their Role as Regulators of

Muscle Protein Degradation. Mar Biotechnol. 2012, 14 (1): 35-45.

58. Zheng X, Torstensen BE, Tocher DR, Dick JR, Henderson RJ, Bell JG: Environmental and dietary influences on highly unsaturated fatty acid biosynthesis and expression of fatty acyl desaturase and elongase genes in liver of Atlantic salmon (Salmo salar). Biochim Biophys Acta Mol Cell Biol Lipids. 2005, 1734 (1): 13-24.
59. Hastings N, Agaba MK, Tocher DR, Zheng X, Dickson CA, Dick JR, Teale AJ: Molecular cloning and functional characterization of fatty acyl desaturase and elongase cDNAs involved in the production of eicosapentaenoic and docosahexaenoic acids from α-linolenic acid in Atlantic salmon (Salmo salar). Mar Biotechnol. 2004, 6 (5): 463-474.
60. Martin SAM, Cash P, Blaney S, Houlihan DF: Proteome analysis of rainbow trout (Oncorhynchus mykiss) liver proteins during short term starvation. Fish Physiol Biochem. 2001, 24 (3): 259-270.
61. Tacchi L, Bickerdike R, Douglas A, Secombes CJ, Martin SAM: Transcriptomic responses to functional feeds in Atlantic salmon (Salmo salar). Fish Shellfish Immunol. 2011, 31 (5): 704-715.
62. Vilhelmson OT, Martin SAM, Médale F, Kaushik SJ, Houlihan DF: Dietary plant-protein substitution affects hepatic metabolism in rainbow trout (Oncorhynchus mykiss). Br J Nutr. 2004, 92 (1): 71-80.
63. Taggart JB, Bron JE, Martin SAM, Seear PJ, Høyheim B, Talbot R, Carmichael SN, Villeneuve LAN, Sweeney GE, Houlihan DF, et al: A description of the origins, design and performance of the TRAITS-SGP Atlantic salmon Salmo salar L. cDNA microarray. J Fish Biol. 2008, 72 (9): 2071-2094.
64. Martin SAM, Douglas A, Houlihan D, Secombes C: Starvation alters the liver transcriptome of the innate immune response in Atlantic salmon (Salmo salar). BMC Genom. 2010, 11 (1): 418-
65. Lucas PJ, Kim SJ, Melby SJ, Gress RE: Disruption of T cell homeostasis in mice expressing a T cell-specific dominant negative transforming growth factor β II receptor. J Exp Med. 2000, 191 (7): 1187-1196.
66. Cazac BB, Roes J: TGF-β receptor controls B cell responsiveness and induction of IgA in vivo. Immunity. 2000, 13 (4): 443-451.
67. Schuster N, Krieglstein K: Mechanisms of TGF-β-mediated apoptosis. Cell Tissue Res. 2002, 307 (1): 1-14.
68. Oka C, Tsujimoto R, Kajikawa M, Koshiba-Takeuchi K, Ina J, Yano M, Tsuchiya A, Ueta Y, Soma A, Kanda H, et al: HtrA1 serine protease inhibits signaling mediated by Tgfβ family proteins. Development. 2004, 131 (5): 1041-1053.
69. Harashima M, Harada K, Ito Y, Hyuga M, Seki T, Ariga T, Yamaguchi T, Niimi S: Annexin A3 expression increases in hepatocytes and is regulated by hepatocyte growth factor in rat liver regeneration. J Biochem. 2008, 143 (4): 537-545.
70. Geay F, Ferraresso S, Zambonino-Infante J, Bargelloni L, Quentel C, Vandeputte M, Kaushik S, Cahu C, Mazurais D: Effects of the total replacement of fish-based diet with plant-based diet on the hepatic transcriptome of two European sea bass (Dicentrarchus labrax) half-sibfamilies showing different growth rates with the plant-based diet. BMC Genom. 2011, 12 (1): 522-
71. Sagstad A, Sanden M, Krogdahl Å, Bakke-Mckellep AM, FrØystad M, Hemre GI: Organs development, gene expression and health of Atlantic salmon (Salmo salar L.) fed genetically modified soybeans compared to the near-isogenic non-modified parental line. Aquac Nutr. 2008, 14 (6): 556-572.
72. Freeman ML, Borrelli MJ, Meredith MJ, Lepock JR: On the path to the heat shock response: Destabilization and formation of partially folded protein intermediates, a consequence of protein thiol modification. Free Radic Biol Med. 1999, 26 (5–6): 737-745.
73. McKnight GL, Reasoner J, Gilbert T, Sundquist KO, Hokland B, McKernan PA, Champagne J, Johnson CJ, Bailey MC, Holly R, et al: Cloning and expression of a cellular high density lipoprotein-binding protein that is up-regulated by cholesterol loading of cells. J Biol Chem. 1992, 267 (17): 12131-12141.
74. Bouraoui L, SÁNchez-Gurmaches J, Cruz-Garcia L, GutiÉRrez J, Benedito-Palos L, PÉRez-SÁNchez J, Navarro I: Effect of dietary fish meal and fish oil replacement on lipogenic and lipoprotein lipase activities and plasma insulin in gilthead sea bream (Sparus aurata). Aquac Nutr. 2011, 17 (1): 54-63.
75. Lamon-Fava S: Genistein Activates Apolipoprotein A-I Gene Expression in the Human Hepatoma Cell Line Hep G2. J Nutr. 2000, 130 (10): 2489-2492.
76. Vandenbrouck Y, Lambert G, Janvier B, Girlich D, Bereziat G, Mangeney-Andreani M: Transcriptional regulation of apolipoprotein A-I expression in Hep G2 cells by phorbol ester. FEBS Lett. 1995, 376 (1–2): 99-102.
77. Houlihan DF, Carter CG, McCarthy ID: Nitrogen Metabolism and Excretion. Protein Turnover in Animals. Edited by: Walsh PJ, Wright P. 1995, CRC Press, Boca Raton, United States of America, 1-32.
78. Martin SAM, Blaney S, Bowman AS, Houlihan DF: Ubiquitin-proteasome-dependent proteolysis in rainbow trout (Oncorhynchus mykiss): Effect of food deprivation. Pflugers Arch Eur J Physiol. 2002, 445 (2): 257-266.
79. Dobly A, Martin SAM, Blaney SC, Houlihan DF: Protein growth rate in rainbow trout (Oncorhynchus mykiss) is negatively correlated to liver 20 S proteasome activity. Comp Biochem Physiol Mol Integr Physiol. 2004, 137 (1): 75-85.
80. Wang K, Yin XM, Chao DT, Milliman CL, Korsmeyer SJ: BID: A novel BH3 domain-only death agonist. Genes Dev. 1996, 10 (22): 2859-2869.
81. Sandri M, Rossini K, Podhorska-Okolow M, Carraro U: Role of Apoptosis in Muscle Disorders. Basic Appl Myol. 1999, 9 (6): 301-310.
82. Laing KJ, Bols N, Secombes CJ: A CXC chemokine sequence isolated from the rainbow trout Oncorhynchus mykissresembles the closely related interferon-γ-inducible chemokines CXCL9, CXCL10 and CXCL11. Eur Cytokine Netw. 2002, 13 (4): 462-473.

83. Jiang J, Zhang Y-B, Li S, Yu F-F, Sun F, Gui J-F: Expression regulation and functional characterization of a novel interferon inducible gene Gig2 and its promoter. Mol Immunol. 2009, 46 (15): 3131-3140.

84. Bange FC, Vogel U, Flohr T, Kiekenbeck M, Denecke B, Böttger EC: IFP 35 is an interferon-induced leucine zipper protein that undergoes interferon-regulated cellular redistribution. J Biol Chem. 1994, 269 (2): 1091-1098.

85. Honda ZI, Ishii S, Shimizu T: Platelet-activating factor receptor. J Biochem. 2002, 131 (6): 773-779.

86. Casadei E, Wang T, Zou J, González Vecino JL, Wadsworth S, Secombes CJ: Characterization of three novel β-defensin antimicrobial peptides in rainbow trout (Oncorhynchus mykiss). Mol Immunol. 2009, 46 (16): 3358-3366.

87. Zheng Q, Wang XJ: GOEAST: a web-based software toolkit for Gene Ontology enrichment analysis. Nucleic Acids Res. 2008, 36: W358-W363. Web Server issue

88. Tacchi L, Bickerdike R, Secombes CJ, Pooley NJ, Urquhart KL, Collet B, Martin SAM: Ubiquitin E3 ligase atrogin-1 (Fbox-32) in Atlantic salmon (Salmo salar): Sequence analysis, genomic structure and modulation of expression. Comp Biochem Physiol Biochem Mol Biol. 2010, 157 (4): 364-373.

89. Bower NI, Johnston IA: Selection of reference genes for expression studies with fish myogenic cell cultures. BMC Mol Biol. 2009, 10: 80-

90. Pfaffl MW: A new mathematical model for relative quantification in real-time RT-PCR. Nucleic Acids Res. 2001, 29 (9): e45-

CHAPTER 10

SUBSTITUTION OF DIETARY FISH OIL WITH PLANT OILS IS ASSOCIATED WITH SHORTENED MID INTESTINAL FOLDS IN ATLANTIC SALMON (SALMO SALAR)

Torfinn Moldal[1], Guro Løkka[2], Jannicke Wiik-Nielsen[1], Lars Austbø[2], Bente E Torstensen[3], Grethe Rosenlund[4], Ole Bendik Dale[1], Magne Kaldhusdal[1] and Erling Olaf Koppang[2]

[1]Norwegian Veterinary Institute, Sentrum, 0106 Oslo, Norway.

[2]Norwegian School of Veterinary Science, Dep, 0033 Oslo,Norway.

[3]National Institute of Nutrition and Seafood Research, Nordnes, 5817 Bergen, Norway.

[4]Skretting ARC, Stavanger, Norway.

ABSTRACT

Background

Fish meal and fish oil are increasingly replaced by ingredients from terrestrial sources in the feeds for farmed salmonids due to expanding production and reduced availability of marine feed raw material. Fish oil that is rich in n-3 polyunsaturated fatty acids is considered beneficial to human health in general and to prevent intestinal inflammation and carcinogenesis in particular. In contrast, n-6 fatty acids that are present in many vegetable oils have been associated with increased risk of colitis and colon cancer in rodents and humans, as well as lowered transcription levels of certain stress and antioxidant-related genes in Atlantic salmon.

The aim of the present study was to investigate the intestinal health in Atlantic salmon fed with different vegetable oils as partial substitutes of fish oil in the diet. A feed trial lasting for 28 weeks included one reference diet containing fish oil as the sole lipid source and three diets where 80% of the fish oil was replaced by a plant oil blend with either olive oil, rapeseed oil or soybean oil as the main lipid source. These plant oils have intermediate or low n-3/n-6-ratios compared to fish oil having a high n-3/n-6-ratio. The protein and carbohydrate fractions were identical in all the feeds.

Results

Morphometric measurements showed significantly shorter folds in the mid intestine in all groups fed vegetable oils compared to the group fed fish oil. In the distal intestine, the complex folds were significantly shorter in the fish fed soybean oil compared to the fish fed rapeseed oil. Histological and immunohistochemical examination did not show clear difference in the degree of inflammation or proliferation of epithelial cells related to dietary groups, which was further confirmed by real-time RT-PCR which revealed only moderate alterations in the mRNA transcript levels of selected immune-related genes.

Citation: Moldal et al.: Substitution of dietary fish oil with plant oils is associated with shortened mid intestinal folds in Atlantic salmon (Salmo salar). BMC Veterinary Research 2014 10:60. doi:10.1186/1746-6148-10-60

Conclusions

Shortened intestinal folds might be associated with reduced intestinal surface and impaired nutrient absorption and growth, but our results suggest that partial substitution of dietary fish oil with vegetable oils does not have any major negative impact on the intestinal health of Atlantic salmon.

BACKGROUND

Salmonids are indigenous carnivores, and fish meal and fish oil have traditionally been the main ingredients in feed for farmed Atlantic salmon (*Salmo salar*) and rainbow trout (*Oncorhynchus mykiss*). Due to limited sources of marine raw materials, a fast growing aquaculture industry and increased focus on sustainability, the salmon farming industry has searched for alternative feeds [1–3]. The inclusion of plant-derived materials in fish feed may be beneficial from an economic and ecological point of view, but used without carefully considering the fish minimum requirements or upper tolerable levels of certain nutrients or anti-nutrients, it can unfortunately also cause adverse effects with regard to fish health, nutritive value and the consumers' acceptance [1].

In the early years of salmonid farming, proteins constituted more than half of the feed content, while the lipid fraction was as low as 10%. These figures have changed during the last decades as the protein level has decreased to less than 40%, while the lipid fraction has increased till about 35% [4]. Fish oil production requires 2-5 times more industrial fish than production of the same weight of fish meal [5], and replacement of fish oil with vegetable oils is hence of growing interest both from an economic and sustainability viewpoint.

Vegetable oils may contain high levels of n-6 polyunsaturated fatty acids (PUFAs) such as linoleic acid (LA, 18:2n-6) that can be further metabolized to arachidonic acid (AA; 20:4n-6). In contrast, fish oils are rich in n-3 PUFAs such as docosahexaenoic acid (DHA, 22:6n-3) and eicosapentaenoic acid (EPA, 20:5n-3) (Figure 1). In mammals, dietary n-3 fatty acids supplant the AA in inflammatory cell membranes [6] and therefore decrease the availability of the major precursor of pro-inflammatory eicosanoids as the same enzymes are involved in the metabolism of n-3 and n-6 fatty acids and further in the synthesis of eicosanoids [7]. Also in Atlantic salmon, dietary lipid has been shown to alter leucocyte phospholipid fatty acid composition and eicosanoid production [8], and increased n-6 levels in the feed gave increased n-6 fatty acids in leucocytes in an *ex vivo* study [9].

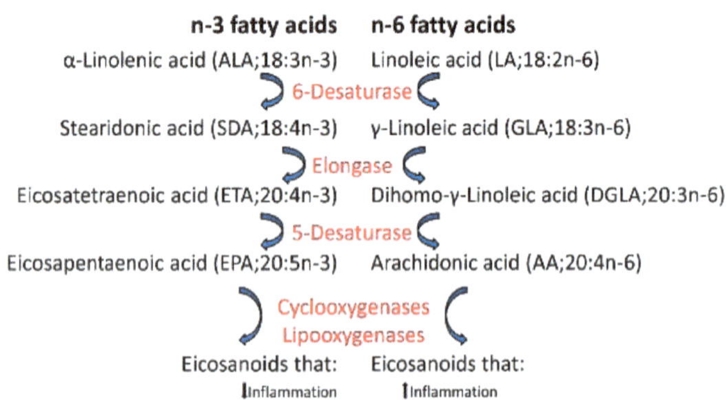

Figure 1 The metabolism of n-3 and n-6 fatty acids and eicosanoids. The same enzymes are involved in the metabolism of n-3 and n-6 fatty acids and synthesis of eicosanoids, but the biological properties of the eicosanoids are different.

In mammals, a diet rich in n-6 fatty acids has been associated with increased risk of ulcerative colitis [10] and promotion of intestinal carcinogenesis [6, 11], while a high intake of n-3 PUFAs is considered to be beneficial for health. Consumption of n-3 fatty acids has been shown to attenuate the dysbiosis and colitis caused by n-6 polyunsaturated fatty acid in mice [12] and to prevent and modulate a wide range of pathological conditions as cardiovascular diseases, diabetes and several inflammatory and neoplastic processes, including inflammatory bowel disease and colon cancer [13]. The n-3 fatty acids also inhibit the prostaglandin synthesizing enzyme cyclooxygenase-2 (COX-2) which is up-regulated during inflammation, the expression of the pro-inflammatory cytokines tumour necrosis factor-α (TNF-α) and interleukin-1 (IL-1) and the proliferation of lymphocytes as shown both *in vitro* and in rodent models [13–15].

Several studies have addressed the effects of vegetable oils as lipid sources in the feed on Atlantic salmon intestinal absorption [16], post-absorptive fates [17], feed uptake, growth rate, metabolism and nutrient content of the fish filet [2, 18–22]. Whereas many studies have addressed the intestinal health of the fish when fish meal is replaced by different plant-derived proteins, and both soybean meal and pea protein concentrate have been shown to induce enteritis [23, 24], there is a knowledge gap regarding the impact on intestinal health when fish oil is replaced by plant oils.

Complete substitution of fish oil with a plant oil blend containing rapeseed oil, palm oil and linseed oil in the feed induced lower transcription levels of certain stress and antioxidant-related genes in the intestine [25]. Another feed trial with the same oil blend partly substituting fish oil in combination with plant proteins at different inclusion levels demonstrated that in response to acute physiological stress, high levels of plant-derived dietary ingredients can enhance COX-2 induction and synthesis of pro-inflammatory eicosanoids in the intestine of salmon [26]. It has also been speculated whether inclusion of plant oils in the feed contributed to intestinal carcinogenesis in brood stock Atlantic salmon [27].

The aim of the present study was to investigate the morphology of the intestinal wall, the presence of antigen presenting cells and T lymphocytes, the proliferation pattern of epithelial cells, and the transcript levels of selected immune-related genes including relevant cytokines, major histocompatibility complex class II (MHC class II), cluster of differentiation 3ζ (CD3ζ), immunoglobulins, the intracellular receptor nucleotide-binding oligomerization domain-containing protein 2 (NOD2) and COX-2a in the intestine of Atlantic salmon when dietary fish oil was partially replaced by different vegetable oil blends with varying n-3/n-6-ratio.

METHODS

Animal Ethics, Fish and Feed

The feed trial was carried out at Skretting ARC Fish Trials Station that is approved by the Norwegian Animal Research Authority and was conducted according to current animal welfare regulations: FOR-1996-01-15-23 (Norway). Six hundred Atlantic salmon (*Salmo salar*) with mean initial weight of 815±28 g were equally distributed into 12 tanks, and triplicate groups of fish were fed a diet with either fish oil as the sole lipid source or a diet where 80% of the fish oil was replaced by one of three vegetable oil blends with olive oil, rapeseed oil or soybean oil as the main lipid source (Figure 2). The n-3/n-6 ratio in the feed with fish oil as the sole lipid source was 5.3, while the n-3/n-6 ratios were 0.7, 0.9 and 0.3 in the diets where fish oil was largely substituted with olive oil, rapeseed oil and soybean oil respectively. The composition of the protein fraction was identical in all diets with 30% fish meal and 70% plant protein. The proximate composition of all diets was similar with 332-341 g kg^{-1} fat and 406-413 g kg^{-1}protein.

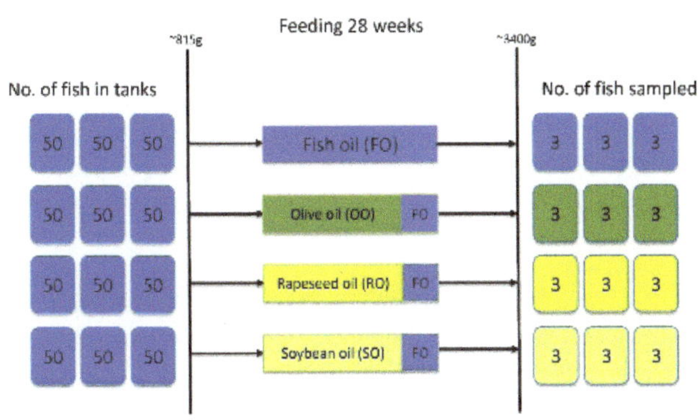

Figure 2 The design of the feed trial. Six hundred Atlantic salmon with mean initial weight of 815±28 g were equally distributed into 12 tanks, and triplicate groups of fish were fed a diet with either fish oil as the sole lipid source or a diet where 80% of the fish oil was replaced by one of three vegetable oil blends with olive oil (OO), rapeseed oil (RO) or soybean oil (SO) as the main lipid source. The protein fraction was identical in all diets, and 70% of the proteins were plant-derived. The trial lasted for 28 weeks, and the mean final weight of all dietary groups was 3399±76 g. Three fish in each tank were sampled by the end of the trial.

The fish were vaccinated intraperitonally (ALPHA JECT micro 6; PHARMAQ, Overhalla, Norway) to give protection towards furunculosis, vibriosis, cold-water vibriosis, winter ulcer disease and infectious pancreas necrosis one month prior to seawater transfer. The trial lasted for 28 weeks, and the mean final weight was 3399±76 g with no significant difference between dietary groups. The fish were anesthetized with MS222 at a concentration of 7 g L^{-1}and euthanized according to regulations (Forskrift om drift av akvakulturanlegg §34. Avlivning av fisk). The design of the feed trial and feed composition is described in detail elsewhere [22].

Histology and Immunohistochemistry

For histological evaluation, tissues from the mid and distal intestine (Figure 3), also called first and second segment of mid intestine, respectively [28], from 24 fish in each dietary group (eight fish per tank) were fixed in 10% phosphate-buffered formalin for 24-48 hours. The tissues were routinely processed, embedded in paraffin and cut in 3 µm thick cross sections for the mid intestine and longitudinal sections (i.e. perpendicular to the macroscopically visible circular folds) for the distal intestine. Sections were routinely deparaffinized in xylene and rehydrated in graded alcohol baths before staining with hematoxylin and eosin (HE).

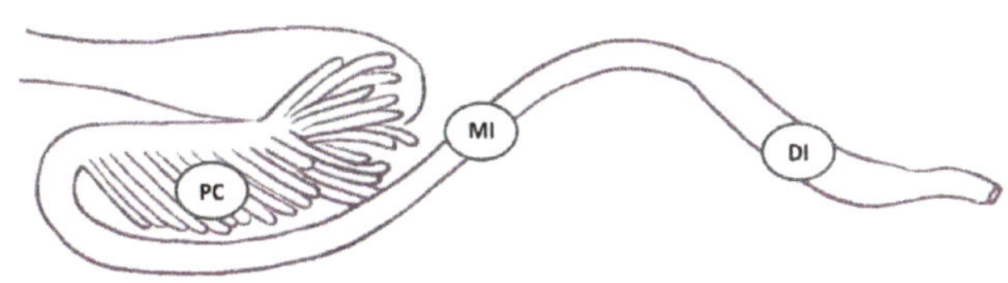

Figure 3 Schematic drawing of the intestinal tract of Atlantic salmon. The sampling sites are indicated as follows: PC=pyloric caeca, MI=mid intestine and DI=distal intestine.

Formalin-fixed and paraffin-embedded tissues from the mid intestine and the distal intestine from nine fish in each dietary groups (three fish in each tank) were prepared for detection of antigen presenting cells expressing MHC class II and T lymphocytes expressing CD3ε by immunohistochemistry with salmon specific polyclonal rabbit antisera as previously described by Koppang *et al.* [29, 30] with some modifications. Proliferating cells were recognized in sections from the mid intestine from nine fish in each dietary group (three fish in each tank) using a monoclonal mouse antibody against proliferating cell nuclear antigen (PCNA; M0879, Dako, Glostrup, Denmark). Sections were routinely deparaffinized in xylene and rehydrated in graded alcohol baths before they were transferred to distilled water. Antigen retrieval was performed by autoclaving the slides in 0.01 M citrate buffer at 121°C for 15 min, and the slides were cooled to room temperature and transferred to phosphate-buffered saline (PBS) before inhibition of endogenous peroxidase with 0.05% phenyl hydrazine (P26252, Sigma-Aldrich, Milwaukee, Wisconsin, US) in PBS at 37°C for 40 min. The slides were then incubated in goat serum diluted 1:50 in 5% bovine serum albumin (BSA) in Tris-buffered saline (TBS) for 20 min to prevent nonspecific binding. Antisera against MHC class II, CD3ε and PCNA were diluted 1:600, 1:400 and 1:3000, respectively, in 1% BSA/TBS before incubation for 30 min. The secondary antibody and substrate chromogen were provided from the EnVision® System kit (K4009, Dako). The sections were counterstained with hematoxylin added acetic acid (Mayer's hematoxylin) for 1 min and mounted with poly-vinyl alcohol mounting media (Ullevål Apotek, Oslo, Norway).

Morphometric Analysis

Micrographs of intestinal sections from nine fish in each dietary group (three fish per tank) were captured and morphometric measurements were performed in the software NIS-Elements D version 3 (Nikon, Tokyo, Japan) using Nikon digital sight camera configured with a Nikon eclipse 80i microscope. The measurements were performed as previously described by Løkka *et al.* [28]. The height of the folds was measured from the fold apex to the bottom of the epithelium at the base of the folds, and both simple and complex folds were measured in the distal intestine. The width of the folds was assessed at two points in every fold, and the thickness of the intestinal wall was measured from beneath the epithelium at the base of the folds (simple folds in distal intestine) to the serosa (Figure 4). Five measurements of the fold height and wall thickness and ten measurements of the fold width in both intestinal segments were recorded for each individual.

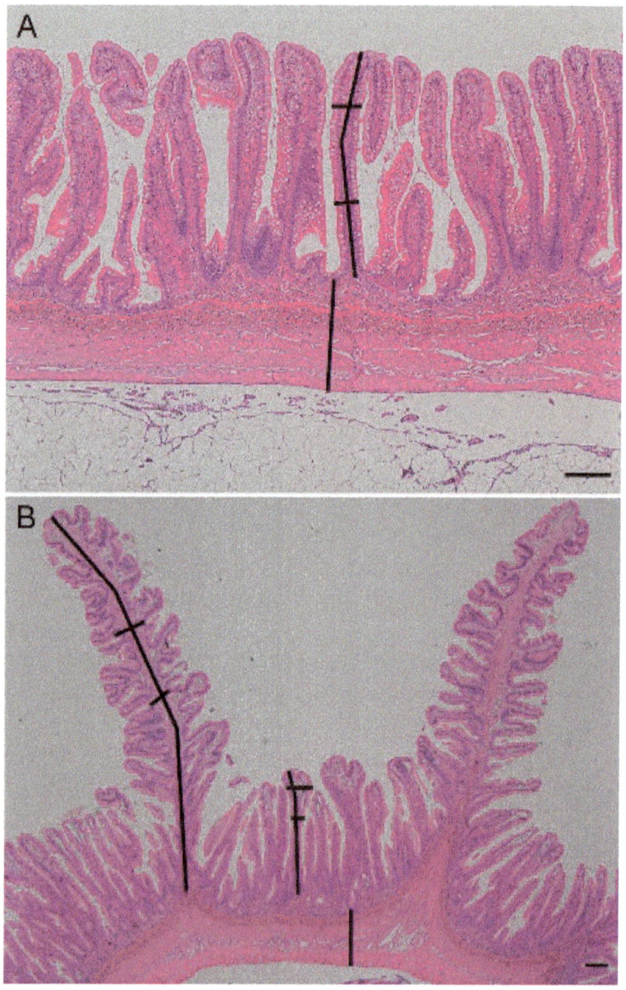

Figure 4 Microphotographs showing measurements in sections of the mid (A) and distal (B) intestine. The lines indicate how the height and width of the folds and the thickness of the wall were measured. Hematoxylin and eosin. Scale bar 200 μm.

Gene Transcription Analysis

Tissues from the pyloric caeca, mid intestine and distal intestine (Figure 3) from nine fish in each dietary group (three fish per tank) were frozen in liquid nitrogen and stored at -80°C. The samples from the mid intestine were homogenized in Trizol using zirconium beads (4 mm) in a Retsch MM 310 homogenizer (Retsch GmbH, Haan, Germany). Subsequent addition of chloroform separated RNA from proteins and DNA, and RNA was then precipitated from the water phase by adding isopropanol. Furthermore, the RNA pellet was cleansed twice in ethanol and dissolved in RNase free water. A DNase treatment with DNA-free™ (Applied biosystems, Foster City, CA, USA) was performed on the RNA extract.

The samples from the pyloric caeca and the distal intestine were homogenized in Buffer RLT added mercaptoetanol using stainless steelbeads (5 mm) in a Retsch MM 300 homogenizer (Retsch GmbH). RNA was extracted with RNeasy Mini kit (QIAGEN, Hilden, Germany) using the protocol "RNeasyMini Animal Tissues and Cells Standard V3" followed by the protocol "Cleanup RNeasyMini RNA Standard V3" in a QIAcube (QIAGEN).

The concentration of RNA was measured using a Biospec-Nano (Shimadzu Corporation, Kyoto, Japan) or Nanodrop ND-1000 UV-Vis Spectrophotometer (NanoDrop Technologies, Wilmington, DE, USA). To verify acceptable quality of the RNA, 24 random samples were selected and tested on an Agilent 2100 Bioanalyzer (Agilent Technologies, Palo Alto, CA, USA). Total RNA was stored at -80°C.

The cDNA synthesis from 1 μg of total RNA was prepared with oligo(dT), random hexamer primers, M-MLV Reverse transcriptase (Promega, Madison, WI, USA) and RNase inhibitor (Promega) to prevent RNA degradation. Real-time PCR was carried out in 13 μl reactions using TaqMan Gene Expression Master Mix (Applied Biosystems, Carlsbad, CA, USA) with cDNA template corresponding to 15 ng of RNA in each reaction in a 7900HT fast real-time PCR system (Applied Biosystems) according to the producer's instructions and running 40 cycles.

The following genes were analyzed by real-time RT-PCR: Cluster of differentiation 3ζ (CD3ζ), cyclooxygenase-2a (COX-2a), interleukin 1β (IL-1β), immunoglobulin M (IgM), immunoglobulin T (IgT), major histocompatibility class II (MHC class II), nucleotide-binding oligomerization domain-containing protein 2 (NOD2), transforming growth factor β (TGF-β), tumour necrosis factor α (TNF-α) and elongation factor $1A_B$ ($EF1A_B$) as the reference gene [31] (Table 1). When possible, primers and probe were designed to span across intron sections. All analyses were performed in triplicates, and a control lacking the template for each master mix was always included in the experiments. The data were analysed using Sequence Detection Systems Software v2.3 (Applied Biosystems).

Table 1 Sequences of primers and probes used in real-time RT-PCR analysis

Gene	Gene Sequence 5′→3′	GenBank accession no.	Pyloric caeca	Mid intestine	Distal intestine
EF1A$_B$	F-TGCCCCTCCAGGATGTCTAC	BG933853	18.19±0.09	19.74±0.28	18.90±0.15
	R-CACGGCCCACAGGTACTG				
	P-FAM-AAATCGGCGGTATTGG-MGB				
CD3ζ	F-AACAGGGATCCAGAGAGTGCTG	BT060238	27.77±0.16	27.74±0.28	27.41±0.20
	R-AAGGGACGTGTAAGTGTCGTCA				
	P-FAM-ACGGCACGCGATAATCGCAGGA-BHQ				
COX-2a	F-CAGATCGCTGGAAGGGTGG	AY848944	33.66±0.31	33.32±0.43	32.83±0.35
	R-TCATGTTGAAGCGTTTCCTGTAG				
	P-FAM-AGCTAAGGCCCTGGAGCACAGC-BHQ				
IgM	F-TGTAAAGAGAGCAGACTGGGACAG	Y12456	25.60±0.55	24.89±0.60	24.63±0.45
	R-GAGACGGGTGCTGCAGATATTC	Y12457			
	P-FAM-TGTTCCACGGCGCATTCAAAGATTT-BHQ				
IgT	F-CAGCAGTCTGCTGAAGGTC	GQ907004	29.91±0.29	30.29±0.50	28.00±0.30
	R-GGTTCTGTTTTGGAGATCG	GQ907003			
	P-FAM-CTGCACCACACAGCTGTACTTGACC-BHQ				
IL-1β	F-GCTACCACAAAGTGCATTTG	AY617117	34.91±0.42	33.91±0.64	31.98±0.33
	R-GAGGTTGGATCCCTTTATGC				
	P-FAM-CCATTGAGACTAAAGCCAGACCTGTAG-BHQ				
MHCII	F-CCACCTGGAGTACACACCCAG	X70165	20.95±0.25	20.34±0.35	20.49±0.18
	R-TTCCTCTCAGCCTCAGGCAG				
	P-FAM-TCCTGCATGGTGGAGCACATCAGC-BHQ				
NOD2	F-GCATCCAGTGTGAGCACTTTCAG	EG915470	32.01 ± 0.19	31.55 ± 0.24	31.86 ± 0.23
	R-TTCATCTTCAGGAGGTGAGCG				
	P-FAM-CAAGCTAACTGATGCCTGCACAGAGTGC-BHQ				
TGF-β	F-TGGAGCTGAGTGAGGAGCAG	EU082211	34.04±0.29	34.24±0.41	33.46±0.32
	R-ACCGCATCTCAGACATGTTG				
	P-FAM-TGTGGACCTCCTTTGCAAAGTATGC-BHQ				
TNF-α	F-GCAGCTTTATGTGCGGCAG	NM_001123589	36.38±0.30	NA	35.47±0.29
	R-TTTTGCACCAATGAGTATCTCCAG	NM_001123590			
	P-FAM-TGGAAGACTGGCAACGATGCAGGA-BHQ				

The mean Ct-values ± SEM for the fish oil group are given for each intestinal segment.

NA, Not applied.

Calculations and Statistical Analysis

Databases for the results for morphometric measurements and real-time RT-PCR were established in Excel® for Windows, and statistical calculations and graphical presentation of real-time PCR results were performed using Prism 6.0 software (GraphPad Software, San Diego, CA, USA). Morphometric data within each dietary group were pooled before calculating the group mean.

Data are given as mean ± standard error of the mean (SEM). The data were analyzed for normality using a Shapiro Wilk's test and for homogeneity of variance using a Brown-Forsythe's test. For data that were non-normal and/or with non-homogeneous residuals, a log transformation was performed prior to one way analysis of variance (ANOVA) followed by Tukey's multiple comparison test. The significance level was set to 0.05.

RESULTS

Histology and Immunohistochemistry

Histological examination of the mid intestine did not show any obvious pathological changes in any fish, while there was shortening, widening and fusion of the simple folds with leucocyte infiltration in the lamina propria in the distal intestine in a few individuals without any obvious association with the different dietary groups. Examination of sections stained with antibodies against MHC class II and CD3ϵ showed positive cells both in the epithelium and in the lamina propria and with similar density and distribution of antigen presenting cells and T lymphocytes regardless of diet (Figure 5). The proliferation pattern of epithelial cells in the mid intestine as demonstrated by immunohistochemical staining with an antibody against PCNA did not differ between the dietary groups.

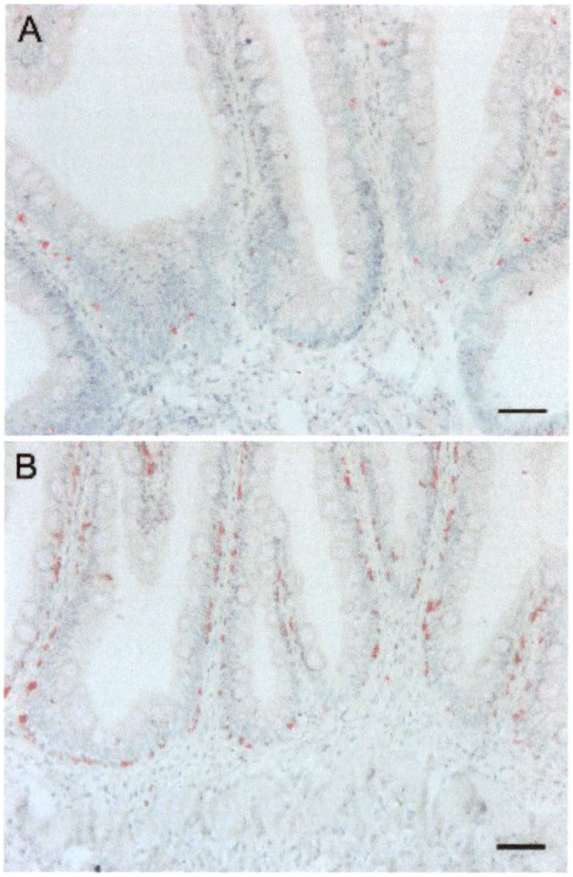

Figure 5 Detection of antigen presenting cells and T lymphocytes in the mid intestine. Microphotographs of mid intestinal sections from fish fed soybean oil stained with antisera against MHC class II **(A)** and CD3ϵ **(B)**. Positive cells visualized with red colour were found in the epithelium and only scattered cells were present in the lamina propria. The density and distribution of positive cells were independent of diet. Scale bar 50 μm.

Morphometric Analysis

Mid intestine: The folds of the mid intestine were tallest in the fish oil group (1393±36.4 μm), intermediate in the olive oil and rapeseed oil groups (1134±31.2 μm and 1131±42.4 μm respectively) and lowest in the soybean oil group (1012±24.8 μm). The differences were highly significant (P<0.0001) between the fish oil group and all vegetable oils groups. The folds of the mid intestine were widest in the soybean oil group (142.0±5.4 μm) and most slender in the olive oil group (122.6±4.3 μm) (P=0.0249). There was no significant difference between the dietary groups regarding the wall thickness. The results from the morphometric analyses of the mid intestine are shown in Table 2.

Table 2 The height and width of folds and wall thickness in mid and distal intestine (**μm**)

Mid intestine				
	FO	**OO**	**RO**	**SO**
Height of folds	**1393±36.4** ^{a}	**1134±31.2** ^{b}	**1131±42.4** ^{b}	**1012±24.8** ^{b}
Width of folds	135.8±4.8^{ab}	**122.6±4.3** ^{a}	129.2±4.5^{ab}	**142.0±5.4** ^{b}
Wall thickness	732.2±21.4^{a}	652.3±24.0^{a}	652.5±21.6^{a}	694.0±20.9^{a}
Distal intestine				
	FO	**OO**	**RO**	**SO**
Height of simple folds	1240±37.8^{a}	1204±39.7^{a}	1221±29.3^{a}	1242±35.7^{a}
Width of simple folds	138.9±5.3^{a}	127.1±5.5^{a}	131.3±5.9^{a}	131.8±5.3^{a}
Height of complex folds	3456±146.9^{ab}	3280±91.7^{ab}	**3598±100.2** ^{a}	**3123±79.7** ^{b}
Width of complex folds	265.2±10.8^{a}	267.1±10.7^{a}	234.1±7.8^{a}	238.7±10.1^{a}
Wall thickness	**705.2±20.4** ^{a}	**616.8±22.8** ^{b}	643.3±21.5^{ab}	**621.4±18.0** ^{b}

FO, fish oil, *OO*, olive oil, *RO*, rapeseed oil, *SO*, soybean oil.

^{a,b}Significant differences (ANOVA, P<0.05) between dietary groups in the same intestinal segment are denoted by different superscript letters. The values that differ significantly are highlighted in bold text.

Distal intestine: The complex folds were tallest in the rapeseed oil group (3598±100.2 μm) and lowest in the soybean oil group (3123±79.7 μm) (P=0.0113), while the wall was thickest in the fish oil group (705.2±20.4 μm) and thinnest in the olive oil group (616.8±22.8 μm) (P=0.0080). There was no significant difference between the dietary groups regarding the height of the simple folds or the width of either simple or complex folds. The results from the morphometric analyses of the distal intestine are shown in Table 2.

Transcript Levels of Immune-Related Genes

Only small differences in the relative transcript levels of the various genes between the dietary groups were detected, and there was generally larger individual variation within a group than between the groups. Neither the transcript levels of the pro-inflammatory cytokines IL-1β and TNF-α, the intracellular receptor NOD2 nor the enzyme COX-2a that is involved in the synthesis of prostaglandins from fatty acids were significantly altered in any dietary group in any of the intestinal segments.

Pyloric caeca: In the pyloric caeca, the olive oil group had significantly higher transcript levels of CD3ζ and MHC class II (P=0.004 and 0.048) while the rapeseed oil group had a significantly higher transcript level of TGF-β (P=0.022) and the soybean oil group had significantly higher transcript levels of CD3ζ and TGF-β (P=0.033 and 0.017) compared to the fish oil group (Figure 6A).

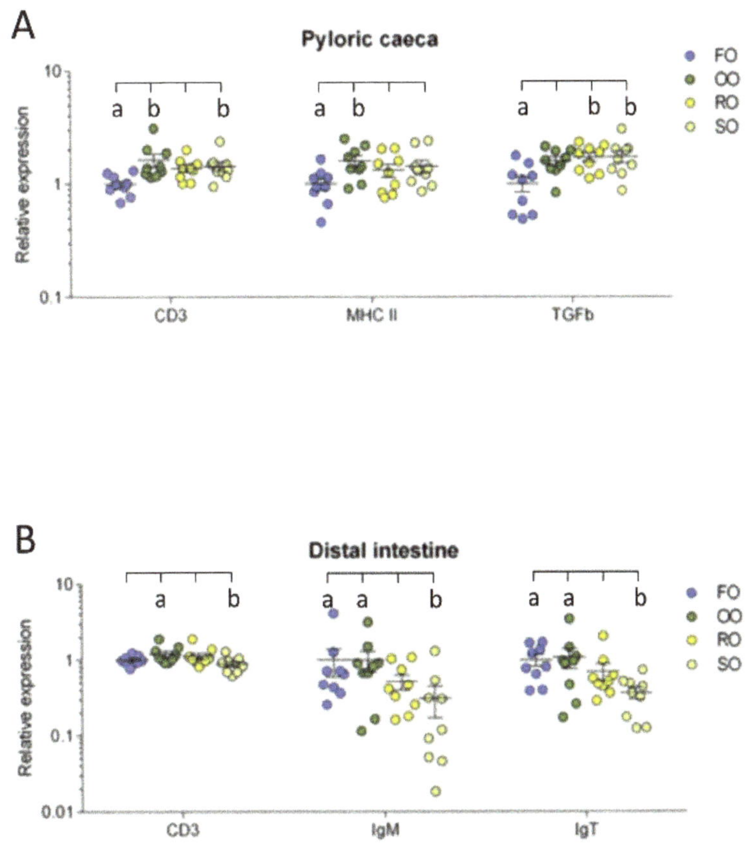

Figure 6 Relative transcription levels of genes with significant differences between dietary groups. The dot plots for CD3ζ, MHC class II and TGF-β in the pyloric caeca **(A)** and CD3ζ, IgM and IgT in the distal intestine **(B)** show the transcript levels for each individual with mean for each group as obtained by real-time RT-PCR analysis. Transcript levels are presented relative to the mean of the fish oil group. Error bars represent standard error of the mean.[a,b]Significant differences (ANOVA, P<0.05) between dietary groups are denoted by different letters above the dot plots. FO=fish oil, OO=olive oil, RO=rapeseed oil and SO=soybean oil.

Mid intestine: The transcript levels of the selected genes appeared to be more uniform in the mid intestine, 7and no significant differences between any dietary groups were detected in this intestinal segment.

Distal intestine: In the distal intestine, the transcript level of CD3ζ was significantly higher in the olive oil group than the soybean oil group (P=0.035). The transcript levels of both IgM and IgT was significantly lower in the soybean oil group than in the fish oil group (P=0.014 and 0.016) and the olive oil group (P=0.015 and 0.038) (Figure 6B).

The underlying data for Figure 6A and B and the normalized transcript levels for all genes and dietary groups relative to the fish oil group are shown in Table 3. A disadvantage of calculating normalized transcription levels is that the differences between genes get lost, while the absolute Ct-values will reveal such differences. The Ct-values for MHC class II were relatively low (~18-20) in all intestinal segments of the fish in all dietary groups, indicative of a high transcript level of mRNA. The Ct-values for CD3ζ and the immunoglobulins were intermediate (~24-30), while the Ct-values for the selected cytokines, NOD2 and COX-2a were in general high (~31-37) and for some individuals beyond the detection limit. The mean Ct-values for all examined genes in the three intestinal segments for the fish oil group were included in Table 1.

Table 3 Relative transcription levels of the immune-related genes normalized to $EF1A_B$

	Pyloric caeca				Mid intestine				Distal intestine			
	FO	**OO**	**RO**	**SO**	**FO**	**OO**	**RO**	**SO**	**FO**	**OO**	**RO**	**SO**
CD3	**1±0.07**^{a}	**1.63±0.21**^{b}	1.38±0.10^{ab}	**1.44±0.13**^{b}	1±0.03^{a}	1.22±0.29^{a}	0.99±0.11^{a}	0.96±0.06^{a}	1±0.05^{ab}	**1.21±0.11**^{a}	1.13±0.11^{ab}	**0.89±0.07**^{b}
COX-2a	1±0.17^{a}	1.56±0.29^{a}	1.36±0.15^{a}	1.28±0.22^{a}	1±0.21^{a}	0.98±0.10^{a}	1.08±0.11^{a}	1.20±0.17^{a}	1±0.33^{a}	0.85±0.17^{a}	1.19±0.21^{a}	0.85±0.24^{a}
IgM	1±0.29^{a}	1.37±0.38^{a}	0.80±0.20^{a}	0.97±0.28^{a}	1±0.30^{a}	0.93±0.28^{a}	0.72±0.20^{a}	0.95±0.17^{a}	**1±0.40**^{a}	**0.99±0.30**^{a}	0.52±0.12^{ab}	**0.31±0.14**^{b}
IgT	1±0.17^{a}	1.97±0.61^{a}	0.85±0.12^{a}	0.94±0.17^{a}	1±0.22^{a}	1.23±0.18^{a}	0.89±0.22^{a}	1.43±0.30^{a}	**1±0.17**^{a}	**1.09±0.33**^{a}	0.71±0.18^{ab}	**0.37±0.07**^{b}
IL-1β	1±0.29^{a}	0.94±0.23^{a}	0.63±0.14^{a}	0.91±0.22^{a}	1±0.29^{a}	1.51±0.68^{a}	0.90±0.20^{a}	0.89±0.29^{a}	1±0.25^{a}	1.26±0.68^{a}	1.07±0.29^{a}	0.92±0.52^{a}
MHCII	**1±0.11**^{a}	**1.59±0.18**^{b}	1.32±0.17^{ab}	1.44±0.19^{ab}	1±0.11^{a}	1.23±0.26^{a}	0.92±0.15^{a}	0.92±0.14^{a}	1±0.14^{a}	1.15±0.13^{a}	1.48±0.18^{a}	1.02±0.15^{a}
NOD2	1±0.09^{a}	1.31±0.12^{a}	1.32±0.10^{a}	1.21±0.09^{a}	1±0.06^{a}	0.98±0.09^{a}	0.92±0.10^{a}	0.87±0.06^{a}	1±0.10^{a}	1.23±0.16^{a}	1.90±0.58^{a}	0.96±0.15^{a}
TGF-β	**1±0.15**^{a}	1.59±0.13^{ab}	**1.71±0.15**^{b}	**1.74±0.21**^{b}	1±0.16^{a}	1.44±0.27^{a}	1.19±0.21^{a}	1.26±0.11^{a}	1±0.16^{a}	1.71±0.41^{a}	1.54±0.27^{a}	0.97±0.33^{a}
TNF-α	1±0.14^{a}	0.93±0.19^{a}	0.68±0.11^{a}	0.93±0.16^{a}	NA	NA	NA	NA	1±0.13^{a}	2.16±1.31^{a}	1.44±0.28^{a}	1.36±0.75^{a}

The data are presented as mean ± SEM relative to the mean of the fish oil group.

FO, fish oil, *OO*, olive oil, *RO*, rapeseed oil, *SO*, soybean oil, *NA*, Not applied.

^{a,b}Significant differences (ANOVA, $P<0.05$) between dietary groups in the same intestinal segment are denoted by different letters. The values that differ significantly are highlighted in bold text.

DISCUSSION

In this study, we have shown that fish fed diets where fish oil was largely replaced by three different vegetable oil blends had significantly shorter folds in the mid intestine compared with fish fed a diet with fish oil as the sole lipid source in a trial lasting for 28 weeks. The fold height decreased to a degree roughly corresponding to a decreasing n-3/n-6 fatty acid ratio of the feed, ie. the fish in the soybean oil group had the shortest folds. Additionally, the mid intestinal folds of the fish in the soybean oil group were significantly wider than of the fish in the olive oil group. In the distal intestine, the complex mucosal folds of the fish in the soybean oil group were significantly shorter than the folds of the fish in the rapeseed oil group, while the wall was significantly thicker for the fish in the fish oil group than the fish in the olive oil and soybean oil groups. Histological and immunohistochemical examination did not reveal any overt signs of inflammation in the lamina propria of the mid intestine. In the distal intestine, however, infiltration of inflammatory cells was observed in some individuals, but this could not be related to the dietary groups as fish in the fish oil group also were affected. Real time RT-PCR revealed only minor alterations in the transcript levels of the selected immune-related genes between dietary groups.

Dietary lipid sources have been reported to affect intestinal morphology in mammals. In weaning pigs, dietary supplementation with fish oil increased villus height in the small intestine combined with a decrease in transcript levels of inflammation related genes compared to a diet with corn oil [32]. Furthermore, dietary fatty acid composition has been reported to affect the height of intestinal villi in ileum in rats, the extent of the reductions increasing with increasing levels of n-6 fatty acids, ie. rats fed with fish oil had higher villi than those fed with olive oil and soybean oil [33]. Interestingly, rats fed with soybean oil had wider villi than the group given olive oil but not the group given fish oil, which is in agreement with our observations. The altered morphology was followed by a corresponding infiltration of mucosal lymphocytes [33].

In Atlantic salmon, shortening and widening of the simple mucosal folds of the distal intestine, in combination with infiltration of inflammatory cells in the lamina propria, has been repeatedly reported when feeding with soybean meal (SBM) and pea protein concentrate [23, 24, 34]. Starvation has also been described to mildly induce similar changes [23]. In humans, shortened intestinal villi and inflammation in the small intestine occurs in patients with coeliac disease caused by reaction to gluten proteins [35]. However, in the present study, histological investigation and immunohistochemical examination with antigen presenting cell and T lymphocyte markers did not show infiltration of inflammatory cells in the intestinal wall corresponding to the fold reduction pattern. Furthermore, there was no significant difference in the transcript levels of the investigated genes between either of the groups in the mid intestinal region. Combined, this indicates that the fold reductions observed in the mid intestinal region in the current study were not connected with a prolonged inflammatory response, but were probably caused by other factors.

In fish, the gastrointestinal microbiota is known to change with different feeding regimes [36, 37] and more specifically with different lipid levels and different vegetable oils [38]. Alterations in intestinal microbiota are hence not to be neglected as a possible explanatory factor for the altered morphology observed.

Significant reductions of mucosal folds in the mid intestine of all vegetable dietary groups were observed, in contrast to the mildly affected distal intestine. This finding might be related to the fact that long chain fatty acids (LCFAs) mainly are absorbed in the pyloric caeca and mid intestine and only to a limited extent in the distal parts of the intestine [16, 39]. Altering the composition of the LCFA in the feed can hence be speculated to cause most pronounced changes in the regions where these fatty acids are mainly absorbed.

Shortening of mucosal folds has been linked to altered proliferation pattern of the intestinal epithelium in Atlantic salmon; decrease in cell proliferation and apoptosis in smolt exposed to sublethal levels of inorganic mercury [40] and increase in cell proliferation in soybean meal induced enteropathy [36]. We did however not observe any differences in the proliferation pattern of epithelial cells in the mid intestine between any of the diet groups, even though lower turn-over in cell proliferation and apoptosis has been previously detected in Atlantic salmon fed with a diet where fish oil was completely replaced by plant oil [25].

Shortening of the mid intestinal folds probably reduces the total surface of the intestine and hence the absorption of nutrients, which may in turn influence the growth of the fish. A substantial proportion of starch and lipids is absorbed in the mid intestine [41]. Fish fed soybean oil were significantly shorter than the fish fed fish oil and somewhat (but not significantly) lighter than the other fish in the trial; results which were linked to reduced feed intake in the soybean oil group [22]. It cannot, however, be ruled out that the pronounced shortening of the folds in the mid intestine of fish fed soybean oil may be an additional factor contributing to the somewhat reduced growth in fish fed soybean oil. Furthermore, the difference in weight between the groups might have been more pronounced if not the fish fed fish oil as the sole lipid source had significantly reduced lipid digestibility due to high levels of dietary saturated fatty acids [22]. It cannot be ruled out that the reduced lipid digestibility in the fish oil group may have affected intestinal morphology as response to saturated fatty acids possibly being above an acceptable threshold level for Atlantic salmon. Overall, the shortened intestinal folds being most pronounced in the soybean oil group suggest that feeding with similar or higher levels of soybean oil as in the present trial might not be unproblematic for the production results.

In the pyloric caeca, the transcript levels of TGF-β and CD3ζ in the soybean oil group, the transcript level of TGF-β in the rapeseed oil group and the transcript levels of MHC class II and CD3ζ in the olive oil group was significantly higher than in the fish oil group. TGF-β is produced by cells of the innate immune system and by regulatory T lymphocytes and is both a pro- and anti-inflammatory cytokine that is involved in cell growth, migration, differentiation and apoptosis including inhibition of lymphocyte proliferation [42]. In SBM-induced inflammation in Atlantic salmon, transcription levels of TGF-β were reported up-regulated by 7-folds after 21 days, combined with a 20-folds up-regulation of IL-1β [43]. In the present trial, although significant, the differences observed in TGF-β transcript levels in the fish fed soybean oil and rapeseed oil compared to the fish fed fish oil was below two-fold, and there was no significant difference in IL-1β transcript levels between the groups; hence it is difficult to ascribe the differences in TGF-β transcript levels to an inflammatory process. The finding that neither the relative transcript levels of TNF-α, NOD2 nor COX-2a did vary significantly between the dietary groups in any of the intestinal segments also suggests that dietary lipids did not affect the degree of inflammation. The transcript levels of TNF-α and IL-1β in leucocytes from Atlantic salmon did not differ significantly between groups that were incubated in plasma with different n-3/n-6 ratio followed by stimulation with LPS, and a relative similar EPA/AA ratio in the cells was launched as an explanation to the lack of influence of fatty acid sources on inflammatory response [9]. A similar EPA/AA ratio in leucocytes may explain the relative stable transcript levels of cytokines regardless of diets in the current study too.

In contrast to mammals harboring mesenteric lymph nodes and distinct lymphoid follicles in the intestinal mucosa, the immune competent cells including antigen presenting cells as well as T and B lymphocytes are more diffusely spread in the intestinal tissue of teleosts like Atlantic salmon [44]. The moderately higher transcript levels of CD3ζ observed in the pyloric caeca of the olive and soybean oil group might indicate a slightly higher number of T lymphocytes as CD3ζ is part of the T cell receptor complex and expressed in all T lymphocytes [45]. MHC class II is in contrast expressed in antigen presenting cells that can present antigenic peptides to T lymphocytes and initiate the adaptive immunity [42]. In mammals, and presumably in teleosts, MHC class II is moreover expressed in intestinal enterocytes [46]. A higher transcript level of MHC class II might indicate a higher level of antigen presentation in the pyloric caeca of the olive oil group compared to the fish oil group. The differences in both CD3ζ and MHC class II transcript levels between the groups were however below two-fold and should be carefully interpreted. The higher transcript levels of certain genes in the plant oil groups compared to the fish oil group in the pyloric caeca could be linked to the high lipid absorption in this region [41]. However, this does not explain why we do not see a corresponding change in the transcript levels for these genes in the mid intestine as absorption rate for LCFAs are reported to be similarly high here [39].

The significantly lower transcript level of CD3ζ in the distal intestine of the soybean oil group compared to the olive oil group might in contrast to the pyloric caeca, indicate a lower number of T lymphocytes. Furthermore, the soybean oil group had significantly lower transcript levels of IgM and IgT than the fish fed both fish oil and olive oil. These immunoglobulins are expressed by different subpopulations of B lymphocytes in the teleost intestine [47] and are present both as membrane bound and secretory forms. Again, the differences were moderate, but summed together they suggest a difference in response to soybean oil between the pyloric cecea and distal intestine.

In mammals, replacement of n-3 fatty acids with n-6 polyunsaturated acids has been associated with intestinal inflammation and promotion of intestinal cancer [6, 10, 11]. Feed-induced intestinal carcinogenesis following inflammation has also been reported in brood stock Atlantic salmon [27], and it was speculated whether this partially could be related to replacement of fish oil with vegetable oils in the feed. Although the brood stock were exposed much longer to the commercial feed than the fish in the current 28 week-long trial, the results of the present study strongly suggest that partial replacement of fish oil with vegetable oils in the feed did not induce prolonged intestinal inflammation in Atlantic salmon.

The amount of fish oil still present in the feed might be of importance for the ability of the fish to cope with the increased amount of n-6 fatty acids. The regular feed composition used in the salmon industry has changed dramatically over the last decades as the lipid fraction has increased from 10% till about 35% [4]. This means that although vegetable oils constitute 80% of the lipid fraction as in the feeds of our study, the feed still contains approximately 1.4% EPA and DHA provided by the fish oil and fish meal included in the feed, which might be enough to sustain general intestinal health. It has been shown previously that 1% EPA and DHA in the feed is essential to attain good growth in fry [48]. A minimum proportion of EPA and DHA is considered to be required also for larger Atlantic salmon [49], however this has not yet been quantified.

CONCLUSIONS

The folds in the mid intestine were significantly shorter in all groups of Atlantic salmon fed vegetable oils compared to the group fed fish oil. In the distal intestine, the complex folds were significantly shorter in the fish fed soybean oil compared to the fish fed rapeseed oil. Histological examination did not reveal clear difference in degree of inflammation related to dietary groups, and this finding was confirmed by real-time RT-PCR that only revealed moderate alterations in the mRNA transcript levels of selected immune-related genes.

The shortening of the intestinal folds was most pronounced in the fish fed soybean oil and might be associated with reduced intestinal surface and impaired nutrient absorption and growth. Kept together with significantly higher transcript levels of TGF-β and CD3ζ in the pyloric caeca and significantly lower transcript levels of IgM and IgT in the distal intestine in the fish fed soybean oil compared to the fish fed fish oil, it can be concluded that inclusion of high levels of soybean oil in the feed for Atlantic salmon should be done with caution.

Notes

Torfinn Moldal, Guro Løkka contributed equally to this work.

ACKNOWLEDGEMENTS

We thank Thommy Holmvåg at Skretting ARC for excellent fish husbandry, PhD student Nina Liland at NIFES for providing RNA samples from the mid intestine and technical staff at the Norwegian Veterinary Institute for cutting sections and stainings. The study was funded by the Research Council of Norway through the projects 'SAFE FEED, SAFE AND HEALTHY SEAFOOD Contaminants, nutrients and health challenges in the novel production of farmed Atlantic salmon' (project number 199626; project leader Bente E. Torstensen) and 'InNoVacc' (grant number 183196/S40), Skretting ARC, the Norwegian Veterinary Institute and the Norwegian School of Veterinary Science.

AUTHORS' CONTRIBUTIONS

TM did the sampling, planned the experiments, performed the histological examination, morphometric measurements, immunohistochemical stainings, parts of the cDNA synthesis and PCR and statistical analyses and drafted the manuscript, GL planned the experiments, performed immunohistochemical stainings, examination of immunohistochemical stainings, parts of the cDNA synthesis and PCR and statistical analyses and drafted the manuscript, JWN performed parts of the RNA isolation, cDNA synthesis and PCR, LA planned the experiments, designed the probes and supervised the PCR work, BT and GR designed and supervised the dietary experiment and samplings, and OBD, EOK and MK planned the experiments and supervised the study. Everyone commented on the manuscript and approved the final version.

REFERENCES

1. Naylor RL, Hardy RW, Bureau DP, Chiu A, Elliott M, Farrell AP, Forster I, Gatlin DM, Goldburg RJ, Hua K, Nichols PD: Feeding aquaculture in an era of finite resources. Proc Natl Acad Sci USA. 2009, 106: 15103-15110. 10.1073/pnas.0905235106.
2. Torstensen BE, Espe M, Sanden M, Stubhaug I, Waagbø R, Hemre GI, Fontanillas R, Nordgarden U, Hevrøy EM, Olsvik P, Berntssen MHG: Novel production of Atlantic salmon (*Salmo salar*) protein based on combined replacement of fish meal and fish oil with plant meal and vegetable oil blends. Aquaculture. 2008, 285: 193-200. 10.1016/j.aquaculture.2008.08.025.
3. Powell K: Eat your veg. Nature. 2003, 426: 378-379. 10.1038/426378a.
4. Obach A: Trends and advances in feed formulation – towards lowest cost fish production. Practical Developments in Aquafeeds - Feed Advances to Boost Profitability: 2013. 2013, Nashville, Tennessee, USA: Nashville Convention Center
5. Sustainibility report 2012. http://www.skretting.com.
6. Chapkin RS, McMurray DN, Lupton JR: Colon cancer, fatty acids and anti-inflammatory compounds. Curr Opin Gastroenterol. 2007, 23: 48-54. 10.1097/MOG.0b013e32801145d7.
7. Calder PC: n-3 polyunsaturated fatty acids, inflammation, and inflammatory diseases. Am J Clin Nutr. 2006, 83: 1505S-1519S.
8. Bell JG, Ashton I, Secombes CJ, Weitzel BR, Dick JR, Sargent JR: Dietary lipid affects phospholipid fatty acid compositions, eicosanoid production and immune function in Atlantic salmon (*Salmo salar*). Prostaglandins Leukot Essent Fatty Acids. 1996, 54: 173-182. 10.1016/S0952-3278(96)90013-7.
9. Seierstad SL, Haugland Ø, Larsen S, Waagbø R, Evensen Ø: Pro-inflammatory cytokine expression and respiratory burst activity following replacement of fish oil with rapeseed oil in the feed for Atlantic salmon (*Salmo salar* L.). Aquaculture. 2009, 289: 212-218. 10.1016/j.aquaculture.2008.12.004.
10. Tjønneland A, Overvad K, Bergmann MM, Nagel G, Linseisen J, Hallmans G, Palmqvist R, Sjödin H, Hägglund G, Berglund G, Lindgren S, Grip O, Palli D, Day NE, Khaw KT, Bingham S, Riboli E, Kennedy H, Hart A: Linoleic acid, a dietary n-6 polyunsaturated fatty acid, and the aetiology of ulcerative colitis: a nested case-control study within a European prospective cohort study. Gut. 2009, 58: 1606-1611.
11. Whelan J, McEntee MF: Dietary (n-6) PUFA and intestinal tumorigenesis. J Nutr. 2004, 134: 3421S-3426S.
12. Ghosh S, DeCoffe D, Brown K, Rajendiran E, Estaki M, Dai CB, Yip A, Gibson DL: Fish oil attenuates omega-6 polyunsaturated fatty acid-induced dysbiosis and infectious colitis but impairs LPS dephosphorylation activity causing sepsis. PLoS One. 2013, 8 (2): e55468-10.1371/journal.pone.0055468. doi:10.1371/journal.pone.0055468
13. Calder PC: Mechanisms of action of (n-3) fatty acids. J Nutr. 2012, 142: 592S-599S. 10.3945/jn.111.155259.
14. Calder PC, Yaqoob P, Thies F, Wallace FA, Miles EA: Fatty acids and lymphocyte functions. Br J Nutr. 2002, 87 (Suppl 1): 31-48.
15. Calder PC: Immunomodulation by omega-3 fatty acids. Prostaglandins Leukot Essent Fatty Acids. 2007, 77: 327-335. 10.1016/j.plefa.2007.10.015.
16. Denstadli V, Vegusdal A, Krogdahl Å, Bakke-McKellep AM, Berge GM, Holm H, Hillestad M, Ruyter B: Lipid absorption in different segments of the gastrointestinal tract of Atlantic salmon (*Salmo salar* L.). Aquaculture. 2004, 240: 385-398. 10.1016/j.aquaculture.2004.06.030.
17. Denstadli V, Bakke AM, Berge GM, Krogdahl Å, Hillestad M, Holm H, Ruyter B: Medium-chain and long-chain fatty acids have different postabsorptive fates in atlantic salmon. J Nutr. 2011, 141: 1618-1625. 10.3945/jn.111.141820.
18. Torstensen BE, Lie Ø, Frøyland L: Lipid metabolism and tissue composition in Atlantic salmon (*Salmo salar* L.) - Effects of capelin oil, palm oil, and oleic acid-enriched sunflower oil as dietary lipid sources. Lipids. 2000, 35: 653-664. 10.1007/s11745-000-0570-6.
19. Torstensen BE, Frøyland L, Lie Ø: Replacing dietary fish oil with increasing levels of rapeseed oil and olive oil - effects on Atlantic salmon (*Salmo salar* L.) tissue and lipoprotein lipid composition and lipogenic enzyme activities. Aquac Nutr. 2004, 10: 175-192. 10.1111/j.1365-2095.2004.00289.x.
20. Torstensen BE, Frøyland L, Ørnsrud R, Lie Ø: Tailoring of a cardioprotective muscle fatty acid composition of Atlantic salmon (*Salmo salar*) fed vegetable oils. Food Chem. 2004, 87: 567-580. 10.1016/j.foodchem.2004.01.009.
21. Torstensen BE, Bell JG, Rosenlund G, Henderson RJ, Graff IE, Tocher DR, Lie Ø, Sargent JR: Tailoring of Atlantic salmon (*Salmo salar* L.) flesh lipid composition and sensory quality by replacing fish oil with a vegetable oil blend. J Agric Food Chem. 2005, 53: 10166-10178. 10.1021/jf051308i.
22. Liland NS, Rosenlund G, Berntssen MHG, Brattelid T, Madsen L, Torstensen BE: Net production of Atlantic salmon (FIFO, Fish in Fish out<1) with dietary plant proteins and vegetable oils. Aquac Nutr. 2013, 19: 289-300. 10.1111/j.1365-2095.2012.00958.x.
23. Bæverfjord G, Krogdahl Å: Development and regression of soybean meal induced enteritis in Atlantic

salmon, *Salmo salar*L., distal intestine: a comparison with the intestines of fasted fish. J Fish Dis. 1996, 19: 375-387. 10.1111/j.1365-2761.1996.tb00376.x.

24. Penn MH, Bendiksen EA, Campbell P, Krogdahl Å: High level of dietary pea protein concentrate induces enteropathy in Atlantic salmon (*Salmo salar* L.). Aquaculture. 2011, 310: 267-273. 10.1016/j.aquaculture.2010.10.040.

25. Olsvik PA, Torstensen BE, Berntssen MHG: Effects of complete replacement of fish oil with plant oil on gastrointestinal cell death, proliferation and transcription of eight genes' encoding proteins responding to cellular stress in Atlantic salmon*Salmo salar* L. J Fish Bio. 2007, 71: 550-568. 10.1111/j.1095-8649.2007.01521.x.

26. Oxley A, Jolly C, Eide T, Jordal AEO, Svardal A, Olsen RE: The combined impact of plant-derived dietary ingredients and acute stress on the intestinal arachidonic acid cascade in Atlantic salmon (*Salmo salar*). Br J Nutr. 2010, 103: 851-861. 10.1017/S0007114509992467.

27. Dale OB, Tørud B, Kvellestad A, Koppang HS, Koppang EO: From chronic feed-induced intestinal inflammation to adenocarcinoma with metastases in salmonid fish. Cancer Res. 2009, 69: 4355-4362. 10.1158/0008-5472.CAN-08-4877.

28. Løkka G, Austbø L, Falk K, Bjerkås I, Koppang EO: Intestinal morphology of the wild Atlantic salmon (*Salmo salar*). J Morphol. 2013, 274: 859-876. 10.1002/jmor.20142.

29. Koppang EO, Hordvik I, Bjerkås I, Torvund J, Aune L, Thevarajan J, Endresen C: Production of rabbit antisera against recombinant MHC class II beta chain and identification of immunoreactive cells in Atlantic salmon (*Salmo salar*). Fish Shellfish Immunol. 2003, 14: 115-132. 10.1006/fsim.2002.0424.

30. Koppang EO, Fischer U, Moore L, Tranulis MA, Dijkstra JM, Köllner B, Aune L, Jirillo E, Hordvik I: Salmonid T cells assemble in the thymus, spleen and in novel interbranchial lymphoid tissue. J Anat. 2010, 217: 728-739. 10.1111/j.1469-7580.2010.01305.x.

31. Olsvik PA, Lie KK, Jordal AEO, Nilsen TO, Hordvik I: Evaluation of potential reference genes in real-time RT-PCR studies of Atlantic salmon. BMC Mol Biol. 2005, 6: doi:10.1186/1471-2199-6-21

32. Liu YL, Chen F, Odle J, Lin X, Jacobi SK, Zhu HL, Wu ZF, Hou YQ: Fish oil enhances intestinal integrity and inhibits TLR4 and NOD2 signaling pathways in weaned pigs after LPS challenge. J Nutr. 2012, 142: 2017-2024. 10.3945/jn.112.164947.

33. Rosa DD, de Sales RL, Moraes LFD, Lourenco FC, Neves CA, Sabarense CM, Ribeiro SMR, Peluzio MDG: Flaxseed, olive and fish oil influence plasmatic lipids, lymphocyte migration and morphometry of the intestinal of Wistar rats. Acta Cir Bras. 2010, 25: 275-280. 10.1590/S0102-86502010000300010.

34. Van den Ingh TSGA, Krogdahl Å, Olli JJ, Hendriks HGCJ, Koninkx JGJF: Effects of soybean-containing diets on the proximal and distal intestine in atlantic salmon (*Salmo salar*) - a morphological study. Aquaculture. 1991, 94: 297-305. 10.1016/0044-8486(91)90174-6.

35. Oberhuber G: Histopathology of celiac disease. Biomed Pharmacother. 2000, 54: 368-372.

36. Bakke-McKellep AM, Penn MH, Salas PM, Refstie S, Sperstad S, Landsverk T, Ringø E, Krogdahl Å: Effects of dietary soyabean meal, inulin and oxytetracycline on intestinal microbiota and epithelial cell stress, apoptosis and proliferation in the teleost Atlantic salmon (*Salmo salar* L.). Br J Nutr. 2007, 97: 699-713. 10.1017/S0007114507381397.

37. Merrifield DL, Dimitroglou A, Bradley G, Baker RTM, Davies SJ: Soybean meal alters autochthonous microbial populations, microvilli morphology and compromises intestinal enterocyte integrity of rainbow trout, *Oncorhynchus mykiss* (Walbaum). J Fish Dis. 2009, 32: 755-766. 10.1111/j.1365-2761.2009.01052.x.

38. Ringø E, Lødemel JB, Myklebust R, Jensen L, Lund V, Mayhew TM, Olsen RE: The effects of soybean, linseed and marine oils on aerobic gut microbiota of Arctic charr *Salvelinus alpinus* L. before and after challenge with *Aeromonas salmonicida*ssp. *salmonicida*. Aquac Res. 2002, 33: 591-606. 10.1046/j.1365-2109.2002.00697.x.

39. Røsjø C, Nordrum S, Olli JJ, Krogdahl Å, Ruyter B, Holm H: Lipid digestibility and metabolism in Atlantic salmon (*Salmo salar*) fed medium-chain triglycerides. Aquaculture. 2000, 190: 65-76. 10.1016/S0044-8486(00)00395-1.

40. Berntssen MHG, Hylland K, Julshamn K, Lundebye AK, Waagbø R: Maximum limits of organic and inorganic mercury in fish feed. Aquac Nutr. 2004, 10: 83-97. 10.1046/j.1365-2095.2003.00282.x.

41. Krogdahl Å, Nordrum S, Sørensen M, Brudeseth L, Røsjø C: Effects of diet composition on apparent nutrient absorption along the intestinal tract and of subsequent fasting on mucosal disaccharidase activities and plasma nutrient concentration in Atlantic salmon *Salmo salar* L. Aquac Nutr. 1999, 5: 121-133. 10.1046/j.1365-2095.1999.00095.x.

42. Zhu LY, Nie L, Zhu G, Xiang LX, Shao JZ: Advances in research of fish immune-relevant genes: A comparative overview of innate and adaptive immunity in teleosts. Dev Comp Immunol. 2013, 39: 39-62. 10.1016/j.dci.2012.04.001.

43. Marjara IS, Chikwati EM, Valen EC, Krogdahl Å, Bakke AM: Transcriptional regulation of IL-17A and other inflammatory markers during the development of soybean meal-induced enteropathy in the distal intestine of Atlantic salmon (*Salmo salar* L.). Cytokine. 2012, 60: 186-196. 10.1016/j.cyto.2012.05.027.

44. Rombout J, Abelli L, Picchietti S, Scapigliati G, Kiron V: Teleost intestinal immunology. Fish Shellfish Immunol. 2011, 31: 616-626. 10.1016/j.fsi.2010.09.001.

45. Liu Y, Moore L, Koppang EO, Hordvik I: Characterization of the CD3 zeta, CD3 gamma delta and CD3 epsilon subunits of the T cell receptor complex in Atlantic salmon. Dev Comp Immunol. 2008, 32: 26-35. 10.1016/j.dci.2007.03.015.

46. Koppang EO, Lundin M, Press CM, Rønningen K, Lie Ø: Differing levels of Mhc class II beta chain expression

in a range of tissues from vaccinated and non-vaccinated Atlantic salmon (*Salmo salar* L.). Fish Shellfish Immunol. 1998, 8: 183-196. 10.1006/fsim.1997.0136.

47. Salinas I, Zhang YA, Sunyer JO: Mucosal immunoglobulins and B cells of teleost fish. Dev Comp Immunol. 2011, 35: 1346-1365. 10.1016/j.dci.2011.11.009.

48. Ruyter B, Røsjø C, Einen O, Thomassen MS: Essential fatty acids in Atlantic salmon: effects of increasing dietary doses of n-6 and n-3 fatty acids on growth, survival and fatty acid composition of liver, blood and carcass. Aquac Nutr. 2000, 6: 119-127. 10.1046/j.1365-2095.2000.00137.x.

49. Sargent J, Bell G, McEvoy L, Tocher D, Estevez A: Recent developments in the essential fatty acid nutrition of fish. Aquaculture. 1999, 177: 191-199. 10.1016/S0044-8486(99)00083-6.

CHAPTER 11

EFFECTS OF GENOTYPE AND DIETARY FISH OIL REPLACEMENT WITH VEGETABLE OIL ON THE INTESTINAL TRANSCRIPTOME AND PROTEOME OF ATLANTIC SALMON (SALMO SALAR)

Sofia Morais[1], Tomé Silva[2], Odete Cordeiro[2], Pedro Rodrigues[2], Derrick R Guy[3], James E Bron[1], John B Taggart[1], J Gordon Bell[1] and Douglas R Tocher[1]

[1] Institute of Aquaculture, University of Stirling, Stirling FK9 4LA, Scotland, UK.

[2] Aquaculture Research Group, Centre of Marine Sciences (CCMAR), University of Algarve, 8005-139 Faro, Portugal.

[3] Landcatch Natural Selection Ltd., Cooperage Way Business Centre, Alloa FK10 3LP, Scotland, UK.

ABSTRACT

Background

Expansion of aquaculture requires alternative feeds and breeding strategies to reduce dependency on fish oil (FO) and better utilization of dietary vegetable oil (VO). Despite the central role of intestine in maintaining body homeostasis and health, its molecular response to replacement of dietary FO by VO has been little investigated. This study employed transcriptomic and proteomic analyses to study effects of dietary VO in two family groups of Atlantic salmon selected for flesh lipid content, 'Lean' or 'Fat'.

Results

Metabolism, particularly of lipid and energy, was the functional category most affected by diet. Important effects were also measured in ribosomal proteins and signalling. The long-chain polyunsaturated fatty acid (LC-PUFA) biosynthesis pathway, assessed by fatty acid composition and gene expression, was influenced by genotype. Intestinal tissue contents of docosahexaenoic acid were equivalent in Lean salmon fed either a FO or VO diet and expression of LC-PUFA biosynthesis genes was up-regulated in VO-fed fish in Fat salmon. Dietary VO increased lipogenesis in Lean fish, assessed by expression of FAS, while no effect was observed on β-oxidation although transcripts of the mitochondrial respiratory chain were down-regulated, suggesting less active energetic metabolism in fish fed VO. In contrast, dietary VO up-regulated genes and proteins involved in detoxification, antioxidant defence and apoptosis, which could be associated with higher levels of polycyclic aromatic hydrocarbons in this diet. Regarding genotype, the following pathways were identified as being differentially affected: proteasomal proteolysis, response to oxidative and cellular stress (xenobiotic and oxidant metabolism and heat shock proteins), apoptosis and structural proteins particularly associated with tissue contractile properties. Genotype effects were accentuated by dietary VO.

Citation: Sofia Morais, Tomé Silva , Odete Cordeiro , Pedro Rodrigues , Derrick R Guy , James E Bron , John B Taggart , J Gordon Bell and Douglas R Tocher, Effects of genotype and dietary fish oil replacement with vegetable oil on the intestinal transcriptome and proteome of Atlantic salmon (Salmo salar), doi:10.1186/1471-2164-13-448.

Conclusions

Intestinal metabolism was affected by diet and genotype. Lean fish may have higher responsiveness to low dietary n-3 LC-PUFA, up-regulating the biosynthetic pathway when fed dietary VO. As global aquaculture searches for alternative oils for feeds, this study alerts to the potential of VO introducing contaminants and demonstrates the detoxifying role of intestine. Finally, data indicate genotype-specific responses in the intestinal transcriptome and proteome to dietary VO, including possibly structural properties of the intestinal layer and defence against cellular stress, with Lean fish being more susceptible to diet-induced oxidative stress.

BACKGROUND

Fish are important components of the human diet, being highly nutritious and valued as the main source of n-3 long-chain polyunsaturated fatty acids (LC-PUFA). These essential fatty acids, mainly eicosapentaenoic acid (EPA) and docosahexaenoic acid (DHA), have well-known health-promoting properties, including protection against a range of cardiovascular and inflammatory diseases, and neurological disorders [1]. With population growth and increasing awareness of the importance of fish consumption as part of a healthy diet, worldwide demand for seafood continues to grow. However, as traditional fisheries are largely in decline, aquaculture must meet this demand. Aquaculture is the fastest-growing food production sector with an average annual growth rate of 6.6%, accounting for 46% of total fish supply [2]. In the European and American continents, aquaculture production is largely dominated by salmonid species, mainly Atlantic salmon, and feeds for such carnivorous species have traditionally relied on fishmeal (FM) and fish oil (FO) from wild stocks. Recent estimates indicated that 88.5% of global production of FO was used by the aquaculture sector, with salmonid culture taking the largest share (56%) [3]. With ever increasing demands for aquafeeds and reduction in fisheries landings, the availability of FO and FM seriously limits the growth of aquaculture production and there is an urgent need to find more sustainable alternatives.

Vegetable oils (VO) can replace FO in salmon feeds without compromising fish growth or condition although, at high levels of replacement, tissue levels of n-3 LC-PUFA are significantly reduced [4, 5]. The effects of FO replacement by VO are becoming well characterized in the hepatic transcriptome of salmonids [6–10], and other species [11]. However, studies on intestinal transcriptome are few and restricted to effects of replacement of FM by plant proteins, particularly soybean meal [12, 13], given its potential to cause enteritis. It is now clear that intestine in salmonids is not simply a site for reacylation and packaging of dietary lipids but it also has important roles in fatty acid metabolism, including LC-PUFA biosynthesis [14, 15]. Furthermore, dietary VO can induce major histological changes in fish enterocytes, originating mostly from supranuclear lipid droplet formation, possibly due to altered reacylation mechanisms and decreased phospholipid synthesis [16–18]. In some cases, these accumulations were large enough to be deemed pathological [19, 20]. A recent study investigating effects of dietary FO replacement by VO on intestinal transcriptome in Atlantic cod indicated potential effects on lipid absorption and transport and suggested morphological and structural changes to the intestinal muscle layer [21]. Furthermore, both this and a previous study on Atlantic salmon showed significant effects on expression of genes involved in cell proliferation and apoptosis [22]. Therefore, there is indication that intestine may be affected by changes in lipid components of feed formulations. Given its crucial roles in nutrient absorption, protection against the entry of pathogens, and immune function [23], further attention is warranted and impacts of FO replacement require investigation in intestine, particularly in salmon where important changes in diet formulation are already being applied. This study is a large-scale analysis of the effects of replacement of dietary FO by VO on the transcriptome and proteome of Atlantic salmon intestine. Furthermore, given recent interest in evaluating genetic selection as a feasible strategy, in conjunction with changes in commercial feed formulation, to meet worldwide demand for farmed fish without compromising animal welfare or nutritional value [9, 10, 24], two groups of Atlantic salmon families, Lean and Fat, were studied to examine the potential effects of genetic background. This experiment was performed in parallel with another microarray study looking at effects in the hepatic transcriptome, analysing samples from the same individuals [9, 10], enabling a global and comprehensive assessment of the physiological and molecular effects of FO replacement by VO in Atlantic salmon, including potential interactions with genotype.

RESULTS

Microarray Analysis

Two way-ANOVA of the cDNA array dataset returned 1409, 1626 and 862 significant genes for the factors diet, genotype and diet × genotype interaction, respectively. Detailed analysis was restricted to the top 100 most significant features, which were categorised according to biological function, based on mammalian homolog genes. Metabolism, particularly of lipid and energy, was the functional category most affected by diet accounting for 39-41% of the top 100 annotated genes (Table 1), and showing highest diet × genotype interaction (Additional file 1). Diet also impacted translation (17%; mostly ribosomal proteins) and signalling (17%). In contrast, genotype affected less markedly metabolism (21%), whereas structural proteins (25%; mostly different types of collagen) and proteins involved in the regulation of transcription (21%) predominated (Table 2).

Table 1 Intestine transcripts corresponding to the top 100 most significant features exhibiting differential expression between diets

Accession no	Gene	VO/FO		p- value
		Lean	Fat	
Metabolism (39%)				
Lipid metabolism (9%)				
DW590668	Fatty aldehyde dehydrogenase	- 1.1	- 1.2	0.0003
CK888998	Ethanolamine kinase 1	1.2	1.2	0.0008
CK884623	Epidermis-type arachidonate lipoxygenase 3	1.4	1.1	0.0010
CK885725	Protein containing a beta-ketoacyl synthase domain	1.4	1.6	0.0017
AF478472	Delta-5 fatty acyl desaturase	1.2	2.5	0.0019
Energy metabolism/generation of precursor metabolites (11%)				
CK885194	Cytochrome c oxidase subunit 5A	- 1.2	- 1.1	0.0000
bra_snb_03E12	NADH dehydrogenase subunit 1	- 1.3	- 1.2	0.0000
EG649067	Ubiquinol-cytochrome c reductase core protein 1	- 1.2	- 1.1	0.0004
CK884638	Uncoupling protein	- 1.3	−1.3	0.0019
CK890974	Mitochondrial calcium-dependent solute carrier SCaMC-2	- 1.5	1.0	0.0021
EG648806	NADH dehydrogenase (ubiquinone) 1 beta subcomplex 9	- 1.1	- 1.1	0.0022
Protein and amino acid metabolism (9%)				
CK877521	Kunitz-type protease inhibitor 2 precursor	1.2	1.1	0.0002
DW590246	Betaine aldehyde dehydrogenase	- 1.4	- 1.0	0.0005
CK885180	Fumarylacetoacetate hydrolase domain-containing protein 2B	- 1.4	- 1.3	0.0007
CK897653	Ubiquitin-conjugating enzyme E2 A	1.2	1.2	0.0012
CK885145	Lysosomal protective protein precursor (Cathepsin A)	1.1	1.1	0.0014
Carbohydrate metabolism (5%)				
EG649263	Endosulfine alpha	- 1.3	- 1.0	0.0005
CK885545	Glucose transporter type 8	1.0	1.5	0.0013
Xenobiotic and oxidant metabolism (5%)				
CK885392	Catalase	1.6	1.1	0.0000
CO472279	Cytochrome P450 1[a]	1.2	1.4	0.0004
CK898307	Selenoprotein H-like protein	1.3	1.1	0.0013
Transport/ intracellular trafficking (5%)				
EG647578	Calcium-activated potassium channel subunit alpha 1	1.3	1.2	0.0005
EG648841	Sodium-coupled neutral amino acid transporter 2	- 1.2	- 1.2	0.0007
EG648303	Mitochondrial import receptor subunit TOM20 homolog	- 1.1	- 1.1	0.0011
Regulation of transcription (13%)				
CK898304	C-Maf	1.6	1.6	0.0000
CK890154	Butyrate response factor 2	1.6	1.2	0.0001
EG647831	Proliferation-associated 2G4 b	- 1.3	- 1.2	0.0003
AM041851	Enhancer trap locus homolog 1	- 1.1	- 1.2	0.0004
CK885179	Steroid receptor RNA activator 1	- 1.3	- 1.3	0.0007
CK884728	Butyrate response factor 1	1.1	1.4	0.0011
CK879187	Enhancer of polycomb homolog 1	1.4	1.1	0.0020
Translation (17%)				
CK878380	40S ribosomal protein S18	- 1.1	- 1.2	0.0003

DW591873	60S acidic ribosomal protein	- 1.3	- 1.2	0.0006
EG648644	60S ribosomal protein L29	- 1.1	- 1.2	0.0011
DW590580	40S ribosomal protein S5	- 1.2	- 1.2	0.0011
CK880046	40S ribosomal protein S28	- 1.1	- 1.2	0.0017
EG649167	60S ribosomal protein L37a	- 1.1	- 1.2	0.0017
EG648512	60S ribosomal protein L41	- 1.1	- 1.2	0.0017
CK886042	60S ribosomal protein L31	- 1.1	- 1.3	0.0021
CK895350	60S ribosomal protein L36	- 1.1	- 1.1	0.0021
Signalling/Signal transduction (17%)				
CK874863	Proto-oncogene serine/threonine-protein kinase	1.1	1.1	0.0002
CK881266	Mitogen-activated protein kinase kinase kinase kinase 1	1.2	1.2	0.0002
CK878590	Serine/threonine protein kinase (Tribbles homolog 2)	- 1.1	- 1.3	0.0004
AJ425351	Tensin-like C1 domain-containing	- 2.0	- 1.2	0.0006
CK885079	Sonic hedgehog-like protein	1.3	1.0	0.0008
CN181161	TGF-beta type II receptor	1.1	1.1	0.0009
EG649257	Serine/threonine-protein kinase VRK3	- 1.1	- 1.2	0.0009
CK886267	Guanylate cyclase 2C	1.1	1.1	0.0011
CK884467	Serine/threonine-protein phosphatase 2A regulatory subunit B	- 1.4	- 1.3	0.0020
Structural proteins (7%)				
mus_snm_13F09	Myosin heavy chain	- 1.2	- 1.1	0.0002
CK885118	Cardiac tropomyosin	1.0	1.1	0.0013
CK883354	Troponin C	1.2	1.2	0.0017
EG647866	Tubulin alpha chain	- 1.2	- 1.2	0.0020
Immune response (2%)				
BM414177	Liver-expressed antimicrobial peptide 2	1.5	1.4	0.0022
Miscellaneous				
CK884634	Ezrin-radixin-moesin binding phosphoprotein 50	1.1	1.1	0.0003
EG648821	Transposase	- 1.1	- 1.2	0.0004
CK883634	Poly [ADP-ribose] polymerase 12	1.3	1.1	0.0006
CK885090	Claudin 15	1.2	1.1	0.0008
DW589291	Transposase	- 1.3	- 1.3	0.0009
CK878888	Splicing factor, arginine/serine-rich 3	- 1.3	- 1.2	0.0013
CK885098	RNA-binding protein with multiple splicing 2	- 1.4	- 1.2	0.0013
CK891747	Occludin	- 1.2	- 1.2	0.0014
CK888908	All-trans-retinol 13,14-reductase precursor	1.1	1.2	0.0017
CK880622	Cyclin G1	- 1.1	- 1.2	0.0019
CK885604	Methylenetetrahydrofolate dehydrogenase (NADP+ dependent) 1-like	- 1.6	- 1.3	0.0020
CK891699	Barrier-to-autointegration factor	- 1.1	- 1.2	0.0020
CK879123	Hemoglobin subunit beta	1.5	1.0	0.0022

Annotated features (67% of all clones) are arranged by categories of biological function (with percentages of distribution of genes within each category being indicated) and, within these, by decreasing significance (assessed by two-way ANOVA). Also indicated are the GenBank accession numbers for each clone (or, when not available, the probe number) and the expression ratio between fish fed VO and those fed FO, for each genotype (Lean and Fat).

Table 2 Intestine transcripts corresponding to the top 100 most significant features exhibiting differential expression between family groups

Accession no	Gene	Lean/Fat		p- value
		FO	VO	
Metabolism (21%)				
Lipid metabolism (5%)				
CK874747	Acylglycerol kinase	- 1.0	- 1.1	0.0003
BI468033	ATP-binding cassette sub-family A member 1	- 1.1	- 1.6	0.0004
Energy metabolism/generation of precursor metabolites (2%)				
CK890974	Mitochondrial calcium-dependent solute carrier SCaMC-2	- 1.1	- 1.6	0.0000
Protein and amino acid metabolism (12%)				
EG647485	Proteasome subunit beta type-8 precursor	- 1.3	- 1.8	0.0000
DW591059	Heat shock protein Hsp-16	1.2	1.4	0.0001
AJ425553	26S proteasome non-ATPase regulatory subunit 9	- 1.2	- 1.1	0.0002
DW592187	26 proteasome complex subunit DSS1	1.5	1.3	0.0008
CK878109	Heat shock protein 60 kDa	1.2	1.4	0.0009
Xenobiotic and oxidant metabolism (2%)				
CO472279	Cytochrome P450 1A	- 1.2	- 1.5	0.0003
Transport/ intracellular trafficking (5%)				
CK886307	Peptide transporter PEPT2	- 1.4	- 1.3	0.0006
DW588251	Post-GPI attachment to proteins factor 2	1.1	1.3	0.0007
Regulation of transcription (21%)				
CK877881	Forkhead box Q1	- 1.4	- 1.7	0.0000
CK884548	Proline-rich nuclear receptor coactivator 2	- 1.1	- 1.2	0.0001
CK885013	Forkhead box protein F1	1.1	1.5	0.0003
CK891975	Y-box binding protein-2	- 1.1	- 1.1	0.0003
DW589341	Signal transducer/activator of transcription Stat1	- 1.2	- 1.2	0.0004
CN181282	Bromodomain containing protein 3	- 1.1	- 1.1	0.0007
CK884407	C-Myc-binding protein	- 1.3	- 1.1	0.0007
CK895950	Transcription factor CP2-like	- 1.5	- 1.2	0.0007
EG648473	BTB/POZ domain-containing protein	- 1.1	- 1.4	0.0009
Translation (7%)				
CN181322	40S ribosomal protein S4	1.1	1.1	0.0002
CK885979	Ribonuclease UK114	- 1.1	- 1.3	0.0005
EG648042	Eukaryotic translation initiation factor 3, subunit J	1.1	1.1	0.0009
Signalling/Signal transduction (16%)				
CO471793	Rho GDP-dissociation inhibitor 1	1.3	1.0	0.0000
CK885560	Protein tyrosine phosphatase non-receptor type 9	1.3	1.5	0.0004
CK885079	Sonic hedgehog-like protein	- 1.3	- 1.0	0.0004
EG648038	Kalirin, RhoGEF kinase	1.2	1.2	0.0005
DW589610	Chimaerin 1	1.2	1.2	0.0005
CK897590	Growth factor receptor-bound protein 2	1.1	1.1	0.0009
EG647545	Protein tyrosine phosphatase receptor type D	1.0	1.2	0.0009
Structural proteins (25%)				

CK882401	Troponin C	- 1.1	- 1.2	0.0000
CK889952	Transgelin 2	1.5	1.7	0.0000
EG648135	Osteonectin	1.3	1.5	0.0000
CK899058	Collagen alpha 3 type VI	1.5	1.5	0.0002
CK879405	Collagen alpha 2 type VI	1.3	1.5	0.0002
EG649361	Collagen alpha 2 type I	1.5	1.7	0.0005
EG649013	Collagen alpha 3 type I	1.4	1.5	0.0005
mus_mfo_15A10	Troponin-I isoform 3	1.4	1.6	0.0007
CK876833	Keratocan	- 1.0	1.6	0.0007
DW590534	Collagen alpha 2 type I	1.5	1.6	0.0010
CK892271	Collagen alpha 2 type V	1.2	1.4	0.00010
CK873441	Collagen alpha1 type VI	1.2	1.4	0.0010
Immune response (5%)				
AJ425527	T-cell immunoglobulin and mucin domain-containing protein 4	1.1	1.1	0.0001
EG649194	Mannose-binding protein C	1.4	1.4	0.0003
Miscellaneous				
AM041770	Preimplantation protein 3	- 1.3	- 1.2	0.0000
EG649106	Ornithine decarboxylase antizyme 2	1.4	1.1	0.0002
CK884355	Guanylin precursor	1.2	1.4	0.0002
CK879185	Aminolevulinate, delta-, synthetase 1	- 2.0	- 1.1	0.0002
CK885814	Similar to p53-associated parkin-like cytoplasmic protein	1.2	1.1	0.0002
CK886214	Septin-7	1.1	1.1	0.0003
BM414079	Tumor protein D52-like 2	- 1.1	- 1.3	0.0003
BM413818	Splicing factor, proline- and glutamine-rich	1.2	1.2	0.0003
CK874550	Follistatin-related protein 1	1.1	1.3	0.0004
DW589104	Nucleoside diphosphate kinase 6	1.1	1.1	0.0005
DW592207	Deoxyhypusine synthase	1.5	1.6	0.0007
CK892359	HCLS1-associated protein X-1	- 1.2	- 1.1	0.0007
BI468158	Nuclear protein 1	- 1.0	- 1.6	0.0009
AJ425003	Lamina-associated polypeptide 2	1.2	1.1	0.0010
CK899283	Small EDRK-rich factor 2	1.3	1.5	0.0010
CK885090	Claudin 15	- 1.2	- 1.1	0.0011

Annotated features (60% of all clones) are arranged by categories of biological function (with percentages of distribution of genes within each category being indicated) and, within these, by decreasing significance (assessed by two-way ANOVA). Also indicated are the GenBank accession numbers for each clone (or, when not available, the probe number) and the expression ratio between Lean and Fat fish fed either FO or VO.

Gene Ontology (GO) enrichment analysis was performed on the complete significant lists, enabling identification of GO terms significantly enriched in the input entity list, in comparison to the whole array, providing clues as to which biological processes might be particularly altered in the experimental conditions being compared. It revealed no significant enrichment of GO terms in the genotype list, while 20 and 7 GO terms were significantly enriched in the diet and interaction lists, respectively. GO terms enriched in the diet list included structural constituents of ribosome, structural molecule activity, cytosolic ribosome, cytosol, ribosomal subunit, translation, cellular biosynthetic process, gene expression, macromolecule and biopolymer biosynthetic process and other related terms. This was explained by the large number of ribosomal proteins, components of both the 40S and 60S subunits, which were down-regulated by dietary VO (Table 1). In contrast, several Δ6-desaturase (Δ6fad) clones showing a diet × genotype interaction (see Additional file 1) caused a significant enrichment of the GO terms oxidoreductase activity, stearoyl-CoA 9-desaturase activity, unsaturated fatty acid biosynthetic activity/metabolic processes and very long chain fatty acid (VLCFA) biosynthetic activity/metabolic processes.

RT-qPCR Analysis of Gene Expression

The expression of several genes significantly affected or related to processes affected by the two factors in the microarray analysis was determined by RT-qPCR (Table 3). For diet, a reasonably good match was found for Δ5 fatty acyl desaturase (Δ5fad), NADH dehydrogenase subunit 1 (NADH1), proliferation-associated 2G4b (PA2G4), 60S acidic ribosomal protein (RP60S), proliferating cell nuclear antigen (PCNA; also known as cyclin) and cytochrome P450 1A (CYP1A), particularly in the Fat group where fold-changes were generally more pronounced and significant. No change in expression of uncoupling protein 2 (UCP2) with diet was measured while, for myosin heavy chain (MYO) and methylenetetrahydrofolate dehydrogenase (NADP+ dependent) 1-like (MTHFD1), RT-qPCR indicated a change opposite to that suggested by microarray. Regarding genotype, a good match was obtained for CYP1A, proteasome subunit beta type-8 precursor (PSMB8) and alpha 2 type I collagen (COL1A2), while transgelin 2 (TAGLN) expression did not differ between family groups, and for ATP-binding cassette sub-family A member 1 (ABCA1) there was an inverse change in expression.

Table 3 Relative expression of genes assayed by RT-qPCR

Genes	VO/FO				Lean/Fat			
	Lean		Fat		**FO**		**VO**	
	Ratio	**p value**	**Ratio**	**p value**	**Ratio**	**p value**	**Ratio**	**p value**
Microarray validation								
Δ5fad (D)	1.72	ns	**4.34**	**0.005**	2.49	ns	−1.01	ns
NADH1 (D)	−1.03	ns	**−1.28**	**0.05**	1.00	ns	1.24	ns
UCP2 (D)	−1.10	ns	1.08	ns	−1.10	ns	**−1.28**	**0.003**
MTHFD1 (D)	1.20	ns	−1.09	ns	−1.08	ns	1.22	ns
PA2G4 (D)	−1.27	ns	**−1.23**	**0.032**	1.11	ns	1.1	ns
RP60S (D)	−1.02	ns	**−1.69**	**0.023**	1.13	ns	**1.9**	**0.006**
MYO (D)	**1.77**	**0.034**	1.22	ns	−1.35	ns	1.08	ns
PCNA (D)	−1.14	ns	−1.27	ns	1.08	ns	1.20	ns
CYP1A (D+G)	**1.71**	**0.004**	**1.81**	**0.023**	**−2.22**	**0.003**	**−2.38**	**0.001**
ABCA1 (G)	1.24	ns	1.17	ns	1.21	ns	**1.28**	**0.015**
PSMB8 (G)	−100.00	ns	**1.87**	**0.033**	**−16.67**	**0.026**	**<−1000**	**0.000**
TAGLN (G)	1.23	ns	−1.02	ns	−1.03	ns	1.22	ns
COL1A2 (G)	1.31	ns	−1.06	ns	1.47	ns	**2.06**	**0.001**
LC-PUFA biosynthesis and regulation								
Δ6fad_a	1.11	ns	**2.96**	**0.008**	1.76	ns	−1.52	ns
elovl5a	1.08	ns	1.2	ns	1.03	ns	−1.08	ns
elovl5b	−1.19	ns	**1.92**	**0.031**	1.54	ns	−1.47	ns
elovl2	1.2	ns	**14.74**	**0.003**	4.89	ns	−2.50	ns
Fatty acid synthesis								
FAS	**1.49**	**0.010**	1.20	ns	1.05	ns	1.31	ns
Fatty acid oxidation								
ACO	1.18	ns	1.28	ns	1.19	ns	1.09	ns
CPT1	1.23	ns	−1.09	ns	−1.16	ns	1.15	ns
Regulation of lipid metabolism								
PPARα	1.14	ns	1.38	ns	1.26	ns	1.04	ns
PPARβ	1.03	ns	1.03	ns	−1.02	ns	−1.02	ns
PPARγ	1.52	ns	**1.60**	**0.036**	1.21	ns	1.15	ns

Xenobiotic and oxidative stress								
GST	−1.30	ns	1.07	ns	1.21	ns	−1.14	ns
MT-A	−1.82	0.051*	−1.05	ns	−1.25	ns	**−2.13**	**0.007**
CAT	**1.57**	**0.036**	1.26	ns	−1.52	ns	−1.22	ns
SOD	−1.20	ns	−1.04	ns	1.21	ns	1.04	ns
Apoptosis								
CASP3B	**1.22**	**0.045**	1.22	ns	−1.04	ns	−1.04	ns
CASP6A/B	1.30	ns	1.37	ns	−1.14	ns	−1.20	ns

fad: Fatty acyl desaturase (Δ5 and Δ6 activities); NADH1: NADH dehydrogenase subunit 1; UCP2: Uncoupling protein 2; elovl: MTHFD1: Methylenetetrahydrofolate dehydrogenase (NADP+ dependent) 1-like; PA2G4: Proliferation-associated 2G4b; RP60S: 60S acidic ribosomal protein; MYO: Myosin heavy chain; PCNA: Proliferating cell nuclear antigen; CYP1A: Cytochrome P4501A; ABCA1: ATP-binding cassette sub-family A member 1; PSMB8: Proteasome subunit beta type-8 precursor; TAGLN: Transgelin 2; COL1A2: Collagen type I alpha 2; elovl2: Elongase 2; FAS: Fatty acid synthase; ACO: Acyl-CoA oxidase; CPT1: Carnitine palmitoyltransferase I; PPARs; Peroxisome proliferator-activated receptors; GST: glutathione S-transferase; MT-A: Metallothionein A; CAT: catalase; SOD: superoxide dismutase; CASP; caspase.

*Only marginally non-significant.

Genes correspond to features significantly affected by either diet (D) or genotype (G) in the microarray analysis or involved in biological processes being analysed in further detail. Values are normalized (by cofilin-2 and elf-1 α) gene expression ratios between fish fed VO in relation to FO for each family group or of Lean fish in relation to the Fat group when fed either one of the diets. Values in bold are significantly different, at $p<0.05$ (REST 2008); ns - non-significant.

As well as validation above, RT-qPCR was used to further analyse genes/pathways identified by microarray and published data as potentially interesting, including lipid metabolism, xenobiotic and oxidative stress, and apoptosis. One was LC-PUFA biosynthesis, given that Δ5fad was significantly affected by diet in the microarray analysis, with a stronger response in Fat fish, whereas Δ6fad showed a significant diet × genotype interaction (supplementary file 1) confirmed by RT-qPCR. The Δ6fad transcript was only significantly up-regulated in Fat fish fed VO, compared to FO, and Lean fish showed higher levels of Δ6fad expression than Fat fish when fed FO, while the opposite trend was noted when fish were fed VO. Fatty acyl elongases (elovl's) were also quantified and their expression (except elovl5a) broadly followed that of fads: significantly up-regulated when dietary VO replaced FO in the Fat group. Additionally, elovl5b and, particularly, elovl2 showed a trend for increased expression (high fold-change) in Lean fish, compared to Fat fish, when fed FO, while an opposite trend was observed when salmon were fed VO. Although genes involved in fatty acid synthesis and oxidation showed few significant differences, expression of fatty acid synthase (FAS) was up-regulated in fish fed VO, but only in Lean fish. The expression of peroxisome proliferator-activated receptors (PPARs), involved in the regulation of multiple lipid metabolism genes, was determined but only PPARγ showed any significant change, being up-regulated in the Fat group when dietary VO replaced FO. Of the xenobiotic and oxidative metabolism genes assayed, apart from CYP1A, only catalase (CAT) was affected by diet (up-regulated in fish fed VO) and only significantly in the Lean family. In contrast, metallothionein A (MT-A) showed higher expression in Fat fish, but only when fed VO, while a marginal down-regulation was observed when comparing VO and FO-fed fish in the Lean group. Of genes related to apoptosis, CASP3B was up-regulated by VO in Lean fish whereas a similar fold-change was marginally non-significant in the Fat fish.

Intestine Fatty Acid Composition

The levels of most fatty acids in pyloric caeca were affected by diet, whereas genotype had no significant effect (Table 4). However, some fatty acids also showed a significant diet × genotype interaction, indicating that the effect of diet depended on the genetic background of the fish. For instance, interactions were observed for some LC-PUFA as a result of higher levels being found in the Lean group, compared to Fat, when fish were fed VO, while the reverse was observed when fed FO. Another unexpected result was that similar levels of DHA in FO- and VO-fed Lean fish meant that, in spite of substantial differences in Fat fish fed the two diets, the effect of diet on DHA was marginally non-significant. Similarly, levels of EPA and 22:5n-3 between FO- and VO-fed fish were noticeably closer in the Lean group.

Table 4 Fatty acid compositions (% of total fatty acids) and lipid contents (% of wet weight) of intestinal tissue of Lean and Fat family groups fed diets containing either FO or VO

	FO		**VO**		**Two-way ANOVA**		
	Fat	**Lean**	**Fat**	**Lean**	**D**	**G**	**D×G**
Fatty acid							
Total saturated	23.70 ± 1.21	22.82 ± 1.09	18.57 ± 0.73	20.08 ± 1.38	0.000	0.742	0.066
Total monoenes	38.25 ± 7.15	44.45 ± 5.31	44.98 ± 2.38	39.34 ± 4.49	0.717	0.928	0.046
18:2n-6	4.19 ± 0.67	4.51 ± 0.24	13.14 ± 0.59	11.28 ± 1.12	0.000	0.139	0.041
18:3n-6	0.10 ± 0.02	0.10 ± 0.02	0.28 ± 0.12	0.25 ± 0.13	0.003	0.611	0.611
20:2n-6	0.57 ± 0.06	0.58 ± 0.07	1.54 ± 0.40	1.65 ± 0.06	0.000	0.199	0.510
20:3n-6	0.17 ± 0.06	0.14 ± 0.02	1.01 ± 0.12	1.47 ± 0.26	0.000	0.056	0.019
20:4n-6	1.20 ± 0.60	0.85 ± 0.31	0.74 ± 0.18	1.40 ± 0.29	0.785	0.419	0.016
22:4n-6	0.08 ± 0.06	0.05 ± 0.06	0.00 ± 0.00	0.09 ± 0.07	0.337	0.337	0.038
22:5n-6	0.21 ± 0.06	0.16 ± 0.07	0.07 ± 0.02	0.12 ± 0.05	0.002	1.000	0.128
Total n-6 PUFA	6.51 ± 0.18	6.38 ± 0.36	16.79 ± 0.86	16.27 ± 0.67	0.000	0.410	0.781
18:3n-3	1.46 ± 0.38	1.67 ± 0.21	4.79 ± 0.53	3.71 ± 0.75	0.000	0.211	0.021
18:4n-3	1.36 ± 0.60	1.67 ± 0.66	0.85 ± 0.31	0.55 ± 0.20	0.002	0.712	0.127
20:3n-3	0.27 ± 0.05	0.33 ± 0.14	0.67 ± 0.18	0.62 ± 0.04	0.000	0.768	0.384
20:4n-3	1.77 ± 0.28	2.16 ± 0.46	0.97 ± 0.13	0.83 ± 0.10	0.000	0.504	0.061
20:5n-3	6.55 ± 0.55	5.79 ± 0.62	3.60 ± 0.90	4.40 ± 0.72	0.000	0.682	0.058
22:5n-3	2.82 ± 0.81	2.30 ± 0.74	1.57 ± 0.36	2.54 ± 0.77	0.157	0.530	0.052
22:6n-3	16.68 ± 6.68	11.65 ± 5.65	6.97 ± 1.58	11.50 ± 3.10	0.057	0.943	0.050
Total n-3 PUFA	30.90 ± 5.94	25.56 ± 5.19	19.41 ± 2.18	24.15 ± 3.76	0.015	0.962	0.042
Total PUFA	38.06 ± 5.96	32.73 ± 5.42	36.45 ± 2.08	40.59 ± 3.16	0.206	0.794	0.059
Total lipids	7.50 ± 3.75	9.56 ± 3.81	6.97 ± 1.07	4.59 ± 1.05	0.081	0.804	0.111

Results are means ± SD (n = 4) and p-values of two-way ANOVA are presented for diet (D), genotype (G) and interaction between both factors.

Proteomic Analysis

Of the protein spots identified as being differentially expressed between diets or genotypes (Additional files 2 and3), only 17 and 29 could be excised and, of these, only 9 and 20, respectively, returned reliable identifications by peptide fragment fingerprinting (Figure 1). Proteins significantly up-regulated by dietary VO are likely implicated in xenobiotic/drug metabolism (epoxide hydrolase 2, EPHX2), protection from oxidative stress (hemopexin-like protein, HPX and peroxiredoxin-1, PRDX1) and induction of apoptosis and inflammatory responses (galectin-2, LGALS2). Those proteins down-regulated by dietary VO included proteins responsible for protein folding and involved in signalling (2-peptidylprolyl isomerase A, PPIA), actin-based motility (myosin light chain, MYL) and DNA replication, repair or transcription (histone H2A, H2A) (Table 5). Proteins affected by genotype encompass a variety of pathways, of which only a few are related to metabolism, namely carbohydrate (alpha-enolase, ENO1; pancreatic alpha-amylase precursor, AMY2; triosephosphate isomerase 1b, TPI1), folate (dihydropteridine reductase, DHPR) or retinol (retinol-binding protein II, RBP2) metabolism (Table 6). Other proteins may have potentially multiple roles but can broadly be assigned roles in response to oxidative and cellular stress (HPX; PRDX; heat shock protein 70, HSP70), oxygen transport (alpha globin, HBA), signal transduction (calreticulin, CALR), transcription/RNA repair (heterogeneous nuclear ribonucleoprotein A0, HNRNPA0; histone H2A, H2A), apoptosis (CASP3), cellular transport, potentially also associated with apoptosis (voltage-dependent anion channel 2–2, VDAC2; annexin A4, ANX4), and proteolysis (proteasome beta 1 subunit, PSMB1). As with the microarray analysis, a few proteins with a more structural function and particularly associated with tissue contractile properties were affected by genotype, showing lower levels in Lean fish. These included calponin-1 (CNN1) and transgelin (TAGLN), the latter which was also found to be significantly affected by microarray, albeit up-regulated in Lean fish. Most proteins significantly affected by genotype showed lower levels of expression in the Lean group, with the exception of ENO1, HSP70, TPI1, H2A and HBA.

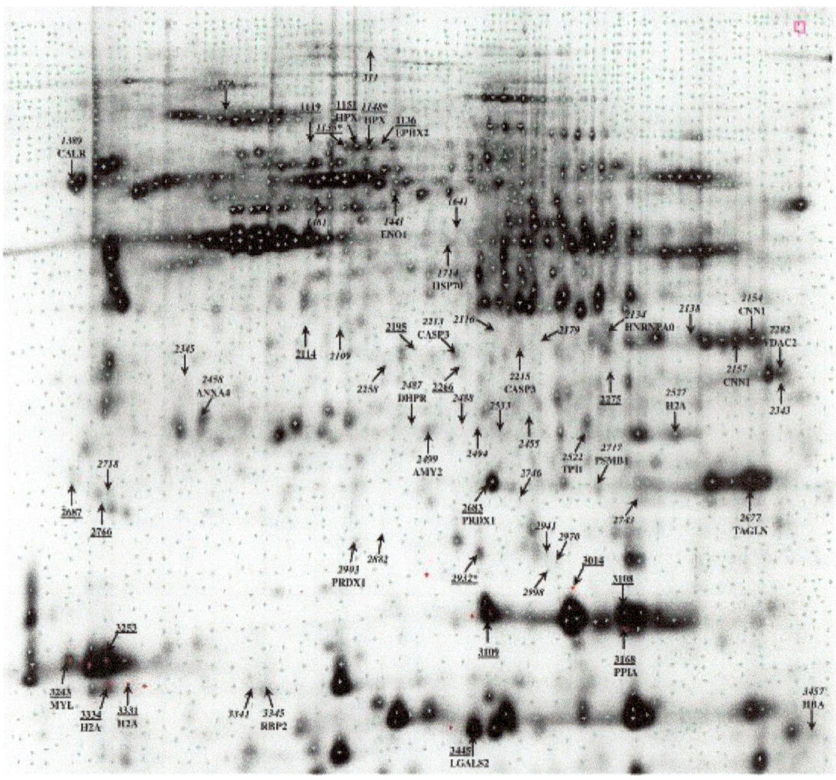

Figure 1 Proteins significantly affected by diet or genotype. Cropped image of a scanned 2-D polyacrylamide gel showing all spots identified as being differentially expressed (P<0.05, two-way ANOVA) in fish fed either FO or VO (underlined spot numbers), or between family groups (remaining spot numbers); the three underlined spot numbers containing an * were differentially affected by both diet and genotype. Spots that returned reliable identifications by peptide fragment fingerprinting are identified by their abbreviation

Table 5 Proteins Differentially Regulated by Diet

Spot No	Protein ID	Accession No. (NCBInr)	Theoretical Mw (KDa)/pI	Protein Score[a]	No. Matched peptides (MS/MS)[b]	Best score peptide	VO/FO Lean	VO/FO Fat	p-value
1136	Epoxide hydrolase 2 (EPHX2)	gb|ACI33129.1|	54.2/5.52	84	1	GGLFVGLPDEIPR	1.23	1.40	0.0190
1148	Hemopexin-like protein* (HPX)	emb|CAA92147.1|	50.4/5.61	169	1	VHLDAITSDDAGNIYAFR	1.32	1.56	0.0002
1151	Hemopexin-like protein* (HPX)	ref|NP_001104617.1|	51.0/6.18	161	1	VHLDAITSDDAGNIYAFR	1.38	1.54	0.0001
2683	Peroxiredoxin 1 (PRDX1)	gb|ACI67145.1|	22.0/6.42	757	8	SISTDYGVLKEDEGIAYR	1.14	1.09	0.0310
3168	2-peptidylprolyl isomerase A (PPIA)	ref|NP_001135161.1|	17.5/7.61	101	1	VYFDITIGDTPAGR	−1.46	−1.18	0.0170
3243	Myosin light chain smooth muscle isoform (MYL)	ref|NP_998803.1|	16.9/4.47	100	2	EAFLLFDR	−1.23	−1.11	0.0490
3331	similar to H2A histone family, member J (H2A)	ref|XP_001521566.1|	13.3/10.84	74	1	AGLQFPVGR	−1.58	−1.18	0.0250
3334	Histone H2A** (H2A)	emb|CAA25528.1|	13.7/10.88	37	1	AGLQFPVGR	−1.48	−1.18	0.0210
3445	Galectin 2 (LGALS2)	gb|ACN10131.1|	14.8/5.93	314	2	SGASSFSINIGHDSDNYALHFNPR	1.17	1.15	0.0170

[a] The protein score probability limit (where $P < 0.05$) is 73.

[b] Peptides with confidence interval above 95% were considered.

* Equivalent to warm-temperature-acclimation-related-65 kDa-protein.

**Same protein identification obtained by MS/PMF (peptide mass fingerprinting): Score = 76; 7 peptides matched.

Only reliable identifications of Actinopterygii, obtained by mass spectrometry (MALDI-TOF-MS/MS) analysis and searches in MASCOT, NCBI and ExPASy (Mw/pI) databases are shown. Expression ratios between fish fed VO and FO diets are given for each family group, as well as p-value for diet (two-way ANOVA, DeCyder V7.0).

Table 6 Proteins differentially regulated by genotype

<table>
<tr><th>Spot No</th><th>Protein ID</th><th>Accession No. (NCBInr)</th><th>Theoretical Mw (KDa)/pI</th><th>Protein Score[a]</th><th>No. Matched peptides (MS/MS)[b]</th><th>Best score peptide</th><th>Lean/Fat FO</th><th>Lean/Fat VO</th><th>p-value</th></tr>
<tr><td>1148</td><td>Hemopexin-like protein* (HPX)</td><td>emb|CAA92147.1|</td><td>50.4/5.61</td><td>169</td><td>1</td><td>VHLDAITSDDAGNIYAFR</td><td>−1.09</td><td>−1.29</td><td>0.0270</td></tr>
<tr><td>1389</td><td>Calreticulin precursor (CALR)</td><td>gb|ACI32936.1|</td><td>47.6/4.33</td><td>81</td><td>2</td><td>FEPFSNEGK</td><td>−1.28</td><td>−1.33</td><td>0.0026</td></tr>
<tr><td>1441</td><td>Alpha-enolase (ENO1)</td><td>ref|NP_001133366.1|</td><td>47.0/5.91</td><td>405</td><td>3</td><td>AAVPSGASTGIYEALELR</td><td>1.28</td><td>1.2</td><td>0.0066</td></tr>
<tr><td>1714</td><td>Heat shock protein 70 (HSP70)</td><td>ref|NP_990334.1|</td><td>70.8/5.47</td><td>111</td><td>1</td><td>IINEPTAAAIAYGLDKK</td><td>1.16</td><td>1.77</td><td>0.0260</td></tr>
<tr><td>2134</td><td>Heterogeneous nuclear ribonucleoprotein A0 (HNRNPA0)</td><td>gb|ACI67551.1|</td><td>28.8/9.10</td><td>235</td><td>2</td><td>LFVGGLNVDTDDDGLRK</td><td>−1.19</td><td>−1.09</td><td>0.0460</td></tr>
<tr><td>2154</td><td>Calponin 1 (CNN1)</td><td>ref|NP_001139857.1|</td><td>33.2/8.56</td><td>565</td><td>4</td><td>KINTSPQNWHQLENIGNFVR</td><td>−1.34</td><td>−1.51</td><td>0.0350</td></tr>
<tr><td>2157</td><td>Calponin 1*** (CNN1)</td><td>ref|NP_001139857.1|</td><td>33.2/8.56</td><td>43</td><td>1</td><td>YDPQKEEELR</td><td>−1.34</td><td>−1.47</td><td>0.0240</td></tr>
<tr><td>2213</td><td>Caspase 3 (CASP3)</td><td>ref|NP_001133393.1|</td><td>31.0/5.97</td><td>254</td><td>2</td><td>IPVEADFLYAYSTAPGYYSWR</td><td>−1.15</td><td>−1.24</td><td>0.0300</td></tr>
<tr><td>2215</td><td>Caspase 3 (CASP3)</td><td>ref|NP_001133393.1|</td><td>31.0/5.97</td><td>279</td><td>2</td><td>VANDQTVQQIQQLLSK</td><td>−1.14</td><td>−1.31</td><td>0.0130</td></tr>
<tr><td>2282</td><td>Voltage-dependent anion channel 2–2 (VDAC2)</td><td>gb|ACH71030.1|</td><td>30.1/8.85</td><td>214</td><td>1</td><td>VNNNSLVGVGYTQTLRPGVK</td><td>−1.24</td><td>−1.19</td><td>0.0160</td></tr>
<tr><td>2458</td><td>Annexin A4 (ANXA4)</td><td>gb|ACI69495.1|</td><td>28.4/5.22</td><td>218</td><td>3</td><td>NHLLQVFK</td><td>−1.14</td><td>−1.11</td><td>0.0300</td></tr>
<tr><td>2487</td><td>Dihydropteridine reductase (DHPR)</td><td>gb|ACI67281.1|</td><td>15.7/8.46</td><td>81</td><td>1</td><td>QSVWTSTISSHLATR</td><td>−1.14</td><td>−1.11</td><td>0.0300</td></tr>
<tr><td>2499</td><td>Pancreatic alpha-amylase precursor (AMY2)</td><td>ref|NP_001036176.1|</td><td>57.4/6.89</td><td>86</td><td>1</td><td>ALVFVDNHDNQR</td><td>−1.37</td><td>−1.27</td><td>0.0021</td></tr>
<tr><td>2522</td><td>Triosephosphate isomerase 1b (TPI1)</td><td>ref|NP_001133174.1|</td><td>26.6/7.63</td><td>336</td><td>2</td><td>LDPNTEVVCGAPSIYLEFAR</td><td>1.26</td><td>1.16</td><td>0.0009</td></tr>
<tr><td>2527</td><td>Histone cluster 1 (H2A)</td><td>ref|NP_001086775.1|</td><td>14.0/10.88</td><td>94</td><td>1</td><td>AGLQFPVGR</td><td>1.12</td><td>1.11</td><td>0.0074</td></tr>
<tr><td>2677</td><td>Transgelin (TAGLN)</td><td>gb|ACM09025.1|</td><td>21.7/7.69</td><td>219</td><td>1</td><td>DGCVLSELINSLHK</td><td>−1.61</td><td>−1.35</td><td>0.0350</td></tr>
<tr><td>2717</td><td>Proteasome beta 1 subunit (PSMB1)</td><td>ref|NP_001003889.1|</td><td>26.1/6.32</td><td>87</td><td>1</td><td>GAVYSFDPVGSYQR</td><td>−1.29</td><td>−1.12</td><td>0.0140</td></tr>
<tr><td>2903</td><td>Peroxiredoxin-1(PRDX1)</td><td>gb|ACI67145.1|</td><td>22.0/6.42</td><td>115</td><td>1</td><td>QITINDLPVGR</td><td>−1.62</td><td>−1.99</td><td>0.0069</td></tr>
<tr><td>3345</td><td>Retinol-binding protein II, cellular (RBP2)</td><td>ref|NP_001139954.1|</td><td>15.6/5.44</td><td>70</td><td>2</td><td>AIDIDFATR</td><td>−1.32</td><td>−1.1</td><td>0.0330</td></tr>
<tr><td>3457</td><td>Alpha globin (HBA)</td><td>emb|CAA65949.1|</td><td>15.1/9.19</td><td>170</td><td>2</td><td>TYFSHWADLSPGSAPVK</td><td>1.03</td><td>2.09</td><td>0.0320</td></tr>
</table>

[a]The protein score probability limit (where $P < 0.05$) is 73.

[b] Peptides with confidence interval above 95% were considered.

* Equivalent to warm-temperature-acclimation-related-65 kDa-protein.

*** Same protein identification obtained by MS/PMF (peptide mass fingerprinting): Score = 193; 27 peptides matched.

Only reliable identifications of Actinopterygii, obtained by mass spectrometry (MALDI-TOF-MS/MS) analysis and searches in MASCOT, NCBI and ExPASy (Mw/pI) databases are shown. Expression ratios between Lean and Fat family groups are given for each dietary treatment, as well as p-value for genotype (two-way ANOVA, DeCyder V7.0).

DISCUSSION

Dietary plant ingredients can induce chronic intestinal inflammatory conditions in salmonids that can ultimately result in carcinogenesis [25]. This extreme reaction is rare and usually associated with soy protein at high levels [26]. Dietary n-3 LC-PUFA have important anti-inflammatory and anti-carcinogenic effects in mammalian intestine [23]. Therefore, use of feeds containing high percentages of plant proteins combined with replacement of FO by VO, as is now prevalent in the industry, requires studies on dietary effects on intestinal transcriptomes and proteomes. However, interpretation of the data was difficult as the effects on dietary treatments and/or family groups were subtle, as also observed in liver transcriptome [9], and is typical of this type of experiment [6, 8, 11,27]. Partly as a consequence, validation of the microarray data gave variable results, from perfect match, to opposite changes in a few, although effects observed in the microarray, with fold changes as low as 1.2 (e.g., NADH1 and PA2G4) were validated by RT-qPCR. In view of the whole genome duplication event that occurred in salmonids [28], gene expression studies are often more challenging due to the presence of highly similar genes which may hybridize with cDNA probes presenting low specificity, further complicated if similar transcripts, corresponding to duplicated genes, are differentially regulated [10]. Nonetheless, the presence of several features related to specific processes in both the transcriptomic and proteomic analysis gave supporting evidence to the pathways likely differentially affected by dietary oil and genetic background related to flesh adiposity.

Intestinal LC-PUFA Biosynthesis Capacity is Differentially Affected by Diet and Genotype

Considering whether genetic selection for fish families showing better adaptation to more sustainable feeds might be a viable approach to develop aquaculture, one outcome of this investigation was to establish if effects of diet on expression of LC-PUFA biosynthesis genes depended on genotype, as shown in the liver transcriptome of these fish [9]. This was not seen in the hepatic transcriptome of European sea bass families showing different growth rates when fed a vegetable diet but, in this, case similar LC-PUFA profiles were also noted in both genotypes in response to the vegetable diet [11]. In both salmon tissues, differences in n-3 LC-PUFA content between fish fed FO or VO were smaller in the Lean family group. This was due to higher levels of n-3 LC-PUFA in Lean salmon, compared to Fat, when fish were fed VO, but higher amounts in the Fat family group when fed FO. However in liver, up-regulation of LC-PUFA biosynthesis when fish were fed VO was much larger in the Lean family group, whereas in intestine the same individuals only showed significant up-regulation in Fat fish. This appears contradictory but can be explained by the differential tissue n-3 LC-PUFA contents. Although the difference was smaller in Lean fish compared to Fat fish in both tissues, in liver there was still a considerable difference in n-3 LC-PUFA levels between fish fed FO or VO, while in intestine levels were similar. PUFA have important activities on transcription factors, either as direct ligands or through effects on membrane composition [29], affecting transcription of many genes involved in lipid metabolism, including desaturases and elongases [30, 31]. In salmon, regulation of genes of LC-PUFA biosynthesis that are known to respond to dietary composition, i.e., Δ5fad, Δ6fad, elovl5b and elovl2 [32–34], appear to show high plasticity and are likely under feedback regulation by tissue n-3 LC-PUFA. Both studies suggest that the Lean family group may show an enhanced response to low dietary n-3 LC-PUFA, with greater up-regulation of biosynthesis when fed VO. In contrast to liver, this response was sufficient in intestine to maintain tissue n-3 LC-PUFA, particularly DHA, at similar levels to fish fed FO. Considering that differences in desaturase expression between the Fat and Lean fish were only significant when FO but not VO, was fed, suggests that the likely mechanism is through negative feedback by high levels of n-3 LC-PUFA rather than positive feedback from low levels of LC-PUFA and/or higher levels of shorter chain precursors.

Other Dietary Effects on Lipid Metabolism

Transcriptional regulation of desaturases and elongases by LC-PUFA may involve both PPARα and sterol regulatory element binding protein 1c (SREBP-1c) [30, 31]. In liver, expression of Δ5fad, Δ6fad, elovl2, PPARα, and possibly PPARβ, appeared co-ordinately regulated by diet depending on genotype, while PPARγ was not affected [9]. In intestine, however, expression of PPARα and PPARβ was not affected by either diet or genotype, while PPARγ was up-regulated by dietary VO, significantly in Fat fish. This suggests that dietary regulation of lipid metabolism genes in fish intestine might differ to mammals, where PPARα showed differential expression in response to dietary EPA and DHA in murine intestine [35]. Reasons for differential regulation of PPARs between salmon liver and intestine are unclear, but may be due to different patterns of tissue expression. In plaice and seabream, there was no nutritional regulation (fasting/feeding) of PPARs in the intestine, where PPARγ was the dominant isotype, in contrast to liver where PPARα was dominant [36]. PPARγ in both mammals and fish is predominantly expressed in adipose tissue and promotes adipocyte differentiation and lipid storage [36–38]. In mammals, PPARγ activates the expression of genes characteristic of mature adipocytes and adipogenesis, including FAS [37, 38] and hence the expression of PPARγ, up-regulated in salmon fed VO, might be related to increased expression of FAS. However, increased PPARγ expression was only significant in Fat fish whereas FAS was significantly up-regulated only in Lean salmon. As fish PPARγ is functionally the most different of the three isotypes compared to mammalian PPARs, and is expressed more widely in fish tissues that in mammals, other mechanisms and functions may underlie the observed regulation [35].

In this study, the hypotriglyceridemic effect of LC-PUFA, well established in mammals [39, 40], was also observed in salmon intestine. Lipogenesis was down-regulated in FO-fed fish, as demonstrated by decreased FAS expression (RT-qPCR) and the presence of a transcript containing a beta-ketoacyl synthase domain, a component of FAS (microarray). The differences in FAS expression were not as marked as in liver [9] and were only significant in Lean fish but, together with the LC-PUFA biosynthesis data, demonstrate the active role of salmon intestine in lipid metabolism [21]. However, despite up-regulation of lipogenesis by dietary VO, lipid accumulation in enterocytes was lower than in fish fed FO, contrary to previous reports of VO promoting lipid accumulation in enterocytes [16–18].

In contrast, the hypotriglyceridemic effect of LC-PUFA did not involve the typical increase in β-oxidation [39], reported in mice intestine [36, 40]. As in liver, no changes were observed in the expression of β-oxidation genes carnitine palmitoyltransferase I (CPT1) and acyl-CoA oxidase (ACO) [10]. Nonetheless, effects of dietary lipid on energy metabolism were observed in intestine. In particular, UCP and transcripts involved in the mitochondrial electron transport chain, including components of cytochrome c oxidase, NADH1 and ubiquinol-cytochrome c reductase complexes, and the mitochondrial metabolite transporter SCaMC-2, were slightly down-regulated by dietary VO, possibly suggesting reduced energetic metabolism. EPA may act as a mitochondrial proliferator in both rat and salmon liver [41, 42], which might also explain this result.

Vegetable oils as Potential Sources of Contaminants in Alternative Diets for Aquaculture?

The use of FO in feeds has also been questioned in relation to levels of persistent organic pollutants, POPs [43], whereas VO are generally considered safer alternatives to dietary FO [44]. However, results suggested that VO had greater impact on xenobiotic metabolism in salmon intestine than FO, with the transcriptome and proteome both showing up-regulation of transcripts and proteins involved in detoxification and protection from oxidative stress in fish fed VO. This was surprising, considering expression of these genes has been associated with FO supplementation and higher levels of organic contaminants and/or increased peroxidative susceptibility of LC-PUFA [22, 45, 46]. In particular, and linked to a detoxification role, up-regulation of CYP1A transcript and epoxide hydrolase 2 (EPHX2) protein was found in fish fed VO. Cytochrome P450 1A metabolizes many exogenous and endogenous molecules, including pollutants (e.g., polychlorinated biphenyls-PCBs, polycyclic aromatic hydrocarbons-PAHs, dioxins and dibenzofurans) as well as several metabolic products (e.g., steroids, bile acids). Its expression is sensitive to low levels of contaminants and therefore it is a commonly used marker [47, 48]. Lipid peroxidation and antioxidant enzymes are also biomarkers for environmental xenobiotic contamination in fish, as the catalytic actions of detoxifying enzymes like CYP1A produce high levels of reactive oxygen species (ROS) [49,50]. We observed up-regulation of CAT and of a selenoprotein transcript, as well as of HPX and PRDX1 proteins in salmon fed VO. CAT and PRDX1 catalyze the decomposition of hydrogen peroxide [51], hemopexin prevents heme-mediated oxidative stress [52], and selenoproteins have diverse roles, including selenium transport and antioxidant defense [53]. Expression of other genes involved in protection from oxidative stress like GST and SOD were not affected, but MT-A was down-regulated in the Lean family group fed VO. MT-A, besides having an antioxidant role, also protects against heavy metal toxicity and maintains physiological zinc homeostasis [54] and hence could be responding to contaminants more abundant in FO (e.g., heavy metals) [55].

Previous analysis of contaminants in comparable feeds using the same standard FO and a similar VO blend, showed that levels of several POPs, including organochlorine pesticides, and heavy metals were substantially lower, but polycyclic aromatic hydrocarbons (PAH) were 10-fold increased in the VO diet compared to the FO diet [55]. PAH derive from incomplete combustion of organic compounds and can be found in high concentrations in fats, including VO, through multiple routes of contamination, before, during or after oil processing [56]. They are metabolized by both CYP1A1 and EPHX2, among other enzymes, and CYP1A1 is induced by PAH in mammals [57]. Therefore, higher PAH levels in the VO diet might explain, at least partly, the results obtained in the present study although, unlike POPs, PAHs are not persistent and are readily eliminated from fish tissues [58].

High doses of PAH result in significant intestinal hyperplasia in fish, with an increase in cell proliferation and faster epithelial turnover [59]. Previous studies on intestinal gene expression in fish indicated a reduction in cell proliferation or differentiation associated with dietary FO replacement by VO [21, 22], possibly due to lower levels of membrane LC-PUFA and reduced oxidative stress [22]. In the present study, no major impact on cell proliferation was apparent in the intestinal transcriptome or proteome data. Two transcripts related to cell proliferation, PA2G4 and cyclin G1, were slightly down-regulated in fish fed VO, but in mammals these have opposing effects [60, 61] and, furthermore, two mammalian PA2G4 isoforms have been shown to have opposite effects in cellular proliferation [62] and hence results are inconclusive.

Previously, expression of caspases, effectors of controlled cell death or apoptosis, was affected by replacement of dietary FO by VO in fish [21, 22]. Apoptosis is particularly important in organs with high rates of cellular turnover such as intestine but, in addition to maintaining normal gut function, apoptosis may be affected by pathological or toxic conditions, including those induced by environmental chemical contaminants [63, 64]. In the present study, expression of CASP3B was up-regulated in salmon fed VO, particularly in the Lean family group and a similar, non-significant trend was observed for CASP6A/B. As ROS are important signalling molecules in apoptotic processes, these results could be linked to a cytotoxic effect causing increased oxidative stress in VO [64]. Relevant to the above was the up-regulation of galectin 2 (LGALS2) in the proteome of salmon fed VO. Galectins are pleiotropic regulators of immune functions and are up-regulated by injury and infectious conditions, have well-recognized modulatory roles in mammalian intestinal inflammatory diseases, and their mode of action involves induction of apoptosis [65, 66]. The lack of major effects on cell proliferation and only slight up-regulation of CASP3 and LGALS2 suggests that any contaminant doses experienced by the fish were unlikely to have caused any serious morphophysiological damage in the intestine. As similar trends were not seen in the hepatic transcriptome of these individuals [9], this may suggest intestine can potentially metabolize and detoxify xenobiotics present in the diet. Furthermore, there were no growth or general performance issues with these fish [5]. Therefore, the data do not imply abnormal gastrointestinal functions or effects on final product quality.

Effect of Genotype in Intestinal Transcriptome and Proteome

Contrary to diet, genotype did not have a major impact on metabolism genes, apart from transcripts related to the proteasomal degradation pathway including a strong down-regulation of PSMB8 in Lean fish, particularly fed VO. This gene has been recently found to have a molecular evolution history that suggests a very strong selective pressure for its functional dimorphism to be maintained in vertebrates [67]. Two different alleles, A-type and F-type, can be found in basal vertebrate species, including Atlantic salmon. The PSMB8F lineage was lost in common ancestors of higher teleosts and tetrapods but was then independently revived de novo through the appearance of F-type alleles within the PSMB8A lineage. In this study we did not find evidence of significant differences between families groups for the A-type allele in the transcriptomic analysis as the probe showing significant variation between families in the microarray corresponded to the PSMB8F allele. Hence, it was also the F-type transcript that was validated by RT-qPCR, using type-F specific primers. However, further studies would be required to confirm this and to assess the functional significance of this result. On the other hand, expression of PSMB1 was down-regulated in the intestine proteome of Lean fish. Proteolysis through this pathway is essential for many cellular processes, including the cell cycle, signalling, cellular defence and responses to oxidative stress [68]. Therefore, this response might be related to defence against cellular stress, as another difference between the two family groups was related to xenobiotic and oxidant metabolism. Apart from lower expression of a CYP1A transcript in Lean fish, two proteins with antioxidant roles, HPX and PRDX, were down-regulated in the proteome of Lean compared to Fat fish. Alpha globin, or haemoglobin alpha (HBA), a major component of blood and potent mediator of oxidative stress, can have both protective and damaging effects depending on complex interactions in H_2O_2-rich environments [69]. However, given its opposite regulation to HPX, whose main role is to scavenge heme and protect from its toxic effects [52], up-regulation of HBA in Lean fish may indicate heme-mediated oxidative stress.

The apoptotic pathway may be differentially affected by genotype, with down-regulation of CASP3, VDAC2 and ANXA4 in the Lean family group, the latter two transport proteins having well-recognized roles in apoptosis [70, 71]. In contrast, heat shock proteins that protect against environmental stresses were increased in the intestine transcriptome and proteome of Lean salmon. This response could be associated with the observed changes in the ubiquitin-proteasome degradation system, as the systems have been functionally coupled in mammals. Thus, moderate exposure to a heat shock can cause a transient increase in intracellular proteolysis by the ubiquitin-proteasome pathway, followed by a phase of slower or even inhibited protein degradation [72, 73]. Furthermore, Pirkkala et al. [74] demonstrated transcriptional induction of heat shock genes when proteasome activity was down-regulated. However, judging by the fold-changes, these effects are only relevant when fish were fed VO, and hence could be more related to dietary changes. Collectively, the data may indicate higher sensitivity of Lean fish to environmental or endogenous stresses due to replacement of dietary FO by VO.

The predominant influence of genotype was in the expression of intestinal transcripts of structural proteins, particularly collagen alpha chains, but also osteonectin, TAGLN, troponin I and keratocan, which were up-regulated in Lean fish, whereas troponin C was down-regulated. Furthermore, CNN1 and TAGLN were down-regulated in the intestinal proteome in Lean fish. Collagen, the main component of connective tissue, helps to maintain the structural integrity of tissues [75], while osteonectin is an extracellular matrix glycoprotein with high affinity towards collagen and whose expression has been associated with remodelling processes in tissues, including human intestine during development/morphogenesis and in diseased mucosa [76]. Troponin, TAGLN and CNN1 are all involved in actin binding, actin-myosin interaction and muscle contraction. The inverse regulation of troponins is not conflicting as they have different roles in actomyosin cross-bridge formation and contraction; binding of troponin C to Ca^{2+} induces conformational changes that counteract the inhibitory action of troponin I [77]. Expression of TAGLN transcript and protein showed opposite effects but a lack of correlation between transcriptomic and proteomic data is not unprecedented [78]. As discussed above, this result might also be explained by the presence of similar duplicated genes in Atlantic salmon that are regulated differently. Transcriptomic results were validated by RT-qPCR for COL1A2, although only significantly when fish were fed the VO diet, for which fold changes were higher. In addition, in the microarray results differences in expression of structural proteins between family groups were consistently more accentuated in fish fed VO which could suggest a cumulative effect of diet. Furthermore, MYO was up-regulated in fish fed VO compared to FO but only in Lean fish (RT-qPCR), and significant diet × genotype interactions were found for alpha-actinin 1, tubulin beta-2 chain and procollagen-lysine 2-oxoglutarate 5-dioxygenase 2 (Additional file 1), which were up-regulated in Lean salmon, compared to Fat, but only when fed VO. In cod, replacement of FO by VO resulted in changes in intestinal expression of structural genes with the potential to alter the structural and mechanical properties of the intestinal muscle layer, including a range of actin-binding transcripts [21].

The present study is the first investigation of the influence of genetic background of families with different flesh adiposity phenotypes on intestinal gene expression of a fish species. Effects were subtle and consequently their potential impacts difficult to fully assess. However, if genetic selection for families better adapted to alternative formulations is an approach taken in the future, the potential for genotype-specific differences being exacerbated when VO replaces dietary FO should be further examined to assess the consequences of these changes in intestinal gene expression.

CONCLUSIONS

Metabolic activity, particularly lipid and energy, of intestinal tissue responded to dietary lipid composition but was also affected by genotype. The LC-PUFA biosynthesis pathway, typically up-regulated when salmon are fed VO, was especially influenced by genetic background. The Lean fish showed an enhanced response to low dietary n-3 LC-PUFA and the expression of Δ5fad, Δ6fad, elovl5b and elovl2 in the intestine showed high plasticity and was reflected in tissue biochemical composition indicating that their transcriptional regulation might be under feedback control by n-3 LC-PUFA, mainly DHA. Lower n-3 LC-PUFA in VO increased lipogenesis in Lean salmon, assessed by expression of FAS, while β-oxidation appeared unaffected, although transcripts involved in mitochondrial respiratory or electron transport chains were down-regulated, suggesting reduced activity in fish fed VO. Higher expression of genes and proteins involved in xenobiotic metabolism (CYP1A and EPHX2), antioxidant defence (CAT, HPX and PRDX1), and apoptosis (Casp3B) were observed in VO-fed fish, suggesting they might be responding to higher levels of contaminants, particularly PAH, in the diet. However, the intestine appeared able to metabolize and detoxify xenobiotic substances potentially present in the diet without major deleterious effects. Nonetheless, the data suggest that further attention should be given to contaminants in VO in the future. On the other hand, the data indicate potential genotype-specific differences in the response of the intestinal transcriptome and proteome to dietary VO. These include potential changes in structural properties of the intestinal layer and defence against cellular stress suggesting the Lean group was more susceptible to diet-induced oxidative stress. Considering the possibility of selecting families better adapted to alternative diet formulations, and the central role of intestine as a major barrier to nutrients, contaminants and pathogens, greater attention should be given to this organ when evaluating the effects of diet and genotype.

METHODS

Feeding trial and sampling

A dietary trial was conducted using two genetically characterized groups of Atlantic salmon (Salmo salar) post-smolts comprising full-sib families selected from a breeding program (Landcatch Natural Selection Ltd., Argyll, Scotland). The choice of the family groups was based on estimated breeding values (EBVs) of the parents for high or low flesh adiposity, assessed by Torry Fatmeter (Distell Industries, West Lothian, UK), a trait that was found to have a heritability ranging from 0.17 to 0.39 in this dataset. The two groups were created from four unrelated full-sib families; two families from the extreme lower end of the EBV distribution for flesh lipid content (Lean) and two families from the extreme upper end of the distribution (Fat). The average EBV for the lipid content of the Fat families was 2.00 percentage units higher than that of the Lean families, representing a standardised selection differential of 2.33 standard deviations. Assessment of the flesh and visceral lipid contents at the end of the trial confirmed differences in adiposity between the groups [5].

Two thousand fish of each group were stocked into eight 12 x 5m^3 net pens at the Ardnish Fish Trials Unit (Marine Harvest Scotland, Lochailort, Scotland; 500 fish pen^{-1}). Duplicate pens of each group were fed one of two experimental diets (Skretting ARC, Stavanger, Norway) containing 25-32% fish meal, 40-45% plant meals and 27.5-30% oil (percentages varying according to pellet size) supplied either as standard northern FO or as a VO blend comprising rapeseed, palm and Camelina oils in a ratio of 5:3:2. Diets were formulated to fully satisfy the nutritional requirements of salmonid fish [79] and contained similar levels of PUFA but different n-3 and n-6 PUFA contents, 25.3% and 4.6% in the FO diet and 13.4% and 17.1% in the VO diet, respectively [5]. After 55 weeks, 25 fish per pen were sampled 24 h after the last meal. Fish were killed by a blow to the head following anaesthesia, and intestinal tissue (pyloric caeca) collected, immediately frozen in liquid nitrogen and stored at −70°C prior to analyses. Further details can be found in Bell et al. [5].

Lipid extraction and fatty acid analyses

Total lipid from 1 g of intestine of four fish per treatment was extracted and determined gravimetrically [80], and fatty acid methyl esters (FAME) prepared by acid-catalysed transesterification of total lipid [81]. FAME were separated and quantified by gas chromatography as described in detail previously [5]. Significant differences in intestinal fatty acid composition were determined by two-way ANOVA ($p<0.05$) using the SPSS 16.0 statistical package (SPSS Inc., Chicago IL, USA).

RNA extraction and purification

Intestinal tissue (0.2 g) from six individuals per experimental group was homogenised in 2mL TRI Reagent and total RNA isolated following manufacturer's instructions (Ambion, Applied Biosystems, Warrington, U.K.). RNA quantity and quality (integrity and purity) were assessed by gel electrophoresis and spectrophotometry (NanoDrop ND-1000, Thermo Scientific, Wilmington, U.S.A.), and 100 μg of total RNA from each sample further cleaned by mini spin-column purification (RNeasy Mini Kit, Qiagen, Crawley, UK).

Microarray hybridizations, image processing and statistical analysis

The TRAITS/SGP (v.2.1) salmon 17k cDNA microarray described by Taggart et al. [33] was used (ArrayExpress accession: A-MEXP-1930). A dual-labelled experimental design was employed, with each sample being competitively hybridised against a common pooled-reference. The experiment comprised 2 genotypes (Lean/Fat) × 2 diets (VO/FO) × 6 biological replicates. Indirect labelling was employed for preparing the microarray targets. Antisense amplified RNA (aRNA) was produced from 500 ng of purified total RNA per sample using the Amino Allyl MessageAmpTM II aRNA Amplification Kit as per manufacturer's instructions (Ambion, Applied Biosystems), followed by Cy3 (for samples) or Cy5 (for pooled reference) fluor (GE HealthCare, Buckinghamshire, U.K.) incorporation through dye-coupling reaction [10]. Microarray hybridizations were performed in a Lucidea semi-automated system (GE Healthcare) without pre-hybridization. For each array, every labelled biological replicate and corresponding pooled reference (40 pmol each dye, c. 150 ng aRNA) were combined and added to the hybridization solution. Two post-hybridization automatic washes followed by six manual washes to a final stringency of 0.1× SSC (EasyDipTM Slide staining system; Canemco Inc., Quebec, Canada) were performed before scanning.

Scanning was performed at 10 μm resolution using an Axon GenePix 4200AL Scanner (MDS Analytical Technologies, Wokingham, Berkshire, U.K.). Laser power was constant (80%) and "auto PMT" was enabled to adjust each channel at less than 0.1% feature saturation and Cy3/Cy5 mean intensity close to one. BlueFuse software (BlueGnome, Cambridge, U.K.) was used to identify features and extract fluorescence intensity values from TIF images. The resulting fluorescence intensity data (BlueFuse proprietary algorithm) and quality annotations for the 17,102 gene features were exported into the GeneSpring GX version 10.0.2 analysis platform (Agilent Technologies, Wokingham, Berkshire, U.K.) after undergoing block Lowess normalization. All control features were excluded from subsequent analyses. Data transformation and quality filtering were as in Morais et al. [9, 10]. This gave a final list of 15,498 genes that were eligible for statistical analysis. Experimental annotations complied fully with minimum information about a microarray experiment (MIAME) guidelines [82] and experimental hybridisations are archived on the EBI ArrayExpress database (http://www.ebi.ac.uk/arrayexpress/) under accession number E-TABM-1173. Hybridization data were analysed in GeneSpring by two-way ANOVA, which examined the explanatory power of the variable diet and genotype and the interaction between the two, followed by Gene Ontology (GO) enrichment analysis of the significant lists of features, at a significance level of 0.05. No multiple test correction was employed, as previous analyses, confirmed by RT-qPCR, indicated that such corrections are over-conservative for this type of data [9, 32].

RT-qPCR gene expression analysis

Expression of selected genes, for microarray validation and to further examine biological processes of interest, was studied by reverse transcription quantitative real time PCR (RT-qPCR), with target qPCR primer sequences given in Additional file 2. In addition, amplification of two reference genes, cofilin-2 and elongation factor-1α (elf-1α), was performed. One μg of column-purified total RNA per sample was reverse transcribed into cDNA using the VersoTM cDNA kit (ABgene, Surrey, U.K.) using a mixture of random hexamers (400ng/μL) and anchored oligo-dT (500ng/μL) at 3:1 (v/v). Negative controls (no enzyme) were performed to check for genomic DNA contamination. A similar amount of cDNA was pooled from all samples and the remaining cDNA diluted 20-fold with water. RT-qPCR analysis used relative quantification with the amplification efficiency of each primer pair assessed by serial dilutions of the cDNA pool. Amplifications were carried out in duplicate using a Quantica machine (Techne, Cambridge, U.K.) in a final volume of 20 μl containing 2–8 μl diluted cDNA, 0.5 μM of each primer and 10 μl AbsoluteTM QPCR SYBR® Green mix (ABgene), with a systematic negative control (NTC-non template control). The qPCR profiles contained an initial activation step at 95°C for 15 min, followed by 30–40 cycles: 15 s at 95°C, 15 s at the specific primer pair annealing temperature (Ta; Additional file 2) and 15 s at 72°C. After amplification, a melt curve was performed confirming a single product in each reaction, RT-qPCR product sizes checked by agarose gel electrophoresis, and identity of amplicons confirmed by sequencing. Gene expression was analysed using the relative expression software tool (REST 2008, http://www.gene-quantification.info/), employing a pair wise fixed reallocation randomisation test (10,000 randomisations) with efficiency correction [83].

Protein extraction and labelling

Six intestine samples (100 mg) per treatment were rapidly disrupted by homogenization and sonication on ice in 1 ml of DIGE lysis/labeling buffer (7M Urea, 2M Thiourea, 4% CHAPS and 30mM Tris, pH 8.5) in the presence of 10 μl of a protease inhibitor cocktail (ref. P8340, Sigma-Aldrich) and 4μl of 250 mM EDTA. After centrifugation at 12,000 × g for 20 min at 4°C, the supernatant was recovered and protein concentration determined (Quick Start Bradford Protein Assay Kit, ref. # 500–0202, Bio-Rad). Protein (500 μg) was purified by precipitation (ReadyPrepTM 2-D Cleanup kit, Bio-Rad) and the pellet re-suspended in DIGE lysis/labeling buffer at ~5μg/μl.

Samples were labelled using CyDye DIGE fluors (5nmol labelling kit; GE Healthcare), following manufacturer's instructions. Three of the experimental replicates of each treatment (50 μg protein with pH corrected to 8.0-8.5) were labelled individually with 400 pmol Cy3 and the remaining three with 400 pmol Cy5. In addition, equal amounts of all experimental samples were pooled and 600 μg of protein (50 μg per 2-D gel) were batch-labelled with Cy2. The three labelled samples, corresponding to two experimental samples and one internal reference pool, were then combined to have in each 2-D gel samples corresponding to fish fed either FO or VO within the same family group.

Two-dimensional (2-D) polyacrylamide gel electrophoresis

Rehydration buffer (ReadyPrep 2-D starter kit, Bio-Rad) containing 0.2% DTT was added to the pooled protein samples to a final volume of 450 μl, which were loaded onto Immobiline DryStrip pH 3–11 NL, 24 cm IPG strips (GE Healthcare) by passive rehydration at room temperature overnight in the dark. Proteins were separated in the first dimension by isoelectric focusing at 20°C (Ettan IPGphor, GE Healthcare), applying increasing voltage until 200 V for 4 h, increasing to 500 V over a period of 3 h, then keeping the applied tension at a constant 1000 V for 1 h, followed by a further increase to 8000 V over 90 min, maintaining this voltage for almost 9 h (total of 79,200 Vh). After isoelectric focusing the strips were equilibrated in two 40 min steps using 50mM Tris–HCl pH 8.8, 6M urea, 30% glycerol, 2% SDS buffer, to which 2 % DTT (w/v) and 2.8% iodoacetamide (GE Healthcare) were added to produce reducing and alkylating buffers, respectively. The strips were loaded onto a 12.5% acrylamide gel cast between low fluorescence glass cassettes (EttanDALTsix gel caster system, GE Healthcare). The strips were overlaid with ReadyPrep Overlay Agarose (Bio-Rad) and the six gel cassettes run in the EttanDALT system in two steps: at 60 mA, 80 V, 6 W (upper limits) for 1 h, and then 240 mA, 500 V, 78 W until the bromophenol blue dye front had run to ~1 cm above the bottom of the gels. Laemmli buffers (250mM Tris, 1.92M glycine, 1% SDS) were used in the lower (1X) and upper (2X) chambers, respectively.

Gel imaging and analysis

Labelled gels were scanned using a Typhoon TRIO (GE Healthcare) and Cy2, Cy3 and Cy5 images acquired using 520BP40, 580BP30 and 670BP30 laser emission filters, respectively, at 500 PMT and 100 μm resolution. Images were cropped to remove extraneous areas prior to analysis (ImageQuantTL, GE Healthcare), and image analysis performed using DeCyder V7.0 (GE Healthcare). The estimated number of spots for each co-detection procedure was set at 10,000 and an exclusion filter was applied to remove spots with a volume lower than 30,000. Differential expression of protein spots was examined by two-way ANOVA at a significance level of 0.05. After verifying that significant spots were well matched across the gels, two pick lists were generated with a total of 22 and 45 spots for the diet and genotype factors, respectively (Figure 1; Additional files 3 and 4).

Spot picking and protein identification by peptide fragment fingerprinting

Four preparative gels were run under the conditions described above but with higher amounts of protein (600 and 1000 μg). They were stained with colloidal Coomassie and, whenever possible, spots were excised and sequenced in the Mass Spectrometry Laboratory ITQB-UNL (Oeiras, Portugal), where in-gel digestion and extraction of the proteins from the gel was performed, followed by micropurification, and peptides identified by mass spectrometry (using an Applied Biosystems (ABI) 4800 MALDI TOF/TOF Analyzer). The search engine MASCOT was then used to identify and confirm protein ID's from the peptide mass fingerprinting (PMF; MS) and peptide fragment fingerprinting (PFF; MS/MS) data.

ACKNOWLEDGEMENTS

This study was funded by the EU FP6 IP "AQUAMAX" (Sustainable Aquafeeds to Maximise the Health Benefits of Farmed Fish for Consumers; 016249–2). SM was supported by a Marie Curie Intra European Fellowship (FP7-PEOPLE-2007-2-1-IEF, Proposal N219667) and by AQUAGENOME Mobility and Resource Exchange Grants in 2008. The authors acknowledge Landcatch Ltd for provision of the Lean and Fat smolts and thank staff of Marine Harvest Ltd, Ardnish FTU, for assistance with fish husbandry and sample collection. Technical assistance from Jacquie Ireland in microarray hybridizations is deeply appreciated. We also acknowledge Mahaut de Vareilles for performing some of the protein identifications by mass spectrometry, and discussions on the use of DeCyder V7.0.

REFERENCES

1. Calder PC, Yaqoob P: Omega-3 polyunsaturated fatty acids and human health outcomes. Biofactors. 2009, 35: 266-272. 10.1002/biof.42.
2. FAO: The State of World Fisheries and Aquaculture (SOFIA) 2010. 2010, Food and Agriculture Organization of the United Nations, Rome
3. Tacon AGJ, Metian M: Global overview on the use of fish meal and fish oil in industrially compounded aquafeeds: Trends and future prospects. Aquaculture. 2008, 285: 146-158. 10.1016/j.aquaculture.2008.08.015.
4. Torstensen BE, Espe M, Sanden M, Stubhaug I, Waagbø R, Hemre G-I, Fontanillas R, Nordgarden U, Hevrøy EM, Olsvik P, Berntssen MHG: Novel production of Atlantic salmon (Salmo salar) protein based on combined replacement of fish meal and fish oil with plant meal and vegetable oil blends. Aquaculture. 2008, 285: 193-200. 10.1016/j.aquaculture.2008.08.025.
5. Bell JG, Pratoomyot J, Strachan F, Henderson RJ, Fontanillas R, Hebard A, Guy DR, Hunter D, Tocher DR: Growth, flesh adiposity and fatty acid composition of Atlantic salmon (Salmo salar) families with contrasting flesh adiposity: Effects of replacement of dietary fish oil with vegetable oils. Aquaculture. 2010, 306: 225-232. 10.1016/j.aquaculture.2010.05.021.
6. Jordal AE, Torstensen BE, Tsoi S, Tocher DR, Lall SP, Douglas SE: Dietary rapeseed oil affects the expression of genes involved in hepatic lipid metabolism in Atlantic salmon (Salmo salar L.). J Nutr. 2005, 135: 2355-2361.
7. Panserat S, Kolditz C, Richard N, Plagnes-Juan E, Piumi F, Esquerré D, Médale F, Corraze G, Kaushik S: Hepatic gene expression profiles in juvenile rainbow trout (Oncorhynchus mykiss) fed fishmeal or fish oil-free diets. Br J Nutr. 2008, 100: 953-967. 10.1017/S0007114508981411.
8. Panserat S, Hortopan GA, Plagnes-Juan E, Kolditz C, Lansard M, Skiba-Cassy S, Esquerre D, Geurden I, Médale F, Kaushik S, Corraze G: Differential gene expression after total replacement of dietary fish meal and fish oil by plant products in rainbow trout (Oncorhynchus mykiss) liver. Aquaculture. 2009, 294: 123-131. 10.1016/j.aquaculture.2009.05.013.
9. Morais S, Pratoomyot J, Taggart JB, Bron JE, Guy DR, Bell JG, Tocher DR: Genotype-specific responses in Atlantic salmon (Salmo salar) subject to dietary fish oil replacement by vegetable oil: a liver transcriptomic analysis. BMC Genomics. 2011, 12: 255-10.1186/1471-2164-12-255.
10. Morais S, Pratoomyot J, Torstensen BE, Taggart JB, Guy DR, Bell JG, Tocher DR: Diet x genotype interactions in hepatic cholesterol and lipoprotein metabolism in Atlantic salmon (Salmo salar) in response to replacement of dietary fish oil with vegetable oil. Br J Nutr. 2011, 106: 1457-1469. 10.1017/S0007114511001954.
11. Geay F, Ferraresso S, Zambonino-Infante JL, Bargelloni L, Quentel C, Vandeputte M, Kaushik S, Cahu CL, Mazurais D: Effects of the total replacement of fish-based diet with plant-based diet on the hepatic transcriptome of two European sea bass (Dicentrarchus labrax) half-sibfamilies showing different growth rates with the plant-based diet. BMC Genomics. 2011, 12: 522-10.1186/1471-2164-12-522.
12. Murray HM, Lall SP, Rajaselvam R, Boutilier LA, Blanchard B, Flight RM, Colombo S, Mohindra V, Douglas SE: A nutrigenomic analysis of intestinal response to partial soybean meal replacement in diets for juvenile Atlantic halibut, Hippoglossus hipploglossus, L. Aquaculture. 2010, 298: 282-293. 10.1016/j.aquaculture.2009.11.001.

13. Skugor S, Grisdale-Helland B, Refstie S, Afanasyev S, Vielma J, Krasnov A: Gene expression responses to restricted feeding and extracted soybean meal in Atlantic salmon (Salmo salar L.). Aquacult Nutr. 2011

14. Bell MV, Dick JR, Porter AEA: Pyloric ceca are a major site of 22:6n-3 synthesis in rainbow trout (Oncorhynchus mykiss). Lipids. 2003, 39: 39-44.

15. Fonseca-Madrigal J, Bell JG, Tocher DR: Nutritional and environmental regulation of the synthesis of highly unsaturated fatty acids and of fatty-acid oxidation in Atlantic salmon (Salmo salar L.) enterocytes and hepatocytes. Fish Physiol Biochem. 2006, 32: 317-328. 10.1007/s10695-006-9109-2.

16. Caballero MJ, Obach A, Rosenlund G, Montero D, Gisvold M, Izquierdo MS: Impact of different dietary lipid sources on growth, lipid digestibility, tissue fatty acid composition and histology of rainbow trout, Oncorhynchus mykiss. Aquaculture. 2002, 214: 253-271. 10.1016/S0044-8486(01)00852-3.

17. Caballero MJ, Izquierdo MS, Kjorsvik E, Montero D, Socorro J, Fernandez AJ, Rosenlund G: Morphological aspects of intestinal cells from gilthead seabream (Sparus aurata) fed diets containing different lipid sources. Aquaculture. 2003, 225: 325-340. 10.1016/S0044-8486(03)00299-0.

18. Olsen RE, Dragsnes BT, Myklebust R, Ringø R: Effect of soybean oil and soybean lecithin on intestinal lipid composition and lipid droplet accumulation of rainbow trout, Oncorhynchus mykiss Walbaum. Fish Physiol Biochem. 2003, 29: 181-192.

19. Olsen RE, Myklebust R, Kaino T, Ringø E: Lipid digestibility and ultrastructural changes in the enterocytes of Arctic char (Salvelinus alpinus L.) fed linseed oil and soybean lecithin. Fish Physiol Biochem. 1999, 21: 35-44. 10.1023/A:1007726615889.

20. Olsen RE, Myklebust R, Ringø E, Mayhew TM: The influences of dietary linseed oil and saturated fatty acids on caecal enterocytes in Arctic char (Salvelinus alpinus L.): a quantitative ultrastructural study. Fish Physiol Biochem. 2000, 22: 207-216. 10.1023/A:1007879127182.

21. Morais S, Edvardsen RB, Tocher DR, Bell JG: Transcriptomic analyses of intestinal gene expression of juvenile Atlantic cod (Gadus morhua) fed diets with Camelina oil as replacement for fish oil. Comp Biochem Physiol B. -

22. Olsvik PA, Torstensen BE, Berntssen MHG: Effects of complete replacement of fish oil with plant oil on gastrointestinal cell death, proliferation and transcription of eight genes' encoding proteins responding to cellular stress in Atlantic salmon Salmo salar L. J Fish Biol. 2007, 71: 550-568. 10.1111/j.1095-8649.2007.01521.x.

23. Teitelbaum JE, Allan Walker W: Review: the role of omega 3 fatty acids in intestinal inflammation. J Nutr Biochem. 2001, 12: 21-32. 10.1016/S0955-2863(00)00141-8.

24. Leaver MJ, Taggart JB, Villeneuve L, Bron JE, Guy DR, Bishop SC, Houston RD, Matika O, Tocher DR: Heritability and mechanisms of n-3 long chain polyunsaturated fatty acid deposition in the flesh of Atlantic salmon. Comp Biochem Physiol Part D Genomics Proteomics. 2011, 6: 62-69. 10.1016/j.cbd.2010.04.002.

25. Dale OB, Tørud B, Kvellestad A, Koppang HS, Koppang EO: From chronic feed-induced intestinal inflammation to adenocarcinoma with metastases in salmonid fish. Cancer Res. 2009, 69: 4355-4362. 10.1158/0008-5472.CAN-08-4877.

26. Burrells C, Williams PD, Southgate PJ, Crampton VO: Immunological, physiological and pathological responses of rainbow trout (Oncorhynchus mykiss) to increasing dietary concentrations of soybean proteins. Vet Immunol Immunopathol. 1999, 72: 277-288. 10.1016/S0165-2427(99)00143-9.

27. Kolditz CI, Paboeuf G, Borthaire M, Esquerré D, SanCristobal M, Lefèvre F, Médale F: Changes induced by dietary energy intake and divergent selection for muscle fat content in rainbow trout (Oncorhynchus mykiss), assessed by transcriptome and proteome analysis of the liver. BMC Genomics. 2008, 9: 506-10.1186/1471-2164-9-506.

28. Allendorf FW, Thorgaard GH: Tetraploidy and the evolution of salmonid fishes. Evolutionary Genetics of Fishes. Edited by: Turner BJ. 1984, Plenum Press, New York, 55-93.

29. Jump DB, Clarke SD, Thelen A, Liimatta M, Ren B, Badin M: Dietary polyunsaturated fatty acid regulation of gene transcription. Prog Lipid Res. 1996, 35: 227-241. 10.1016/S0163-7827(96)00007-0.

30. Nakamura MT, Nara TY: Gene regulation of mammalian desaturases. Biochem Soc Trans. 2002, 30: 1076-1079.

31. Wang Y, Botolin D, Christian B, Busik J, Xu J, Jump DB: Tissue-specific, nutritional, and developmental regulation of rat fatty acid elongases. J Lipid Res. 2005, 46: 706-715. 10.1194/jlr.M400335-JLR200.

32. Leaver MJ, Villeneuve LA, Obach A, Jensen L, Bron JE, Tocher DR, Taggart JB: Functional genomics reveals increases in cholesterol biosynthetic genes and highly unsaturated fatty acid biosynthesis after dietary substitution of fish oil with vegetable oils in Atlantic salmon (Salmo salar). BMC Genomics. 2008, 9: 299-10.1186/1471-2164-9-299.

33. Taggart JB, Bron JE, Martin SAM, Seear PJ, Høyheim B, Talbot R, Carmichael SN, Villeneuve LAN, Sweeney GE, Houlihan DF, Secombes CJ, Tocher DR, Teale AJ: A description of the origins, designs and performance of the TRAITS-SGP Atlantic salmon Salmo salar L. cDNA microarray. J Fish Biol. 2008, 72: 2071-2094. 10.1111/j.1095-8649.2008.01876.x.

34. Morais S, Monroig O, Zheng X, Leaver MJ, Tocher DR: Highly unsaturated fatty acid synthesis in Atlantic salmon: characterization of ELOVL5- and ELOVL2-like elongases. Mar Biotechnol. 2009, 1: 627-639.

35. van Schothorst EM, Flachs P, Franssen-van Hal NL, Kuda O, Bunschoten A, Molthoff J, Vink C, Hooiveld GJ, Kopecky J, Keijer J: Induction of lipid oxidation by polyunsaturated fatty acids of marine origin in small intestine of mice fed a high-fat diet. BMC Genomics. 2009, 10: 110-10.1186/1471-2164-10-110.

36. Leaver MJ, Boukouvala E, Antonopoulou E, Diez A, Favre-Krey L, Ezaz MT, Bautista JM, Tocher DR, Krey G: Three peroxisome proliferator-activated receptor isotypes from each of two species of marine fish. Endocrinology. 2005, 146: 3150-3162. 10.1210/en.2004-1638.

37. Rosen ED, Walkey CJ, Puigserver P, Spiegelman BM: Transcriptional regulation of adipogenesis. Genes Dev. 2000, 14: 1293-1307.

38. Schmid B, Rippmann JF, Tadayyon M, Hamilton BS: Inhibition of fatty acid synthase prevents preadipocyte differentiation. Biochem Biophys Res Commun. 2005, 328: 1073-1082. 10.1016/j.bbrc.2005.01.067.

39. Davidson MH: Mechanisms for the hypotriglyceridemic effect of marine omega-3 fatty acids. Am J Cardiol. 2006, 98: 27i-33i.

40. Mori T, Kondo H, Hase T, Tokimitsu I, Murase T: Dietary fish oil upregulates intestinal lipid metabolism and reduces body weight gain in C57BL/6J mice. J Nutr. 2007, 137: 2629-2634.

41. Madsen L, Frøyland L, Kryvi H, Auwerx J, Staels B, Berge RK: The mitochondrion is the principal target for nutritional and pharmacological control of plasma triglyceride. Lipids. 1999, 34 (Suppl): S167-

42. Kjær MA, Vegusdal A, Gjøen T, Rustan AC, Todorcević M, Ruyter B: Effect of rapeseed oil and dietary n-3 fatty acids on triacylglycerol synthesis and secretion in Atlantic salmon hepatocytes. Biochim Biophys Acta. 2008, 178: 112-122.

43. Jacobs MN, Covaci A, Schepens P: Investigation of selected persistent organic pollutants in farmed Atlantic salmon (Salmo salar), salmon aquaculture feed, and fish oil components of the feed. Environ Sci Technol. 2002, 36: 2797-2805. 10.1021/es011287i.

44. Bell JG, McGhee F, Dick JR, Tocher DR: Dioxin and dioxin-like polychlorinated biphenyls (PCBs) in Scottish farmed salmon (Salmo salar): effects of replacement of dietary marine fish oil with vegetable oils. Aquaculture. 2005, 243: 305-314. 10.1016/j.aquaculture.2004.10.016.

45. Takahashi M, Tsuboyama-Kasaoka N, Nakatani T, Ishii M, Tsutsumi S, Aburatani H, Ezaki O: Fish oil feeding alters liver gene expressions to defend against PPARα activation and ROS production. Am J Physiol Gastrointest Liver Physiol. 2002, 282: G338-G348.

46. Saera-Vila A, Benedito-Palos L, Sitjà-Bobadilla A, Nácher-Mestre J, Serrano R, Kaushik S, Pérez-Sánchez J: Assessment of the health and antioxidant trade-off in gilthead sea bream (Sparus aurata L.) fed alternative diets with low levels of contaminants. Aquaculture. 2009, 296: 87-95. 10.1016/j.aquaculture.2009.07.028.

47. Moore MJ, Mitrofanov IV, Valentini SS, Volkov VV, Kurbskiy AV, Zhimbey EN, Eglinton LB, Stegeman JJ: Cytochrome P4501A expression, chemical contaminants and histopathology in roach, goby and sturgeon and chemical contaminants in sediments from the Caspian Sea, Lake Balkhash and the Ily River Delta, Kazakhstan. Mar Pollut Bull. 2003, 46: 107-119. 10.1016/S0025-326X(02)00325-9.

48. Lee RF, Anderson JW: Significance of cytochrome P450 system responses and levels of bile fluorescent aromatic compounds in marine wildlife following oil spills. Mar Pollut Bull. 2005, 50: 705-723. 10.1016/j.marpolbul.2005.04.036.

49. Bacanskas LR, Whitaker J, Di Giulio RT: Oxidative stress in two populations of killifish (Fundulus heteroclitus) with differing contaminant exposure histories. Mar Environ Res. 2004, 58: 597-601. 10.1016/j.marenvres.2004.03.048.

50. Galar-Martínez M, Gomez-Olivan LM, Amaya-Chavez A, Razo-Estrada C, Garcia-Medina S: Oxidative stress induced on Cyprinus carpio by contaminants present in the water and sediment of Madín Reservoir. J Environ Sci Health, Part A: Tox Hazard Subst Environ Eng. 2010, 45: 155-160. 10.1080/10934520903425780.

51. Wood ZA, Schröder E, Robin Harris J, Poole LB: Structure, mechanism and regulation of peroxiredoxins. Trends Biochem Sci. 2003, 28: 32-40. 10.1016/S0968-0004(02)00003-8.

52. Tolosano E, Altruda F: Hemopexin: structure, function, and regulation. DNA Cell Biol. 2002, 21: 297-306. 10.1089/104454902753759717.

53. Burk RF, Hill KE, Motley AK: Selenoprotein metabolism and function: evidence for more than one function for selenoprotein P. J Nutr. 2003, 133 (5 Suppl 1): 1517S-1520S.

54. Kang YJ: Metallothionein redox cycle and function. Exp Biol Med. 2006, 231: 1459-1467.

55. Berntssen MH, Julshamn K, Lundebye AK: Chemical contaminants in aquafeeds and Atlantic salmon (Salmo salar) following the use of traditional- versus alternative feed ingredients. Chemosphere. 2010, 78: 637-646. 10.1016/j.chemosphere.2009.12.021.

56. Moret S, Conte LS: Polycyclic aromatic hydrocarbons in edible fats and oils: occurrence and analytical methods. J Chromatogr A. 2000, 882: 245-253. 10.1016/S0021-9673(00)00079-0.

57. Shimada T: Xenobiotic-metabolizing enzymes involved in activation and detoxification of carcinogenic polycyclic aromatic hydrocarbons. Drug Metab Pharmacokinet. 2006, 21: 257-276. 10.2133/dmpk.21.257.

58. Hellou J, Leonard J, Anstey C: Dietary exposure of finfish to aromatic contaminants and tissue distribution. Arch Environ Contam Toxicol. 2002, 42: 470-476. 10.1007/s00244-001-0038-x.

59. Yuen BB, Wong CK, Woo NY, Au DW: Induction and recovery of morphofunctional changes in the intestine of juvenile carnivorous fish (Epinephelus coioides) upon exposure to foodborne benzo[a]pyrene. Aquat Toxicol. 2007, 82: 181-194. 10.1016/j.aquatox.2007.02.010.

60. Zhang Y, Lu Y, Zhou H, Lee M, Liu Z, Hassel BA, Hamburger AW: Alterations in cell growth and signaling in ErbB3 binding protein-1 (Ebp1) deficient mice. BMC Cell Biol. 2008, 9: 69-10.1186/1471-2121-9-69.

61. Celis JE, Madsen P, Celis A, Nielsen HV, Gesser B: Cyclin (PCNA, auxiliary protein of DNA polymerase delta) is a central component of the pathway(s) leading

to DNA replication and cell division. FEBS Lett. 1987, 220: 1-7. 10.1016/0014-5793(87)80865-7.

62. Liu Z, Ahn JY, Liu X, Ye K: Ebp1 isoforms distinctively regulate cell survival and differentiation. Proc Natl Acad Sci USA. 2006, 103: 10917-10922. 10.1073/pnas.0602923103.

63. Sweet LI, Passino-Reader DR, Meier PG, Omann GM: Xenobiotic-induced apoptosis: significance and potential application as a general biomarker of response. Biomarkers. 1999, 4: 237-253. 10.1080/135475099230778.

64. Ramachandran A, Madesh M, Balasubramanian KA: Apoptosis in the intestinal epithelium: its relevance in normal and pathophysiological conditions. J Gastroenterol Hepatol. 2000, 15: 109-120. 10.1046/j.1440-1746.2000.02059.x.

65. Paclik D, Lohse K, Wiedenmann B, Dignass AU, Sturm A: Galectin-2 and −4, but not galectin-1, promote intestinal epithelial wound healing in vitro through a TGF-beta-independent mechanism. Inflamm Bowel Dis. 2008, 14: 1366-1372. 10.1002/ibd.20499.

66. Loser K, Sturm A, Voskort M, Kupas V, Balkow S, Auriemma M, Sternemann C, Dignass AU, Luger TA, Beissert S: Galectin-2 suppresses contact allergy by inducing apoptosis in activated CD8+ T cells. J Immunol. 2009, 182: 5419-5429. 10.4049/jimmunol.0802308.

67. Tsukamoto K, Miura F, Fujito NT, Yoshizaki G, Nonaka M: Long-Lived Dichotomous Lineages of the Proteasome Subunit Beta Type 8 (PSMB8) Gene Surviving More than 500 Million Years as Alleles or Paralogs. Mol Biol Evol. 2012, May 2. [Epub ahead of print]

68. Nandi D, Tahiliani P, Kumar A, Chandu D: The ubiquitin-proteasome system. J Biosci. 2006, 31: 137-155. 10.1007/BF02705243.

69. Widmer CC, Pereira CP, Gehrig P, Vallelian F, Schoedon G, Buehler PW, Schaer DJ: Hemoglobin can attenuate hydrogen peroxide-induced oxidative stress by acting as an antioxidative peroxidase. Antioxid Redox Signal. 2010, 12: 185-198. 10.1089/ars.2009.2826.

70. Shimizu S, Matsuoka Y, Shinohara Y, Yoneda Y, Tsujimoto Y: Essential role of voltage-dependent anion channel in various forms of apoptosis in mammalian cells. J Cell Biol. 2001, 152: 237-250. 10.1083/jcb.152.2.237.

71. Jeon YJ, Kim DH, Jung H, Chung SJ, Chi SW, Cho S, Lee SC, Park BC, Park SG, Bae KH: Annexin A4 interacts with the NF-kappaB p50 subunit and modulates NF-kappaB transcriptional activity in a Ca2+−dependent manner. Cell Mol Life Sci. 2010, 67: 2271-2281. 10.1007/s00018-010-0331-9.

72. Parag HA, Raboy B, Kulka RG: Effect of heat shock on protein degradation in mammalian cells: involvement of the ubiquitin system. EMBO J. 1987, 6: 55-61.

73. Raboy B, Sharon G, Parag HA, Shochat Y, Kulka RG: Effect of stress on protein degradation: role of the ubiquitin system. Acta Biol Hung. 1991, 42: 3-20.

74. Pirkkala L, Alastalo TP, Zuo X, Benjamin IJ, Sistonen L: Disruption of heat shock factor 1 reveals an essential role in the ubiquitin proteolytic pathway. Mol Cell Biol. 2000, 20: 2670-2675. 10.1128/MCB.20.8.2670-2675.2000.

75. Di Lullo GA, Sweeney SM, Korkko J, Ala-Kokko L, San Antonio JD: Mapping the ligand-binding sites and disease-associated mutations on the most abundant protein in the human, type I collagen. J Biol Chem. 2002, 277: 4223-4231. 10.1074/jbc.M110709200.

76. Lussier C, Sodek J, Beaulieu JF: Expression of SPARC/osteonectin/BM4O in the human gut: predominance in the stroma of the remodeling distal intestine. J Cell Biochem. 2001, 81: 463-476. 10.1002/1097-4644(20010601)81:3<463::AID-JCB1060>3.0.CO;2-L.

77. Layland J, Solaro RJ, Shah AM: Regulation of cardiac contractile function by troponin I phosphorylation. Cardiovasc Res. 2005, 66: 12-21. 10.1016/j.cardiores.2004.12.022.

78. Olivares-Hernández R, Usaite R, Nielsen J: Integrative analysis using proteome and transcriptome data from yeast to unravel regulatory patterns at post-transcriptional level. Biotechnol Bioeng. 2010, 107: 865-875. 10.1002/bit.22868.

79. National Research Council (NRC): Nutrient requirements of fish and shrimp. 2011, National Academy Press, Washington

80. Folch J, Lees M, Sloane-Stanley GH: A simple method for the isolation and purification of total lipides from animal tissues. J Biol Chem. 1957, 226: 497-509.

81. Christie WW: Lipid analysis. 2003, The Oily Press, Bridgewater, UK, 3

82. Brazma A, Hingamp P, Quackenbush J, Sherlock G, Spellman P, Stoeckert C, Aach J, Ansorge W, Ball CA, Causton HC, Gaasterland T, Glenisson P, Holstege FC, Kim IF, Markowitz V, Matese JC, Parkinson H, Robinson A, Sarkans U, Schulze-Kremer S, Stewart J, Taylor R, Vilo J, Vingron M: Minimum information about a microarray experiment (MIAME)-toward standards for microarray data. Nat Genet. 2001, 29: 365-371. 10.1038/ng1201-365.

83. Pfaffl MW, Horgan GW, Dempfle L: Relative expression software tool (REST) for group-wise comparison and statistical analysis of relative expression results in real-time PCR. Nucleic Acids Res. 2002, 30: e36-10.1093/nar/30.9.e36.

84. Zheng X, Tocher DR, Dickson CA, Bell JG, Teale AJ: Highly unsaturated fatty acid synthesis in vertebrates: new insights with the cloning and characterization of a Δ6 desaturase of Atlantic salmon. Lipids. 2005, 40: 13-24. 10.1007/s11745-005-1355-7.

85. Torstensen BE, Nanton DA, Olsvik PA, Sundvold H, Stubhaug I: Gene expression of fatty acid-binding proteins, fatty acid transport proteins (cd36 and FATP) and β-oxidation-related genes in Atlantic salmon (Salmo salar L.) fed fish oil or vegetable oil. Aquacult Nutr. 2009, 15: 440-451. 10.1111/j.1365-2095.2008.00609.x.

86. Rees CB, Li W: Development and application of a real-time quantitative PCR assay for determining CYP1A

transcripts in three genera of salmonids. Aquat Toxicol. 2004, 66: 357-368. 10.1016/j.aquatox.2003.10.004.

87. Kleveland EJ, Ruyter B, Vegusdal A, Sundvold H, Berge RK, Gjøen T: Effects of 3-thia fatty acids on expression of some lipid related genes in Atlantic salmon (Salmo salar L.). Comp Biochem Physiol B Biochem Mol Biol. 2006, 145: 239-248. 10.1016/j.cbpb.2006.07.012.

88. Olsvik PA, Kristensen T, Waagbø R, Rosseland BO, Tollefsen KE, Baeverfjord G, Berntssen MHG: mRNA expression of antioxidant enzymes (SOD, CAT and GSH-Px) and lipid peroxidative stress in liver of Atlantic salmon (Salmo salar) exposed to hyperoxic water during smoltification. Comp Biochem Physiol C Toxicol Pharmacol. 2005, 141: 314-323. 10.1016/j.cbpc.2005.07.009.

89. Takle H, McLeod A, Andersen O: Cloning and characterization of the executioner caspases 3, 6, 7 and Hsp70 in hyperthermic Atlantic salmon (Salmo salar) embryos. Comp Biochem Physiol B Biochem Mol Biol. 2006, 144: 188-198. 10.1016/j.cbpb.2006.02.006.

CHAPTER 12

FUNCTIONAL GENOMIC ANALYSIS OF THE IMPACT OF CAMELINA (CAMELINA SATIVA) MEAL ON ATLANTIC SALMON (SALMO SALAR) DISTAL INTESTINE GENE EXPRESSION AND PHYSIOLOGY

Tyler D. Brown[1], Tiago S. Hori[1], Xi Xue [1], Chang Lin Ye [2], Derek M. Anderson[2], Matthew L. Rise[1]

[1]Department of Ocean Sciences, Memorial University of Newfoundland, 1 Marine Lab Road, St. John's, NL A1C 5S7, Canada

[2] Department of Plant and Animal Sciences, Faculty of Agriculture, Dalhousie University, Truro, NS, Canada B2N 5E3

ABSTRACT

The inclusion of plant meals in diets of farmed Atlantic salmon can elicit inflammatory responses in the distal intestine (DI). For the present work, fish were fed a standard fish meal (FM) diet or a diet with partial replacement of FM with solvent-extracted camelina meal (CM) (8, 16, or 24 % CM inclusion) during a 16-week feeding trial. A significant decrease in growth performance was seen in fish fed all CM inclusion diets (Hixson et al. in Aquacult Nutr 22:615–630, 2016). A 4x44K oligonucleotide microarray experiment was carried out and significance analysis of microarrays (SAM) and rank products (RP) methods were used to identify differentially expressed genes between the DIs of fish fed the 24 % CM diet and those fed the FM diet. Twelve features representing six known transcripts and two unknowns were identified as CM responsive by both SAM and RP. The six known transcripts (including thioredoxin and ependymin), in addition to tgfb, mmp13, and GILT, were studied using qPCR with RNA templates from all four experimental diet groups. All six microarray-identified genes were confirmed to be CM responsive, as was tgfb and mmp13. Histopathological analyses identified signs of inflammation in the DI of salmon fed CM-containing diets, including lamina propria and sub-epithelial mucosa thickening, infiltration of eosinophilic granule cells, increased goblet cells and decreased enterocyte vacuolization. All of these were significantly altered in 24 % CM compared to all other diets, with the latter two also altered in 16 % CM compared with 8 % CM and control diet groups. Significant correlation was seen between histological parameters as well as between five of the qPCR analyzed genes and histological parameters. These molecular biomarkers of inflammation arising from long-term dietary CM exposure will be useful in the development of CM-containing diets that do not have deleterious effects on salmon growth or physiology

---

Citation: Brown TD, Hori TS, Xue X, Ye CL, Anderson DM, Rise ML. Functional Genomic Analysis of the Impact of Camelina (Camelina sativa) Meal on Atlantic Salmon (Salmo salar) Distal Intestine Gene Expression and Physiology. Marine Biotechnology (New York, N.y). 2016;18:418-435. doi:10.1007/s10126-016-9704-x.

INTRODUCTION

Fish products are an important component of many human diets, and the demand for marine products is increasing as the global population has grown past seven billion (FAO 2009). The increasing demand for fish products for human consumption has fueled a rapid increase in the production of the aquaculture industry and its continued growth is required to meet the global demand. In 2003, the global production from aquaculture of aquatic animals was approximately 39 million tonnes, while in 2012, this production had reached approximately 67 million tonnes; in this same time, the global production from capture fisheries remained relatively unchanged, at around 90 million tonnes per year (FAO 2012).

Atlantic salmon (*Salmo salar*) is a carnivorous species and, as such, aquaculture feeds for farmed Atlantic salmon typically include marine animal-based protein and oil sources (such as fish and crustacean) that rely heavily on the inputs of unsustainable fishery resources (Tacon et al. 2009). The supplies of these raw marine products are limited and include fish oil (FO) and fish meal (FM) from natural stocks of lower trophic level fish species such as anchovy and herring (Tacon et al. 2009). The global output of FO and FM is restricted as the populations of some natural fish stocks are declining due to over-exploitation (Tacon and Metian 2009). Therefore, sustainable, cost-effective substitutes for FM are currently being sought with many plant-based alternatives being investigated (Tacon and Metian 2008). Plant-based products are not typically part of the diet of wild Atlantic salmon, and the supplementation of feeds for Atlantic salmon with plant meals (PM) may lead to inflammation of the distal intestine (DI) (enteritis) (Chikwati et al. 2013; Marjara et al. 2012; Moldal et al. 2014; Overland et al. 2009; Sahlmann et al. 2013). This inflammation is typically characterized by the shortening of intestinal villi, a decrease in supranuclear vacuoles (SNV) in the intestinal epithelial cells, significant infiltration of inflammatory cells in the lamina propria (LP), an increase in the number of goblet cells (GC), and thickening of the sub-epithelial mucosa (SEM) (Baeverfjord and Krogdahl 1996). The presence of inflammation is preceded by changes in the expression of genes that are responsible for mediating inflammatory responses. This inflammatory response, paired with many other factors, leads to decreased digestibility of essential nutrients and a decrease in the overall growth performance of the fish (Chikwati et al. 2013; Marjara et al. 2012; Moldal et al. 2014; Overland et al. 2009; Sahlmann et al. 2013). The specific components of PMs that are responsible for this inflammatory response are not known for certain. However, many plant products contain antinutritional factors (Sahlmann et al. 2013), and these components may be responsible for the inflammatory response seen in the intestine. In camelina, the main antinutrients are glucosinolates, phytic acid, sinapine, and tannins (Russo and Reggiani 2012). Glucosinolates and their metabolites are known to cause impairment of growth and irritation of the gastrointestinal tract. Phytic acid and tannins have been shown to render essential minerals insoluble and inhibit digestive enzymes, respectively, which may also inhibit growth due to decreased utilization of vitamins and minerals (Amarowicz et al. 2010; Schlemmer et al. 2009). In Atlantic salmon, the hindgut is the main site of amino acid absorption in the digestive tract and, as such, any loss of function that may be caused by inflammation can lead to a decline in the growth performance of the fish (Lokka et al. 2013; Sire et al. 1981).

A novel source of plant-based protein that is currently being considered as a supplementation to FM is the meal from camelina (*Camelina sativa*). Camelina is a flowering oilseed crop of the Brassicaceae family with many characteristics that are favorable to agriculture. These include low fertilizer and nitrogen requirements and high resistance to biotic and abiotic stress (Ghamkhar et al. 2010). CM contains up to 45 % crude protein and is comparable in this regard to other oilseed crops such as canola and rapeseed (Acamovic et al. 1999; Frame et al. 2007; Hixson et al. 2016). CM also contains adequate levels of at least 18 amino acids, with nine of them being essential, and only requires supplementation of some limiting amino acids such as methionine and/or lysine (Cherian 2012; Zubr 2003).

Recently, a feeding trial was conducted to examine the growth performance of Atlantic salmon-fed diets with partial replacement of FM with CM. The current study uses some of the same diets in the same feeding trial as Hixson et al. (2016), which focused on tissue lipid and amino acid analyses. The current study examines the effect of CM-supplemented diets on DI transcript expression and histology. Functional genomics analyses of hindgut tissue samples were carried out using DNA microarrays and real-time quantitative polymerase chain reaction (qPCR). These techniques have been used in previous studies for nutrigenomic analyses to assess the impact of other PM inclusion on the hindgut of Atlantic salmon (Chikwati et al. 2013; Marjara et al. 2012; Moldal et al. 2014; Overland et al. 2009; Sahlmann et al. 2013). The current functional genomics study aimed to identify and validate putative inflammatory biomarker genes that are responsive to long-term exposure to CM-containing aquafeeds. The gene expression data were paired with histological data of the hindgut tissue to allow for a more thorough understanding of the effect of CM on the Atlantic salmon hindgut. Molecular biomarkers of chronic inflammation arising from this study will be useful in future research aimed at determining the maximum inclusion of CM that does not cause deleterious effects in the Atlantic salmon DI.

A study (Ye 2014) analyzing the effect of high oil residue CM, rather than solvent-extracted CM (SECM, which was used in the current study), on growth performance, carcass composition, and DI histology of Atlantic salmon smolts was recently carried out; in addition to using a different CM product, Ye (2014) is different from the current study in that it did not include gene expression assays, and the two studies used distinct methods for histological analyses. Another recent study (Ye et al. 2016) evaluated the effects of SECM-containing diets on Atlantic salmon parr using carcass composition analysis and the same semi-quantitative histological methods as in Ye (2014).

MATERIALS AND METHODS

Diet Formulation

Camelina (Calena cultivar) was grown and harvested by the staff of the Dalhousie Agricultural Campus in Canning, NS, Canada. The seeds were cold pressed using a KEK 0500 screw press at Atlantic Oilseed Processing (Summerside, PEI,

Canada). The resulting meal was ground with a hammer mill (screen size 8 mm) into a pre-pressed meal cake at Atlantic Oilseed Processing, and solvent extracted with petroleum ether at a concentration of 3 mL g^{-1} at the Faculty of Agriculture Campus, Dalhousie University (Truro, NS, Canada). The solvent-extracted CM used in this study contained 39 % crude protein and had a gross energy content of 4,320 kcal/kg (Hixson et al. 2016; Ye 2014).

All diets in this trial were formulated at the Dalhousie Agricultural Campus to be iso-nitrogenous, iso-energetic practical diets and were formulated to meet the nutritional requirements of Atlantic salmon (National Research Council 2011). The experimental diets included a control diet with 34.9 % FM; and CM-containing diets with partial replacement of FM with 8 % CM (8CM); 16 % CM (16CM), or 24 % CM (24CM). The diet formulations are outlined in Table 1.

Table 1 Formulation and composition of control and experimental diets

Ingredients	0CM	8CM	16CM	24CM
Wheat gluten meal	15	15	15	15
Empyreal 75®	5.0	5.0	5.0	5.0
D/L methionine	0.17	0.17	0.17	0.17
Vitamin mineral premix[a]	0.2	0.2	0.2	0.2
Antioxidant/pigment premix	0.25	0.25	0.25	0.25
Choline chloride	0.5	0.5	0.5	0.5
Whey	5.0	5.0	5.0	5.0
Pregelatinized starch	2.5	2.5	2.5	2.5
Fish meal	34.9	34.2	29.9	27.4
Fish oil	14.0	15.7	17.3	18.9
Wheat middlings	22.4	15.3	8.2	1.0
Camelina meal	0	8.0	16	24

Experimental Animals

A feeding trial experiment was conducted in accordance with the regulations set out by the Canadian Council of Animal Care for the ethical treatment of animals (Memorial University animal care protocol: 12-50-MR). This feeding trial was carried out in the Dr. Joe Brown Aquatic Research Building (JBARB) at the Ocean Sciences Centre (OSC) of Memorial University of Newfoundland (Canada) using Atlantic salmon smolts (initial weight 242.1±46.0 g; initial length, 27±1.8 cm) from the Saint John River stock. The fish were received from Cooke Aquaculture (St. Alban's, NL, Canada), where they were held in freshwater, and were transferred to seawater at the JBARB to undergo smoltification. Six hundred fish were randomly distributed into twelve 500-L tanks (50 fish per tank) with flow-through water supply at a rate of 12 L min^{-1}(temperature, 14 °C; dissolved oxygen, 10 mg L^{-1}) and all fish were kept on a photoperiod of 12-h light and 12-h dark. Fish were fed a commercial diet (Nutra Transfer NP, 3 mm, Skretting Canada, St. Andrews, NB, Canada) during a 1-week acclimation period. Once the fish were acclimated, they were gradually moved onto the control diet over a period of 3 days. Initial sampling was done 1 week after the fish were completely weaned from the commercial diet. Subsequently, fish were weaned onto their assigned experimental diet (control, 8CM, 16CM, or 24CM) over another period of 3 days. Fish were fed the experimental diets to apparent satiety twice daily in triplicate tanks over a period of 16 weeks.

Tissue Sampling

At each sampling time point fish were removed from their tank using a net and over-anesthetized in 400 mg L^{-1} tricaine-methane-sulfonate bath (TMS; Syndel Laboratories, Vancouver, BC, Canada). Three fish per tank were dissected by removing the total viscera, and the DI was removed. For each fish, an ~2-cm-long section was removed from the middle of the DI; the distal-most ~0.5-cm portion was flash-frozen for RNA isolation (i.e., placed in RNase-free 1.5-mL microcentrifuge tubes, flash-frozen in liquid nitrogen and placed on dry ice until they could be moved to a −80 °C freezer for long-term storage), and the remainder was fixed for histological analyses (i.e., stored in 10 % formalin at room temperature). Further details on the rearing conditions and sampling can be found in Hixson et al. (2014, 2016), which focused on lipid analyses of diets and tissues and did not include gene expression analyses.

RNA Extraction, DNAse-I Treatment, and Column Purification

Frozen tissue samples were homogenized using ceramic mortars and pestles, which were previously cleaned with bleach and baked at 220 °C for 5 h to inactivate RNases. Homogenization was done under liquid nitrogen; after the addition of TRIzol reagent (Invitrogen, Carlsbad, CA, USA), samples were further homogenized using QIAshredders (QIAGEN, Mississauga, ON, Canada) and RNA was extracted following the manufacturers' instructions. Twenty micrograms of total RNA were treated with 6.8 Kunitz units DNaseI (RNase-free DNase kit, QIAGEN) to remove genomic DNA. DNaseI-treated RNA samples were column-purified using the RNeasy MinElute Cleanup Kit (QIAGEN) following the manufacturer's protocol. RNA integrity and purity were assessed using 1 % agarose gel electrophoresis, and A260/230 and A260/280 NanoDrop spectrophotometry, respectively.

Microarray Hybridization and Scanning

The experimental microarray design can be seen in Figure. 1. Three individuals from each of the three replicate tanks ($n=9$) from the control diet (no CM inclusion) and from the 24CM (high level CM inclusion diet) were used for microarray analysis with the 4x44K Atlantic salmon oligonucleotide array (Agilent Technologies, Mississauga, ON, Canada; Jantzen et al. 2011).

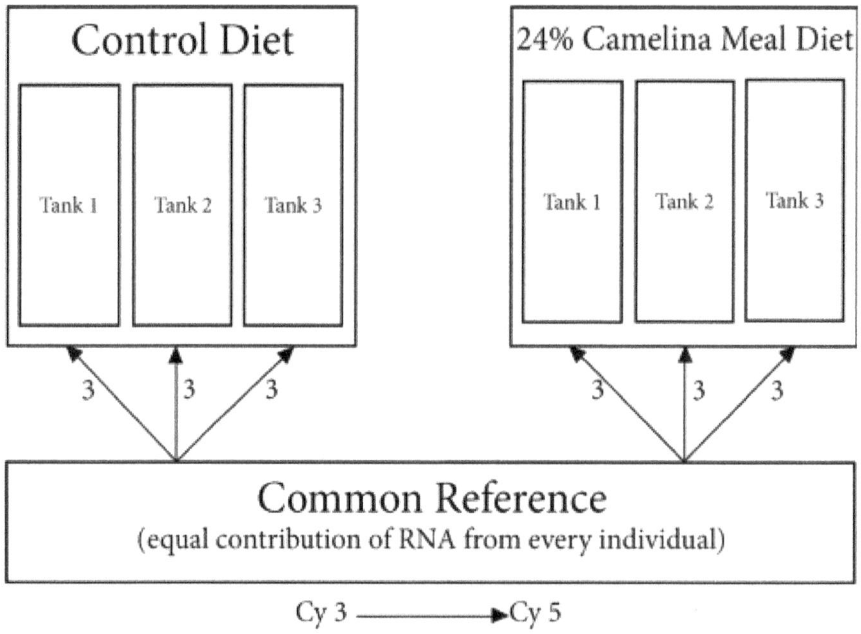

Figure. 1 Microarray experimental design. *Arrows* indicate arrays with number of biological replicates indicated by *numbers next to the arrows*. Eighteen arrays were used in total.

Microarray hybridization was carried out following the protocols outlined in Xue et al. (2015). Microarrays were hybridized using labeled antisense amplified RNA (aRNA) synthesized from 1 µg of DNaseI-treated, column-purified RNA that had satisfactory purity measurements (i.e., A260/230 and A260/280>1.8) and showed tight 18S and 28S ribosomal RNA bands (i.e., high integrity). A "common reference" sample was generated by pooling 5 µg of RNA from each of the 18 samples so that it included RNA from all of the samples analyzed. The concentration, purity, and integrity of the pooled RNA sample were checked as previously described. For each of the experimental samples or the common reference pool, 1 µg of DNaseI-treated, column-purified RNA was used to synthesize aRNA using the Amino Allyl MessageAmp II aRNA Amplification Kit (Life Technologies, Burlington, ON, Canada) following the manufacturer's instructions. Briefly, 1 µg of DNaseI-treated, column-purified RNA was reverse transcribed to cDNA which was used as a template for an in vitro transcription reaction using amino allyl modified dUTP. Twenty micrograms of aRNA were precipitated overnight, re-suspended in coupling buffer, and labeled with either Cy5 dye (experimental samples) or Cy3 dye (common reference) (GE Healthcare, Baied'Urfe, QC, Canada) following the manufacturer's instructions. Labeled aRNA from one experimental sample was pooled together with the common reference and, for each target, 825 ng of labeled aRNA was fragmented using 25X fragmentation buffer as well as 10X blocking agent (Agilent Technologies) and RNAse-free dH_2O. These were co-hybridized to a consortium for Genomic Research on All Salmonids Project (cGRASP)-designed Agilent 4x44K salmonid oligonucleotide array (GEO accession: GPL11299) (Jantzen et al. 2011; Sahlmann et al.2013) as per the manufacturer's instructions. Hybridizations were carried out at 65 °C for approximately 16 h with 10 rpm rotation in an Agilent hybridization oven (Jantzen et al. 2011). The arrays were washed immediately in Agilent Gene Expression Wash buffer I on a rocker platform for 5 min at room temperature, then washed in Agilent Gene Expression Wash Buffer II on a rocker platform for 5 min at 37 °C. The arrays were dried by centrifuging in 50-mL tubes at 200 rpm for 5 min at room temperature.

The arrays were immediately scanned using a ScanArrayGx Plus scanner (Perkin Elmer, Waltham, MA, USA) at a resolution of 5 µm. Photomultiplier tube (PMT) settings were adjusted for each array individually with the laser power set at 90 % for both channels (Cy3 and Cy5). Scanned array images were saved in TIFF format to be used for data extraction.

Scanned images were processed in the Imagene v9.0 software (BioDiscovery, El Segundo, CA, USA) using the cGRASP 4x44K GAL file to extract probe intensities. The extracted data were imported into the R statistical computing software where background correction, data transformation (Log_2), print-tip Loess normalization and removal of control and flagged (low-quality) spots were carried out. This was performed using the Bioconductor and mArray packages for R, using scripts adapted from those reported in Booman et al. (2011). After spot quality filtering, spots absent on more than 30 % of the arrays (i.e., 5 out of 18 arrays) were rejected, which resulted in a final list of 14,897 probes for analysis. The microarray data has been submitted to the Gene Expression Omnibus (GEO) under the accession number GSE68732.

Microarray Data Analyses

Prior to significance analysis of microarrays (SAM) and rank products (RP) analyses, missing data points for the 14,897 probes were imputed using the EM_array method from LSimpute (Bo et al. 2004; Chiu et al.2013). A reference design comparison was carried out between the control group and the 24CM group using the SAM algorithm (Tusher et al. 2001) to identify any genes that were significantly differentially expressed between the groups. SAM analysis was performed at a false discovery rate (FDR) threshold of 6.25 % using Bioconductor in the siggenes package for R. The resulting gene list was re-annotated using the expressed sequence tag (EST) information or contiguous sequence (contig) from which the 60mer oligonucleotide array probes were designed. Annotation was done manually with a BLASTX alignment against the NCBI non-redundant (nr) database using an E value threshold of 10^{-5}. To functionally annotate the microarray-identified transcripts, gene ontology (GO) terms were manually obtained using the putative ortholog from *Homo sapiens* (i.e., best human BLASTX hit) from the UniProt Knowledgebase (http://www.uniprot.org).

In order to confirm the results obtained using SAM and to potentially identify additional transcripts that were differentially expressed between 24CM and control groups, the microarray data were also analyzed using the RP method (Breitling et al. 2004). RP is a statistical approach whose ability to identify differentially expressed genes is less affected by high variability and low biological replication (Breitling et al. 2004; Jeffery et al. 2006). RP analysis was performed at a percentage of false-positives (PFP) threshold of 5 % using the Bioconductor package, RankProd (Hong et al. 2006). The resulting gene lists were annotated in Blast2GO (Conesa et al. 2005) using the BLASTX algorithm against the nr protein database of NCBI (2015.09.30). The best BLASTX hit that had an E value $< 10^{-5}$ and an informatively named protein product was chosen. Gene Ontology (GO) terms mapped to each microarray probe in Blast2GO were collected.

Real-Time Quantitative Polymerase Chain Reaction (qPCR)

Complimentary DNA (cDNA) was synthesized from 1 µg of column-purified total RNA in 20 µl reactions using 250 ng of random hexamers (Life Technologies), 1 µL of dNTPs (10 mM each), and Moloney-Murine Leukemia Virus (M-MLV) reverse transcriptase (Life Technologies) with the first strand buffer and Dithiothreitol (DTT) supplied by the manufacturer. Samples were incubated as follows: 65 °C for 5 min after the addition of the random hexamers and dNTPs, then 25 °C for 10 min and 37 °C for 50 min after the addition of the first-strand buffer, DTT, and M-MLV. cDNA was diluted 1:10 using nuclease-free water (Gibco, Life Technologies). Negative controls were performed by omitting the reverse transcriptase.

Candidate genes from the microarray analysis were subjected to qPCR to validate transcript levels. cDNA was synthesized from nine individuals (three from each replicate tank) from each of the four experimental diets (control, 8CM, 16CM and 24CM) for a total of 36 individuals. Primers used for qPCR analysis can be found in Table.2. All primers were quality checked (QC) before use. qPCR primer QC included running five-point 1:3 dilutions using pooled cDNA from all samples representing a starting quantity of 10 ng of input total RNA. This was done to determine the amplification efficiency of each primer pair ($E = 10^{[-1/\text{slope}]}$; Pfaffl 2001). Melt curves were also performed (+1 % increases over 30 s from 60 to 95 °C) to verify that only a single product was being amplified and that there was no dimerization of primers. This was also done to verify that there was no amplification in the no-template controls (i.e., to confirm the absence of contamination).

qPCR analysis was carried out using the ViiA7 real-time PCR system and SYBR Green I dye chemistry (Applied Biosystems, Foster City, CA, USA) with the following conditions: 50.0 °C for 5 min, pre-incubation at 95.0 °C for 10 min, and 40 cycles of 95 °C for 15 s and 60 °C for 1 min. Each amplification was performed using 13 µl reactions with 4 µl of input cDNA (representing 10 ng of input RNA), 1.46 µl of PCR-grade water (Life Technologies), 6.5 µl of 2x SYBR Green Mastermix (Applied Biosystems) and 0.52 µl each of the forward and reverse primers (1.25 µM concentration). Expression levels were normalized to *elongation factor 1 alpha* (*ef1a*) and *60S ribosomal rna subunit L32*. Five candidate normalizer genes were tested across 18 samples using the GeNorm software (Vandesompele et al. 2002) (BioGazelle, Zwijnaarde, Belgium) and *ef1a* and *60S ribosomal rna subunit L32* were selected as normalizer genes based on their high expression stability across all samples (GeNorm V<0.15, average GeNorm M<0.05). Expression analyses were carried out in 384-well format with each sample run in triplicate. The respective C_T values were determined using the ViiA7 Gene Expression Study Application (Applied Biosystems) with automated threshold determination and walking baseline. The relative starting quantity (RQ) was determined for each transcript using the $2^{-\Delta\Delta CT}$ method (Livak and Schmittgen 2001) using the sample with the lowest expression for each gene of interest as a calibrator. Amplification efficiencies (Table 2) were used to calculate each RQ value.

Histology

DI tissue was prepared from fish from the week-16 time point using standard histological techniques. Briefly, samples were dehydrated in ethanol, cleared using xylene and infiltrated with paraffin. Samples were embedded in paraffin blocks and 4 µm cross-sections were taken (i.e., across the lumen of the intestine). All slides were stained using hematoxylin and eosin, and blinded microscopic evaluations were carried out (i.e., fish were evaluated without knowing from which experimental group they came).

Table 2 Primers used for qPCR analysis of microarray-identified transcripts responsive to camelina meal-containing diets in Atlantic salmon

Gene name (symbol)		Primer sequence (5′–3′)	Sequence used for primer design	Amplification efficiency (%)	Amplicon size (bp)
ES1 protein homolog (*es1*)	F	ACTATAAAGGGAACAGGGTGCC	EB085110	89.3	167
	R	CAAGCTGGTCACCTCCTCTG			
transmembrane protease serine 9-like (*tmprss9*; alias: polyserase-I)	F	TAGTGCCGTTAGTGCTGGTG	EG876684	95.8	194
	R	CCAGACCTCTGTGGTGGAAT			
thioredoxin (*txn*)	F	GGATTCCTTCTTCAGTGCCC	BT049355	90.1	196
	R	GATGTCACAGTGTTTGGCCA			
pirin (*pir*; alias: probable quercetin 2,3-dioxygenase)	F	GAGGTCTAAGATCAGGGATTCC	NM_001141576	102.6	200
	R	GGAGGCACAGTGACAAAACA			
ependymin (*epd*)	F	TGACTGGAGCCATGTCAGTG	BT058166	99.4	125
	R	CAGGCCGAATGTCTTGTTCT			
N-acyl-phosphatidylethanolamine-hydrolyzing phospholipase D (*napepld*)	F	AAGACATCCAGGCCAGACAC	DY730770	99.9	126
	R	GCTCCACATTCAACCCATTC			
gamma-interferon-inducible lysosomal thiol reductase (*GILT*)	F	AAATGGGGAGCACACAGAAG	BT047373	100.2	105
	R	GCCGTACTACAGGCTTCAGG			
matrix metalloproteinase 13 (*mmp13*)	F	ACGATGACGAGTCCTTCAGC	DW539943	102.2	153
	R	AGGTGCTGGGGTTTGTGTAG			
transforming growth factor β3 (*tgfb*)	F	AGTTGCCTTGTGATTGTGGGA	EU082211	108.0	191
	R	CTCTTCAGTAGTGGTTTGTCG			
elongation factor 1 alpha-2 (*ef1a2*; normalizer)	F	TCGAGTGAGCGCACAGTAAC	GO064943	103.6	123
	R	CCATCTCAGCTGCTTCCTTC			
60S ribosomal subunit L32 (*rpl32*; normalizer)	F	AGGCGGTTTAAGGGTCAGAT	BT043656	102.4	119
	R	TCGAGCTCCTTGATGTTGTG			

Quantitative evaluations were performed to determine physiological changes in mucosal fold height, LP width, presence of GCs, infiltration of EGCs in the LP and SEM thickness. For mucosal fold height, measurements were taken from the tip of the mucosal fold to the bottom of the epithelial cells at the base of the fold. The LP width was measured at the top of the fold at the end of the LP, at the midpoint of the mucosal fold, and at the base of the mucosal fold, and an average was taken for each villus. GCs and EGCs were counted and normalized to the area of the mucosal fold and LP, respectively. Three fish were assessed from each tank with 10 simple villi from a single section being examined from each fish. SEM width was measured at three points per fish: 0°, 120°, and 240° around the lumen (three fish per tank, $n=9$). All quantitative measurements were made using the ImageJ image analysis software (Schneider et al. 2012). A semi-quantitative scoring system, described in Uran et al. (2008), was used to measure the degree of supranuclear vacuolization. Briefly, a scoring system ranging from 1 to 5 was used, with 1 being a normal intestinal state (i.e., high level of vacuolization in the enterocytes) and 5 being the most extreme inflammation state (i.e., complete extinction of vacuolization in the enterocytes).

Statistical Analyses

Statistical analysis of growth and histology data was carried out using Minitab 16 (Minitab Inc., State College, PA, USA). A two-way nested ANOVA was performed using the general linear model with a term that accounts for the effect of diet on the response variable and a term that nests tanks within diets to negate any tank effect or individual variance that may be present. Tukey's post hoc tests for multiple comparisons at a significance of 5 % (i.e., $p<0.05$) were carried out for pairwise comparisons.

RQ values obtained from the Applied Biosystems Gene Expression Study Application were statistically analyzed using GraphPad Prism version 6.0 (La Jolla, CA, USA). Data were tested for normality using the Shapiro–Wilks test and analyzed using a one-way ANOVA and Tukey's multi-comparison pairwise post hoc test. Statistically significant difference in RQ values between test groups was taken at 5 %.

Correlation between histopathological parameters (except villus height), or between histopathological parameters and transcript expression data, was evaluated using Pearson's correlation coefficients using MiniTab 16.

RESULTS

Growth Performance

The growth performance of the fish involved in this feeding trial was previously reported in Hixson et al. (2016), and the growth performance of fish fed the control diet (0CM) was also reported in Hixson et al. (2014); however, the results for the subset of diets that were used in this study are briefly described and reported herein as they are directly pertinent to the current study. The Atlantic salmon grew from an initial weight of 230–246 g to a final weight of 565–591 g. The fish groups fed any of the CM-containing diets in this study gained less weight and had a lower final weight than the fish group fed the control diet after the 16-week feeding trial was completed ($p=0.003$ and $p=0.001$, respectively). The apparent feed intake was decreased by the CM-containing diets compared to the FM control ($p=0.001$). However, fish fed all diets had comparable feed conversion ratios ($p=0.299$). The visceral somatic index (VSI) of fish fed the CM-containing diets also increased when compared to the fish fed the control diet ($p=0.001$) but there was no significant difference between the CM diet groups. The growth performance of the fish, as measured by the specific growth rate, remained unchanged by the CM-containing diets ($p=0.107$). Full growth performance results are reported in Table 3 and additional details about growth performance for the fish in this trial, as well as the other groups involved in this feeding trial (i.e., those that were not analyzed in this study) are reported in Hixson et al. (2014, 2016)

Table 3 Week 16 growth performance of Atlantic salmon fed a control diet or camelina meal (CM)-containing diets

Growth parameter	0CM	8CM	16CM	24CM	*p* value
Initial weight (g)[a]	230 ± 41	246 ± 62	250 ± 42	241 ± 49	0.637
Final weight (g)[a]	691 ± 153a	576 ± 152b	560 ± 129b	565 ± 117b	0.001
Weight gain (g)[b]	471 ± 39a	329 ± 72b	309 ± 45b	327 ± 17b	0.003
Initial length (cm)[a]	26.2 ± 2.4	26.9 ± 1.9	27.5 ± 1.5	26.8 ± 1.6	0.424
Final length (cm)[a]	35.0 ± 4.1	33.8 ± 2.8	33.4 ± 2.6	33.3 ± 2.9	0.081
SGR (%/day)[b]	0.99 ± 0.1	0.77 ± 0.2	0.72 ± 0.1	0.76 ± 0.01	0.107
CF[a]	1.53 ± 0.1	1.46 ± 0.1	1.48 ± 0.1	1.54 ± 0.5	0.106
VSI (%)[c]	9.8 ± 1.1a	10.8 ± 1.0b	11.2 ± 1.2b	11.4 ± 0.9b	0.001
AFI[d]	515 ± 7.6a	420 ± 57b	384 ± 33b	391 ± 20b	0.001
FCR[d]	1.01 ± 0.1	1.20 ± 0.2	1.16 ± 0.1	1.10 ± 0.1	0.299

Intestine Transcriptome Analysis

A microarray experiment utilizing nine biological replicates was carried out to identify transcripts responsive to a CM-containing diet (24CM) compared to a control diet in the DI of Atlantic salmon. SAM analysis (FDR 6.25 %) identified 16 differentially expressed features, with all being more highly expressed in the fish fed the 24CM diet; the fold change values of these microarray-identified genes ranged from 1.57 to 3.94. In total, eight different transcripts were represented by the SAM-identified features and six were annotated through BLASTx searches against the NCBI nr protein database (Table.4). RP detected a total of 136 differentially expressed features with a PFP threshold of 5 % (82 more highly expressed in the salmon fed 24CM diet and 54 more highly expressed in the control diet) (Electronic Supplementary Material (ESM), Supplemental Tables 1 and 2). Twelve of the 16 SAM-identified features were also represented in the RP lists (Table 4). The remainder of the Results, as well as the Discussion, will focus mainly on the genes that were identified as CM responsive using both SAM and RP (with reported fold change values calculated using SAM).

Table 4 Microarray-identified transcripts that were significantly upregulated in salmon fed the 24CM diet compared to salmon fed the control diet

Probe ID[a]	Fold change[b]	BLASTx identification[c]				Gene ontology (GO) of putative human ortholog[d]	
			Best named BLASTx hit [species]	**Accession no.**	***Evalue***		**% ID (AA)**
C011R096	3.33	ES1 protein like protein (Es1) [*Chelonia mydas*]	EMP32012	3e-12	27/39 (69 %)	Mitochondrion (CC)	
C215R002	2.59	PREDICTED transmembrane protease serine 9-like (Tmprss9) [*Stegastes partitus*]	XP_008289151	3e-29	71/133 (53 %)	Proteolysis (BP), catalytic activity (MF) serine-type endopeptidase activity (MF), hydrolase activity (MF), integral component of plasma membrane (CC)	
C227R083	2.46	Thioredoxin (Txn) [*Salmo salar*]	ACM09155	2e-68	107/108 (99 %)	Negative regulation of transcription from RNA polymerase II promoter (BP), Response to reactive oxygen species (BP), Glycerol ether metabolic process (BP), Movement of cell or subcellular component (BP), Signal transduction (BP), Cell proliferation (BP), Response to radiation (BP), Nucleobase-containing small molecule interconversion (BP), Activation of protein kinase B activity (BP), Positive regulation of peptidyl-serine phosphorylation (BP), Translocation (BP), Nucleotide-binding domain (BP), Leucine rich repeat containing receptor signaling pathway (BP), Positive regulation of DNA binding (BP), Innate immune response (BP), Cell redox homeostasis (BP), Negative regulation of hydrogen peroxide-induced cell death (BP), Protein binding (MF), Protein disulfide oxidoreductase activity (MF), Poly(A) RNA binding (MF), Extracellular region (CC), Nucleus (CC), Cytoplasm (CC), Mitochondrion (CC), Extracellular vesicular exosome (CC).	
C075R139	2.25						
C203R158	2.27						
C040R065	2.04						
C182R133	1.97						
C050R109	1.89						
C110R097	1.92						

Probe ID[a]	Fold change[b]	BLASTx identification[c]				Gene ontology (GO) of putative human ortholog[d]	
			Best named BLASTx hit [species]	**Accession no.**	***E*value**		**% ID (AA)**
C251R109	3.94	Pirin (Pir) [*Salmo salar*]	NP_001135048	7e-69	107/115 (93 %)	Transcription, DNA-templated (BP), myeloid cell differentiation (BP), monocyte differentiation (BP), oxidation–reduction process (BP), transcription cofactor activity (MF), protein binding (MF), quercetin 2,3-dioxygenase activity (MF), oxidoreductase activity (MF), metal ion binding (MF), nucleus (CC), cytoplasm (CC)	
C230R082	1.89	Ependymin (Epd) precursor [*Salmo salar*]	ACM09231	1e-62	111/112 (99 %)	Cell-matrix adhesion (BP), calcium ion binding (MF), extracellular region (CC), lysosome (CC), extracellular vesicular exosome (CC)	
C210R115	2.00						
C005R058	1.57						
C231R006	2.81	PREDICTED: N-acyl-phosphatidyl ethanolamine-hydrolyzing phospholipase D-like (Napepld) [*Lepisosteus oculatus*]	XP_006633527	2e-75	108/143 (76 %)	Lipid metabolic process (BP), phospholipase activity (MF), zinc ion binding (MF), hydrolase activity (MF), N-acylphosphatidylethanolamine-specific phospholipase D activity (MF), membrane (CC), photoreceptor outer segment membrane (CC), extracellular vesicular exosome (CC)	
C034R078	2.02	Unknown	N/A	N/A	N/A	N/A	
C202R142	3.88	Unknown	N/A	N/A	N/A	N/A	

The differentially expressed genes identified by the microarray included *ES1 protein homolog* (*es1*), which was identified by one common (i.e., identified by both SAM and RP analyses) feature that was 3.3-fold upregulated (Table 4). The only GO term associated with this gene was "mitochondrion". *Transmembrane protease serine 9* (*tmprss9*) was identified by one common differentially expressed feature (2.6-fold upregulated) and GO annotation included "proteolysis" and "integral component of plasma membrane".*Thioredoxin* (*txn*) was represented by five common upregulated features (1.9- to 2.5-fold) and had many functional annotations, including "cell proliferation", "innate immune responses", "cell redox homeostasis", and "negative regulation of hydrogen peroxide induced cell death". *Pirin (pir)* was represented by one common upregulated feature (3.9-fold) and was associated with GO terms including "myeloid cell differentiation", "monocyte differentiation" and "oxidoreductase activity". *Ependymin* (*epd*) was represented by one common upregulated feature (2.0-fold). The functional annotation for this gene included the GO terms "cell matrix adhesion" and "calcium ion binding". The last gene identified by both SAM and RP with a significant BLASTX hit was *N-acyl-phosphatidylethanolamine-hydrolyzing phospholipase D* (*napepld*), which was represented by one common upregulated feature (2.8-fold) and included the GO terms "phospholipase activity" and "zinc ion binding". Two features were identified by both SAM and RP that did not have significant BLASTX hits, meaning that their identities could not be determined for this study. These genes were represented by one common feature each and had fold changes of 2.0 and 3.9 (Table 4).

qPCR Validation

The expression of the six transcripts identified by both SAM and RP, and with significant BLASTX hits, was quantified using qPCR (Figure. 2a–f). This was done to validate the upregulation of the genes in the 24CM group and also to measure the expression of these genes in the fish fed the 8CM and 16CM diets. Five of the six genes (i.e., all except *napepld)* were validated to be responsive to the 24CM diet. All six genes were significantly upregulated in the fish fed the 16CM diet compared to the fish fed the control diet, and five out of the six genes (i.e., all except *tmprss9*) were upregulated in the fish fed the 8CM diet compared to those fed the control diet.

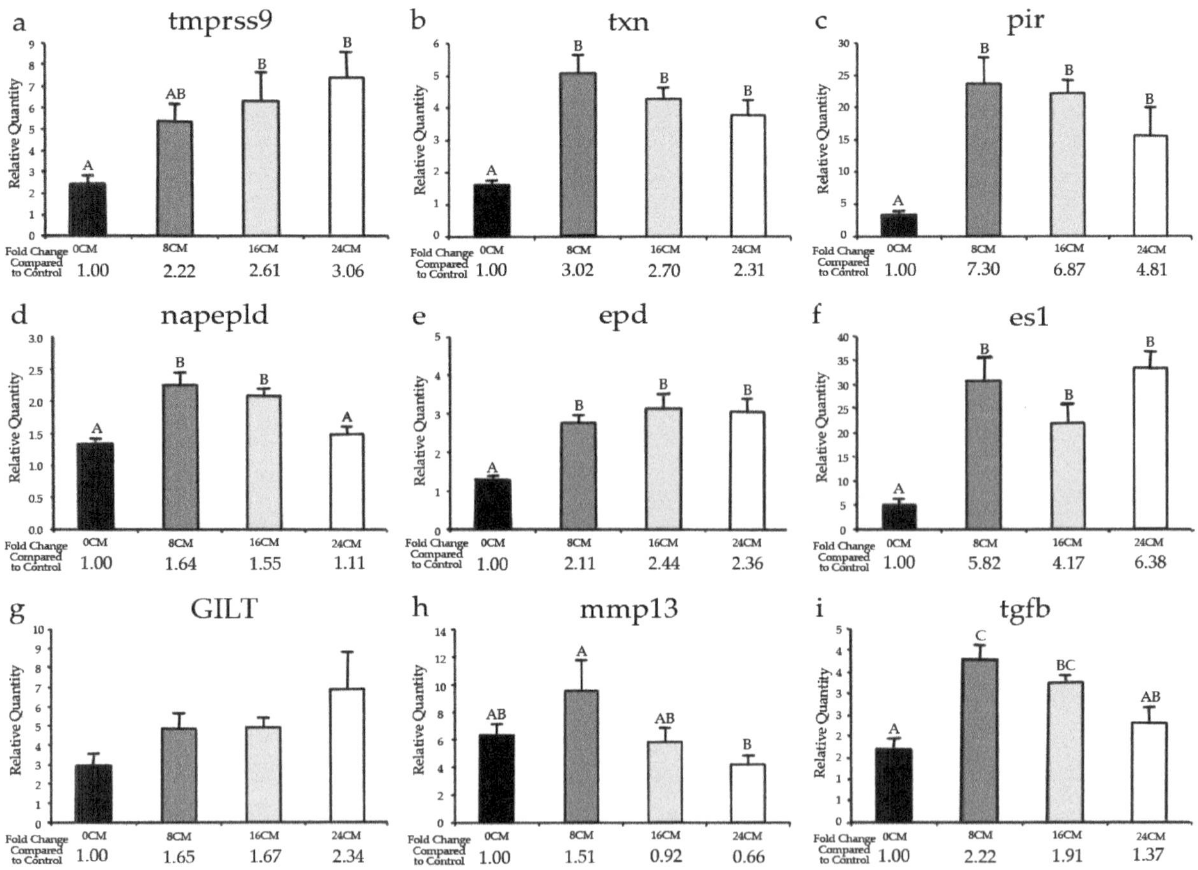

Figure. 2 qPCR validation of microarray-identified transcripts. Relative transcript levels were measured in the hindguts of fish fed the control diet and the camelina meal-containing test diets (8CM, 16CM, and 24CM) at week 16. Transcript relative quantity (RQ) ...

Additional qPCR assays were designed (see Table 2 for primers) for three inflammatory biomarker genes that are typically responsive during acute food-mediated enteritis in Atlantic salmon (Marjara et al. 2012; Sahlmann et al. 2013) and were not identified by the microarray in the current study. These genes include*gamma-interferon-inducible lysosomal thiol reductase* (*GILT*), *matrix metalloproteinase 13* (*mmp13*), and*transforming growth factor beta* (*tgfb*) (Figure. 2g–i). Of these additional genes, *tgfb* was significantly upregulated by the 8CM and 16CM diets compared with fish fed the control diet, and *mmp13* was significantly downregulated by the 24CM diet (p=0.050) compared with fish fed the 8CM diet (Figure. 2h, i).

Histological Analysis

Morphological changes in the Atlantic salmon DI were observed at the week 16 time point as a result of the dietary inclusion of CM and are reported in Table.5. Quantitative analyses indicate no significant change in the height of the villi; however, a significant widening of the LP was seen in the fish fed the 24CM diet compared to all other diets with no significant change in the LP width in the fish fed the 8CM or 16CM diets compared to control (p=0.001). There was a significant increase in the number of GCs per square millimeter in the DI of the fish fed the 24CM diet compared to all other diets and in fish fed 16CM diets compared to the 8CM diet and control (p=0.027). There was no significant change in the number of GCs per square millimeter seen in the fish fed the 8CM diet compared to the control diet. The counts of EGCs that had infiltrated into the LP showed a significant increase seen in the fish fed the 24CM diet compared to all other diet groups (p=0.004); however, there was no significant change in the number of EGCs that had infiltrated into the LP seen in the fish fed the 8CM or 16CM diet compared to the control group. Images showing increased EGCs and GCs can be seen in Figure. 3. It was also noted that the population of EGCs in the SEM appeared to increase in density with an increase in dietary CM inclusion, as seen in Figure. 4. In the fish fed the 16CM diet, there was a significantly decreased amount of SNV when compared to fish fed the control and 8CM diets and a significant decrease seen in SNV in fish fed the 24CM diet when compared to all other diets (p=0.001). This is illustrated with representative images in Figure. 3. The SEM was significantly thickened in fish fed the 24CM diet when compared to all other diets (p=0.046), although there was no notable change in the fish fed the other CM-containing diets compared to the control (Table,5, Figure. 4). It was also noted that many fused villi were seen in the fish fed the 24CM diet; although not quantified, villus fusion appeared to be much reduced in the fish fed the other experimental diets. All of these morphological changes are consistent with the presence of intestinal inflammation in Atlantic salmon.

Table 5 Quantitative[a] histopathological analysis of the DI of Atlantic salmon fed CM-containing diets for 16 weeks

	0CM	8CM	16CM	24CM	*F* stat	*p* value
Villus height (µm)	482.73 ± 193.12	544.79 ± 281.67	397.53 ± 174.77	454.51 ± 215.61	2.71	0.115
Lamina propria width (µm)	13.94 ± 8.05bc	17.14 ± 14.83b	11.89 ± 6.86c	24.56 ± 23.36a	29.00	0.001
SEM width (µm)	11.94 ± 4.45b	16.23 ± 6.08b	20.66 ± 10.34b	39.95 ± 25.10a	4.20	0.046
Eosinophilic granular cells (cells/mm^2)	42.26 ± 118.41b	75.52 ± 156.25b	204.73 ± 361.91b	550.89 ± 594.81a	9.95	0.004
Goblet cells (cells/mm^2)	347.95 ± 235.83c	306.41 ± 128.69c	495.70 ± 311.76b	793.23 ± 409.65a	5.24	0.027
Supranuclear vacuolization[a]	1.22 ± 0.42c	1.28 ± 0.45c	1.72 ± 0.60b	2.91 ± 0.65a	113.96	0.001

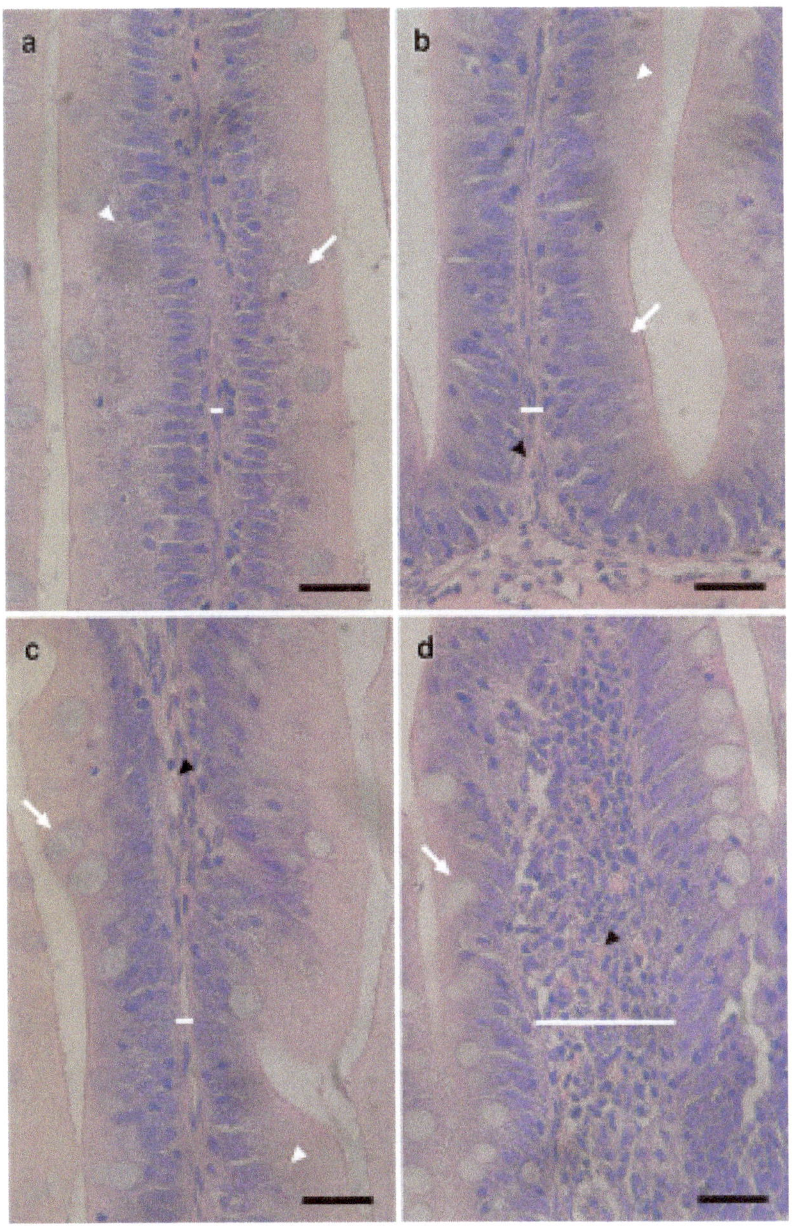

Figure. 3 Light micrographs of villi from the DI of Atlantic salmon showing histopathological changes in the DI of Atlantic salmon fed either a control FM-containing diet or one of an 8CM, 16CM, or 24CM diet. **a**0CM, **b** 8CM, **c** 16CM, **d** 24CM. *White lines* indicate lamina ...

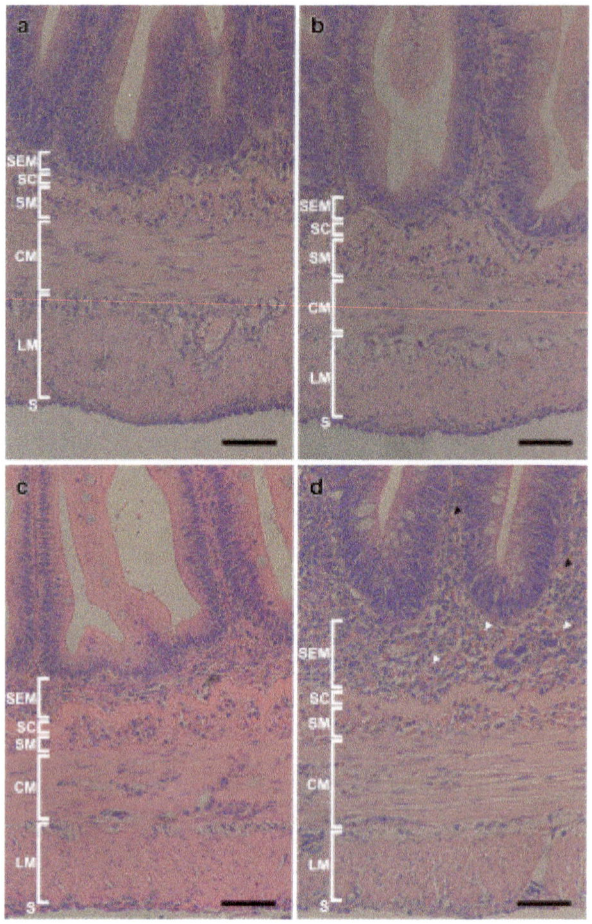

Figure. 4 Light micrographs of DI walls of Atlantic salmon fed either a control FM-containing diet or one of an 8CM, 16CM, or 24CM diet. **a** 0CM,**b** 8CM, **c** 16CM, **d** 24CM. *M*=mucosa, *SEM*=sub-epithelial mucosa,*SC*=stratum ...

Correlation between Tissue Gene Expression and Histopathological Parameters

Pearson's correlation analyses revealed significant ($p<0.05$) correlation between many of the histological parameters (Table 6). LP width correlated with SEM width, GCs and SNV; in addition, SEM width correlated with GCs and SNV, EGCs correlated with GCs and SNV, and GCs correlated with SNV (Table 6). Significant correlation was also seen between five of the qPCR analyzed genes and the histological data, four of which correlated with two different histological parameters (Table 7). *Es1* showed significant correlation with SEM width, *epd* correlated with SEM width and SNV ($p=0.050$), *tmprss9* correlated with LP width and SNV, *GILT* correlated with LP width and SNV, and *mmp13* correlated with GCs and SNV.

Table 6 Correlations among histopathological parameters

Histological parameter	Histological parameter	Pearson's r^a	p value
LP width	SEM width	0.350	0.036
LP width	EGC	0.259	0.127
LP width	GC	0.466	0.004
LP width	SNV	0.479	0.003
SEM width	EGC	0.259	0.127
SEM width	GC	0.400	0.016
SEM width	SNV	0.565	<0.001
EGC	GC	0.600	<0.001
EGC	SNV	0.613	<0.001
GC	SNV	0.652	<0.001

Table 7 Correlations among gene expression (RQ data) with histopathological parameters

Gene		LP width	SEM width	EGC	GC	SNV
pir	Pearson's r[a]	−0.059	−0.011	0.036	0.063	0.033
	p value	0.733	0.951	0.834	0.716	0.851
epd	Pearson's r	0.153	0.364	−0.024	0.207	0.329
	p value	0.374	**0.029**	0.888	0.226	**0.050**
tmprss9	Pearson's r	0.425	0.224	0.230	0.283	0.456
	p value	**0.010**	0.189	0.178	0.095	**0.005**
napepld	Pearson's r	−0.058	−0.153	−0.263	−0.216	−0.274
	p value	0.736	0.374	0.121	0.205	0.106
es1	Pearson's r	0.182	0.355	0.124	0.016	0.255
	p value	0.280	**0.034**	0.472	0.928	0.134
txn	Pearson's r	−0.079	0.183	−0.011	−0.051	0.151
	p value	0.645	0.286	0.948	0.769	0.378
GILT	Pearson's r	0.401	0.160	0.047	0.009	0.459
	p value	**0.015**	0.351	0.784	0.959	**0.005**
mmp13	Pearson's r	−0.061	−0.155	−0.165	−0.413	−0.331
	p value	0.722	0.366	0.335	**0.012**	**0.049**
tgfb	Pearson's r	−0.114	−0.017	−0.138	−0.300	−0.112
	p value	0.509	0.923	0.423	0.075	0.516

DISCUSSION

Growth Performance of Salmon

The epithelium of the DI is an important site of nutrient absorption in Atlantic salmon and plays a critical role in growth performance and the maintenance of inflammatory responses against food antigens (Chikwati et al.2013; Marjara et al. 2012; Moldal et al. 2014; Overland et al. 2009; Sahlmann et al. 2013). As previously reported (Hixson et al. 2016), Atlantic salmon fed CM-containing diets showed a significantly lower weight gain and higher VSI than the fish fed the control diet; however, the CM-fed fish also showed a decreased AFI. This may be a direct result of the inflammation in the intestine, and/or may be due to an aversion of the salmon to the diets containing CM. In this study, the lowest dietary inclusion of CM (8 %) was enough to cause a reduction in growth performance, with similar performance seen in the fish fed the 16CM and 24CM diets (Table 3). Other studies have shown that the inclusion of some PMs (such as lupin meal and wheat gluten) in feeds for Atlantic salmon led to decreased growth performance (Gu et al. 2014), while others showed that some PMs (such as pea meal) could be successfully included up to ~23 % without any apparent detriment to growth performance (Overland et al. 2009). Collectively, these data indicate that Atlantic salmon show different growth performance when fed different PMs. This may be due to the differences in antinutritional factors found in different types of PMs, with some of the main ANFs in CM being glucosinolates (Colombini et al. 2013). PMs from different sources also have different levels of digestibility, which affect the level of nutrient absorption and availability of amino acids needed for growth (Sugiura et al.1998).

Unlike the current study, Ye et al. (2016) reported no influence of SECM (up to 20 % of the diet) on feed consumption or weight gain of Atlantic salmon. However, Ye et al. (2016) used earlier life stage salmon (parr, ~ 8 g starting weight) compared with the current study (post-smolt, ~240 g starting weight). Collectively, these results suggest that earlier juvenile stages of Atlantic salmon may have a higher tolerance to dietary SECM compared with later juvenile stages.

The Transcriptional Effect of Dietary CM in the DI

Molecular Biomarkers of Chronic Inflammation

The microarray experiment carried out in the current study indicated that the dietary inclusion of 24 % CM resulted in the modulation of the expression of a relatively small number of transcripts related to inflammatory responses, and all of the BLASTX-annotated genes identified by both SAM and RP in the microarray experiment were qPCR validated. The three microarray-identified and qPCR validated genes that were positively correlated with physical signs of inflammation (*tmprss9*, *epd*, and *es1*) are discussed herein to elucidate the transcriptional effect of chronic exposure to dietary CM on inflammatory processes in the DI of Atlantic salmon.

Tmprss9 was significantly upregulated in the fish fed the 16CM and 24CM diets when compared to fish fed the control diet; this response appeared to be dose-dependent, with the highest degree of upregulation in fish fed the 24CM diet. This gene also showed significant positive correlation with LP width and SNV. *Tmprss9*, also commonly known as *polyserase-I*, was first characterized in Cal et al. (2003). It has been identified as a pro-tumor protease in epithelioid carcinomas (cell line PANC-1) and adenocarcinomas (cell line SK-PC-1) of the pancreas (Fontanil et al. 2014). Cancerous cells demonstrating *tmprss9* expression were found to have significant activation of the urokinase plasminogen activator (uPA) system, which is heavily involved in remodeling of the extracellular matrix (ECM) for the promotion of tumor progression and metastasis (Okumura et al. 2006; Ulisse et al. 2009; Fontanil et al. 2014). This gene and gene product may aid in ECM remodeling to promote a proinflammatory environment in response to the CM-containing diets.

Epd was upregulated by all of the CM-containing diets and with no significant difference among CM-fed groups. This gene was positively correlated with SEM width and SNV. *Epd* has been widely studied in the cerebrospinal fluid and central nervous system of teleost fish (Hoffman and Schwarz 1996; Shashoua 1977) and has been linked to neuroplasticity in many species (Pradel et al. 1999; Rother et al. 1995; Shashoua1991). The physiological role of *epd* outside of the nervous system is largely unknown, but studies have shown that a homolog of this gene is overexpressed during the regeneration of intestinal tissue in the gastrointestinal tract of some invertebrates (Suarez-Castillo et al. 2004; Zheng et al. 2006). There is also evidence that *epd* in teleost fish plays a role in the maintenance of the ECM in response to changing calcium levels (Ganss and Hoffmann 1993). This is significant for the study at hand as the physical signs of inflammation that were seen in the histological analyses are indicative of increased cell turnover and regeneration of damaged intestinal tissue. To our knowledge, this is the first study showing that *epd* is a molecular biomarker of chronic inflammation in the intestine of Atlantic salmon.

Es1 was significantly upregulated in all fish fed the CM-containing diets with no significant difference between those fed the 8CM, 16CM, or 24CM diets. Additionally, this gene was positively correlated with SEM width. The role of *es1* is not yet understood in humans, but it is proposed that it serves a basic function in the mitochondria (Shin et al. 2004). *Es1* has been shown to be upregulated in the caecum of chicken in response to infection with *Salmonella* (Rychlik et al. 2014), which suggests that *es1* may play a role in the development of immune responses in the vertebrate gut. While the role of *es1* remains largely undetermined, these results, paired with the positive correlation between the gene expression and histology results, collectively suggest it may have an important physiological role in the intestine involved in regulation of inflammatory responses. As with *tmprss9* and *epd*, *es1* appears to be a novel molecular biomarker of chronic inflammation in the salmon intestine.

Two of the qPCR analyzed genes that were not identified by the microarray were also shown to be responsive to CM inclusion. The first gene, *mmp13*, also known as *collagenase-3*, was significantly downregulated by the 24CM diet when compared to the 8CM diet. This gene has been shown to be responsive to numerous inflammatory conditions, including rheumatoid arthritis, wound healing, and inflammatory bowel disease, to name a few (Toriseva et al. 2012; Vizoso et al. 2006). The main function of*mmp13* is in the remodeling of the ECM during proliferation and during events involving high cell turnover, with chronic mucosal inflammation being one of the main conditions in which this gene is responsive (Uitto et al. 1998). The expression profile of *mmp13* (i.e., highest in 8CM-fed fish) and negative correlation of*mmp13* transcript expression with GCs and SNV, were not expected. In Sahlmann et al. (2013), *mmp13* was among the microarray-identified, upregulated differentially expressed genes (DEGs) at day 5 in a 20 % soybean meal (SBM) feeding trial, but was not differentially expressed at any other time points in the 1-week trial. Collectively, these results suggest that *mmp13* expression response to PMs may be dynamic and time-dependent. The potential roles of *mmp13* in PM-associated acute versus chronic inflammation of salmon DI warrant further investigation.

GILT transcript expression displayed significant positive correlation with LP width and SNV but did not show any differential expression between the fish fed the different diets. *GILT* has been shown to be heavily involved in many aspects of immunity; it is constitutively expressed in professional antigen presenting cells and can be induced in many other cells by inflammatory cytokines (Phan et al. 2000). GILT aids in the processing of antigenic proteins by reducing disulfide bonds to promote protein unfolding and presentation on MHC proteins for immune recognition (Maric et al. 2001; West and Cresswell 2013). In addition to this, increased *GILT* expression has been linked to decreased T cell activation in mouse models (Barjaktarevic et al. 2006). Although GILT was not significantly differentially expressed between fish fed the different diets, we hypothesize that it may be playing a role in the presentation of food antigens for immune recognition.

The RP analysis revealed four genes with GO terms directly related to "inflammatory response" that were not identified by SAM analysis (ESM, Supplemental Table 1). These genes were *dual oxidase 1* (4.7-fold upregulated), *tumor necrosis factor alpha induced protein 8-like protein 2* (2.4-fold upregulated), *b-cell lymphoma 6 protein homolog* (1.5-fold upregulated), and *c-c motif chemokine 19*, which was represented by two upregulated transcripts (both 2.4-fold upregulated).

Interestingly, RP analysis also identified 20 downregulated immunoglobulin (Ig) chain transcripts (ranging from 1.5-fold to 3.3-fold downregulated). Igs are present in many mucosal surfaces and are necessary for adaptive mucosal immunity (reviewed by Salinas et al. 2011). In fish, the mucosa-associated lymphoid tissue (MALT) contains B cells and various Igs, which are essential for the regulation of response to pathogens in the mucosa (reviewed by Brandtzaeg 2009). The downregulation of Ig encoding transcripts in the 24CM group was surprising, as one might expect Ig transcripts to be upregulated during an inflammatory response. Xue et al. (2015) reported potential immune suppression in Atlantic salmon fed a diet with 100 % replacement of fish oil with camelina oil, solvent-extracted FM, and 10 % CM. This conclusion was based on immune-relevant gene expression results. The downregulation of numerous Ig transcripts may suggest immune suppression in the fish fed the 24CM diet in the current study. Secreted Igs are involved in the maintenance of mucosal immunity (Salinas et al. 2011), and the downregulation of these genes may have influenced the homeostasis of the mucosa in the DI.

Other Genes Responsive to Long-Term Dietary CM Exposure

There were some transcripts that were identified by both SAM and RP to be responsive to CM and were qPCR validated, but did not correlate with any of the histological parameters (i.e., *txn*, *napepld*, and *pir*). It is likely that other cellular processes were altered in the CM fed fish, and that these genes may play roles in the modulation of these processes. Those genes are briefly discussed herein to elucidate possible further effects of CM outside of an inflammatory response.

Txn was upregulated in all of the fish fed CM-containing diets and there was no significant difference between the fish fed the 8CM, 16CM, or 24CM diets. Txn serves as a key reductase in one of the major reducing pathways in the intracellular environment (Collet and Messens 2010). The thioredoxin system is ubiquitous in biology and plays key roles in resistance to oxidative stress through control of the aggregation of reactive oxygen species (ROS), control of growth, and regulation of apoptosis (Arnér and Holmgren2000). It is possible that *txn* was upregulated as a protective response to oxidative stress resulting from the long-term dietary exposure to CM.

Pir was significantly upregulated in the fish fed all of the CM-containing diets when compared to those fed the control diet but did not correlate with any of the histological parameters. *Pir* is a highly conserved protein found in cell nuclei (Wendler et al. 1997). Its tight association with nuclear factor I has led to suggestions that it may serve as a potential cofactor and could be involved in biological redox reactions through its interaction with iron (Pang et al. 2004; Wendler et al. 1997). It is possible that, as with *txn*, *pir* was upregulated in response to oxidative stress induced by the long-term dietary exposure to CM.

Napepld was upregulated in the fish fed the 8CM and 16CM diets when compared to those fed the control diet, but no difference was seen in the fish fed the 24CM diet when compared to the control. *Napepld*encodes a key enzyme in the hydrolysis of many important lipid aldehydes, which generate numerous signaling molecules (Guo et al. 2013). These lipid mediators are involved in myriad cellular responses, including, most notably, innate immunity, energy balance, and response to cellular stress (Magotti et al.2015). The upregulation of *napepld* in this study may be a result of increased cellular stress because of the CM in the DI.

The gene expression analyses carried out in this study present a snapshot of the Atlantic salmon DI transcriptome response to long-term exposure to dietary CM. The dynamic transcriptome response of salmon intestine to PM-containing diets has been documented in various studies, including Sahlmann et al. (2013), which showed 48 DEGs in response to dietary SBM after 5 days of feeding, but only 5 DEGs after 7 days of feeding. Another study by Sahlmann et al. (2015) showed variable responses to dietary SBM, which peaked at 67 days post-hatching with lower responses from that point onward. This variability was also shown in Marjara et al. (2012), which looked at genomic responses to SBM and demonstrated significant changes in the expression of genes from one time point to another (**i.e.**, from days 3–7 to days 10–21). As shown in the aforementioned studies, the response to dietary PMs is very dynamic and changes significantly over time. In the current study, a high level of variability was also seen between fish, which may have limited our ability to identify DEGs. As a result, it is possible that other cellular processes were altered in response to long-term exposure to dietary CM that we were not able to detect in our analyses.

Histopathological Changes in the DI in Response to Dietary CM

Transcriptional modulation was paired with histopathological changes in the DI of salmon fed the CM-containing diets that are evidence of the manifestation of an inflammatory response. Quantitative measurements were taken in place of a conventional semi-quantitative scoring system for most histological parameters analyzed to increase the accuracy and sensitivity of the analyses carried out in detecting pathophysiological changes. Notable changes detected include an increase in the population of GCs in the epithelium of fish fed the 16CM and 24CM diets. GCs are responsible for the production and maintenance of the mucus layer that protects the epithelium of the intestine (Kim and Ho 2010). Increased mucus production caused by proliferation of GCs in the intestine is a protective response that can be triggered by food antigens, and an increased population of GCs is a direct indication of this protective response to inflammation (van den Ingh et al. 1991). Increased infiltration of EGCs was also observed in the villi of salmon fed the 24CM diet. EGCs are immune cells found in teleost fish that are functionally homologous to mammalian mast cells (Reite and Evensen 2006). EGCs, which are normally present in relatively high numbers in fish sub-epithelial tissues such as in the digestive tract, gills and skin (Reite 1997; Vergnolle 2004), mediate inflammatory responses through secreting a variety of immuno-active compounds that directly regulate the inflammatory response (Reite 1997). These compounds, which include granular proteases (e.g. serine proteases) and glycoaminoglycans (e.g. heparin) (Reite 1997), are released during degranulation and an increase in resident populations of EGCs has been observed during persistent inflammatory reactions in the intestines of salmonids (Bullock 1963; Sharp et al. 1989). The significant increase in the population of EGCs in the villi of fish fed the 24CM diet suggests that food antigens in the intestinal lumen are eliciting an inflammatory response that is mediated through EGCs.

Other histopathological changes seen in the intestine include a significant thickening of the SEM in fish fed the 24CM diet, which is consistent with the presence of inflammation and has been described in other studies investigating the effect of PMs on the physiology of the DI in salmon (Uran et al. 2008). There was also a widening of the LP seen in the fish fed the 24CM diet. The LP is a layer of connective tissue beneath the epithelium of the intestine and serves as an immune effector site in the intestine (Biorivant et al. 1999; Maxwell 1994). Stimulated T lymphocytes in the LP contribute to the maintenance of proinflammatory processes, and increased T lymphocyte populations are seen in LP widening resulting from inflammatory processes (Circu and Aw 2012). While not directly tested for, our transcriptional evidence suggests the presence of T lymphocytes in the LP through the elevated expression of *thioredoxin*, which is associated with the maintenance of T lymphocyte function

(Bennett and Griffiths 2013; Mougiakakos et al. 2011; Sido et al.2005).

A decrease in SNVs was observed in the fish fed the 16CM and 24CM diets which also provides evidence for an inflammatory response. SNVs are involved in the endocytosis of food nutrients and a decrease in SNVs may be triggered as a protective response against food antigens and ANFs that can cause inflammation (Baeverfjord and Krogdahl 1996). Interestingly, there were no changes observed in the height of the intestinal villi. Typically, inflammation in the DI is associated with shortening of the villi (Baeverfjord and Krogdahl1996).

Ye (2014) reported similar growth performance results as described in this study: decreased growth performance in the Atlantic salmon smolt fed the high oil residue CM diets with a significant decrease in the final weight and weight gain; the fish fed the CM diets also had significantly lower feed consumption but showed a significantly higher FCR. Ye (2014) also reported a significant increase in GCs in the DI of fish fed the high oil residue CM-containing diets, similar to GC results in the current study (i.e., higher numbers of GCs in fish fed the 16 and 24 % solvent-extracted CM-containing diets). Ye et al. (2016) examined the effect of solvent-extracted CM on the DI of Atlantic salmon parr and reported a significant widening of the LP in response to dietary exposure to 15 % CM- and 20 % CM-containing diets; this study, however, found no significant effect on SNV, GCs or SEM with up to 20 % CM inclusion. Collectively, these results support our previously mentioned hypothesis that Atlantic salmon parr may be more tolerant of CM-containing diets than Atlantic salmon post-smolts.

Many other studies have examined the effects of dietary PMs in the DI of Atlantic salmon. Sahlmann et al. (2013) reported that many immune-relevant transcripts were responsive to SBM inclusion, including *mmp13*and *NF B*. Sahlmann et al. (2013) also identified several other cellular processes that were responsive to SBM, including lipid metabolism, and cytoskeleton and ECM remodeling. This is similar to what was seen in Marjara et al. (2012), which presented several immune-relevant transcripts that were responsive to SBM. Some notable transcripts include *trypsin, PAR-2A/B, interleukin 17A*, and *GILT* (which was found to be non-responsive in the current study). Both of these papers reported histology results that are congruous with what is reported here. Other studies have also reported differential transcript expression and decreased growth performance in Atlantic salmon fed SBM (Skugor et al. 2011; Uran et al. 2008).

Other PMs and plant proteins have been evaluated in Atlantic salmon; decreased growth performance was reported for fish fed narrow leafed lupin and field pea meal (Carter and Hauler 2000), and Tacchi et al. (2012) showed mid-intestine transcriptional changes in response to dietary plant proteins. Overland et al. (2009), however, showed no significant change in growth performance in fish fed SBM or pea protein concentrate when compared to a FM diet.

CONCLUSIONS

The data presented in the current study demonstrate an inflammatory reaction in response to long-term dietary CM exposure in the DI of Atlantic salmon post-smolts. This conclusion was established based on decreased overall growth performance in response to dietary CM inclusion, which was paired with the modulation of the expression of some novel inflammation-relevant genes and histopathological changes associated with inflammation (i.e., LP and SEM widening, increased CGs and EGCs, and decreased SNV). Some of the identified genes correlated significantly with histopathological parameters and allowed for the identification of novel molecular biomarkers of inflammation (i.e., *tmprss9, epd*, and *es1*) that are responsive to long-term dietary CM exposure. While high levels of biological variability may have limited out ability to detect all of the genes that were responding to the CM-containing diets, we believe that this study provides valuable new information for understanding the long-term effects of dietary CM exposure in the DI of Atlantic salmon.

ACKNOWLEDGMENTS

The authors would like to thank the staff of the Dr. Joe Brown Aquatic Research Building (JBARB) for their assistance with animal holding and sampling; Dr. Marije Booman and Qingheng Xu for assistance with qPCR assay development, Charles Yu Feng for fish sampling, Dr. Christopher Parrish for assistance with experimental design, and Dr. Stefanie Hixson for assistance with experimental design and analysis of growth performance data. This study was supported by Genome Atlantic, Atlantic Canada Opportunities Agency (ACOA) Atlantic Innovation Fund (AIF), and a Research and Development Corporation of Newfoundland (RDC) Ocean Industries Student Research Award. Funding was also provided by a Natural Sciences and Engineering Research Council of Canada (NSERC) Discovery Grant and a Canada Research Chair to MLR.

REFERENCES

1. Acamovic T, Gilbert C, Lamb K, Walker KC. Nutritive value of *Camelina sativa* meal for poultry. Br Poult Sci. 1999;40:S27–S41. doi: 10.1080/00071669986657.
2. Amarowicz R, Estrella I, Hernandez T, Robredo S, Troszynska A, Kosinska A, Pegg RB. Free radical-scavenging capacity, antioxidant activity, and phenolic composition of green lentils (*Lens culinaris*) Food Chem. 2010;121:705–711. doi: 10.1016/j.foodchem.2010.01.009.
3. Arnér ES, Holmgren A (2000) Physiological functions of thioredoxin and thioredoxin reductase. Eur J Biochem 267:6102–6109
4. Baeverfjord G, Krogdahl A. Development and regression of soybean meal induced enteritis in Atlantic salmon, *Salmo salar* L., distal intestine: a comparison with the intestine of fasted fish. J Fish Dis.1996;19:375–387. doi: 10.1111/j.1365-2761.1996.tb00376.x.
5. Barjaktarevic I, Rahman A, Radoja S, Bogunovic B, Vollmer A, Vukmanovic S, Maric M. Inhibitory role of IFN-gamma-inducible lysosomal thiol reductase in T cell activation. J Immunol.2006;177:4369–4375. doi: 10.4049/jimmunol.177.7.4369.
6. Bennett SJ, Griffiths HR. Regulation of T-cell functions by oxidative stress. In: Alcaraz MJ, Gualillo O, Sánchez-Pernaute O, editors. Oxidative stress in applied basic

research and clinical practice.Heidelberg: Springer Science and Business Media; 2013.

7. Biorivant M, Marini M, Di Felice G, Pronio AM, Montesani C, Tersigni R, Strober W. Lamina propria T cells in Crohn's disease and other gastrointestinal inflammation show defective CD2 pathway-induced apoptosis. Gastroenterology. 1999;116:557–565. doi: 10.1016/S0016-5085(99)70177-0.

8. Bo TH, Dysvik B, Jonassen I. LSimpute: accurate estimation of missing values in microarray data with least squares methods. Nucleic Acids Res. 2004;32:e34. doi: 10.1093/nar/gnh026.

9. Booman M, Borza T, Feng CY, Hori TS, Higgins B, Culf A, Leger D, Chute IC, Belkaid A, Rise M, Gamperl AK, Hubert S, Kimball J, Ouellette RJ, Johnson SC, Bowman S, Rise ML. Development and experimental validation of a 20K Atlantic cod (*Gadus morhua*) oligonucleotide microarray based on a collection of over 150,000 ESTs. Mar Biotechnol. 2011;13:733–750. doi: 10.1007/s10126-010-9335-6.

10. Brandtzaeg P. Mucosal immunity: induction, dissemination, and effector functions. Scand J Immunol.2009;70:505–515. doi: 10.1111/j.1365-3083.2009.02319.x.

11. Breitling R, Armengaud P, Amtmann A, Herzyk P. Rank products: a simple, yet powerful, new method to detect differentially regulated genes in replicated microarray experiments. FEBS Lett.2004;573:83–92. doi: 10.1016/j.febslet.2004.07.055.

12. Bullock WL. Intestinal histology of some salmonid fishes with particular reference to the histopathology of acanthocephalan infections. J Morphol. 1963;112:23–43. doi: 10.1002/jmor.1051120104.

13. Cal S, Quesada V, Garabaya C, Lopez-Otin C. Polyserase-I, a human polyprotease with the ability to generate independent serine protease domains from a single translation product. Proc Natl Acad Sci U S A.2003;100:9185–9190.doi: 10.1073/pnas.1633392100.

14. Carter CG, Hauler RC. Fish meal replacement by plant meals in extruded feeds for Atlantic salmon,*Salmo salar* L. Aquaculture. 2000;185:299–311. doi: 10.1016/S0044-8486(99)00353-1.

15. Cherian G. *Camelina sativa* in poultry diets: oppotunities and challenges. In: Makkar HPS, editor. Biofuel co-products as livestock feed: opportunities and challenges. Rome: FAO; 2012.

16. Chikwati EM, Sahlmann C, Holm H, Penn MH, Krogdahl A, Bakke AM. Alterations in digestive enzyme activities during the development of diet-induced enteritis in Atlantic salmon, *Salmo salar* L.Aquaculture. 2013;402–403:28–37. doi: 10.1016/j.aquaculture.2013.03.023.

17. Chiu CC, Chan SY, Wang CC, Wu WS. Missing value imputation for microarray data: a comprehensive comparison study and a web tool. BMC Syst Biol. 2013;7(Suppl 6):S12. doi: 10.1186/1752-0509-7-S6-S12.

18. Circu ML, Aw TY. Intestinal redox biology and oxidative stress. Semin Cell Dev Biol. 2012;23:729–737. doi: 10.1016/j.semcdb.2012.03.014.

19. Collet JF, Messens J. Structure, function, and mechanism of thioredoxin proteins. Antioxid Redox Signal. 2010;13:1205–1216. doi: 10.1089/ars.2010.3114.

20. Colombini S, Broderick GA, Galasso I, Martinello T, Rapetti L, Russo R, Reggiani R. Evaluation of*Camelina sativa* (L.) Crantz meal as an alternative protein source in ruminant rations. J Sci Food Agric. 2013;94:736–743. doi: 10.1002/jsfa.6408.

21. Conesa A, Götz S, García-Gómez JM, Terol J, Talón M, Robles M. Blast2GO: a universal tool for annotation, visualization and analysis in functional genomics research. Bioinformatics. 2005;21:3674–3676. doi: 10.1093/bioinformatics/bti610.

22. FAO . The state of world fisheries and aquaculture 2008. Rome: Food and Agriculture Organization of the United Nations; 2009.

23. FAO (2012) Fisheries and aquaculture statistics. Food and Aquaculture Organization of the United Nations, Rome

24. Fontanil T, Mohamedi Y, Esteban MM, Obaya AJ, Cal S. Polyserase-1/TMPRSS9 induces pro-tumor effects in pancreatic cancer cells by activation of pro-uPA. Oncol Rep. 2014;31:2792–2796.

25. Frame DD, Palmer M, Peterson B. Use of *Camelina sativa* in the diets of young turkeys. J Appl Poult Res. 2007;16:381–386. doi: 10.1093/japr/16.3.381.

26. Ganss B, Hoffmann W. Calcium binding to sialic acids and its effect on the conformation of ependymins. Eur J Biochem. 1993;217:275–280. doi: 10.1111/j.1432-1033.1993.tb18243.x.

27. Ghamkhar K, Croser J, Aryamanesh N, Campbell M, Kon'kova N, Francis C. Camelina (*Camelina sativa* (L.) Crantz) as an alternative oilseed: molecular and ecogeographic analyses. Genome.2010;55:558–567.

28. Gu M, Kortner TM, Penn M, Hansen AK, Krogdahl A. Effects of dietary plant meal and soya-saponin supplementation on intestinal and hepatic lipid droplet accumulation and lipoprotein and sterol metabolism in Atlantic salmon (*Salmo salar* L.) Br J Nutr. 2014;111:432–444. doi: 10.1017/S0007114513002717.

29. Guo L, Gragg SD, Chen Z, Zhang Y, Amarnath V, Davies SS. Isolevuglandin-modified phosphatidylethanolamine is metabolized by NAPE-hydrolyzing phospholipase D. J Lipid Res.2013;54:3151–3157. doi: 10.1194/jlr.M042556.

30. Hixson SM, Parrish CC, Anderson DM. Full substitution of fish oil with camelina (*Camelina sativa*) oil, with partial substitution of fish meal with camelina meal, in diets for farmed Atlantic salmon (*Salmo salar*) and its effect on tissue lipids and sensory quality. Food Chem. 2014;157:51–61. doi: 10.1016/j.foodchem.2014.02.026.

31. Hixson SM, Parrish CC, Wells JS, Winkowski EM, Anderson DM, Bullerwell CN. Inclusion of camelina meal as a protein source in diets for farmed salmonids. Aquac Nutr. 2016;22:615–630. doi: 10.1111/anu.12276.

32. Hoffman W, Schwarz H. Ependymins: meningeal-derived extracellular matrix proteins at the blood–brain barrier. Int Rev Cytol. 1996;165:121–158. doi: 10.1016/S0074-7696(08)62221-4.

33. Hong F, Breitling R, McEntee CW, Wittner BS, Nemhauser JL, Chory J. RankProd: a bioconductor package for detecting differentially expressed genes in meta-analysis. Bioinformatics. 2006;22:2825–2827. doi: 10.1093/bioinformatics/btl476.

34. Jantzen SG, Sanderson DS, Von Schalburg KR, Yasuike M, Marass F, Koop BF. A 44K microarray dataset of the changing transcriptome in developing Atlantic salmon (*Salmo salar* L.) BMC Res Notes.2011;4:88. doi: 10.1186/1756-0500-4-88.

35. Jeffery IB, Desmond GH, Culhane AC. Comparison and evaluation of methods for generating differentially expressed gene lists from microarray data. BMC Bioinformatics. 2006;7:359. doi: 10.1186/1471-2105-7-359.

36. Kim YS, Ho SB. Intestinal goblet cells and mucins in health and disease: recent insights and progress.Curr Gastroenterol Rep. 2010;12:319–330. doi: 10.1007/s11894-010-0131-2.

37. Livak KJ, Schmittgen TD. Analysis of relative gene expression data using real-time quantitative PCR and the 2(−delta delta C(T)) method. Methods. 2001;25:402–408. doi: 10.1006/meth.2001.1262.

38. Lokka G, Austbo L, Falk K, Bjerkas I, Koppang EO. Intestinal morphology of the wild Atlantic salmon (*Salmo salar*) J Morphol. 2013;274:859–876. doi: 10.1002/jmor.20142.

39. Magotti P, Bauer I, Igarashi M, Babagoli M, Marotta R, Piomelli D, Garau G. Structure of human N-acylphosphatidylethanolamine-hydrolyzing phospholipase D: regulation of fatty acid ethanolamide biosynthesis by bile acids. Structure. 2015;23:598–604. doi: 10.1016/j.str.2014.12.018.

40. Maric M, Arunachalam B, Phan UT, Dong C, Garrett WS, Cannon KS, Alfonso C, Karlsson L, Flavell RA, Cresswell P. Defective antigen processing in GILT-free mice. Science. 2001;294:1361–1365. doi: 10.1126/science.1065500.

41. Marjara IS, Chikwati EM, Valen EC, Krogdahl A, Bakke AM. Transcriptional regulation of IL-17A and other inflammatory markers during the development of soybean meal-induced enteropathy in the distal intestine of Atlantic salmon (*Salmo salar* L.) Cytokine. 2012;60:186–196. doi: 10.1016/j.cyto.2012.05.027.

42. Maxwell WL. Wheater's functional histology. Edinburgh: Churchill Livingston; 1994.

43. Moldal T, Løkka G, Wiik-Nielsen J, Austbø L, Torstensen BE, Rosenlund G, Dale OB, Kaldhusdal M, Koppang OA. Substitution of dietary fish oil with plant oils is associated with shortened mid intestinal folds in Atlantic salmon (*Salmo salar*) BMC Vet Res. 2014;10:60. doi: 10.1186/1746-6148-10-60.

44. Mougiakakos D, Johansson CC, Jitschin R, Bottcher M, Kiessling R. Increased thioredoxin-1 production in human naturally occurring regulatory T cells confers enhanced tolerance to oxidative stress. Blood. 2011;117:857–861. doi: 10.1182/blood-2010-09-307041.

45. National Research Council . Nutrient requirements of fish and shrimp. Washington, DC: The National Academies Press; 2011.

46. Okumura Y, Hayama M, Takahashi E, Fujiuchi M, Shimabukuro A, Yano M, Kido H. Serase-1B, a new splice variant of polyserase-1/TMPRSS9, activates urokinase-type plasminogen activator and the proteolytic activation is negatively regulated by glycosaminoglycans. Biochem J. 2006;400:551–561. doi: 10.1042/BJ20060212.

47. Overland M, Sorensen M, Storebakken T, Penn M, Krogdahl A, Skrede A. Pea protein concentrate substituting fish meal or soybean meal in diets for Atlantic salmon (*Salmo salar*)—effect on growth performance, nutrient digestibility, carcass composition, gut health, and physical feed quality.Aquaculture. 2009;288:305–311. doi: 10.1016/j.aquaculture.2008.12.012.

48. Pang H, Bartlam M, Zeng Q, Miyatake H, Hisano T, Miki K, Wong LL, Gao GF, Rao Z. Crystal structure of human pirin: an iron-binding nuclear protein and transcription cofactor. J Biol Chem.2004;279:1491–1498. doi: 10.1074/jbc.M310022200.

49. Pfaffl MW. A new mathematical model for relative quantification in real-time RT-PCR. Nucleic Acids Res. 2001;29:e45. doi: 10.1093/nar/29.9.e45.

50. Phan UT, Arunachalam B, Cresswell P. Gamma-interferon-inducible lysosomal thiol reductase (GILT). Maturation, activity, and mechanism of action. J Biol Chem. 2000;275:25907–25914. doi: 10.1074/jbc.M003459200.

51. Pradel G, Schachner M, Schmidt R. Inhibition of memory consolidation by antibodies against cell adhesion molecules after active avoidance conditioning in zebrafish. J Neurobiol. 1999;39:197–206. doi: 10.1002/(SICI)1097-4695(199905)39:2<197::AID-NEU4>3.0.CO;2-9.

52. Reite OB. Mast cells/eosinophilic granule cells of salmonids: staining properties and responses to noxious agents. Fish Shellfish Immunol. 1997;7:567–584. doi: 10.1006/fsim.1997.0108.

53. Reite OB, Evensen O. Inflammatory cells of teleostean fish: a review focusing on mast cells/eosinophilic granule cells and rodlet cells. Fish Shellfish Immunol. 2006;2:192–208. doi: 10.1016/j.fsi.2005.01.012.

54. Rother S, Schmidt R, Brysch W, Schlingensiepen KH. Learning-induced expression of meningeal ependymin mRNA and demonstration of ependymin in neurons and glial cells. J Neurochem.1995;65:1456–1464. doi: 10.1046/j.1471-4159.1995.65041456.x.

55. Russo R, Reggiani R. Antinutritive compounds in twelve *Camelina sativa* genotypes. Am J Plant Sci.2012;3:1408–1412. doi: 10.4236/ajps.2012.310170.

56. Rychlik I, Elsheimer-Matulova M, Kyrova K. Gene expression in the chicken caecum in response to infections with non-typhoid Salmonella. Vet Res. 2014;45:119. doi: 10.1186/s13567-014-0119-2.

57. Sahlmann C, Sutherland BJ, Kortner TM, Koop BF, Krogdahl A, Bakke AM. Early response of gene expression in the distal intestine of Atlantic salmon (*Salmo salar* L.) during the development of soybean meal induced enteritis. Fish Shellfish Immunol. 2013;34:599–609. doi: 10.1016/j.fsi.2012.11.031.

58. Sahlmann C, Gu J, Kortner TM, Lein I, Krogdahl A, Bakke AM. Ontogeny of the digestive system of Atlantic salmon (*Salmo salar* L.) and effects of soybean meal from start-feeding. PLoS One.2015;10:e0124179. doi: 10.1371/journal.pone.0124179.

59. Salinas I, Zhang Y-A, Sunyer LO. Mucosal immunoglobulins and B cells of teleost fish. Dev Comp Immunol. 2011;35:1346–1365. doi: 10.1016/j.dci.2011.11.009.

60. Schlemmer U, Frolich W, Prieto RM, Grases F. Phytate in foods and significance for humans: food sources, intake, processing, bioavailability, protective role and analysis. Mol Nutr Food Res.2009;53(Suppl 2):S330–S375. doi: 10.1002/mnfr.200900099.

61. Schneider CA, Rasband WS, Eliceiri KW. NIH Image to ImageJ: 25 years of image analysis. Nat Methods. 2012;9:671–675. doi: 10.1038/nmeth.2089.

62. Sharp GJE, Pike AW, Secombes CJ. The immune response of wild rainbow trout, *Salmo gairdneri*Richardson, to naturally acquired plerocercoid infections of *Diphyllobothrium dendriticum* (Nitzsch, 1824) and *D. ditremum* (Creplin, 1825) J Fish Biol. 1989;35:781–793. doi: 10.1111/j.1095-8649.1989.tb03029.x.

63. Shashoua VE. Brain protein metabolism and the acquisition of new patterns of behavior. Proc Natl Acad Sci U S A. 1977;74:1743–1747. doi: 10.1073/pnas.74.4.1743.

64. Shashoua VE. Ependymin, a brain extracellular glycoprotein, and CNS plasticity. Ann N Y Acad Sci.1991;627:94–114. doi: 10.1111/j.1749-6632.1991.tb25916.x.

65. Shin JH, Weitzdoerfer R, Fountoulakis M, Lubec G. Expression of cystathionine beta-synthase, pyridoxal kinase, and ES1 protein homolog (mitochondrial precursor) in fetal Down syndrome brain.Neurochem Int. 2004;45:73–79. doi: 10.1016/j.neuint.2003.12.004.

66. Sido B, Giese T, Autschbach F, Lasitschka F, Braunstein J, Meuer SC. Potential role of thioredoxin in immune responses in intestinal lamina propria T lymphocytes. Eur J Immunol. 2005;35:408–417. doi: 10.1002/eji.200424500.

67. Sire MF, Lutton C, Vernier JM. New views on intestinal absorption of lipids in teleostean fishes: an ultrastructural and biochemical study in the rainbow trout. J Lipid Res. 1981;22:81–94.

68. Skugor S, Grisdale-Helland B, Afanasyev S, Vielma J, Krasnov A. Gene expression responses to restricted feeding and extracted soybean meal in Atlantic salmon (*Salmo salar* L.) Aquac Nutr.2011;17:505–517. doi: 10.1111/j.1365-2095.2010.00832.x.

69. Suarez-Castillo EC, Medina-Ortiz WE, Roig-Lopez JL, Garcia-Arraras J. Ependymin, a gene involved in regeneration and neuroplasticity in vertebrates, is overexpressed during regeneration in the echinoderm *Holothuria glaberrima*. Gene. 2004;9:133–143. doi: 10.1016/j.gene.2004.03.023.

70. Sugiura SH, Dong FM, Rathbone CK, Hardy RW. Apparent protein digestibility and mineral availabilities in various feed ingredients for salmonid feeds. Aquaculture. 1998;159:177–202. doi: 10.1016/S0044-8486(97)00177-4.

71. Tacchi L, Secombes CJ, Bickerdike R, Adler MA, Venegas C, Takle H, Martin SA. Transcriptomic and physiological responses to fishmeal substitution with plant proteins in formulated feed in farmed Atlantic salmon (*Salmo salar*) BMC Genomics. 2012;13:363. doi: 10.1186/1471-2164-13-363.

72. Tacon AJ, Metian M. Global overview on the use of fish meal and fish oil in industrially compounded aquafeeds: Trends and future prospects. Aquaculture. 2008;285:146–158. doi: 10.1016/j.aquaculture.2008.08.015.

73. Tacon AJ, Metian MM. Fishing for aquaculture: non-food use of small pelagic forage fish—a global perspective. Rev Fish Sci. 2009;17:305–317. doi: 10.1080/10641260802677074.

74. Tacon AJ, Metian M, Turchini GM, De Silva SS. Responsible aquaculture and trophic level implications to global fish supply. Rev Fish Sci. 2009;18:94–105. doi: 10.1080/10641260903325680.

75. Toriseva M, Laato M, Carpen O, Ruohonen ST, Savontaus E, Inada M, Krane SM, Kahari VM. MMP-13 regulates growth of wound granulation tissue and modulates gene expression signatures involved in inflammation, proteolysis, and cell viability. PLoS One. 2012;7:e42596. doi: 10.1371/journal.pone.0042596.

76. Tusher VG, Tibshirani R, Chu G. Significance analysis of microarrays applied to the ionizing radiation response. Proc Natl Acad Sci U S A. 2001;98:5116–5121. doi: 10.1073/pnas.091062498.

77. Uitto VJ, Airola K, Vaalamo M, Johansson N, Putnins EE, Firth JD, Salonen J, Lopez-Otin C, Saarialho-Kere U, Kahari VM. Collagenase-3 (matrix metalloproteinase-13) expression is induced in oral mucosal epithelium during chronic inflammation. Am J Pathol. 1998;152(6):1489–1499.

78. Ulisse S, Baldini E, Sorrenti S, D'armiento M. The urokinase plasminogen activator system: a target for anti-cancer therapy. Curr Cancer Drug Targets. 2009;9:32–71. doi: 10.2174/156800909787314002.

79. Uran PA, Schrama JW, Rombout JHWM, Obach A, Jensen L, Koppe W, Verreth JAJ. Soybean meal-induced enteritis in Atlantic salmon (*Salmo salar* L.) at different temperatures. Aquac Nutr.2008;14:324–330. doi: 10.1111/j.1365-2095.2007.00534.x.

80. Van Den Ingh TSGAM, Krogdahl A, Olli JJ, Hendriks HGCJM, Koninkx JGJF. Effects of soybean-containing diets on the proximal and distal intestine in Atlantic salmon (*Salmo salar*): a morphological study. Aquaculture. 1991;94:297–305. doi: 10.1016/0044-8486(91)90174-6.

81. Vandesompele J, De Preter K, Pattyn F, Poppe B, Van Roy N, De Paepe A, Speleman F. Accurate normalization

of real-time quantitative RT-PCR data by geometric averaging of multiple internal control genes. Genome Biol. 2002;3:RESEARCH0034. doi: 10.1186/gb-2002-3-7-research0034.

82. Vergnolle N. Modulation of visceral pain and inflammation by protease-activated receptors. Br J Pharmacol. 2004;141:1264–1274. doi: 10.1038/sj.bjp.0705750.

83. Vizoso FJ, Gonzalez LO, Corte MD, Corte MG, Bongera M, Martinez A, Martin A, Andicoechea A, Gava RR. Collagenase-3 (MMP-13) expression by inflamed mucosa in inflammatory bowel disease. Scand J Gastroenterol. 2006;41:1050–1055. doi: 10.1080/00365520600554667.

84. Wendler WM, Kremmer E, Forster R, Winnacker EL. Identification of pirin, a novel highly conserved nuclear protein. J Biol Chem. 1997;272:8482–8489. doi: 10.1074/jbc.272.13.8482.

85. West LC, Cresswell P. Expanding roles for GILT in immunity. Curr Opin Immunol. 2013;25:103–108. doi: 10.1016/j.coi.2012.11.006.

86. Xue X, Hixson SM, Hori TS, Booman M, Parrish CC, Anderson DM, Rise ML. Atlantic salmon (*Salmo salar*) liver transcriptome response to diets containing *Camelina sativa* products. Comp Biochem Physiol Part D Genomics Proteomics. 2015;14:1–15. doi: 10.1016/j.cbd.2015.01.005.

87. Ye C (2014) Evaluation of camelina (*Camelina sativa*) byproducts fed to Atlantic salmon (*Salmo salar*) in practical diets. M.Sc. Thesis. Dalhousie University, Truro, NS, Canada

88. Ye C, Anderson DM, Lall SP. The effects of camelina oil and solvent extracted camelina meal on the growth, carcass composition and hindgut histology of Atlantic salmon (*Salmo salar*) parr in freshwater. Aquaculture. 2016;450:397–404. doi: 10.1016/j.aquaculture.2015.08.019.

89. Zheng FX, Sun XQ, Fang BH, Hong XG, Zhang JX. Comparative analysis of genes expressed in regenerating intestine and non-eviscerated intestine of *Apostichopus japonicus* Selenka (Aspidochirotida: Stichopodidae) and cloning of ependymin gene. Hydrobiologia. 2006;571:102–122. doi: 10.1007/s10750-006-0231-z.

90. Zubr J. Dietary fatty acids and amino acids of *Camelina sativa* seed. J Food Qual. 2003;26:451–462. doi: 10.1111/j.1745-4557.2003.tb00260.x.

CHAPTER 13

GENOMIC PREDICTION OF HOST RESISTANCE TO SEA LICE IN FARMED ATLANTIC SALMON POPULATIONS

HsinYuan Tsai[1], Alastair Hamilton[2] , Alan E. Tinch[2] , Derrick R. Guy[2] , James E. Bron[3] , John B. Taggart[3] , Karim Gharbi[4] , Michael Stear[5] , Oswald Matika[1] , Ricardo PongWong[1] , Steve C. Bishop[1] and Ross D. Houston[1]

[1]The Roslin Institute and Royal (Dick) School of Veterinary Studies, The Univer sity of Edinburgh, Midlothian EH25 9RG, UK.

2 Landcatch Natural Selection Ltd., 15 Beta Centre, Stirling University Innovation Park, Stirling FK9 4NF, UK.

3 Insti tute of Aquaculture, University of Stirling, Stirling FK9 4LA, UK.

4 Edinburgh Genomics, Ashworth Laboratories, King's Buildings, University of Edinburgh, Edinburgh EH9 3JT, UK.

5 Institute of Biodiversity, Animal Health and Compara tive Medicine, University of Glasgow, Bearsden Road, Glasgow G61 1QH, UK.

ABSTRACT

Background

Sea lice have significant negative economic and welfare impacts on marine Atlantic salmon farming. Since host resistance to sea lice has a substantial genetic component, selective breeding can contribute to control of lice. Genomic selection uses genome-wide marker information to predict breeding values, and can achieve markedly higher accuracy than pedigree-based methods. Our aim was to assess the genetic architecture of host resistance to sea lice, and test the utility of genomic prediction of breeding values. Individual lice counts were measured in challenge experiments using two large Atlantic salmon post-smolt populations from a commercial breeding programme, which had genotypes for ~33 K single nucleotide polymorphisms (SNPs). The specific objectives were to: (i) estimate the herit ability of host resistance; (ii) assess its genetic architecture by performing a genome-wide association study (GWAS); (iii) assess the accuracy of predicted breeding values using varying SNP densities (0.5 to 33 K) and compare it to that of pedigree-based prediction; and (iv) evaluate the accuracy of prediction in closely and distantly related animals.

Results

Heritability of host resistance was significant (0.22 to 0.33) in both populations using either pedigree or genomic relationship matrices. The GWAS suggested that lice resistance is a polygenic trait, and no genome-wide significant quantitative trait loci were identified. Based on cross-validation analysis, genomic predictions were more accurate than pedigree-based

Citation: HsinYuan Tsai, Alastair Hamilton, Alan E. Tinch, Derrick R. Guy, James E. Bron, John B. Taggart, Karim Gharbi, Michael Stear, Oswald Matika, Ricardo PongWong, Steve C. Bishop and Ross D. Houston, Genomic prediction of host resistance to sea lice in farmed Atlantic salmon populations, DOI 10.1186/s12711-016-0226-9

predictions for both populations. Although prediction accuracies were highest when closely-related animals were used in the training and validation sets, the benefit of having genomic-versus pedigreebased predictions within a population increased as the relationships between training and validation sets decreased. Prediction accuracy reached an asymptote with a SNP density of ~5 K within populations, although higher SNP den sity was advantageous for cross-population prediction.

Conclusions

Host resistance to sea lice in farmed Atlantic salmon has a significant genetic component. Phenotypes relating to host resistance can be predicted with moderate to high accuracy within populations, with a major advan tage of genomic over pedigree-based methods, even at relatively sparse SNP densities. Prediction accuracies across populations were low, but improved with higher marker densities. Genomic selection can contribute to lice control in salmon farming.

BACKGROUND

Genomic selection (GS) involves the prediction of individual breeding values for complex traits by combining statistical methods with genome-wide single nucleotide polymorphism (SNP) data. Relationships between SNPs and traits of interest are first determined within a reference (or training) population, and then they are used to identify selection candidates with high genetic merit in the absence of phenotype records [1,2]. The feasibility of GS schemes depends on the availability of a high-quality SNP genotyping platform and on extensive trait records collected in the reference populations. Due to the increased availability of high-density SNP chips and the development of genotyping-by-sequencing for several economically important livestock and aquaculture species (e.g. [3–7]), GS has become a widely used approach, particularly for traits of economic and welfare importance (e.g. disease resistance). The accuracy of predicted breeding values based on genomic data is expected to be substantially higher than that based on pedigree records alone, but depends on many variables, including the genetic architecture of the trait, SNP density, sample size, and the degree of relationship between the reference and validation sets [8, 9].

In Atlantic salmon farming, ectoparasitic copepods, commonly known as sea lice (specifically*Lepeophtheirus salmonis* in Europe and *Caligus rogercresseyi* in Chile), are the primary threat to sustainable production, and have a negative economic, animal welfare, and environmental impact. Symptoms of *L. salmonis* infection include skin lesions, osmotic imbalance, and increased susceptibility to other infections as a result of host immune suppression and skin damage [10]. Frequent chemical treatments are required to control louse infections on commercial farms and result in large annual costs, potential environmental damage, and a high prevalence of drug-resistant lice [10, 11]. However, there is encouraging evidence from challenge trials that revealed heritabilities of approximately 0.2 to 0.3 for lice resistance, as measured by counts of lice on the fish (e.g. [11–14]), which highlights host genetic variation in resistance to lice. Therefore, selective breeding to improve host resistance to lice in farmed salmon populations is an increasingly important component of disease control [9, 11]. Given the importance of the sea lice issue to the salmon industry, this trait is also a high priority candidate for GS to accelerate the production of stocks with increased resistance.

The quantitative genetic models that underpin GS can be broadly split into two categories based on the assumptions that underlie the genetic architecture of the trait. The first category assumes an even distribution of the genetic variance across the genome and includes genomic best linear unbiased prediction (GBLUP) methods. The second category allows for heterogeneity in the contribution of the markers to the genetic variance, which is typically modelled using Bayesian methods (e.g. [15]). While the Bayesian methods (e.g. Bayes B) are generally more accurate than GBLUP on simulated data, particularly when the number of quantitative trait loci (QTL) that underlie the genetic variance is small [8], prediction accuracy using 'experimental' data in livestock breeding schemes is often very similar with either of these two methods [16]. Genomic prediction using these models relies both on capturing linkage disequilibrium (LD) between SNPs and QTL and on accurate estimates of realised genetic relationships between individuals [9, 17]. In typical farm animal populations, prediction accuracy depends largely on the latter [18], but the persistency of prediction accuracy across generations and between unrelated populations depends on the LD between SNPs and QTL [2, 9, 17]. For most commercial aquaculture breeding programmes, the availability of large full-sib families facilitates extensive trait measurements on individuals that are closely related to the selection candidates. Therefore, within-population genomic prediction will capitalise on realised genetic relationships, and the role of LD between SNPs and QTL may be less crucial [9, 18]. However, for salmon with a discrete 3-or 4-year generation interval, accuracy of prediction across adjacent year groups with limited genetic connectivity between them will depend more on LD, and is likely more challenging.

Family-based selective breeding programmes for Atlantic salmon have traditionally focused on economically important traits that can be easily measured on the selection candidates (e.g. growth) and on traits that can be measured on close relatives (e.g. full and half siblings), such as disease resistance and processing traits. Studies of GS in aquaculture using both simulated and 'experimental' data have suggested that genomic prediction can result in more accurate breeding values than traditional pedigree-based approaches (e.g. [9,19–21]). However, the cost-efficiency of GS is critical; both high-density SNP arrays and extensive collection of trait data can be prohibitively expensive for routine genomic evaluations. Therefore, knowledge of the optimal design of reference populations and of the required SNP density is important, as well as quantification of the benefit that can be expected from the implementation of GS.

The objectives of this study were to (i) estimate the heritability of host resistance to sea lice using both genomic and pedigree-based methods, (ii) analyse the genetic architecture of host resistance by performing a GWAS, (iii) assess the accuracy of genomic prediction using various SNP densities up to 33 K SNPs and compare it to that of pedigree-based prediction, and (iv) test genomic prediction accuracies in closely and more distantly related reference and validation populations.

METHODS

Animals and challenge experiment

The animals used in the study originated from a commercial Atlantic salmon breeding programme (Landcatch, UK). Due to the 4-year generation interval, the breeding program consists of four sub-populations (referred to as year groups), two of which were studied. Full details for population I (2007 year group, n = 624) were previously described in Tsai et al. [21]. Briefly, this population consisted of 531 genotyped offspring with complete phenotype and genotype information, derived from 61 nucleus families (30 sires and 59 dams). The families in population I were reared in separate tanks until approximately 9 months post-hatch, at which time they were mixed. Population II (2010 year group, n = 874) comprised 151 families (98 sires and 188 dams), with 588 offspring that were phenotyped and genotyped. The families in population II were mixed at first feeding and reared in a single common tank. The lice challenge trials were conducted at the Marine Environmental Research Laboratory (Machrihanish, UK) in 2007 and 2010, respectively. The challenge protocols were similar for both populations; the fish (approximately 1 year post-hatching) were challenged in a single tank with a moderate dose of copepodid larvae (90 to 96 larvae per fish) and then monitored daily until most lice had moulted into chalimus I. Sampling and measurements began on day 7 post-challenge and lasted 1 and 4.5 days for populations I and II, respectively (for population I, lice counts were shown to be stable between 7 and 17 days post-challenge [11]). Prior to lice counting, fish were euthanized with benzocaine as described in Gharbi et al. [11]. Phenotypes including weight (g), length (mm), and sea lice count [number of sea lice per fish, measured using a stereo-microscope (Olympus SZ-40)] were recorded for each fish. An adipose fin clip was collected and stored in ethanol for DNA extraction. For population I, pedigree information for each individual was traced by using passive integrated transponder (PIT) tags. For population II, a standard parentage assignment panel of 108 SNPs was screened on a Sequenom platform (DNA LandMarks Inc., Canada) to construct the pedigree.

All animals were reared in accordance with relevant national and EU legislation concerning health and welfare. The challenge experiment was performed by the Marine Environmental Research Laboratory (Machrihanish, UK) under approval of the ethics review committee of the University of Stirling (Stirling, UK) and according to Home Office license requirements. Landcatch are accredited participants in the RSPCA Freedom Foods standard, the Scottish Salmon Producers Organization Code of Good Practice, and the EU Code-EFABAR Code of Good Practice for Farm Animal Breeding and Reproduction Organizations.

SNP genotyping

DNA was extracted from fin tissue samples using the DNeasy 96 tissue DNA extraction kit (Qiagen, UK). Population I was genotyped with an Affymetrix Axiom SNP array that included ~132 K SNPs [22] and population II was genotyped with the custom Affymetrix Axiom ~35 K array described in Tsai et al. [21]. This 35 K array is used for routine genomic evaluations and includes a subset of high-quality SNPs of the 132 K array that were selected based on having a good distribution throughout the genome and minimal LD between pairs of SNPs [21]. Sex of the fish was predicted by using the Y-specific probes on the 132 K array, as described by Houston et al. [22]. Filtering of SNP data was performed using the Plink software [23], excluding SNPs with Mendelian errors, SNPs with a minor allele frequency (MAF) lower than 0.1 and SNPs with a proportion of missing genotypes greater than 0.03. Finally, approximately 33 K SNPs were retained for analyses in both populations.

Genetic parameters for lice resistance

Data normalization

The raw data for lice counts showed a positively skewed distribution (See Additional file 1: Figure S1), thus to normalize this distribution, we transformed the data using a previously applied approach that also accounts for an approximation of the surface area of the fish [13]:

$$\log_e LD = \log_e((LC + 1)/(BW)^{2/3}), \quad (1)$$

where LC is the number of lice counted on the fish (plus 1 to avoid a computation error since some fish may have zero lice), $(BW)^{2/3}$ is an approximation of the whole surface of the skin of each individual, where BWrepresents the body weight (g) at the time of the sea lice challenges. A moderate correlation of 0.35 was found between body surface and lice count.

Estimation of genetic parameters

The heritability of host resistance to sea lice count (and of weight and length traits) was estimated using both genomic and pedigree-based analyses for the two populations. Only fish with complete phenotype and genotype records were included, resulting in 531 and 588 fish in populations I and II, respectively. Heritabilities were estimated by ASReml 3.0 [24] using genomic and pedigree-based relationship matrices (**G**-matrix and **A**-matrix, respectively) with the following mixed model:

$$\mathbf{y} = \mu + \mathbf{Xb} + \mathbf{Za} + \mathbf{e}, \quad (2)$$

where **y** is a vector of observed phenotypes, μ is the overall mean of phenotype records, **b** is the vector of fixed effects, **a** is a vector of additive genetic effects distributed as ~N(0,Gσ2a) or N(0,Aσ2a) where σ2a is the additive (genetic) variance, **G** and **A** are the genomic and pedigree relationship matrices, respectively. **X** and **Z** are the corresponding incidence matrices for fixed and additive effects, respectively, and **e** is a vector of residuals. If the SNPs applying sex as the fixed effect did not surpass the genome-wide significance threshold (Bonferroni correction (0.05/N), where N represents the number of QC-filtered SNPs across the entire genome), it was omitted from subsequent analyses. The genomic relationship matrix was constructed by the Genabel R package [25] using the method of VanRaden [26] and then inverted by applying a standard R function (https://www.r-project.org/). Narrow sense heritability was estimated as the ratio of additive genetic variance to total phenotypic variance. Phenotypic correlations between traits were estimated using ASReml 3.0 [24] and genetic correlations were estimated using bivariate analyses implemented in ASReml 3.0 [24] as well.

Genome-wide association study

The two-step 'GRAMMAR' approach was used to perform the GWAS using the GenABEL R Package [25]. The GWAS was performed in each population separately, and on the two populations combined. First, model (2) was applied to adjust the lice count data based on fixed (year group in the combined analysis) and polygenic effects (relationships between animals as measured by the genomic relationship matrix). Subsequently, the mmscore method [27] of GenABEL was applied to measure the association between individual SNPs and the residuals from model (2) (which are corrected for family relatedness). Significance thresholds were calculated using a Bonferroni correction to obtain genome-wide (0.05/ number of all quality-control filtered SNPs, ~33 K) and chromosome-wide (0.05/number of SNPs on the corresponding chromosome) thresholds, respectively. For the SNPs that were closest to chromosome-wide significance (i.e. those with the lowest P values), allele substitution effects were estimated using model (2) in ASReml 3.0 [24] by including the fixed effects of SNP genotype and population. The additive effect (*a*) of the SNP was calculated as half the difference between the predicted phenotypic means of the two homozygotes, i.e. (AA - BB)/2, and the dominance effect (*d*) was calculated as AB - [(AA + BB)/2], where the AB represents the predicted phenotypic mean of the heterozygote. The additive genetic SNP variance (σ2SNP) was estimated using the following equation:

$$\sigma 2SNP=2pq(a+d(q-p))2, \quad (3)$$

where p and q are the frequency of the major and minor alleles at the SNP, respectively. The proportion of variance explained by the SNP is then given by:

$$\sigma 2SNP/\sigma 2a, \quad (4)$$

where σ2a is the total additive genetic variance of the trait when no SNP effects are included in the model.

Assessment of genomic prediction

The utility of genomic prediction for resistance to lice was assessed by cross-validation analyses under various scenarios (see below) in which (i) varying SNP densities (0.5, 1, 5, 10, 20 K (all chosen at random), and 33 K (full dataset)) and (ii) varying degrees of relationships between training and validation sets were applied.

Scenario (i): Random selection

Within each population (which correspond to discrete 'year groups' of a commercial Atlantic salmon breeding programme), cross-validation analysis was performed by selecting five random non-overlapping training and validation sets as described previously [21]. At each SNP density (0.5 to 33 K SNPs), GBLUP was applied to predict the masked phenotypes of the validation sets and the resulting prediction accuracy was compared to that of pedigree-based BLUP (PBLUP), as described above. The average accuracy across the five cross-validation replicates for each SNP density was computed.

Scenario (ii): Sibling

Within each population, training and validation sets were established such that both sets contained representatives of each family. The same cross-validation analyses were performed as for Scenario (i).

Scenario (iii): Non-sibling

Within each population, training and validation sets were established such that full siblings were not included in either set (i.e. different full-sibling families were used for training and validation sets). The resulting training and validation sets were more distantly related than for Scenarios (i) and (ii), although they did contain some half-sibs. The same cross-validation analyses were performed as for Scenarios (i) and (ii).

Scenario (iv): Across populations

To assess prediction accuracy across populations per year group, population I was used as the training set and population II as the validation set, and vice versa. The same genomic prediction and cross-validation analyses were performed as for Scenarios (i) to (iii), but pedigree-based prediction was not possible since genetic links between the two populations were absent from the available pedigree.

Cross-validation

The five-fold cross-validation analyses for each scenario described above were performed using the methods described in Tsai et al. [21]. Briefly, for the within-population analyses, populations I and II were each divided into a training (80 %) and validation (20 %) set. Phenotypes (i.e. lice counts) of the samples in the validation sets were then masked and GBLUP or pedigree-based BLUP (PBLUP) was applied to predict the phenotypes of the masked individuals using model (2) implemented in ASReml 3.0 [24]. The Pearson correlation coefficient of the estimated breeding values (EBV) [either genomic EBV (GEBV) or pedigree-based EBV (PEBV)] with the adjusted phenotype of the masked validation set. Accuracy was calculated as the correlation divided by the square root of the heritability using all individuals, and then averaged across the five replicates (Figs. 2, ,33).

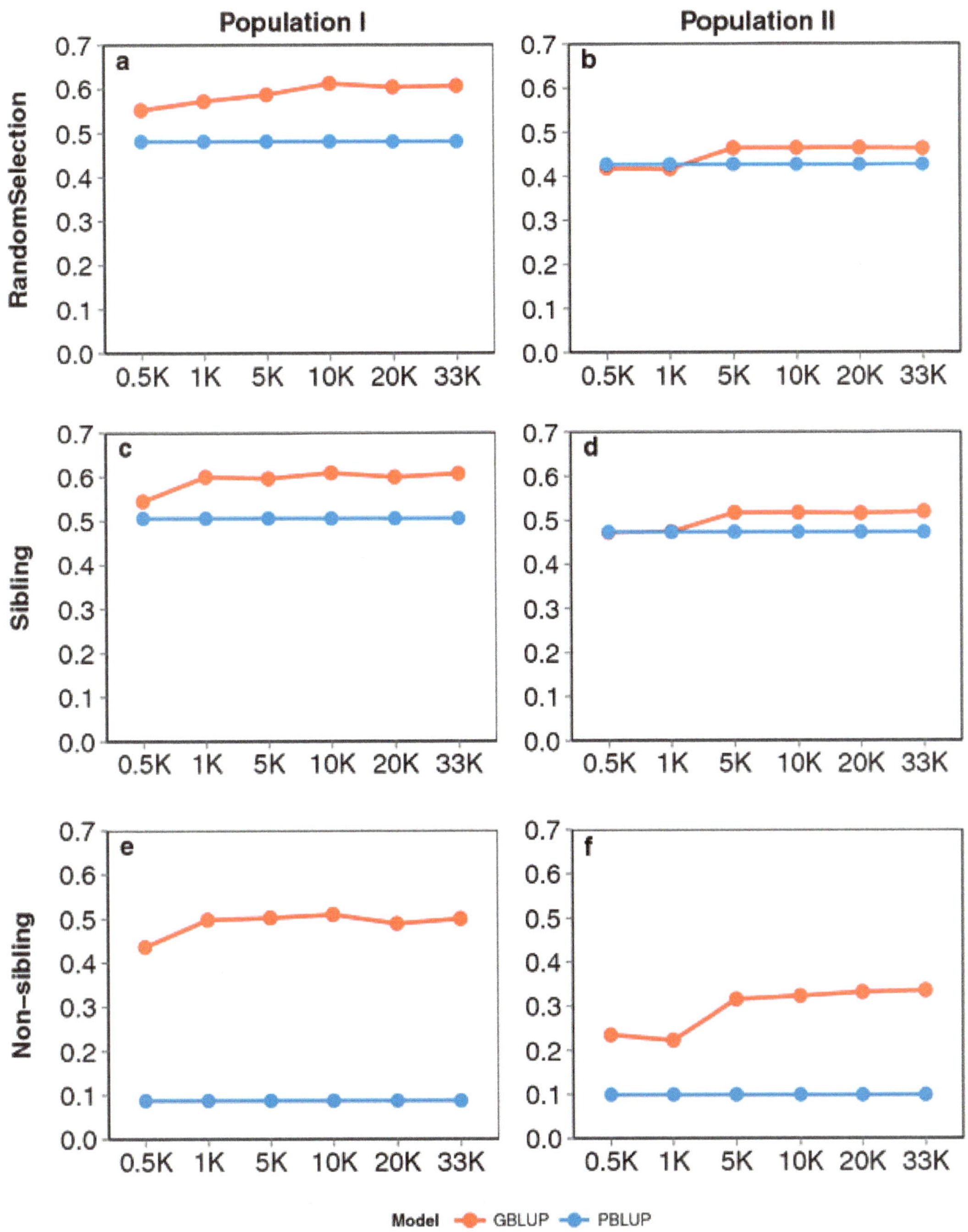

Figure. 2 Accuracy of genomic and pedigree-based prediction within populations. Comparison of prediction accuracy (Y-axis) of two populations using increasing SNP densities from 0.5 to 33 K (X-axis) assessed by cross-validation analyses. "Random ...

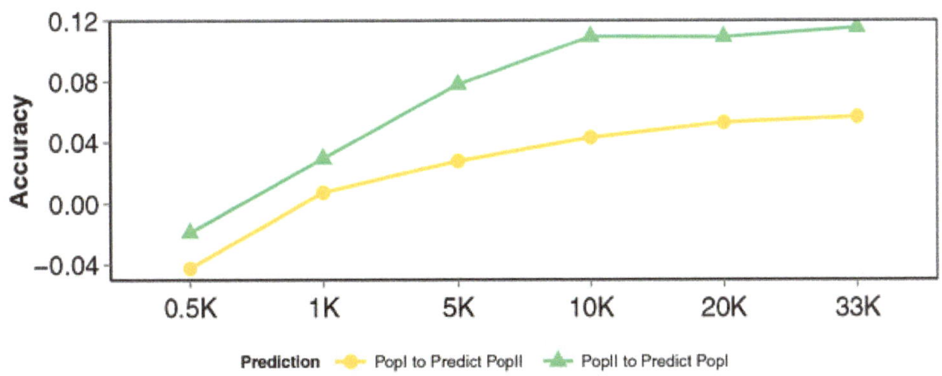

Figure. 3 Accuracy of genomic prediction across populations. Based on setting population I as the training set and population II as the validation set and vice versa. Accuracy of prediction (Y-axis) for the two populations was estimated using increasing SNP density ...

RESULTS

General statistics and genetic parameters of resistance to lice and growth

Estimated heritability for lice count was moderate (~0.3) and relatively consistent when using a pedigree relationship matrix (Table 1). Estimates of heritability for the growth-related traits (weight and length) were higher (~0.6), in line with previous estimates [21]. The two growth traits had high positive phenotypic and genetic correlations with each other (~0.93 to 0.96), and correlations of the growth traits with lice count were either equal to zero or slightly negative (Table 2).

Table 1 General statistics and heritability estimates for lice count and growth traits

	Population I		Population II	
	Mean (SD)	Heritability[a] (SE)	Mean (SD)	Heritability[a] (SE)
Lice[b]	25.8 (12.3)	0.33 (0.08)/0.27 (0.08)	18.3 (9.1)	0.22 (0.06)/0.27 (0.08)
Length	214.2 (16.1)[c]	0.61 (0.07)/0.51 (0.11)[c]	206.2 (14.3)	0.51 (0.07)/0.50 (0.10)
Weight	112.0 (21.0)[c]	0.61 (0.07)/0.49 (0.10)[c]	89.9 (19.9)	0.50 (0.07)/0.50 (0.10)

SD is the standard deviation and SE is the standard error

[a] Heritability was estimated based on the G-matrix/A-matrix

[b] The lice count data (number of lice per fish) used here was without data adjustment

[c] The results are from Tsai et al. [21]

Table 2 Estimates of genetic and phenotypic correlations between lice count and growth traits in populations I and II

Genetic correlation	Phenotypic correlation		
	Lice	Length	Weight
Population I			
Lice	–	−0.04	−0.06
Length	0.10	–	0.96
Weight	0.11	0.96	–
Population II			
Lice	–	−0.1	−0.1
Length	−0.3	–	0.93
Weight	−0.3	0.95	–

Genome-wide association study

The results of the GWAS suggest that lice resistance is a polygenic trait, with no SNPs surpassing the Bonferroni-corrected significance thresholds (Figure. 1). Indeed, when each population was analysed separately, there was little overlap between regions that had the lowest P values (Figure. 1a, b). When the two populations were combined (Figure. 1c), SNPs with the lowest P values were located on chromosomes 1, 3, 9 and 23. The estimated proportion of additive genetic variance explained by these SNPs ranged from ~2 to 6 % each. The quantile–quantile (Q–Q) plots for each GWA analysis are in Figure S2 (See Additional file 2: Figure S2).

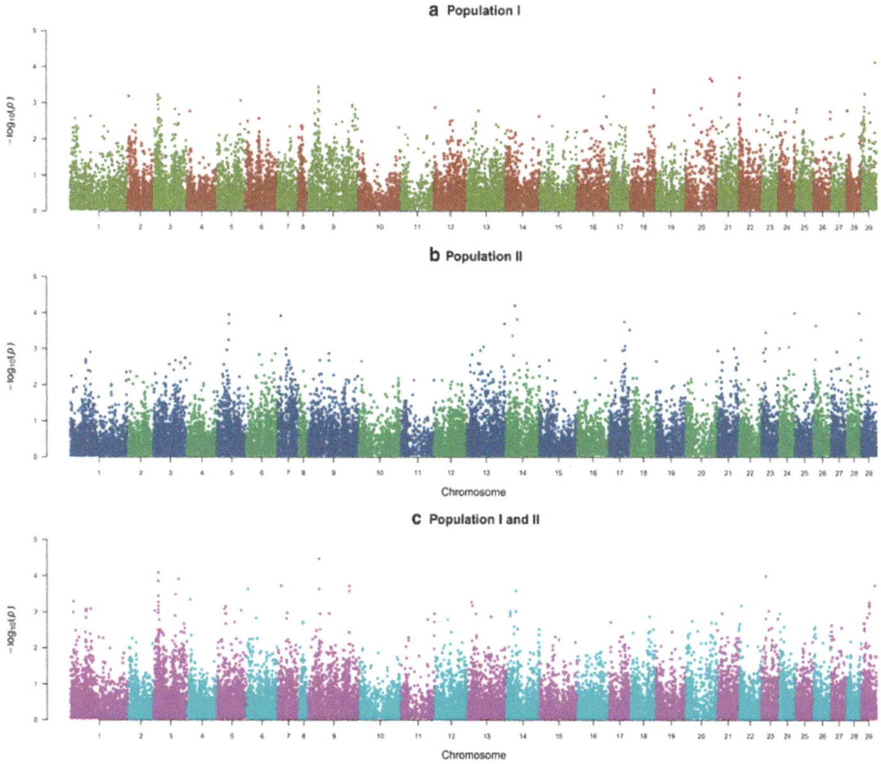

Figure. 1 Manhattan plots of the genome-wide association study for populations I (**a**), II (**b**), and I and II combined (**c**). *Top markers* are close to chromosome-wide significance ($\alpha < 0.05$) but do not pass the threshold

Accuracy of predicted breeding values

The putative polygenic architecture of lice resistance in these populations means that genomic prediction may be a practical and effective method of predicting breeding values for lice resistance, which was tested using cross-validation analyses under different scenarios in which varying SNP densities and varying levels of relatedness between training and validation sets were applied (see "Methods" for details). Accuracy of prediction using the genomic relationship matrix (GBLUP) was generally higher than that using the pedigree relationship matrix (PBLUP). Greater SNP density tended to improve prediction accuracy, but the asymptote was generally reached at ~5 K SNPs for both populations (Figure. 2).

The results of genomic prediction under the "random selection" (where training and validation sets were chosen at random), and "sibling" (where full siblings from each family were deliberately included in both the training and validation sets) scenarios were very similar for both populations (Figure. 2a–d). Therefore, including animals that share close relationships did not improve the accuracy of genomic predictions for these populations, which indicates that "random selection" will result in the presence of several closely-related fish across the training and validation datasets by chance. In both cases, GBLUP resulted in more accurate predictions of lice count in the validation data than PBLUP, with a relative advantage of approximately 27 % for population I and 10 % for population II (Figure. 2a–d). Increasing marker density to more than ~5 K randomly chosen SNPs had little impact on prediction accuracy, which may be expected when the training and validation sets are closely related [9].

When the training and validation sets were less related, predictions of both pedigree- and genomic-based methods were less accurate, as expected. In the "non-sibling" scenario (where no full-siblings were included in both the training and validation sets), accuracies of prediction obtained with both GBLUP and PBLUP were substantially lower than those in the previous two scenarios. However, the benefit of genomic prediction was greatest under this scenario, with prediction accuracies increasing fivefold (population I) and 2.5-fold (population II) relative to pedigree-based prediction accuracies. Perhaps surprisingly, there was little benefit from increasing SNP density above ~5 K SNPs under this scenario as well (Figure. 2e, f). When the accuracy of genomic prediction was assessed across the two populations (corresponding to 2 year groups of the Landcatch broodstock), accuracies were markedly lower (0.05–0.11) than with all of the within-population scenarios (0.34–0.61). Increasing SNP density did seem to yield incremental (albeit small) increases in prediction accuracies when predicting across populations (Figure. 3), which suggested that this scenario was likely to benefit most from a high-density SNP array. However, these two populations were probably too small to achieve high prediction accuracy for these distantly-related animals, and a more thorough test of across-population prediction in salmon should use larger sample sizes.

Discussion

Genomic selection is an increasingly important component of modern aquaculture breeding schemes, with simulated and applied studies highlighting its benefits over pedigree-based selection [9, 28]. However, the substantial cost of genome-wide genotyping means that the traits targeted by GS are likely to be those of high economic value, particularly those that cannot be easily measured on the selection candidates themselves. Currently, sea lice present the largest threat to the sustainability of salmon farming, which relies heavily on expensive and potentially environmentally-damaging chemical treatments [10]. Host resistance to sea lice has consistently been shown to have a substantial genetic component [11]. Therefore, resistance to lice is an ideal candidate trait for the application of GS. In our study, lice count data and genome-wide SNP genotypes were collected for two pedigreed salmon populations from a commercial breeding programme to assess the utility of genomic prediction of host resistance to sea lice under different scenarios, including a comparison to predictions based on pedigree records alone.

The heritability of resistance to lice was estimated at ~0.3 and 0.2 in populations I and II, respectively, which is similar to the findings of Gharbi et al. [11] (~0.3) and Gjerde et al. [29] (~0.2 to 0.3), and slightly higher than those of Ødegård et al. [9] (~0.13 to 0.14). However, it should be noted that the challenge experiments that are reported in Gharbi et al. [11], Gjerde et al. [13], and in our study, were all conducted in controlled tanks conditions, whereas the study of Ødegård et al. [9] was based on challenges in a sea-cage environment, which may display greater environmental variation. Furthermore, it should be noted that the higher heritability estimates for all traits in population I may be due in part to confounding between genetic and common environmental effects due to the family-specific rearing of the fry (compared to population II, for which individuals were mixed into a single tank as first feeding fry).

The GWAS indicated that host resistance to lice likely has a polygenic architecture, with no major QTL segregating in these populations (Figure. 1). Therefore, it is likely that individual QTL for lice resistance explain only a small percentage of the genetic variance, and a proportion of the QTL may be population-specific. As such, GBLUP and similar methods of genomic prediction are likely to be suitable for predicting breeding values for host resistance to lice, particularly within populations.

The degree of the genetic relationships between training and validation sets is critical for the efficacy of genomic prediction. In our study, genomic prediction was found to be highly effective and showed a significant advantage in terms of accuracy over pedigree-based methods within populations (which correspond to year groups of a salmon breeding programme, Figure. 2). The accuracy of prediction and the relative advantage of genomic prediction were lower for population II than for population I (Figure. 2), which may reflect the lower estimated heritability in this population because a low heritability can contribute to low prediction accuracy [20, 30]. Also, the family structure of population II was potentially less amenable to accurate prediction since it comprised a larger number of smaller families, which decreased the chance of having useful numbers of full siblings in the training and validation sets. Prediction accuracies were highest when training and validation sets were closely related, as was shown with the "Random selection" and "Sibling" scenarios. In addition, these results showed that deliberately including highly-related animals (i.e. full siblings) in the training and validation sets yielded little advantage over random assignment. This likely reflects the typical family structure of commercial salmon breeding populations, which consist of large full sibling families (thousands of fish per family) that result in close relationships between selection candidates and test individuals. However, the benefit of using genomic prediction over pedigree-based prediction was largest under the "Non-sibling" scenario, in which training and validation sets were established such that no full-siblings were included (i.e. the sets were less related than would be expected by chance, Figure. 2). Prediction across populations or year groups (for which genetic relationships are more distant) was substantially less effective, with relatively low prediction accuracies (Figure. 3). This may reflect, in part, inadequate sample size of the populations, or possibly differences in the experimental procedures between the two studies. However, our findings imply that either the GBLUP analyses did not efficiently capture short range LD between SNPs and QTL for resistance to sea lice, and/or that the QTL were population-specific. Therefore, in commercial salmon breeding schemes, regular phenotype data collection on animals that are closely-related to the selection candidates, combined with medium- or low-density (and cost) SNP panel genotyping appears to be the most effective means of using genomic prediction for resistance to lice. This strategy is supported by results from previous simulation studies (e.g. [28]).

Using data collected from a challenge trial performed in a sea cage environment, Ødegård et al. [9] also showed that genomic prediction of breeding values for lice resistance was more accurate compared to pedigree-based prediction. As in our study, the observed improvements depended partly on SNP density with ~32 (1 K SNPs) and 51 % (220 K SNPs) higher reliabilities than those obtained from predictions based on pedigree records alone [9]. Interestingly, increasing SNP density above a threshold of around 5 K SNPs had little impact on accuracy of prediction in both studies (Figure. 2, [9]). This may reflect the relatively close relationships between the training and validation sets, since higher SNP density did slightly improve the accuracy of cross-population predictions, as shown in our study, up to ~33 K SNPs (the highest density tested) (Figure. 3). However, it seems unlikely that linkage alone is underpinning the predictions, since predictions with low SNP densities (<1 K) and predictions based on an IBD (identity-by-descent) genomic relationship matrix were less accurate [9]. Therefore, short or long range LD between SNP and QTL alleles may be an important component of prediction. Obviously, such LD can be captured by a relatively sparse SNP set, a finding that may be related to the relatively close relationships between training and validation sets, recent population admixture [9], or slower decay of LD due to the lack of male recombination in male salmon across much of the genome [31, 32].

A difference between simulation studies and those performed on experimental data is often observed in genomic prediction studies. Previous simulation studies indicated that the accuracy of breeding value prediction can reach values of 0.8 to 1.0 if the reference population size is sufficiently large (e.g. more than 100,000) [2, 33]. However, in practice, due to financial and practical limitations, research programs that use 'experimental' data usually involve the analysis of relatively small reference populations [9, 21, 34]. It is likely that if we had used larger population sizes, higher accuracies of prediction would have been obtained, particularly for predictions across the two distantly-related populations (subject to sufficient SNP density). As such, cost-effective means of generating high-density SNP data remain a relevant goal, and genotype imputation is likely to be increasingly important, particularly now that the majority of the Atlantic salmon reference genome has been assembled and ordered onto chromosomes (Genbank assembly accession GCA_000233375.4, [35]). Genotyping-by-sequencing may be crucial for reaching such high SNP density at moderate cost and its potential for genomic prediction in livestock has already been reported [36]. With a high SNP density across large sample sizes, one may expect to capture LD between SNPs and QTL, and co-segregation of chromosome segments among related individuals, although the resolution of mapping causative variants may be hampered by the strong relationship structure in the population. Within populations/year groups, the requirement in terms of SNP density for accurate prediction is clearly lower and as few as 1 to 5 K informative SNPs are sufficient. However, while this points to the potential utility of cheaper and lower density genotyping platforms in aquaculture breeding, it is important to keep in mind that SNP informativeness can vary between populations.

CONCLUSIONS

Genomic prediction is an effective method for predicting breeding values for host resistance to sea lice in Atlantic salmon populations from a commercial breeding programme. The GWAS results suggested that lice resistance is a polygenic trait. Cross-validation tests of genomic prediction highlighted the substantial improvements in prediction accuracy compared to that of pedigree-based prediction. The accuracy of GBLUP was highest when training and validation sets were closely related but the relative advantage over pedigree-based prediction within a population was largest when relationships were more distant. Relatively low SNP densities (from 1 to 5 K SNPs) were sufficient for accuracy to reach the asymptote under most of the scenarios tested. Prediction accuracy is generally much lower across distantly-related populations, although a trend was evident that increased marker density was advantageous in such situations. Therefore, larger population sample sizes and high-density SNP genotypes are probably necessary to improve across-population prediction. Given the economic importance of resistance to sea lice, and the efficacy of genomic prediction, it is likely that selective breeding for this trait using genomic data will become an important component of sea lice control.

AUTHORS' CONTRIBUTIONS

RDH and SCB (the author, Stephen C. Bishop, unfortunately passed away before completion of the manuscript) were responsible for the overall experimental design; JEB, JBT, MJS and KG designed experiments and collected data; AET, AH, and DRG provided phenotype and pedigree data; HYT, OM, RPW and RDH analysed data; HYT and RDH prepared the manuscript. All authors read and approved the final manuscript.

ACKNOWLEDGEMENTS

The authors gratefully thank the Edinburgh Genomics facility for SNP array genotyping, and Bill Roy and Matt Tinsley at the Marine Environmental Research Laboratory for assistance with trait data collection. This project received funding from the Technology Strategy Board (TP 5771-40299), Innovate UK (45266-329178), and Biotechnology and Biological Sciences Research Council (BBSRC) Grants (BB/H022007/1, BB/J004235/1 and BB/J004324/1). JBT and JEB were partly supported by the MASTS pooling initiative (The Marine Alliance for Science and Technology for Scotland) for the completion of this study. MASTS is funded by the Scottish Funding Council (Grant reference HR09011) and contributing institutions. HYT is supported by funding from the Ministry of Education, Taiwan.

REFERENCES

1. Habier D, Fernando RL, Garrick DJ. Genomic BLUP decoded: a look into the black box of genomic prediction. Genetics. 2013;194:597–607. doi: 10.1534/genetics.113.152207.
2. Meuwissen THE, Hayes BJ, Goddard ME. Prediction of total genetic value using genome-wide dense marker maps. Genetics. 2001;157:1819–1829.
3. Wilkinson S, Wiener P, Archibald AL, Law A, Schnabel RD, McKay SD, et al. Evaluation of approaches for identifying population informative markers from high density SNP chips. BMC Genet. 2011;12:45. doi: 10.1186/1471-2156-12-45.
4. Riggio V, Matika O, Pong-Wong R, Stear MJ, Bishop SC. Genome-wide association and regional heritability mapping to identify loci underlying variation in nematode resistance and body weight in Scottish Blackface lambs. Heredity (Edinb) 2013;110:420–429. doi: 10.1038/hdy.2012.90.
5. Bermingham ML, Bishop SC, Woolliams JA, Pong-Wong R, Allen AR, McBride SH, et al. Genome-wide association study identifies novel loci associated with resistance to bovine tuberculosis. Heredity (Edinb)2014;112:543–551. doi: 10.1038/hdy.2013.137.
6. Gutierrez AP, Yáñez JM, Fukui S, Swift B, Davidson WS. Genome-wide association study (GWAS) for growth rate and age at sexual maturation in Atlantic salmon (*Salmo salar*) PLoS One. 2015;10:e0119730. doi: 10.1371/journal.pone.0119730.
7. Yáñez JM, Newman S, Houston RD. Genomics in aquaculture to better understand species biology and accelerate genetic progress. Front Genet. 2015;6:128.
8. Daetwyler HD, Pong-Wong R, Villanueva B, Woolliams JA. The impact of genetic architecture on genome-wide evaluation methods. Genetics. 2010;185:1021–1031. doi: 10.1534/genetics.110.116855
9. Odegård J, Moen T, Santi N, Korsvoll SA, Kjøglum S, Meuwissen THE. Genomic prediction in an admixed population of Atlantic salmon (*Salmo salar*) Front Genet. 2014;5:402.
10. Frazer LN, Morton A, Krkosek M. Critical thresholds in sea lice epidemics: evidence, sensitivity and subcritical estimation. Proc Biol Sci. 2012;279:1950–1958. doi: 10.1098/rspb.2011.2210.
11. Gharbi K, Matthews L, Bron J, Roberts R, Tinch A, Stear MJ, et al. The control of sea lice in Atlantic salmon by

selective breeding. J R Soc Interface. 2015;12:0574. doi: 10.1098/rsif.2015.0574

12. Kolstad K, Heuch PA, Gjerde B, Gjedrem T, Salte R. Genetic variation in resistance of Atlantic salmon (*Salmo salar*) to the salmon louse *Lepeophtheirus salmonis*. Aquaculture. 2005;247:145–151. doi: 10.1016/j.aquaculture.2005.02.009.
13. Gjerde B, Ødegård J, Thorland I. Estimates of genetic variation in the susceptibility of Atlantic salmon (*Salmo salar*) to the salmon louse *Lepeophtheirus salmonis*. Aquaculture. 2011;314:66–72. doi: 10.1016/j.aquaculture.2011.01.026.
14. Houston RD, Bishop SC, Guy DR, Tinch AE, Taggart JB, Bron JE, et al. Genome wide association analysis for resistance to sea lice in Atlantic salmon: application of a dense SNP array. In: Proceedings of the 10th World congress of genetics applied to livestock production: 17–22 August 2014; Vancouver. 2014.https://asas.org/docs/default-source/wcgalp-proceedings-oral/265_paper_9597_manuscript_751_0.pdf?sfvrsn=2.
15. Habier D, Fernando RL, Kizilkaya K, Garrick DJ. Extension of the bayesian alphabet for genomic selection. BMC Bioinformatics. 2011;12:186. doi: 10.1186/1471-2105-12-186.
16. Daetwyler HD, Calus MPL, Pong-Wong R, de los Campos G, Hickey JM. Genomic prediction in animals and plants: Simulation of data, validation, reporting, and benchmarking. Genetics. 2013;193:347–365. doi: 10.1534/genetics.112.147983.
17. Habier D, Fernando RL, Dekkers JCM. The impact of genetic relationship information on genome-assisted breeding values. Genetics. 2007;177:2389–2397.
18. Wientjes YCJ, Veerkamp RF, Calus MPL. The effect of linkage disequilibrium and family relationships on the reliability of genomic prediction. Genetics. 2013;193:621–631. doi: 10.1534/genetics.112.146290
19. Hickey JM, Dreisigacker S, Crossa J, Hearne S, Babu R, Prasanna BM, et al. Evaluation of genomic selection training population designs and genotyping strategies in plant breeding programs using simulation. Crop Sci. 2014;54:1476–1488. doi: 10.2135/cropsci2013.03.0195.
20. Vela-Avitúa S, Meuwissen THE, Luan T, Ødegård J. Accuracy of genomic selection for a sib-evaluated trait using identity-by-state and identity-by-descent relationships. Genet Sel Evol. 2015;47:9. doi: 10.1186/s12711-014-0084-2.
21. Tsai HY, Hamilton A, Tinch AE, Guy DR, Gharbi K, Stear MJ, et al. Genome wide association and genomic prediction for growth traits in juvenile farmed Atlantic salmon using a high density SNP array.BMC Genomics. 2015;16:969. doi: 10.1186/s12864-015-2117-9.
22. Houston RD, Taggart JB, Cézard T, Bekaert M, Lowe NR, Downing A, et al. Development and validation of a high density SNP genotyping array for Atlantic salmon (*Salmo salar*) BMC Genomics.2014;15:90. doi: 10.1186/1471-2164-15-90.
23. Purcell S, Neale B, Todd-Brown K, Thomas L, Ferreira MAR, Bender D, et al. PLINK: a tool set for whole-genome association and population-based linkage analyses. Am J Hum Genet. 2007;81:559–575. doi: 10.1086/519795.
24. Gilmour AR, Gogel BJ, Cullis BR, Thompson R. ASReml user guide. 4th ed. Hemel Hempstead: VSN International Ltd; 2014.
25. Aulchenko YS, Ripke S, Isaacs A, van Duijn CM. GenABEL: an R library for genome-wide association analysis. Bioinformatics. 2007;23:1294–1296. doi: 10.1093/bioinformatics/btm108.
26. VanRaden PM. Efficient methods to compute genomic predictions. J Dairy Sci. 2008;91:4414–4423. doi: 10.3168/jds.2007-0980.
27. Chen WM, Abecasis GR. Family-based association tests for genomewide association scans. Am J Hum Genet. 2007;81:913–926. doi: 10.1086/521580.
28. Sonesson AK, Meuwissen THE. Testing strategies for genomic selection in aquaculture breeding programs. Genet Sel Evol. 2009;41:37. doi: 10.1186/1297-9686-41-37.
29. Gjerde B, Pante MJR, Baeverfjord G. Genetic variation for a vertebral deformity in Atlantic salmon (*Salmo salar*) Aquaculture. 2005;244:77–87. doi: 10.1016/j.aquaculture.2004.12.002.
30. Nirea KG, Sonesson AK, Woolliams JA, Meuwissen THE. Strategies for implementing genomic selection in family-based aquaculture breeding schemes: double haploid sib test populations. Genet Sel Evol.2012;44:30. doi: 10.1186/1297-9686-44-30.
31. Lien S, Gidskehaug L, Moen T, Hayes BJ, Berg PR, Davidson WS, et al. A dense SNP-based linkage map for Atlantic salmon (*Salmo salar*) reveals extended chromosome homeologies and striking differences in sex-specific recombination patterns. BMC Genomics. 2011;12:615. doi: 10.1186/1471-2164-12-615
32. Gonen S, Lowe NR, Cezard T, Gharbi K, Bishop SC, Houston RD. Linkage maps of the Atlantic salmon (*Salmo salar*) genome derived from RAD sequencing. BMC Genomics. 2014;15:166. doi: 10.1186/1471-2164-15-166.
33. Daetwyler HD, Villanueva B, Woolliams JA. Accuracy of predicting the genetic risk of disease using a genome-wide approach. PLoS One. 2008;3:e3395. doi: 10.1371/journal.pone.0003395.
34. Calus MPL. Genomic breeding value prediction: methods and procedures. Animal. 2010;4:157–164. doi: 10.1017/S1751731109991352.
35. Davidson WS, Koop BF, Jones SJM, Iturra P, Vidal R, Maass A, et al. Sequencing the genome of the Atlantic salmon (*Salmo salar*) Genome Biol. 2010;11:403.
36. Gorjanc G, Cleveland MA, Houston RD, Hickey JM. Potential of genotyping-by-sequencing for genomic selection in livestock populations. Genet Sel Evol. 2015;47:12. doi: 10.1186/s12711-015-0102-z

CHAPTER 14

GENETIC HETEROGENEITY OF WITHIN-FAMILY VARIANCE OF BODY WEIGHT IN ATLANTIC SALMON (SALMO SALAR)

Anna K Sonesson[1]*, Jørgen Ødegård[1] and Lars Rönnegård[2]

[1] Nofima AS, Ås, Norway.

[2] Dalarna University, 791 88 Falun, Sweden.

ABSTRACT

Background

Canalization is defined as the stability of a genotype against minor variations in both environment and genetics. Genetic variation in degree of canalization causes heterogeneity of within-family variance. The aims of this study are twofold: (1) quantify genetic heterogeneity of (within-family) residual variance in Atlantic salmon and (2) test whether the observed heterogeneity of (within-family) residual variance can be explained by simple scaling effects.

Results

Analysis of body weight in Atlantic salmon using a double hierarchical generalized linear model (DHGLM) revealed substantial heterogeneity of within-family variance. The 95% prediction interval for within-family variance ranged from ~0.4 to 1.2 kg^2, implying that the within-family variance of the most extreme high families is expected to be approximately three times larger than the extreme low families. For cross-sectional data, DHGLM with an animal mean sub-model resulted in severe bias, while a corresponding sire-dam model was appropriate. Heterogeneity of variance was not sensitive to Box-Cox transformations of phenotypes, which implies that heterogeneity of variance exists beyond what would be expected from simple scaling effects.

Conclusions

Substantial heterogeneity of within-family variance was found for body weight in Atlantic salmon. A tendency towards higher variance with higher means (scaling effects) was observed, but heterogeneity of within-family variance existed beyond what could be explained by simple scaling effects. For cross-sectional data, using the animal mean sub-model in the DHGLM resulted in biased estimates of variance components, which differed substantially both from a standard

Citation: Sonesson et al.: Genetic heterogeneity of withinfamily variance of body weight in Atlantic salmon (Salmo salar). Genetics Selection Evolution 2013 45:41. doi:10.1186/1297-9686-45-41

linear mean animal model and a sire-dam DHGLM model. Although genetic differences in canalization were observed, selection for increased canalization is difficult, because there is limited individual information for the variance sub-model, especially when based on cross-sectional data. Furthermore, potential macro-environmental changes (diet, climatic region, etc.) may make genetic heterogeneity of variance a less stable trait over time and space.

BACKGROUND

In livestock and aquaculture production systems, individuals that can perform well in different environments and against different background genetics are preferred. In evolutionary genetics, the term canalization describes the stability or buffering ability of a genotype against minor variations, not only in the environment but also in the genetic make-up [1].

In animal breeding, genetic differences in phenotypic stability can be detected as genetic heterogeneity of (within-family) residual variance, where a low variance indicates stable performance across environmental factors. Genetic heterogeneity of residual variance has been estimated for different traits in a number of land-based species, e.g., litter size in sheep [2], pigs [3], mice [4] and rabbits [5] and weight in piglets [6] and broilers [7] (review of Hill and Mulder [8]) and more recently somatic cell scores and milk yield in dairy cattle [9] and body weight in rainbow trout [10]. Overall, these studies show substantial genetic heterogeneity of (within-family) residual variance but heritability of individual dispersion as defined in [11] is usually low (<0.05). This is due to the fact that the dispersion parameters of an individual cannot be measured directly but are inferred through estimated residual(s) (i.e., deviations from the expected mean), which are expected to be distributed around zero regardless of the underlying variance. Hence, a single individual phenotype does not convey much information about the underlying variance, although differences between family groups may be substantial and estimable.

The presence of genetic heterogeneity of residual variance suggests that selection to change residual variance is possible, based mainly on family information. Thus, large families will be an advantage to study the genetic heterogeneity of variance. In fish species, both sexes are usually highly fecund, with a potential for large full- and half-sib families. The aims of this study were twofold: (1) quantify the genetic heterogeneity of (within-family) residual variance for body weight in Atlantic salmon and (2) test whether the observed heterogeneity of (within-family) residual variance can be explained by simple scaling effects. We also discuss the data structures that are best suited for estimating variance heterogeneity and individual breeding values for residual variance.

METHODS

Phenotypic Data

The analysis was restricted to year-class 2008 of Atlantic salmon from the AquaGen breeding program (http://www.aquagen.no/). Descriptive statistics of the dataset are in Table 1. Data included body weight records until harvest for 13 633 fish. Fish that were sexually mature or had visible deformities were excluded from the analysis. Potential outliers were not removed from the raw data, because this could remove the real effects that we wanted to study, e.g., occurrence of runts (undersized animals), which is a known phenomenon in Atlantic salmon.

Table 1 Descriptive statistics of the data

Item	
Number of observations (fish)	13633
Number of full-sib families	502
Number of sires	346
Number of dams	324
Number of full-sib families with paternal and/or maternal half sibs	440
Average weight, kg (SD)	3.70 (0.90)
Skewness	0.31
Kurtosis	1.25

For the production of families, both dams and sires were mated with a maximum of two partners. The final dataset included 502 full-sib families from 346 sires and 324 dams. Of these parents, 156 sires and 178 dams were represented with two full-sib offspring groups, a structure that facilitates proper separation of common (non-additive) family effects, e.g., effects due to common rearing, maternal effects and potential dominance genetic effects.

Statistical Models

The theory behind the data analysis was previously described in detail by Rönnegård et al. [12]. Both untransformed, log- and square root-transformed data were analyzed. Transformation was applied to study the potential relationships between mean and variance (scaling effects). Sire-dam models are often used in aquaculture breeding [13, 14, 15], especially for non-normal (e.g. binary) traits, and were used here to model heterogeneity of variance. In animal models, the full additive genetic effect (mid-parent mean + Mendelian sampling deviation) of each animal is fitted, whereas sire-dam models only fit the additive genetic mid-parent means (sire + dam effects) and the Mendelian sampling deviations of the offspring are included in the residuals of the model. When parents do not have own phenotypes, an animal mean model allowing for heterogeneous residual variance is equivalent to a sire-dam model in which the residual of an individual i is defined as [16]:

$$e_i^* = m_i + e_i,$$

where m_i and e_i are, respectively, the Mendelian sampling deviation and the micro-environmental residual of individual i, and the variance of e∗i is thus equal to:

$$\sigma^2_{e_i^*} = \frac{1}{2}\sigma^2_a + \sigma^2_{e_i},$$

where σ2a is the (homogeneous) additive genetic variance and σ2ei is the (heterogeneous) micro-environmental residual variance for animal i in the mean animal model. Hence, fitting an animal mean model with heterogeneous residual variance necessarily implies that the Mendelian sampling variance is assumed identical for all animals, i.e., that variance heterogeneity is strictly micro-environmental within each individual.

Because a sire-dam mean model includes the Mendelian sampling deviations into the residuals, the variance model is used to model the full within-family variance. Thus, a sire-dam variance model implies that genetic heterogeneity of residual variance in the mean model is completely explained by sire and dam effects. However, if genetic heterogeneity of variance exists, Mendelian sampling deviations are also expected to affect the level of residual variance and an animal model for the variance would be theoretically more correct. However, in our cross-sectional data (i.e. one observation per individual), these Mendelian sampling deviations only affect the variance of a single observation, i.e., some residuals will be more or less extreme than expected, but the overall effect of within-family variance on genetic heterogeneity will probably be small.

Against this background, the following models were used and compared in this study.

Linear Mean Animal Model

This is a classical linear mean animal model for body weight at a given age.

$$\mathbf{y} = \mu + \mathbf{e},$$

(1)

$$\mu = \mathbf{Xb} + \mathbf{Za} + \mathbf{Qc},$$

where **y** is the vector of recorded body weights (untransformed, log-, or square root-transformed) for all fish,**b** is a vector of fixed effects (includes the effect of group of rearing at early life stages (2 groups)), $\mathbf{a} \sim N(0,A\sigma^2_a)$ is a vector of additive genetic effects, $\mathbf{c} \sim N(0,I\sigma^2_c)$ is a vector of common environmental family effects, $\mathbf{e} \sim N(0,I\sigma^2_e)$ is a vector of random residuals, the matrices **X**, **Z** and **Q** are appropriate incidence matrices and **I** is an identity matrix of appropriate size.

Double hierarchical generalized linear models (DHGLM)

DHGLM1 -The mean sub-model of the DHGLM is a classical linear mean animal model for body weight at a given age, as given above. However, the residual variance is modeled using a gamma sire-dam model, where:

(2)

$$E\left(e_i^2\right) = \phi_i,$$

$$Var\left(e_i^2\right) = 2\phi_i^2,$$

$$\log\left(\phi_i\right) = \mathbf{X_i b_D} + \left(\mathbf{Z_{Pi}} + \mathbf{Z_{Mi}}\right)\mathbf{u_D},$$

where $\mathbf{b_D}$ is a vector of fixed effects on dispersion (equal to those for the mean model) and $\mathbf{u_D}$ is a vector of additive genetic sire and dam effects on dispersion with distribution $\mathbf{u_D} \sim N(0,A\sigma^2_u)$, where $\sigma 2u=14\sigma^2_a$. The matrices $\mathbf{X_i}$, $\mathbf{Z_{Pi}}$ and $\mathbf{Z_{Mi}}$ are row incidence vectors associated with observation *i*, for the fixed, sire, and dam effects, respectively. Since the residuals are unobserved, but rather estimated as e^i=yi−μ^i, equation (2) is reformulated as:

$$E\left(\hat{e}_i^2 / \left(1 - h_i\right)\right) = \phi_i,$$

and can be fitted using a gamma model with weights $(1-h_i)/2$, where h_i is the ith diagonal element in the hat matrix of the mean sub-model; the hat matrix is described in Rönnegård et al. [12].

DHGLM2-This model is similar to the DHGLM1 described above, except that the additive genetic animal effect in the mean model is replaced by additive genetic effects of the sire and the dam (as in the variance model):

$$\mu = \mathbf{X_b} + \left(\mathbf{Z_P} + \mathbf{Z_m}\right)\mathbf{u} + \mathbf{Qc},$$

where **u** is a vector of additive genetic effects (on the mean) of the sire and the dam.

Restricted maximum likelihood methods were used to estimate variance components with the ASREML software [17]. For the DHGLM models, an iterative approach within the R environment (http://www.r-project.org), alternating between weighted analyses of the mean and variance sub-models was used, with weights being updated for each round, as described in Rönnegård et al. [12].

RESULTS

Variance components estimated based on the different models are in Table 2 for the untransformed data, in Table 3 for the log-transformed data and in Table 4 for the square root-transformed data. The linear mean animal model revealed considerable amounts of additive genetic variance for body weight in Atlantic salmon, with a genetic variance of 0.30 kg^2 and a heritability of 0.36 ± 0.04 (0.32 ± 0.04 and 0.35 ± 0.04 for log- and square root-transformed data, respectively). However, for both the original and transformed data, the two DHGLM models deviated considerably from each other. The DHGLM2 (sire-dam model) and the classical linear mean animal model gave similar estimates of genetic (after appropriate rescaling) and common environmental variances on the mean, whereas the DHGLM1 (animal/sire-dam model) produced a considerably lower (~ −70%) estimate of genetic variance and a much larger (~ +300%) estimate of common environmental variance on the mean. With the linear mean animal model (ignoring heterogeneity of within-family variance), genetic variance is estimated based on covariance between relatives. Hence, even in the presence of genetic heterogeneity of within-family variance, the mean model is expected to produce unbiased estimates of the genetic variance on the mean level, which was also confirmed in a preliminary simulation study (results not shown). Since the linear mean animal model is expected to give unbiased estimates of genetic variance (on the mean), the discrepancy between estimates from the linear mean animal model and the DHGLM1 indicates that the latter model is substantially biased, while the DHGLM2 appears appropriate.

Table 2 Variance components of the classical linear mean animal model and two double hierarchical generalized linear models DHGLM1 and DHGLM2 (untransformed data)

Sub-model	Variance component	Classical	DHGLM1	DHGLM2
Mean	Sire-dam	-	-	0.071±0.010
	Genetic[1]	0.303±0.041	0.099±0.023	0.283±0.039
	Common environment	0.013±0.008	0.063±0.010	0.016±0.008
	Residual	0.521±0.022	-	-
	Heritability	0.362±0.043	-	-
Variance	Sire-dam	-	0.051±0.008	0.043±0.008
	Genetic[1]	-	0.204±0.033	0.174±0.031

[1]Genetic variance is defined as 4*sire-dam variance.

Table 3 Variance components of the classical linear mean model and two double hierarchical generalized linear models DHGLM1 and DHGLM2 (log-transformed data)

Sub-model	Variance component	Classical	DHGLM1	DHGLM2
Mean	Sire-dam	-	-	0.005±0.001
	Genetic[1]	0.023±0.003	0.007±0.002	0.022±0.003
	Common environment	0.001±0.001	0.005±0.001	0.001±0.001
	Residual	0.047±0.002	-	-
	Heritability	0.321±0.040	-	-
Variance	Sire-dam	-	0.042±0.011	0.034±0.009
	Genetic[1]	-	0.166±0.042	0.138±0.037

[1]Genetic variance is defined as 4*sire-dam variance.

Table 4 Variance components of the classical linear mean model and two double hierarchical generalized linear models DHGLM1 and DHGLM2 (root-transformed data)

Sub-model	Variance component	Classical	DHGLM1	DHGLM2
Mean	Sire-dam	-	-	0.019±0.003
	Genetic[1]	0.082±0.011	0.030±0.007	0.078±0.011
	Common environment	0.003±0.002	0.016±0.003	0.004±0.002
	Residual	0.149±0.006	-	-
	Heritability	0.350±0.042	-	-
Variance	Sire-dam	-	0.040±0.008	0.033±0.007
	Genetic[1]	-	0.160±0.032	0.133±0.028

[1]Genetic variance is defined as 4*sire-dam variance.

For the variance sub-model, the estimated sire-dam genetic variances from DHGLM2 for the untransformed and transformed data were 0.04±0.01 and 0.03±0.01 (log- and square root), respectively. The sire-dam variance component is ¼ of the total genetic variance (σ^2_a) for the trait (sire and dam combined explain ½ σ^2_a). Thus, the results obtained with DHGLM2 for all transformations indicate that genetic variance in canalization exists, which implies that families differed in their sensitivity to environmental (and potentially genetic) changes. The importance of the genetic heterogeneity of within-family variance can be illustrated by the following example. Assuming a population with an average within-family variance for body weight of 0.67 (as in the current untransformed dataset), equation (2) implies an expectation on the log scale of $\mu=-0.40$. A log-scale 95% prediction interval for mid-parent variance breeding values, i.e. the interval for which 95% of future families is expected to fall in absence of selection, can then be defined as: $\left[\mu-1.96\sqrt{\tfrac{1}{2}\sigma^2_a},\mu+1.96\sqrt{\tfrac{1}{2}\sigma^2_a}\right]=[-0.98,-0.18]$, which corresponds to within-family variances of body weight in the range $[e^{-0.98},e^{-0.18}]=[0.38,1.18]$. This shows that the variation of the most extreme high families is expected to be approximately three times larger than the extreme low families.

Pearson correlation coefficients between estimated breeding values for the mean and variance sub-models using DHGLM2 were all significantly different from zero, but differed substantially between transformations (0.42, -0.17 and 0.14 for untransformed, log-transformed and square root transformed data, respectively). Thus, transformation of the data makes estimated breeding values for the mean and variance more independent. This may also explain the somewhat reduced genetic variance of the variance model after transformation.

DISCUSSION

The main result of this study is that substantial heterogeneity of within-family variance exists for body weight in Atlantic salmon. Furthermore, the iterative approach revealed that the DHGLM animal mean model (DHGLM1) resulted in severely biased estimates of variance components when employed with cross-sectional data, both when compared with a linear mean animal model and a sire-dam DHGLM (DHGLM2) model.

Statistical Models

Since genetic heterogeneity of variance exists between families, an equivalent genetic heterogeneity must also exist within families. Hence, an animal model should ideally be fitted in both the mean and the variance sub-models of the DHGLM. This would be of particular importance for datasets with repeated records on individuals, and such data structures would allow proper estimation of variances on an individual level. However, the current dataset contained cross-sectional data (one observation per individual), which implies that the variance cannot be estimated on the individual level but only for groups of individuals (families). This may explain why any attempt to model the variance using an animal model did not converge, possibly due to the iterative approach used, where one iterates between the mean and variance sub-models and the models are thus allowed to interact. In earlier studies, more approximate non-iterative methods have been used [7,10, 18], where a linear mean animal model is fitted first and the variance model is subsequently applied to the (log of) squared residuals. For these non-iterative methods (with fixed variance "phenotypes" coming from a linear mean model), animal variance models could be fitted. For the more advanced iterative approaches, more research is needed on the optimization of data structures with respect to proper identification of individual breeding values for the variance.

Since genetic variance in the classical linear mean animal model is estimated based on between-family variation in observed mean phenotypes, the estimated genetic variance is expected to be unbiased. Likewise, the estimated residual variance is expected to reflect the typical level of within-family variation. This would be the case even if the true underlying model includes genetic heterogeneity of residual variance, which was confirmed in a simulation study with 200 families of 40 individuals resulting from a perfect 2 sires $\times$ 2 dams mating system of 100 sires and 100 dams (results not shown). In principle, the DHGLM models are expected to give similar genetic variance estimates on the mean as the linear mean animal model, because these estimates are also based on observed between-family variation. Hence, in the real data analysis, the mean sub-models of the two DHGLM models are expected to give similar results as a linear mean animal model. This was indeed the case for the DHGLM2 model but the DHGLM1 model showed substantial deviations. Hence, the results of this study indicate that the DHGLM1 model is severely biased when the data does not contain repeated records. The two DHGLM models differed in the flexibility of the variance sub-model; in the DHGLM1 model, the Mendelian sampling variance is assumed homogeneous (and restricted to $\frac{1}{2}\sigma_a^2$), while DHGLM2 includes the Mendelian sampling effects in the residuals, and the entire within-family variance is modeled by the variance sub-model, allowing for heterogeneity. The latter is consistent with the concept of canalization, which leads to genetic heterogeneity of variance due to varying sensitivities to potentially both environmental and genetic perturbations [1]. If sensitivities to both environmental and genetic perturbations contribute to the observed heterogeneity of variance, the underlying assumption of DHGLM1 is violated while DHGLM2 would still be appropriate. Our initial hypothesis was therefore that the bias of the DHGLM1 model was due to an incorrect assumption of homogeneous Mendelian sampling variance. To test this, we fitted the DHGLM1 and DHGLM2 models to simulated cross-sectional data (as described above), using either homogeneous (as in DHGLM1) or heterogeneous Mendelian sampling variances (appropriate for DHGLM2); the residual (micro-environmental) variance was always heterogeneous, which is consistent with both models. The results were striking: the DHGLM1 model always resulted in biased estimates of the variance components (similar to the real data analysis), even when the Mendelian sampling variance was homogeneous, while DHGLM2 was unbiased in both cases (results not shown). Hence, DHGLM1 was biased even when its assumptions were true. Furthermore, we found that the bias was restricted to statistical models in which both animal additive genetic and common environmental family effects were fitted in the mean sub-model, i.e., when between-family variance on the mean can potentially be explained by common environmental factors. Thus, it appears that DHGLM models that include random family-specific effects (genetic and/or common environment) in the variance sub-model will use these to describe as much of the within-family variance as possible. This would be appropriate for sire-dam mean sub-models (since the entire within-family variance is included in the residual variance) but is not appropriate for animal mean sub-models (since Mendelian deviations are included in the additive genetic effects). Again, such bias is more likely observed in iterative methods such as those of Rönnegård et al. [12], which allow the mean and variance sub-models to interact with each other. However, a non-iterative procedure, like the one applied in [7, 10, 18], should not be used without critically assessing the assumptions [19], because there is a risk of drawing conclusions from data that contains no (or very little) information to infer genetic heterogeneity of the residual variance.

We have chosen to quantify the genetic heterogeneity in terms of variance components rather than heritability because variance components in the variance sub-model are easier to interpret and are comparable across experiments and species (when an exponential model for the variance is used). For instance, a genetic variance of 0.17 in the variance model, as found for DHGLM2, can be expressed as a standard deviation of 0.41 (i.e. the square-root of 0.17), which implies that a unit standard deviation increase/decrease of the breeding value for the variance increases/decreases the residual variance by 51% and 36%, respectively, because the model is multiplicative and can be shown to be [9]:

$$\sigma^2_{e-new} = \sigma^2_{e-old} \cdot \exp\left(\Delta G_v\right),$$

Where σ^2_{e-old} is the residual variance prior to selection, σ^2_{e-new} is the residual variance after selection and G_v is the genetic gain on the underlying log scale of the variance. Hence, the genetic standard deviation of residual variance is an informative measure for two reasons: (1) it gives an idea of how large the proportional change in the residual variance would be for a unit standard deviation increase/decrease of the residual variance breeding values and (2) this proportional change does not depend on the phenotypic variance and can therefore be compared across experiments and species. The value of using the genetic standard deviation of residual variance was emphasized already in Mulder et al. [11] where they refer to it as the genetic coefficient of variation. An alternative would be to use the Mulder-Hill heritability for the residual variance, i.e. the regression of the variance breeding values on squared phenotype values [11], but this measure is model-specific and not as easy to interpret.

Common environmental effects were not included in the variance sub-models. The main reason for this is that genetic and common environmental effects are difficult to disentangle, and more so with respect to the variance model, since there is only one variance group per family. In contrast, data with repeated records per individual will have one variance group per animal rather than one per family. In our case, the data structure makes the separation between genetic and common environment factors on the variance scale difficult, which was confirmed in the simulation study described above, where the DHGLM2 model was extended with a common environmental family effect in the variance sub-model. Although all the simulated heterogeneity of variance was due to genetics, the model distributed the dispersion effects more or less randomly over genetic and common environmental factors (varying between replicates). This confirms the low power of separating such effects in the variance model, despite a perfect mating system of 2 sires × 2 dams. Furthermore, a preliminary analysis of the real data also revealed that partitioning additive genetic and common environmental effects in the variance model was sensitive to transformation of data, while the mean model was rather robust in this regard. Judging by the results of the linear animal mean model, the common environmental effects were small and not very significant (likelihood ratio test, $P>0.05$, results not shown). Still, we cannot rule out the existence of common environmental effects acting on the variance, and if they exist, the estimated genetic variance of the variance sub-model is probably inflated.

It may be argued that heterogeneity of within-family variances may, to some extent, be explained by simple scaling effects, i.e., variance may be a function of the expected mean. If this is the case, larger genetic and environmental variances should be expected for families with higher mean body weight. Yang et al. [20] studied the effect of (Box-Cox) transformation of the data to stabilize variances. In one dataset (rabbits), evidence for genetic heterogeneity was weaker after transformation, while the opposite was true for other datasets (pigs). Based on the moderately positive correlation between estimated parental breeding values from the mean and variance models for untransformed data (DHGLM2) of the current dataset, scaling effects may explain some of the genetic heterogeneity of within-family variance. The phenotypic data was therefore also log-transformed to remove a potential relationship between mean and variance. This reduced but did not remove heterogeneity of within-family variances (Table 3). Both log and untransformed data are extremes of the Box-Cox transformation (corresponding to $\lambda=0$ and $\lambda=1$, respectively). Therefore, an intermediate transformation (square root, equivalent to $\lambda=0.5$) was also tested but gave similar results to those of the log-transformation (Table 3). Hence, our results indicate substantial heterogeneity of (within-family) variance for body weight in Atlantic salmon, apart from what is expected from simple scaling effects.

Implications for Aquaculture Breeding

The results of this study indicate that there is genetic heterogeneity of (within-family) variance for body weight in Atlantic salmon. Selection for decreased within-family variance may give more homogeneous stocks, which would be an advantage in commercial fish farming. Increased homogeneity is consistent with the concept of canalization, but according to theory, canalization may be due to varying sensitivity to both environmental and genetic perturbations, i.e., the Mendelian sampling variance may also differ between families. Our study does not prove the presence of heterogeneity of Mendelian sampling variance, nor does it rule out the potential existence of such effects. A complication for selection for increased homogeneity based on cross-sectional data, which is by far the most common data structure in aquaculture, is that the Mendelian sampling and micro-environmental residual effects are completely confounded, which implies that pure selection against micro-environmental sensitivity cannot be guaranteed.

Genetic heterogeneity in sensitivity to micro-environmental change (on the individual scale) is one form of genotype by environment interaction. However, genotype by environment interactions may also be present on the macro-environmental scale (diet, climatic region, etc.), potentially causing re-ranking among families. For example, some families may have highly variable phenotypes for one diet but more homogeneous phenotypes for another diet. Hence, one may expect substantial re-ranking of estimated breeding values for variance among cohorts. Such factors would complicate meaningful selection for altered environmental sensitivity in practical breeding programs aimed at providing genetic material for production systems over a wide range of macro-environments.

CONCLUSIONS

Substantial heterogeneity of within-family variance was found for body weight in Atlantic salmon, with a 95% prediction interval for within-family variance ranging from ~0.4 to ~1.2 kg^2, showing that the variation of the most extreme high families is approximately three times larger than the extreme low families. A tendency towards higher variance with higher means (scaling effects) was observed, but heterogeneity of within-family variance existed beyond what could be explained by simple scaling effects. Since there were no repeated observations, the fitting algorithm biased the estimates of the variance components of the DHGLM animal mean model (DHGLM1), but not of a classical linear mean animal model and a pure sire-dam DHGLM (DHGLM2) model. However, selection for increased homogeneity of variance may still prove difficult, especially when based on cross-sectional data. Furthermore, potential macro-environmental changes may make genetic heterogeneity of variance a less stable trait over time and space.

ACKNOWLEDGEMENTS

AquaGen AS is gratefully acknowledged for use of the data. Calculations were done on the TITAN/ABEL computer cluster at University of Oslo, Norway.

AUTHORS' CONTRIBUTIONS

AKS initiated the study. JØ analyzed the data. The manuscript was drafted by AKS and JØ. LR provided the necessary R scripts for the analyses. All authors read and approved the final manuscript.

REFERENCES

1. Waddington CH: Canalization of development and the inheritance of acquired characters. Nature. 1942, 150: 563-565. 10.1038/150563a0.
2. SanCristobal-Gaudy M, Bodin L, Elsen JM, Chevalet C: Genetic components of litter size variability in sheep. Genet Sel Evol. 2001, 33: 249-271. 10.1186/1297-9686-33-3-249.
3. Sorensen D, Waagepetersen R: Normal linear models with genetically structured residual variance heterogeneity: a case study. Genet Res. 2003, 82: 207-222. 10.1017/S0016672303006426.
4. Gutiérrez J, Nieto B, Piqueras P, Ibáñez N, Salgado C: Genetic parameters for canalisation analysis of litter size and litter weight traits at birth in mice. Genet Sel Evol. 2006, 38: 445-462. 10.1186/1297-9686-38-5-445.
5. Ibáñez-Escriche N, Sorensen D, Waagepetersen R, Blasco A: Selection for environmental variation: A statistical analysis and power calculations to detect response. Genetics. 2008, 180: 2209-2226. 10.1534/genetics.108.091678.
6. Damgaard LH, Rydhmer L, Løvendahl P, Grandinson K: Genetic parameters for within-litter variation in piglet birth weight and change in within-litter variation during suckling. J Anim Sci. 2003, 81: 604-610.
7. Mulder HA, Hill WG, Vereijken A, Veerkamp RF: Estimation of genetic variation in residual variance in female and male broiler chickens. Animal. 2009, 3: 1673-1680. 10.1017/S1751731109990668.
8. Hill WG, Mulder HA: Genetic analysis of environmental variation. Genet Res. 2010, 92: 381-395. 10.1017/S0016672310000546.
9. Rönnegård L, Felleki M, Fikse WF, Mulder HA, Strandberg E: Variance component and breeding value estimation for genetic heterogeneity of residual variance in Swedish Holstein dairy cattle. J Dairy Sci. 2013, 96: 2627-2636. 10.3168/jds.2012-6198.
10. Janhunen M, Kause A, Vehviläinen H, Järvisalo O: Genetics of microenvironmental sensitivity of body weight in rainbow trout (*Oncorhynchus mykiss*) selected for improved growth. PLoS One. 2012, 7: e38766-10.1371/journal.pone.0038766.
11. Mulder HA, Bijma P, Hill WG: Prediction of breeding values and selection responses with genetic heterogeneity of environmental variance. Genetics. 2007, 175: 1895-1910. 10.1534/genetics.106.063743.
12. Rönnegård L, Felleki M, Fikse F, Mulder HA, Strandberg E: Genetic heterogeneity of residual variance - estimation of variance components using double hierarchical generalized linear models. Genet Sel Evol. 2010, 42: 8-10.1186/1297-9686-42-8.
13. Ødegård J, Olesen I, Dixon P, Jeney Z, Nielsen H-M, Way K, Joiner C, Jeney G, Ardó L, Rónyai A, Gjerde B: Genetic analysis of common carp (*Cyprinus carpio*) strains. II: Resistance to koi herpesvirus and Aeromonas hydrophila and their relationship with pond survival. Aquaculture. 2010, 304: 7-13. 10.1016/j.aquaculture.2010.03.017.
14. Ødegård J, Olesen I, Gjerde B, Klemetsdal G: Evaluation of statistical models for genetic analysis of challenge test data on furunculosis resistance in Atlantic salmon (*Salmo salar*): Prediction of field survival. Aquaculture. 2006, 259: 116-123. 10.1016/j.aquaculture.2006.05.034.
15. Drangsholt TM, Gjerde B, Ødegård J, Finne-Fridell F, Evensen O, Bentsen HB: Quantitative genetics of disease resistance in vaccinated and unvaccinated Atlantic salmon (*Salmo sala* r *L.*). Heredity. 2001, 107: 471-477.
16. Ødegård J, Meuwissen THE, Heringstad B, Madsen P: A simple algorithm to estimate genetic variance in an animal threshold model using Bayesian inference. Genet Sel Evol. 2010, 42: 29-10.1186/1297-9686-42-29.
17. Gilmour AR, Gogel BJ, Cullis BR, Thompson R: ASReml user guide release 3.0. 2009, VSN International Ltd: Hemel Hempstead
18. Neves HR, Carvalheiro R, Queiroz SA: Genetic and environmental heterogeneity of residual variance of weight traits in Nellore beef cattle. Genet Sel Evol. 2012, 44: 19-10.1186/1297-9686-44-19.
19. Rönnegård L, Valdar W: Recent developments in statistical methods for detecting genetic loci affecting phenotypic variability. BMC Genet. 2012, 13: 63-
20. Yang Y, Christensen OF, Sorensen D: Analysis of a genetically structured variance heterogeneity model using the Box–Cox transformation. Genet Res. 2011, 93: 33-46. 10.1017/S0016672310000418.

CHAPTER 15

HEPATIC TRANSCRIPTOME ANALYSIS OF INTER-FAMILY VARIABILITY IN FLESH N-3 LONG-CHAIN POLYUNSATURATED FATTY ACID CONTENT IN ATLANTIC SALMON

Sofia Morais[1], John B Taggart[1], Derrick R Guy[2], J Gordon Bell[1] and Douglas R Tocher[1]

[1]Institute of Aquaculture, University of Stirling, Stirling, FK9 4LA Scotland, UK.

[2] Landcatch Natural Selection Ltd, Cooperage Way, Alloa, FK10 3LP Scotland, UK.

ABSTRACT

Background

Genetic selection of Atlantic salmon families better adapted to alternative feed formulations containing high levels of vegetable ingredients has been suggested to ensure sustainable growth of aquaculture. The present study aimed to identify molecular pathways that could underlie phenotypic differences in flesh n-3 long-chain polyunsaturated fatty acid (LC-PUFA) levels when fish are fed vegetable oil diets. Liver transcriptome was analyzed and compared in four families presenting higher or lower n-3 LC-PUFA contents at two contrasting flesh total lipid levels.

Results

The main effect of n-3 LC-PUFA contents was in the expression of immune response genes (38% of all significantly affected genes), broadly implicated in the modulation of inflammatory processes and innate immune response. Although genetic evaluations of traits used in the breeding program revealed that the chosen families were not balanced for viral disease resistance, this did not fully explain the preponderance of immune response genes in the transcriptomic analysis. Employing stringent statistical analysis no lipid metabolism genes were detected as being significantly altered in liver when comparing families with high and low n-3 LC-PUFA flesh contents. However, relaxing the statistical analysis enabled identification of potentially relevant effects, further studied by RT-qPCR, in cholesterol biosynthesis, lipoprotein metabolism and lipid transport, as well as eicosanoid metabolism particularly affecting the lipoxygenase pathway. Total lipid level in flesh also showed an important effect on immune response and 8% of significantly affected genes related to lipid metabolism, including a fatty acyl elongase (elovl2), an acyl carrier protein and stearoyl-CoA desaturase.

Citation: Sofia Morais, John B Taggart, Derrick R Guy, J Gordon Bell, Douglas R Tocher, Hepatic transcriptome analysis of inter-family variability in flesh n-3 long-chain polyunsaturated fatty acid content in Atlantic salmon, DOI: 10.1186/1471-2164-13-410

Conclusions

Inter-family differences in n-3 LC-PUFA content could not be related to effects on lipid metabolism, including transcriptional modulation of the LC-PUFA biosynthesis pathway. An association was found between flesh adiposity and n-3 LC-PUFA in regulation of cholesterol biosynthesis, which was most likely explained by variation in tissue n-3 LC-PUFA levels regulating transcription of cholesterol metabolism genes throughsrebp2. **A preponderance of immune response genes significantly affected by n-3 LC-PUFA contents could be potentially associated with disease resistance, possibly involving anti-inflammatory actions of tissue n-3 LC-PUFA through eicosanoid metabolism. This association may have been fortuitous, but it is important to clarify if this trait is included in future salmon breeding programmes.**

BACKGROUND

Aquaculture is the fastest growing animal production activity worldwide, supplying an increasing proportion of fish for human consumption, estimated at around 50% of total supply in 2008 [1]. However, the growth of marine aquaculture is threatened by its excessive reliance on fishmeal (FM) and fish oil (FO) from wild stocks for the production of fish feeds, which is also an ecologically unsound practice. Almost 89% of the total global production of FO is currently used by aquaculture [2] and the future of this activity strongly depends on the reduction of dependency on FM and FO and its replacement with alternative ingredients, such as vegetable oils (VO) and plant meals, while maintaining fish welfare and health benefits for the human consumer. Fish are highly nutritious components of the human diet and the main source of essential n-3 long-chain polyunsaturated fatty acids (LC-PUFA). The beneficial effects of fatty acids, such as eicosapentaenoic acid (EPA) and docosahexaenoic acid (DHA), are numerous and important, including protection against a range of cardiovascular and inflammatory diseases, as well as neurological disorders [3]. Atlantic salmon (Salmo salar) can grow well on diets where FO has been completely replaced by VO but this results in lower levels of n-3 LC-PUFA in their flesh, compromising their nutritional value and health-promoting effects to the human consumer [4].

The use of selective breeding programs to enhance traits of commercial importance is becoming increasingly common in aquaculture [5]. It has been suggested that combining genetic selection for fish that are more efficient in retaining and/or biosynthesising n-3 LC-PUFA with changes in commercial diet formulations (i.e., higher levels of inclusion of VO) might be a viable strategy to meet growing worldwide demands for aquaculture products, without loss of nutritional value. Previous studies have shown wide individual variability in the capacity of Atlantic salmon to retain or synthesize n-3 LC-PUFA when fed VO diets [6]. Following this, Leaver et al. [7] demonstrated that deposition and/or retention in flesh of dietary n-3 LC-PUFA, EPA and DHA, is a highly heritable trait ($h^2=0.77$) in salmon. These results have prompted further interest in large-scale in-depth studies exploring genotype × nutrient interactions in salmon, analysing whether the genetic background of the fish could affect the physiological response to complete dietary replacement of FO by VO [8, 9]. In the present study we investigated this further by analyzing the transcriptome from liver, the primary site of synthesis and export of lipids to extra-hepatic tissues including flesh, from four Atlantic salmon families phenotyped for different levels of flesh n-3 LC-PUFA content in response to a VO diet. The objective was to identify gene pathways and molecular mechanisms that might underlie differences in flesh n-3 LC-PUFA contents when salmon families were fed the same low LC-PUFA diet. Furthermore, because n-3 LC-PUFA level is a component of, and associated with total lipid content in a tissue, a factorial design was chosen in which families containing higher and lower proportions of flesh n-3 LC-PUFA were compared at similar flesh total lipid contents.

RESULTS

Family Lipid Contrasts

Lipid analysis of fifty Atlantic salmon families showed flesh lipid levels ranging from 2.3 to 5.7% of wet weight, with relative and absolute n-3 LC-PUFA contents varying from 71 to 136 (μg/mg lipid) and 314 to 554 (mg/100 g flesh), respectively. As expected, high correlations between lipid level and n-3 LC-PUFA content were observed ($r=-0.65$ or 0.70 for relative and absolute contents, respectively), indicating that only families with near identical lipid levels should be compared to avoid confounding effects associated with the lipid level factor (additional file 1). Using these results, four families were identified; two with high (H) levels of lipid (5 g/100 g flesh), and two with low (L) levels of lipid (3.5 g/100 g flesh) and, within each level of total lipid, the two families had significantly contrasting relative n-3 LC-PUFA contents (similarly termed H and L). Therefore, the four families constituted a 2 x 2 factorial design, labelling each family by the total lipid/n-3 LC-PUFA contrasts as LL, LH, HL and HH, respectively (Table 1), which allowed comparisons of n-3 LC-PUFA contents at a constant lipid level and, similarly, comparisons of total lipid at constant n-3 LC-PUFA levels.

Table 1 Lipid phenotypes of families chosen for molecular analysis

Family	Total Lipid (g/100 g flesh)	n-3 LC-PUFA		ARA	
		Relative (μg/mg lipid)	Absolute (mg/100 g flesh)	Relative (μg/mg lipid)	Absolute (mg/100 g flesh)
LL	3.5±0.4	105.1±3.8 *	363.0±30.3	3.45±0.13 *	11.93±1.00
LH	3.5±0.7	133.8±4.8 *	468.0±92.9	4.25±0.06 *	14.87±2.94
HL	5.1±0.8	83.7±14.0 *	426.9±103.2	2.70±0.53	13.81±3.80
HH	5.0±0.7	112.0±7.9 *	554.3±50.7	3.67±0.63	18.04±2.09

Indicated are levels of total lipid (g/100 g flesh, wet weight), and relative and absolute contents of total n-3 LC-PUFA and of the n-6 LC-PUFA arachidonic acid (ARA) in the flesh (n=3 pools) of the 4 Atlantic salmon families used in the transcriptomic analysis. Asterisks signify significant differences between the two families with the same total lipid content.

Microarray analysis

A two-way ANOVA analysis employing the Benjamini-Hochberg multiple testing correction (at a significance level of 0.05 and fold change cut-off of 1.2) was performed to assess significant effects of the factors 'n-3 LC-PUFA' and 'total lipid', which returned lists with 43, 109 and 66 entities for each factor and their interaction, respectively. These significant lists were then analyzed in detail and genes were categorized according to their biological function, in some cases inferred from mammalian homolog genes (Tables 2 and 3). Because the focus of this work was to identify genes that are specifically affected by the trait n-3 LC-PUFA content without the interference of total lipid level, the interaction between the two factors is not presented. Distribution of genes by categories of biological function (excluding 12-18% non-annotated probes, those representing the same gene or with a miscellaneous function) revealed that there was a preponderance of immune response genes significantly affected by both factors: 38% by 'n-3 LC-PUFA' and 29% by 'total lipid'. Gene Ontology (GO) enrichment analysis, which enables the identification of GO terms significantly enriched in the input entity list when compared to the whole array dataset, revealed that this is a true over-representation in the list of genes significantly affected by the 'total lipid' factor (Additional file 2). In contrast, genes involved in the broad category of metabolism only corresponded to 21% of genes significantly affected by n-3 LC-PUFA content and 30% by the 'total lipid' factor. Surprisingly, no lipid metabolism genes were significantly altered in liver when comparing families with higher and lower contents of n-3 LC-PUFA in their flesh, while about 8% were significantly affected by flesh lipid level. Within these, noteworthy was the down-regulation of fatty acyl elongase (elovl2) and of acyl carrier protein transcripts in salmon having a higher lipid level in their flesh, independent of LC-PUFA content. On the other hand, stearoyl-CoA desaturase was significantly up-regulated in fish with higher lipid levels in their flesh. The interaction between both factors is not presented but it did not substantially affect lipid metabolism genes. Finally, and in general, genes involved in regulation of transcription and signalling were also prevalent, 17% in response to 'n-3 LC-PUFA' and 12-13% to 'total lipid'.

Table 2 Liver transcripts differentially expressed when examining the explanatory power of the factor 'n-3 LC-PUFA' content in flesh of four families of Atlantic salmon fed the same low FM/high VO diet

Probe name	Gene	High/Low LC-PUFA		p-value
		L Lipid (LH/LL)	H Lipid (HH/HL)	
Metabolism (21%)				
Energy metabolism (4%)				
Ssa#S31995754	Cytochrome c oxidase subunit 2	7.1	- 1.0	0.0001
Protein and amino acid metabolism (13%)				
Ssa#CB502423	N-acetylated alpha-linked acidic dipeptidase-like 1	3.7	79.8	0.0011
Ssa#STIR03710	Proteasome subunit beta type-9 precursor	- 14.9	- 1.1	0.0026
Ssa#S31993738_S	Ubiquitin-conjugating enzyme E2	- 3.3	- 1.4	0.0188
Xenobiotic and oxidant metabolism (4%)				
Ssa#S18892279	Cytochrome P450 1A	1.8	1.4	0.0096
Ssa#STIR00161_2	Cytochrome P450 1A	1.9	1.4	0.0160
Ssa#STIR00161_3	Cytochrome P450 1A	2.5	1.8	0.0213
Con_CANDS_13	Cytochrome P450 1A	2.0	1.4	0.0494

Translation (8%)				
Ssa#STIR26031	Mitochondrial 28 S ribosomal protein S34	- 11.1	1.5	0.0017
Ssa#S18867312	Ribonuclease UK114	1.4	1.4	0.0450
Regulation of transcription (17%)				
Ssa#S35510106	Zinc finger protein 367	- 1.4	6.5	0.0026
Omy#S18104058	Zinc finger protein 235	4.1	2.3	0.0058
Ssa#TC111702	Reverse transcriptase-like protein	- 1.3	4.6	0.0104
Ssa#TC112002	Retinoid X receptor beta	1.0	- 19.1	0.0134
Signalling and protein modification (17%)				
Ssa#STIR15776	Dolichyl-diphosphooligosaccharide-protein glycosyltransferase subunit 4	- 1.7	123.3	0.0000
Ssa#STIR23530	Dolichyl-diphosphooligosaccharide-protein glycosyltransferase subunit 4	- 1.9	115.6	0.0000
Ssa#STIR03642	Dolichyl-diphosphooligosaccharide-protein glycosyltransferase subunit 4	- 1.7	68.1	0.0000
Ssa#STIR01857	Dolichyl-diphosphooligosaccharide-protein glycosyltransferase subunit 4	- 1.5	125.5	0.0001
Ssa#STIR31840	Sphingomyelin phosphodiesterase acid transcript variant 1	2.1	2.3	0.0303
Ssa#STIR07369	RAF1 proto-oncogene serine/threonine-protein kinase	1.8	1.3	0.0343
Ssa#S35552908	Tyrosine 3-monooxygenase/tryptophan 5-monooxygenase activation protein, epsilon polypeptide	- 1.6	- 2.1	0.0455
Immune response (38%)				
Ssa#S35536179	similar to novel NACHT domain containing protein	2.4	1.8	0.0026
Ssa#S35516341	Tripartite motif-containing protein 25 (trim25)	5.2	6.7	0.0066
Ssa#S30241035	MHC class I	- 1.3	- 26.0	0.0079
Ssa#STIR02298	c-c motif chemokine 13 precursor (ccl13)	4.4	3.0	0.0100
Ssa#S35581943	Myelin and lymphocyte protein (mal)	- 22.2	- 25.0	0.0100
Ssa#KSS3969	Leukocyte cell-derived chemotaxin 2 precursor (lect2)	4.6	7.1	0.0134
Ssa#STIR15577	Tissue factor pathway inhibitor a	22.3	1.7	0.0135
Omy#S15332652	Pentraxin	- 1.1	199.4	0.0172
Ssa#TC70262	Cathepsin K	1.4	2.3	0.0279
Ssa#S35558945	Tripartite motif-containing protein 25 (trim25)	4.8	6.5	0.0347
Miscellaneous/unknown function				
Ssa#STIR09736	Transmembrane protein 42	2.3	1.9	0.0009
Ssa#S35519407	Tetratricopeptide repeat protein 23	3.7	1.0	0.0011
Ssa#STIR02307	Family with sequence similarity member a (FAM36A)	1.3	3.6	0.0026
Omy#BX309274	X-ray repair complementing defective repair in Chinese hamster cells 3	- 1.0	- 8.0	0.0100
Ssa#STIR21287	Solute carrier family 30 (zinc transporter) member 7	2.1	1.5	0.0100
Ssa#S35521859_S	Family with sequence similarity member a (FAM36A)	- 1.1	4.8	0.0358
Ssa#S18842295	Alveolin	1.7	4.6	0.0422

Results were obtained by two-way ANOVA analysis ($p < 0.05$; fold change cut-off of 1.2) with Benjamini-Hochberg multiple testing correction. Transcripts (43 probes; 12% unknowns) are arranged by categories of biological function and, within these, by p-value. Indicated are also the probe names and the expression ratio determined separately between families with high vs low n-3 LC-PUFA contents, for each total lipid level. Percentages of distribution of genes within each category do not include non-annotated probes, those representing the same gene or with a miscellaneous function.

Table 3 Liver transcripts differentially expressed when examining the explanatory power of the variable 'total lipid' level in flesh of four families of Atlantic salmon fed the same low FM/high VO diet

Probe name	Gene	High/Low Lipid		p-value
		Lo LC-PUFA (HL/LL)	Hi LC-PUFA (HH/LH)	
Metabolism (30%)				
Lipid metabolism (8%)				
Ssa#STIR02479	Butyrophilin subfamily 2 member A2 precursor	7.76	1.23	0.0001
Ssa#STIR03356	Acyl carrier protein, mitochondrial precursor	- 5.17	- 1.55	0.0004
Ssa#STIR00151_3	Polyunsaturated fatty acid elongase (elovl2)	- 1.40	- 1.87	0.0190
Ssa#STIR00151_2	Polyunsaturated fatty acid elongase (elovl2)	- 1.28	- 1.97	0.0238
Con_CANDS_03	Polyunsaturated fatty acid elongase (elovl2)	- 1.33	- 1.82	0.0310
Ssa#STIR24266	Acyl carrier protein, mitochondrial precursor	- 2.88	- 1.52	0.0312
Ssa#STIR21802	Stearoyl-CoA desaturase	3.53	3.17	0.0324
Ssa#STIR00151_4	Polyunsaturated fatty acid elongase (elovl2)	- 1.41	- 1.82	0.0393
Ssa#S35501441_S	Acyl carrier protein, mitochondrial precursor	- 2.65	- 1.44	0.0457
Ssa#KSS4634	Stearoyl-CoA desaturase	3.52	2.82	0.0486
Energy metabolism and generation of precursor metabolites (11%)				
Ssa#S31995754	Cytochrome c oxidase subunit 2	6.57	- 1.12	0.0002
Ssa#STIR03496	LYR motif-containing protein 5	1.20	1.48	0.0233
Ssa#STIR03740	6-Phosphogluconate dehydrogenase, decarboxylating	1.60	1.93	0.0238
Ssa#TC106663	Creatine kinase, testis isozyme	1.76	1.46	0.0380
Ssa#STIR19155	NADH dehydrogenase iron-sulfur protein 7	1.21	2.30	0.0409
Ssa#STIR12872	V-type ATPase B subunit	- 1.19	- 1.24	0.0419
Protein and amino acid metabolism (4%)				
Ssa#STIR03710	Proteasome subunit beta type-9 precursor	1.94	25.52	0.0001
Ssa#S30294677	Serine protease HTRA1 precursor (Serine protease 11)	3.42	2.44	0.0089
Carbohydrate metabolism (4%)				
Ssa#TC106766	Glycogenin	8.64	1.67	0.0093
Ssa#S30290426	Serine dehydratase-like	1.60	1.72	0.0393
Xenobiotic and oxidant metabolism (4%)				
Ssa#S35671757	Extracellular superoxide dismutase	2.18	1.09	0.0061
Ssa#STIR25620	Microsomal glutathione S-transferase 1	2.45	2.72	0.0096
Omy#S18159333	Microsomal glutathione S-transferase 1	2.47	3.15	0.0233
Transport (2%)				
Ssa#S35599996	ATPase, H+transporting, lysosomal, V1 subunit H	45.10	15.16	0.0000
Translation (4%)				

Ssa#STIR26031	Mitochondrial 28 S ribosomal protein S34	1.19	19.22	0.0000
Ssa#S30241612	39 S ribosomal protein L16	1.30	1.51	0.0324
Regulation of transcription (12%)				
Ssa#STIR06878	Cytosolic iron-sulfur protein assembly 1	- 1.16	- 1.56	0.0106
Ssa#TC112002	Retinoid X receptor beta	1.20	- 16.00	0.0254
Omy#S15320037	SWI/SNF-related matrix-associated actin-dependent regulator of chromatin a4	- 1.97	- 1.53	0.0324
Ssa#CN181280	alpha thalassaemia mental retardation X-linked protein	- 3.05	- 6.90	0.0419
Ssa#S35697153	YLP motif containing 1	- 1.49	- 1.32	0.0428
Ssa#S35486480	Zinc finger protein 492	- 1.08	- 2.29	0.0452
Signalling and protein modification (13%)				
Ssa#STIR15776	Dolichyl-diphosphooligosaccharide-protein glycosyl-transferase subunit 4	- 84.27	2.44	0.0000
Ssa#STIR23530	Dolichyl-diphosphooligosaccharide-protein glycosyl-transferase subunit 4	- 83.34	2.60	0.0000
Ssa#STIR03642	Dolichyl-diphosphooligosaccharide-protein glycosyl-transferase subunit 4	- 49.76	2.34	0.0001
Ssa#STIR01857	Dolichyl-diphosphooligosaccharide-protein glycosyl-transferase subunit 4	- 74.19	2.51	0.0011
Ssa#S35516167	14 kDa phosphohistidine phosphatase	1.57	1.34	0.0019
Ssa#STIR11086	Lunatic fringe	- 5.47	- 2.00	0.0154
Ssa#S30263209	HtrA serine peptidase 3	3.71	2.33	0.0154
Ssa#STIR22920	HCLS1-associated protein X-1	1.73	2.33	0.0390
Ssa#S35701148	Phosphatase and actin regulator 3	- 1.52	- 65.53	0.0404
Omy#S15290792	Serine/threonine-protein kinase PLK2	1.70	2.03	0.0468
Immune response (29%)				
Ssa#STIR00130_4	Complement factor H precursor (cfh)	- 1.43	- 1.67	0.0012
Ssa#S35536179	novel NACHT domain containing protein	- 1.89	- 2.41	0.0013
Ssa#S30241035	MHC class I	- 1.55	- 30.59	0.0024
Ssa#S35516341	Tripartite motif-containing protein 25 (trim25)	- 5.64	- 4.34	0.0125
Ssa#S18834140	Complement factor H precursor (cfh)	- 1.34	- 1.57	0.0131
Omy#utu04b09	Complement factor H precursor (cfh)	- 1.56	- 1.70	0.0134
Ssa#STIR15577	Tissue factor pathway inhibitor a	21.03	1.59	0.0151
Ssa#S35558236	C-type lectin domain family 16, member A	4.59	2.25	0.0190
Ssa#STIR10409	CD83 antigen precursor	1.98	1.68	0.0238
Omy#S15332652	Putative pentraxin	- 1.33	159.84	0.0246
Ssa#S35685271	GTPase IMAP family member 7	−168.44	- 4.09	0.0263
Ssa#S35551959	Duodenase-1	- 2.94	- 2.77	0.0349
Ssa#S35558945	Tripartite motif-containing protein 25 (trim25)	- 5.74	- 4.28	0.0373
Ssa#S35685273	Lactose-binding lectin l-2 precursor putative	1.56	4.04	0.0380

Ssa#S31977617	Scavenger receptor cysteine-rich type 1 protein m130	- 1.31	- 2.05	0.0404
Ssa#STIR04893	Lactose-binding lectin l-2 precursor putative	1.57	4.15	0.0419
Ssa#S30264865	Indoleamine-pyrrole 2,3-dioxygenase	2.83	1.06	0.0444
Ssa#S31981622	Granzyme A	- 5.31	- 4.33	0.0444
Ssa#S35685718	CD83 antigen precursor	2.03	1.58	0.0462
Ssa#KSS3969	Leukocyte cell-derived chemotaxin 2 precursor (lect2)	3.16	4.86	0.0486
Structural proteins (10%)				
Ssa#STIR03004	Troponin I, slow skeletal muscle	- 202.28	- 135.03	0.0000
Ssa#CK891024	Vitelline envelope protein gamma	1.07	- 15.43	0.0002
Ssa#STIR02053	Troponin I, slow skeletal muscle	- 70.21	- 10.07	0.0005
Omy#S34312003	similar to Titin (Connectin)	- 44.29	- 29.86	0.0013
Omy#S15317515	Type I keratin E7	5.08	2.43	0.0167
Ssa#STIR05140	Troponin I, slow skeletal muscle	- 13.16	- 14.20	0.0124
Ssa#STIR08802	Myosin regulatory light chain 2, smooth muscle isoform	1.55	1.34	0.0324
Miscellaneous/unknown function				
Ssa#S35519407	Tetratricopeptide repeat protein 23	3.64	1.02	0.0012
Ssa#S35584894	NCK-associated protein 1-like	- 1.82	- 1.70	0.0019
Ssa#S35521859_S	FAM36A	1.11	5.75	0.0047
Ssa#STIR02307	FAM36A	1.16	3.31	0.0058
Ssa#TC87798	Envelope polyprotein	- 1.83	- 1.03	0.0233
Ssa#STIR20319	TPA-induced transmembrane protein	- 1.09	1.75	0.0233
Ssa#TC110493_S	Beta-3-galactosyltransferase	6.80	1.80	0.0233
Ssa#S30270166	Transmembrane protein 37	2.13	1.90	0.0254
Ssa#STIR08658	Tetraspanin-3 putative	- 3.20	- 1.47	0.0312
Ssa#S30293470	Deoxyribonuclease gamma precursor	2.19	2.85	0.0366
Ssa#S35667723	Dynein, cytoplasmic 1, light intermediate chain 2	- 1.90	- 1.48	0.0366
Ssa#CX354464	Retinol dehydrogenase 12	2.37	2.35	0.0373
Ssa#S35582016	Type I iodothyronine deiodinase	1.51	1.74	0.0380
Ssa#S35515630	C7orf57	1.32	1.13	0.0380
Ssa#STIR15617	FAM36A	1.02	4.60	0.0403
Ssa#STIR31448	osteopontin-like	- 4.84	- 1.58	0.0419
Ssa#STIR26263	Anterior gradient protein 2 homolog precursor	3.72	1.68	0.0419
Ssa#TC65497_S	Adipocyte plasma membrane-associated protein	2.50	1.64	0.0419
Ssa#TC93681	Ring finger protein 44	- 1.23	- 1.17	0.0455
Ssa#EG819142	Glutaminyl-peptide cyclotransferase-like	- 1.11	- 1.49	0.0455

Results were obtained by two-way ANOVA analysis ($p<0.05$; fold change cut-off of 1.2) with Benjamini-Hochberg multiple testing correction. Transcripts (109 probes; 18% unknowns) are arranged by categories of biological function and, within these, by p-value. Indicated are also the probe names and the expression ratio determined separately between families with high vs low total lipid level, for each n-3 LC-PUFA grouping. Percentages of distribution of genes within each category do not include non-annotated probes, those representing the same gene or with a miscellaneous function.

Therefore, the results did not identify lipid metabolism pathways that might underlie differences in flesh n-3 LC-PUFA composition between families. However, previous studies demonstrated that hepatic metabolism genes typically show only low fold changes, even when comparing highly contrasting nutritional compositions (e.g., inclusion of 100% FO versus 100% VO in diets [8, 9]), compared to immune response genes that tend to be regulated with higher magnitudes of change [10]. Hence, nutritional data such as the present data have been analysed previously without multiple testing correction and this was found to result in relevant biological interpretations, when validated by reverse transcription real time quantitative PCR (RT-qPCR) [9, 11]. For this reason, we examined the significant effects of ‹n-3 LC-PUFA› without the correction, and from within the list containing 1951 features (significance level, 0.05; fold change cut-off, 1.2), we identified and categorized all 48 lipid metabolism transcripts present (Table 4). An effect on cholesterol metabolism was apparent for the factor ‹n-3 LC-PUFA›, with several genes of the biosynthesis pathway and its regulation being down-regulated in fish with a high n-3 LC-PUFA phenotype. In addition, glycerophospholipid synthesis, lipid hydrolysis and eicosanoid synthesis and metabolism were also affected, while other genes were associated with lipid and fatty acid transport, fatty acid synthesis and regulation of lipid metabolism.

Table 4 Lipid metabolism genes differentially expressed in the liver of fish having high or low flesh n-3 LC-PUFA contents

Probe name	Gene	High/Low LC-PUFA		p- value
		L Lipid (LH/LL)	H Lipid (HH/HL)	
Cholesterol biosynthesis, metabolism and regulation				
Omy#S15288895	Sterol-C4-methyl oxidase-like	1.84	1.32	0.0033
Ssa#STIR00031_3	7-dehydrocholesterol reductase (7dchr)	- 1.55	- 1.48	0.0036
Ssa#S30286041	7-dehydrocholesterol reductase (7dchr)	- 1.48	- 1.42	0.0120
Omy#CF752841	Sterol regulatory element-binding transcription factor 2 (srebp2)	- 1.71	- 1.31	0.0170
Ssa#TC102141	Cytochrome P450, family 27, subfamily A, polypeptide 1	- 3.97	- 1.04	0.0192
Ssa#STIR16974	7-dehydrocholesterol reductase (7dchr)	- 1.46	- 1.31	0.0195
Ssa#AM402497	Hydroxymethylglutaryl-CoA synthase 1	- 2.39	1.21	0.0199
Ssa#STIR00031_4	7-dehydrocholesterol reductase (7dchr)	- 1.41	- 1.46	0.0212
Ssa#STIR00098_4	Isopentenyl-diphosphate delta isomerise (ipi)	- 2.12	- 1.15	0.0228
Ssa#S18867829	7-dehydrocholesterol reductase (7dchr)	- 1.33	- 1.25	0.0228
Ssa#DY741343	Lanosterol 14-alpha demethylase	- 1.42	- 1.28	0.0346
Omy#S22913656	Acetoacetyl-CoA synthetase	- 1.77	- 2.52	0.0358
Ssa#STIR00033_3	Mevalonate kinase (mev)	- 1.54	- 1.08	0.0393
Ssa#CA064135	Vigilin	2.80	- 1.22	0.0423
Ssa#DW582478	Cytochrome P450, family 8, subfamily B, polypeptide 1	- 1.26	- 1.50	0.0449
Ssa#STIR00102_3	Squalene epoxidase	- 2.11	- 1.36	0.0468
Glycerophospholipid synthesis				
Ssa#STIR39152_S	Lipid phosphate phosphohydrolase 2 (lpp2)	1.20	1.19	0.0177
Ssa#KSS4003	Chka protein	1.18	1.53	0.0264
Ssa#S35538062_S	Monoacylglycerol O-acyltransferase 1 (mgat)	1.20	1.51	0.0303
Ssa#S31963704	Diacylglycerol O-acyltransferase homolog 2	1.41	1.02	0.0358
Ssa#S48418830	Phosphatidylglycerophosphate synthase 1	- 1.52	- 1.72	0.0369

Fatty acid synthesis				
Ssa#KSS4155	Trans-2-enoyl-CoA reductase, mitochondrial precursor	- 1.26	- 1.25	0.0440
Intracellular fatty acid transport				
Ssa#S35585414	Acyl-CoA-binding protein	10.47	1.65	0.0081
Ssa#DY703528	Fatty acid-binding protein, intestinal	- 1.25	- 1.46	0.0331
Ssa#CB509140	Fatty acid-binding protein, intestinal	- 1.20	- 1.35	0.0375
Ssa#STIR04578	Fatty acid-binding protein, heart	1.22	1.47	0.0401
Lipid transport				
Ssa#CK898816	Low density lipoprotein receptor-related protein 1 (lrp1)	- 1.14	- 1.39	0.0341
Ssa#S32008850	Apolipoprotein A-IV precursor (apoa4a)	- 1.43	- 1.01	0.0377
Omy#BX318293	Low density lipoprotein receptor-related protein 1 (lrp1)	- 3.61	- 1.67	0.0404
Ssa#S18866963	Apolipoprotein A-IV precursor (apoa4b)	- 1.39	- 1.43	0.0428
Lipid hydrolysis				
Ssa#S18887340	Acyl-coenzyme A thioesterase 5	1.64	1.18	0.0007
Ssa#STIR02708	Isoamyl acetate-hydrolyzing esterase 1 homolog	1.54	1.12	0.0013
Ssa#DW007099	Similar to patatin-like phospholipase domain containing 7	- 3.75	- 2.37	0.0070
Ssa#S31963297	Acyl-CoA thioesterase 11	1.35	1.38	0.0097
Ssa#STIR07750	N-acylsphingosine amidohydrolase 1	- 1.14	- 1.34	0.0122
Ssa#STIR05034	Isoamyl acetate-hydrolyzing esterase 1 homolog	1.74	1.01	0.0208
Omy#CA366823	Acyl-coenzyme A thioesterase 3	1.33	2.17	0.0248
Ssa#STIR22551	Lipoprotein lipase (lpl)	1.31	1.31	0.0346
Eicosanoid synthesis and metabolism				
Ssa#TC68569	Thromboxane-A synthase (thas)	- 1.33	- 1.67	0.0269
Ssa#TC110080	Phospholipase A2 (pla2g4)	1.26	1.08	0.0296
Omy#TC147730	Prostaglandin I2 (prostacyclin) synthase (ptgis)	- 2.53	- 4.25	0.0349
Ssa#S35581706	15-hydroxyprostaglandin dehydrogenase	- 1.51	- 1.09	0.0400
Ssa#EG930234	Arachidonate 5-lipoxygenase (alox5)	1.91	1.55	0.0449
Regulation of lipid metabolism				
Ssa#TC112002	Retinoid X receptor beta	1.00	- 19.09	0.0000
Ssa#KSS2129	Adiponectin receptor protein 1	- 1.30	- 1.31	0.0098
Ssa#CA056493	Angiopoietin-like 6	- 1.63	- 1.07	0.0290
Ssa#S35490606	Adiponectin, C1Q and collagen domain containing, like	1.34	1.66	0.0441
Ssa#S18888608	Adiponectin receptor protein 1	- 1.38	- 1.40	0.0451

Results were obtained by two-way ANOVA analysis ($p<0.05$; fold change cut-off of 1.2) without multiple testing correction ($n=1951$ total features). Transcripts are arranged by functional categories and, within these, by p-value. Indicated are also the probe names and the expression ratio between families determined separately for each total lipid level.

Validation of results by RT-qPCR

To validate the microarray analysis results, expression of selected genes was quantified by RT-qPCR. These genes were chosen from lipid metabolism pathways that were more highly affected by the factor 'n-3 LC-PUFA', and also included immune response genes, which was the category most highly affected by both 'n-3 LC-PUFA' and 'total lipid' factors. In addition, the expression of two fatty acyl desaturases (Δ5fad and Δ6fad) and one elongase (elovl2), which are typically responsive to dietary levels of n-3 LC-PUFA were also determined. The LC-PUFA biosynthesis pathway was not identified by the microarray analysis as being differentially expressed in families with different n-3 LC-PUFA flesh contents but, given the potential importance of this pathway in determining n-3 PUFA phenotypes, we specifically aimed to verify this result. The RT-qPCR results confirmed that genes involved in LC-PUFA biosynthesis were not differentially expressed in families with higher and lower levels of n-3 LC-PUFA (Table 5). Furthermore, the RT-qPCR results confirmed significant down-regulation of genes involved in hepatic cholesterol biosynthesis, such as isopentenyl-diphosphate isomerase (ipi), 7-dehydrocholesterol reductase (7dchr) and sterol regulatory element-binding protein 2 (srebp2) in families containing higher levels of n-3 LC-PUFA in their flesh although this was only observed when this phenotype was also associated with low lipid level, except for 7dchr, which was significantly down-regulated irrespective of lipid level. With regards to lipoprotein metabolism (lipid transport) genes, general trends such as the magnitude and direction of change were broadly similar between the microarray and the RT-qPCR analysis for the high versus low n-3 LC-PUFA comparison at low lipid contents, although RT-qPCR results were not significant. In the case of high lipid contents, the match between microarray and RT-qPCR data was less consistent, except for lipoprotein lipase (lpl), which was similarly up-regulated albeit non-significantly. Up-regulation of the glycerophospholipid biosynthesis pathway in fish with higher n-3 LC-PUFA contents was also indicated when associated with high lipid levels, significant for monoacylglycerol O-acyltransferase 1 (mgat). With regards to the eicosanoid biosynthesis pathway, the microarray results could only be confirmed for arachidonic 5-lipoxygenase (alox5). Validation of lipid metabolism genes affected by the 'total lipid' factor (Table 6) confirmed the lower expression of elovl2 in salmon presenting higher lipid levels in their flesh, independent of LC-PUFA content. Finally, good agreement was found between the microarray and RT-qPCR results for immune response genes in response to both ‹n-3 LC-PUFA› (Table 5) and ‹total lipid› (Table 6) factors.

Table 5 Validation of microarray results and expression of genes of interest in relation to the factor 'n-3 LC-PUFA level'

Gene	Low Lipid LH/LL		High Lipid HH/HL	
	Microarray	**RT-qPCR**	**Microarray**	**RT-qPCR**
LC-PUFA biosynthesis				
Δ5fad		−1.19		1.17
Δ6fad		1.13		−1.21
elovl2		1.14		1.06
Cholesterol biosynthesis				
ipi	−2.13	**−3.92**	−1.15	1.32
mev	−1.54	−1.51	−1.08	1.06
7dchr	−1.33 to −1.54	**−1.47**	−1.25 to −1.47	**−1.34**
srebp2	−1.72	**−1.68**	−1.30	1.60
Lipid transport and lipoprotein metabolism				
lrp1	−1.14 to −3.57	−1.36	−1.39 to −1.67	1.24
apoA4a	−1.43	−1.09	1.00	1.32
apoA4b	−1.39	−1.48	−1.43	−1.10
lpl	1.31	1.23	1.31	1.38
Glycerophospholipid synthesis				
lpp2	1.20	−1.19	1.19	1.30

mgat	1.20	1.04	1.51	1.78
Eicosanoid biosynthesis				
alox5	1.91	**1.48**	1.55	**1.62**
pla2g4	1.26	−1.08	1.08	1.06
thas	−1.33	−1.08	−1.67	**1.34**
ptgis	−2.53	−1.25	−4.25	1.27
Immune response				
mal	−22.20	**−3.70**	−25.00	**−5.00**
ccl13	4.40	**5.98**	3.00	2.14
trim25	5.20	**2.80**	6.70	**2.51**
lect2	4.60	1.92	7.10	**7.57**

Values represent the expression ratios between high PUFA / low PUFA, for fish containing either low or high total lipid levels in their flesh, obtained by microarray analysis or RT-qPCR. Expression ratios in bold were significant by REST2008 analysis of RT-qPCR results.

Delta5 and 6 fatty acyl desaturases (Δ5fad and Δ6fad); fatty acyl elongase (elovl2); isopentenyl-diphosphate isomerase (ipi); mevalonate kinase (mev); 7-dehydrocholesterol reductase (7dchr); sterol regulatory element-binding protein 2 (srebp2); low density lipoprotein receptor-related protein 1 (lrp1); apolipoprotein A-IV (apoa4a and apoa4b); lipoprotein lipase (lpl); lipid phosphate phosphohydrolase 2 (lpp2); monoacylglycerol O-acyltransferase 1 (mgat); arachidonate 5-lipoxygenase (alox5); phospholipase A2 (pla2g4); thromboxane-A synthase (thas); prostaglandin I2 (prostacyclin) synthase (ptgis); myelin and lymphocyte protein (mal); c-c motif chemokine 13 precursor (ccl13); tripartite motif-containing protein 25 (trim25); leukocyte cell-derived chemotaxin 2 precursor (lect2).

Table 6 Validation of microarray results and expression of genes of interest in relation to the factor 'Lipid level'

Gene	**Low n-3 LC-PUFA HL/LL**		**High n-3 LC-PUFA HH/LH**	
	Microarray	**RT-qPCR**	**Microarray**	**RT-qPCR**
LC-PUFA biosynthesis				
Δ5fad		−1.03		1.35
Δ6fad		1.04		−1.32
elovl2	−1.28 to −1.41	**−1.51**	−1.82 to −1.97	**−1.62**
Cholesterol biosynthesis				
ipi		**−3.95**		1.31
mev		−1.40		1.14
7dchr		−1.01		1.09
srebp2		**−2.09**		1.29
Lipid transport and lipoprotein metabolism				
lrp1		**−1.82**		−1.08
apoA4a		**−1.37**		−0.95
apoA4b		**−2.43**		**−1.80**
lpl		−1.18		−1.05
Immune response				
cfh	−1.33 to −1.56	−1.24	−1.56 to −1.69	**−1.31**
trim25	−5.64	**−2.09**	−4.34	−2.33
lect2	3.16	1.23	4.86	**4.84**

Values represent the expression ratios between high lipid / low lipid, for fish containing either low or high n-3 LC-PUFA levels in their flesh, obtained by microarray analysis or RT-qPCR. Expression ratios in bold were significant by REST2008 analysis of RT-qPCR results.

Delta5 and 6 fatty acyl desaturases (Δ5fad and Δ6fad); fatty acyl elongase (elovl2); isopentenyl-diphosphate isomerase (ipi); mevalonate kinase (mev); 7-dehydrocholesterol reductase (7dchr); sterol regulatory element-binding protein 2 (srebp2); low density lipoprotein receptor-related protein 1 (lrp1); apolipoprotein A-IV (apoa4a and apoa4b); lipoprotein lipase (lpl); complement factor H precursor (cfh); tripartite motif-containing protein 25 (trim25); leukocyte cell-derived chemotaxin 2 precursor (lect2).

Genetic Evaluations

Subsequent to the dietary trial and microarray analyses, genetic evaluations (estimated breeding values, EBVs) became available for a range of traits upon which the families are under active selection in the breeding program. Given the unexpectedly high preponderance of immune response genes identified by transcriptomic analysis, we investigated associations with traits that could potentially explain the gene expression data. In this respect, one of the most relevant traits was 'survival to infectious pancreatic necrosis (IPN) virus', known to be almost entirely controlled by a major QTL [12]. Genetic evaluations included data collected from a freshwater experimental IPN challenge on full-sibs from the same families as the trial fish. Examining the families, selected on their lipid phenotypes, used for transcriptomic analysis it was seen that family HH, containing both high total lipid and high n-3 LC-PUFA flesh contents, also showed a high EBV for survival to IPN (selection differential on a standardized normal distribution = 1.86 standard deviations), contrasting with −0.83 (LL) -0.99 (LH) and −1.28 (HL) for the other families, that could introduce a potential for bias in interpretation of the transcriptomic responses. However, no such imbalance was present in the lower lipid grouping, comparing families LL and LH (additional file 3).

DISCUSSION

The present study which ascertained lipid profiles of 50 Atlantic salmon families confirmed previous results showing important inter-family variation in the ability to retain n-3 LC-PUFA in the flesh when fish are fed diets with low levels of these fatty acids [7]. Furthermore, even though a high correlation was found between flesh lipid levels and n-3 LC-PUFA contents, families with the same total lipid level varied significantly in n-3 LC-PUFA contents. In the present study we did not examine whether these differences have a genetic basis, as this was established previously [7], but instead aimed to identify molecular pathways whose transcriptional regulation might underlie the phenotypic differences, independent of lipid content.

LC-PUFA biosynthesis

Differences in flesh n-3 LC-PUFA content in individuals fed the same diet is likely to arise from either selective incorporation or retention of fatty acids supplied by the diet or from biosynthesis from precursors in tissues such as the liver. In the present study we performed a transcriptomic study to identify molecular mechanisms potentially underlying flesh n-3 LC-PUFA phenotypes. Expression of candidate genes of the LC-PUFA biosynthesis pathway were also quantified as there was good evidence that these genes are transcriptionally regulated and that mRNA levels correlate with enzymatic activity of this pathway [13, 14], and so this appeared a likely mechanism that required specific investigation. Flesh was the target tissue for analysis of the n-3 LC-PUFA retention trait because salmon accumulate lipid reserves in muscle and this is the main product for human consumption, and so its composition will affect the health-promoting properties of salmon. However, hepatic tissue was analyzed for effects on gene expression since the production of both LC-PUFA and the lipoproteins that transport them to the tissues takes place mainly in the liver [15].

The transcriptomic analysis revealed few effects of the n-3 LC-PUFA factor on metabolism in general and, in particular, a lack of effect on lipid metabolism genes, when the statistical analysis employed multiple testing correction. However, this correction is typically not used when examining effects of diet and genetic background on metabolic genes, which tend to show subtle, but physiologically relevant, changes [9, 11, 16]. Without multiple testing correction we were able to identify pathways of lipid metabolism that might be altered in response to this factor, although a clear mechanism for the observed inter-family differences in n-3 LC-PUFA content was not identified. Potential effects on lipid transport and lipoprotein metabolism were indicated by the presence of two apolipoprotein A4 transcripts (apoa4a and b), a low density lipoprotein (LDL) receptor-related protein (lrp1) and a lipoprotein lipase (lpl) transcript in the microarray analysis, albeit these were not validated by RT-qPCR. In contrast, the RT-qPCR results clearly confirmed that the flesh n-3 LC-PUFA phenotype cannot be explained by transcriptional modulation of genes of LC-PUFA biosynthesis and so other mechanisms must be in operation. One hypothesis might be that phenotypic differences between families originates from the presence of different alleles of fatty acyl desaturases and/or elongases encoding proteins with altered biological activity or specificity, as described for the nematode Caenorhabditis elegans[17].

Effects of n-3 LC-PUFA Flesh Contents on Hepatic Cholesterol Biosynthesis

Within the lipid metabolism genes that were differentially expressed in the liver between fish showing higher or lower n-3 LC-PUFA contents in flesh, the category of cholesterol biosynthesis and its regulation was the most apparent, based on the number of probes for interrelated genes present in this list, all with coordinated regulation indicating reduced cholesterol biosynthesis in salmon having higher flesh n-3 LC-PUFA. In addition, and inferred by the magnitude of change (i.e., fold-changes), effects were more pronounced in fish containing lower flesh lipid levels. These results were confirmed by quantifying the expression of three enzymes catalyzing steps in cholesterol biosynthesis (mev, ipi and 7dchr) as well as srebp2, a transcription factor that regulates cholesterol synthesis [18]. Furthermore, the RT-qPCR analysis indicated that this regulation was only associated with lower flesh lipid levels given that in the high lipid group only 7dchr was down-regulated. Therefore, this experiment confirmed previous studies suggesting an association between flesh adiposity and n-3 LC-PUFA in the regulation of cholesterol biosynthesis in Atlantic salmon families, with lean fish showing a higher responsiveness to n-3 LC-PUFA [8]. However, an important novel outcome of the present study was the demonstration that the previous results were not solely a consequence of a higher dietary intake of cholesterol supplied by a FO diet in contrast to a VO diet [11] but also resulted from higher incorporation and increased tissue levels of n-3 LC-PUFA. The likely explanation for these results is the role of n-3 LC-PUFA as regulators of gene transcription, including some implicated in cholesterol biosynthesis, mediated bysrebp2[18, 19, 20]. Nonetheless, the mechanism for why this response was only observed when associated with low flesh lipid levels requires clarification. Recent studies showed that lean humans are also more responsive, in terms of plasma lipid and lipoprotein composition, to cholesterol-reducing diets containing lower levels of saturated fatty acids and cholesterol than obese individuals, and several mechanisms have been proposed to explain this [21]. In the present case, the absolute, rather than the relative, level of n-3 LC-PUFA may be the determinant factor affecting gene transcription and, in the high lipid group, absolute levels of these fatty acids might have been sufficiently high to repress cholesterol biosynthesis genes, even at lower relative n-3 LC-PUFA contents (i.e., group HL). This hypothesis is supported by the RT-qPCR analysis comparing the families with regards to lipid level, HL/LL and HH/LH. In the HL/LL comparison, contrasting absolute n-3 LC-PUFA levels of 427 versus 363 mg/100 g flesh, there was down-regulation of both ipi andsrebp2 (−3.95 and −2.09, respectively), whereas comparison of the families HH/LH, containing 554 versus 468 mg/100 g flesh, showed no difference in the expression of the genes. Similarly, genes involved in lipoprotein metabolism, which are also regulated by LC-PUFA through different mechanisms [20], also showed more significant changes when comparing fatter and leaner salmon with lower LC-PUFA levels, indicating that a similar regulatory mechanism might occur. Therefore, the present study is consistent with previous work identifying cholesterol and lipoprotein metabolism as pathways significantly and differentially affected by n-3 LC-PUFA depending on flesh adiposity [8].

Effects of Total Lipid Level on Lipid Metabolism

Lipid level significantly affected expression of lipid metabolism genes, although effects were still relatively small (8% of all genes assigned to a biological function category). A noteworthy result was the down-regulation of elovl2 (confirmed by RT-qPCR) in salmon presenting higher flesh lipid, independent of LC-PUFA content. Elovl2 has substrate specificity towards LC-PUFA and is highly responsive to dietary n-3 LC-PUFA levels in salmon [22]. However, the expression of this gene is often co-ordinately regulated with other genes of LC-PUFA biosynthesis, such as Δ5fad and Δ6fad[9], which was not the case here. Hence, the biological significance of this result is not clear and may indicate other roles of elovl2 in lipid metabolism. For instance, an association between overexpression of elovl2 and enhanced triacylglycerol synthesis and lipid droplet accumulation, as well as induction of PPARγ target genes, was shown in mouse preadipocyte cell lines [23]. In addition, elovl2 was up-regulated in the liver transcriptome of rats with nephrotic syndrome, a condition characterized by hyperlipidemia [24]. Elovl2 was only recently characterized in salmon [22], and this is the first indication of an association between its expression and lipid accumulation in a non-mammalian vertebrate, with results suggesting that increased lipid level in salmon flesh repressed elovl2 expression independent of n-3 LC-PUFA level although this requires further investigation. Another gene down-regulated at higher lipid levels was a mitochondrial acyl carrier protein, involved in acyl transfer steps, including roles in fatty acid synthesis and functioning of the electron transport chain [25], which could conceptually be responding to similar regulatory mechanisms affecting elovl2. In contrast, stearoyl-CoA desaturase, responsible for the synthesis of monounsaturated fatty acids from saturated precursors, was up-regulated in salmon with higher flesh lipid levels. This gene was positively correlated with fat accumulation in bovine skeletal muscle [26], consistent with up-regulation in salmon families with increased fat stores.

Possible Association between Flesh N-3 Lc-Pufa Contents and Immune Response

The predominance of immune response genes responding to total lipid level and, particularly, n-3 LC-PUFA contents in salmon flesh was unexpected. This was a true over-representation as GO enrichment analysis enabled identification of several GO terms related to regulation of immune and inflammatory responses in relation to the total lipid factor. However, as mentioned above, the transcriptomic comparison, although balanced for total lipid, was not balanced for viral disease resistance (specifically IPN in this case) and, as a consequence, higher contrast between families was imposed on the high lipid group (families HL and HH) due to the fortuitous selection of family HH presenting a much higher viral resistance EBV. Nonetheless, if family HH biased the results of the two-way ANOVA we would expect a preponderance of immune-related genes to occur only when comparing these two families, presenting higher and lower flesh n-3 LC-PUFA contents at the higher lipid level. In order to assess this, t-tests were performed comparing separately the higher versus lower n-3 LC-PUFA families at each total lipid level, i.e., LH/LL and HH/HL. A Venn diagram contrasting the two t-test significant lists was then performed and when analyzing the genes that were similarly affected by n-3 LC-PUFA contents at both higher and lower total lipid level, a similar preponderance (33%) of immune response genes was observed (Additional file 4). Finally, examination of the fold changes of immune-related genes, indicating magnitude of effects, between families with higher and lower contents of n-3 LC-PUFA at either higher or lower total lipid levels (Tables 2 and 5), showed no clear evidence of the effect being more marked for the high lipid comparison, which is what would be expected if results were caused simply by inclusion of family HH in the transcriptomic analysis.

Hence, there is evidence to suggest that there may be some correlation between flesh n-3 LC-PUFA contents and immune response in the families analysed. An anti-inflammatory role of n-3 LC-PUFA is well established in mammals and fish [27, 28, 29]. Immune cells are typically rich in arachidonic acid (ARA), the precursor for eicosanoids with a pro-inflammatory action, whereas EPA and DHA give rise to eicosanoids that are less biologically active, as well as to resolvins and protectins presenting anti-inflammatory properties [30]. Higher incorporation of n-3 LC-PUFA in biological membranes of immune cells can modulate immune responses in several ways [reviewed in [15, 30, 31, 32, 33]. They alter the production of inflammatory eicosanoid mediators of which they are precursors, directly affect the organization and properties of the immune cell membranes with effects on signalling pathways, phagocytic capacity and antigen presenting capability, and activate transcription of various genes involved in inflammatory responses. Therefore, families with higher tissue levels of n-3 LC-PUFA may show differential expression of immune response and inflammation-related genes, as well as of genes involved in signalling and regulation of transcription (as observed in the present study). Furthermore, although liver is chiefly a metabolic organ, it has other physiological functions including removal of pathogens and antigens from the blood and modulation of immune responses, as well as the production of inflammatory mediators [34, 35].

Related to the above, microarray analysis revealed the presence of several genes that intervene in eicosanoid synthesis and metabolism including phospholipase A_2 (pla2), arachidonate 5-lipoxygenase (alox5), thromboxane-A synthase (thas), prostaglandin I_2 synthase (ptgis) and 15-hydroxyprostaglandin dehydrogenase [36]. However, RT-qPCR only confirmed up-regulation of hepatic alox5 in families presenting higher flesh n-3 LC-PUFA and, given that alox5 acts on LC-PUFA of both n-3 and n-6 series and that ARA levels generally accompanied the n-3 LC-PUFA phenotype (Table 1), it cannot be ascertained whether this transcript was responding to higher levels of membrane ARA or EPA and hence if it would result in increased pro-inflammatory 4-series, or less potent 5-series, leukotrienes [37].

The immune response genes whose expression was correlated with ‹n-3 LC-PUFA› are mainly involved in the modulation of inflammatory processes and innate immune response to pathogens, which are particularly important in fish species and that can be easily compromised in aquaculture conditions [38]. We could speculate that the changes in expression may give enhanced protection from inflammation or pathological conditions in fish with higher n-3 LC-PUFA in their tissues. Up-regulation associated with high flesh n-3 LC-PUFA was noted in expression of NACHT domain containing protein, tripartite motif-containing protein 25 (trim25), c-c motif chemokine 13 precursor (ccl13), leukocyte cell-derived chemotaxin 2 precursor (lect2), tissue factor pathway inhibitor a, pentraxin and cathepsin K. In contrast, down-regulation in the high n-3 LC-PUFA families was observed for MHC class I (mostly in the high total lipid group), and for myelin and lymphocyte protein (mal). NACHT domain containing proteins are pathogen-sensing molecules (recognizing intracellular pathogen-associated molecular patterns – PAMPs) implicated in early host defence, inflammation and innate immune signalling pathways in mammals [39], by activating transcription of MHC class II and the apoptotic pathway. The trim25 protein is involved in antiviral innate immune responses through activation of signalling pathways leading to production of interferons and in teleost cells TRIM genes are induced in response to viral infections [40, 41]. The ccl13 (also known as monocyte chemotactic protein 4) and lect2proteins are both involved in inflammation, having roles in attracting monocytes and T lymphocytes in tissues exposed to exogenous pathogens, and have neutrophil chemotactic function [42, 43]. Expression of lect2 was increased in fish liver and spleen after bacterial infections [43]. Tissue factor pathway inhibitor inhibits the initial reactions of the blood coagulation cascade and modulates cell proliferation, and may protect vascular tissue in inflammatory conditions in mammals [44]. Cathepsin K mediates immune responses in cells, having a critical role in signalling events proximal to the Toll-like receptor 9 (TLR9) that has a fundamental role in pathogen recognition (recognizing PAMPs) and activation of mammalian innate immunity [45]. Finally, pentraxins are pattern recognition proteins of the innate immune system that play a role in the acute phase response, activating complement pathways to clear pathogens in both mammals and fish [46, 47]. In this case, up-regulation of pentraxin in salmon with higher n-3 LC-PUFA in their flesh was only observed with high lipid levels. Similarly, down-regulation of the MHC class I transcript was observed only in the high lipid group. In mammalian studies, high LC-PUFA contents (EPA, DHA and ARA) reduced cell surface expression of MHC I, decreasing antigen presentation and altering T-cell signalling [34, 35]. Therefore, the high IPN resistance genotype observed in family HH in later genetic evaluations of the families could potentially involve effects on both the complement pathway and T-cell mediated immunity, and involve co- or post-translational modification of proteins by N-linked glycosylation through up-regulation of dolichyl-diphosphooligosaccharide-protein glycosyltransferase subunit 4 (Table 2; [48]). Given the high economic impact of IPN in salmonid culture, identification of genes potentially involved in the progression of the disease using transcriptomic approaches is already in progress [49]. Finally, down-regulation of mal, associated with T-cell differentiation and signal transduction [50], was observed at higher n-3 LC-PUFA levels.

As mentioned above, several immune response-related genes were also affected by the total lipid factor with results validated by RT-qPCR. However, we cannot exclude the possibility that this results from the strong correlation between total lipid levels and absolute LC-PUFA contents, which makes it difficult to dissociate both factors.

CONCLUSIONS

It has been demonstrated earlier that LC-PUFA flesh content is a highly heritable trait [7], but the present study has shown that the underlying mechanisms do not appear to involve changes in the expression of lipid metabolism genes, including the LC-PUFA biosynthesis pathway. Other possible mechanisms, such as alleles with different biological activity, require investigation. The present study revealed an association between flesh adiposity and n-3 LC-PUFA in the regulation of cholesterol biosynthesis, which was down-regulated by higher n-3 LC-PUFA levels but only in the lean families. This response was not caused by dietary factors, given that the fish were all fed the same VO-based diet, and is most likely explained by variation in tissue n-3 LC-PUFA levels, regulating transcription of cholesterol metabolism genes through srebp2. Furthermore, the transcriptional repression of these genes may be sensitive to the absolute levels of these fatty acids in the tissues, which could explain the lack of regulation when comparing the families containing higher flesh lipid levels. It is likely that n-3 LC-PUFA exert similar roles in regulation of gene expression in fish as in mammals and, furthermore, fish might be a useful model to study important relationships between genetics, diet, adiposity/

obesity and lipoprotein/cholesterol metabolism. However, unexpected differences were found in the expression of genes implicated in the modulation of inflammatory processes and innate immune response between families differing in lipid composition, both in terms of total lipid level and, particularly, n-3 LC-PUFA contents. Although the evidence is generally circumstantial it is important to clarify this association if flesh n-3 LC-PUFA level is included as a trait for genetic selection in Atlantic salmon breeding programmes. If such a relationship is confirmed, the data suggest that the underlying mechanism might involve anti-inflammatory actions of tissue n-3 LC-PUFA on the eicosanoid biosynthesis pathway (particularly affecting the lipoxygenase pathway), although direct effects through regulation of transcription of immune genes or more indirectly through changes in architecture and properties of immune cell membranes are also possible.

METHODS

Feeding trial and Sampling

Fifty full-sib families selected from the 200 broodstock families of the Landcatch Natural Selection (LNS) Atlantic salmon breeding program (2005-strip year-group) were specifically selected for the feeding trial. On the basis of parental genetic evaluations, 25 high flesh lipid contrasting with 25 low flesh lipid families were identified, and 35 fish (initial weight, ~100 g) from each family were transferred and grown in communal seawater pens (Marine Harvest, Ardnish, Scotland). All fish were tagged with electronic transponders (PIT tags) to allow family identification while rearing in a common environment. After acclimation, the fish were grown for 12 weeks on the same low FM/high VO diet (Nutreco ARC, Stavanger, Norway) containing 25% FM and 44% plant meals and a VO blend including rapeseed oil/palm oil/camelina oil (2.5:1.5:1). At the end of the trial (378 g average weight), flesh samples (Norwegian Quality Cut) were collected, frozen on dry ice and stored at −20°C until lipid analysis. Liver samples were also taken and stored at −70°C for subsequent molecular analyses.

Lipid Analysis and Choice of Families for Transcriptomic Comparisons

The 50 selected families were screened for their ability to retain and/or synthesize n-3 LC-PUFA when fed a low FM/high VO diet. De-boned and skinned flesh samples were combined into 3 pools per family for lipid analysis. Total lipids were extracted and determined gravimetrically from 1–2 g of pooled flesh [51]. Fatty acid methyl esters (FAME) were prepared by acid-catalyzed transesterification of total lipids [52]. Following purification, FAME were separated and quantified by gas–liquid chromatography as described in [9]. These data were used to select four families for transcriptomic analysis: two with equivalent high levels of lipid 'H', and two with equivalent low levels of lipid 'L'. Within each level of total lipid, two families with significantly contrasting ($p<0.05$ on Student's t-test; Graphpad Prism™, version 4.0, Graphpad Software, San Diego, CA) relative n-3 LC-PUFA levels (similarly termed H and L) were identified (Table 1).

RNA Extraction and Purification

Hepatic tissue (200 mg) from ten individuals per family was rapidly homogenized in 2 ml TRI Reagent (Ambion, Applied Biosystems, Warrington, U.K.). Total RNA was isolated, following manufacturer's instructions, and RNA quality and quantity was assessed by gel electrophoresis and spectrophotometry (NanoDrop ND-1000, Thermo Scientific, Wilmington, U.S.A.), respectively. Equal amounts (50 µg) of total RNA were pooled from two individuals to produce five biological replicates per family, which were further purified by mini spin-column purification (RNeasy Mini Kit, Qiagen, Crawly, U.K.).

Microarray Hybridization and Analysis

A custom-made Atlantic salmon oligoarray with 44 K features per array on a four-array-per-slide format (Agilent Technologies, Cheshire, U.K.), with experimental features printed singly was used [described more fully in [53]. The probes were co-designed at the Institute of Aquaculture, University of Stirling, U.K. and Nofima, Norway, with array design available in the EBI ArrayExpress database (http://www.ebi.ac.uk/arrayexpress/arrays/browse.html) under accession number A-MEXP-2065. The features were mainly derived from a core set of Atlantic salmon Unigenes (NCBI) supplemented with other unique cDNAs derived from Genbank and the Atlantic Salmon Gene Index (http://compbio.dfci.harvard.edu/tgi/tgipage.html). Probe annotations were derived from Blastx comparisons across four protein databases, as detailed elsewhere [54]. The entire experiment comprised 20 hybridizations (5 slides): 4 groups (families) × 5 biological replicates (pools of 2 individuals each).

Indirect labelling was employed in preparing the microarray targets, as described in detail previously [8]. Antisense amplified RNA (aRNA) was produced from 500 ng of each total RNA purification reaction using the Amino Allyl MessageAmpTM II aRNA Amplification Kit (Ambion, Applied Biosystems), following the manufacturer's methodology followed by Cy3 or Cy5 fluor (PA23001 or PA25001, GE HealthCare) incorporation through a dye-coupling reaction.

The hybridizations were performed using SureHyb hybridisation chambers (Agilent) in a DNA Microarray Hybridisation Oven (Agilent). Sample order was semi-randomized, with one replicate per experimental group being loaded into each slide. Each biological replicate pool was co-hybridized in a two-dye experiment with a single pooled reference sample. This pooled reference comprised equal quantitites of aRNA from all 20 biological replicate pools. Microarry manufacturer's instructions were followed. Briefly, for each hybridization, 825 ng of Cy3-labelled experimental biological replicate and Cy5-labelled reference pool were combined. A fragmentation master mix containing 10× blocking agent (Agilent), 25× fragmentation buffer (Agilent) and nuclease-free water, was dispensed into the Cy-dyes mix. After incubating in the dark at 60°C for 30 mins, 2× GE Hybridization buffer (pre-heated to 37°C; Agilent) was added, contents gently mixed, spun at 16 K g for 1 min and finally kept on ice until loaded onto the microarray slides. Hybridization was carried out in the oven rotator (Agilent) at 65°C and 10 rpm for 17 h. Post-hybridization washes were carried out in EasyDipTM Slide staining containers (Canemco Inc., Quebec,

Canada). After disassembling the array-gasket sandwiches submersed in wash buffer 1 (Agilent) at room temperature, the microarray slides were incubated in wash buffer 1 for 1 min at 31°C in a Stuart Orbital Incubator S150 rotating at 150 rpm, and then a further 1 min at 31°C at 150 rpm in wash buffer 2 (Agilent). A final dip in wash buffer 2 at room temperature was performed, after which the slides were dried by centrifugation (500 xg for 6 mins) and kept in a desiccator and in the dark until scanned, the same day.

Scanning was performed at 5 μm resolution using an Axon GenePix 4200AL Scanner (MDS Analytical Technologies, Wokingham, Berkshire, U.K.). Laser power was kept constant (50%) and the "auto PMT" function within the acquisition software (v.4) was enabled to adjust PMT for each channel such that less than 0.1% of features were saturated and that the mean intensity ratio of the Cy3 and Cy5 signals was close to one. Agilent Feature Extraction Software (v 9.5) was used to identify features and extract fluorescence intensity values from the resultant TIF images. Analysis of the intensity values was performed in the GeneSpring GX version 11 analysis platform (Agilent Technologies, Wokingham, Berkshire, U.K.). All intensity values <0.1 were set to equal 0.1 followed by a Lowess normalization. After removing control features, four quality filtering steps were carried out sequentially using a range of quality control metrics produced by the Agilent Feature Extraction software to remove features that were saturated, non uniform, population outliers and spots non-significantly different from background. This gave a final list of 32,566 probes that were eligible for statistical analysis. Experimental annotation complied fully with minimum information about a microarray experiment (MIAME) guidelines [55]. The experimental hybridizations and further methodological details are archived on the EBI ArrayExpress database under accession number E-TABM-1204.

Normalized and quality-filtered fluorescence intensity data was analysed in GeneSpring GX v11 by two-way ANOVA, which examined the explanatory power of the variables 'total lipid' and 'n-3 LC-PUFA' and the interaction between the two, at a significance level of 0.05 and expression ratio (i.e., fold change) cut-off of 1.2. Two sets of analysis were performed, with or without Benjamini-Hochberg multiple testing correction. In the set with multiple testing correction, GO enrichment analysis was performed at a significance level of 0.05.

RT-qPCR

Expression of selected genes found by microarray analysis to be significantly affected by either 'total lipid' or 'n-3 LC-PUFA' content was quantified by RT-qPCR. In addition, the expression of two fatty acyl desaturases (Δ5fad and Δ6fad) and one elongase (elovl2) that are typically responsive to dietary n-3 LC-PUFA was determined. Primers were designed using Primer3 software (http://biotools.umassmed.edu/bioapps/primer3_www.cgi) (Table 7). Two reference genes, elongation factor-1α (elf-1α) and β-actin, were also quantified.

Table 7 Primers used for RT-qPCR analyses

Transcript	Primer sequence (5'-3')	Fragment	Ta	Efficiency	Accession No.	Source
Δ5fad	GTGAATGGGGATCCATAGCA	192 bp	56°C	0,945	AF478472 [1]	[9]
	AAACGAACGGACAACCAGA					
Δ6fad_a	CCCCAGACGTTTGTGTCAG	181 bp	56°C	0,928	AY458652 [1]	[9]
	CCTGGATTGTTGCTTTGGAT					
elovl2	CGGGTACAAAATGTGCTGGT	145 bp	60°C	0,926	TC91192 [2]	[24]
	TCTGTTTGCCGATAGCCATT					
ipi	ACAGCCCTATGGTTATGTGTCATCTC	230 bp	60°C	0,985	CK875291 [1]	[11]
	CAAGGTGAGGCGAATGTTTGAAC					
mev	CCCTTAATCAGGGTCCCAAT	247 bp	60°C	0,910	DW005667 [1]	[11]
	GGTGCTGGTTGATGTCAATG					
7dchr	CTTCTGGAATGAGGCATGGT	230 bp	60°C	0,977	TC99602 [2]	[11]
	ACAGGTCCTTCTGGTGGTTG					
srebp2	GACAGGCACAACACAAGGTG	215 bp	60°C	0,887	DY733476 [1]	[11]
	CAGCAGGGGTAAGGGTAGGT					
lrp1	ACCAACCGCATCTACTGGAC	204 bp	60°C	0,996	CK898816 [1]	New design
	CAGATTACCAGCCACCCAGT					
apoA4a	CCCAAACCAACACCACTCCT	150 bp	60°C	0,997	BT047465 [1]	New design
	GGTTTATATTTCTCACCCTGCAC					
apoA4b	CTCTTGCCCTCTTGATGACTG	154 bp	60°C	0,918	BT047267 [1]	New design
	TGACTCATCAGAGCCAATTCA					
lpl	AGGGCGTTAATCCATGTCAG	223 bp	60°C	0,917	TC84899 [2]	[8]
	GACCTTTCAAAAGGGCATGA					
lpp2	TCCGGAAGAACTCGCAATAC	174 bp	60°C	0,926	NM_001140716 [1]	[9]
	ACATCACGTCCACCAAGACA					

mgat	TTAACCCAAAGATGCTGCAA	157 bp	60°C	0,977	EG824440 [1]	New design
	CACGCAGTTGTCAGTGGTTT					
alox5	TATCTCCCTCTCCCTCAGTCC	155 bp	56°C	0,987	CX727592 [1]	[57]
	GGTCAGCAGTGCCATCA					
pla2g4	GTCGCTGGCTGGAGCTGTGG	138 bp	60°C	0,998	NM_001141333 [1]	New design
	AGCCCTATGGGCCCTGGTCA					
thas	TGTTCACACGGACCTGATTC	150 bp	60°C	0,986	NM_001165312 [1]	New design
	GACCGGATCGTCATTCTGTT					
ptgis	GCGTGTTTGTGGTCATTACG	247 bp	60°C	0,836	GE778709 [1]	New design
	TTCCCTTAGCAAGGTCTGGA					
mal	GGCCTCAGTCAAAGAGGAGA	156 bp	60°C	0,946	NM_001141320 [1]	New design
	GGGGAGTGCACACTTTAGGA					
ccl13	CGAGGATCCCTCTTCAACAA	178 bp	60°C	0,996	EG831431 [1]	New design
	ATCGTCGACTAGGCAGCAGT					
trim25	GCAGGGTCCTATCTCATCCA	215 bp	60°C	0,951	BT048046 [1]	New design
	GGACTGGACCTTTTTATTCTCTCA					
lect2	CTGTGTTGTCAGAGTGCGAGATGGT	150 bp	60°C	0,996	BT050009 [1]	[58]
	TACACACAATGTCCAGGCCCTGA					
cfh	TGTGATGATGGAGAGATGCAG	193 bp	60°C	0,966	TC141997 [2]	New design
	CAAGCGACAAAGAAACCACA					
Reference genes*						
elf-1α	CTGCCCCTCCAGGACGTTTACAA	175 bp	60°C	1.000	AF321836 [1]	[11]
	CACCGGGCATAGCCGATTCC					
β-actin	ACATCAAGGAGAAGCTGTGC	141 bp	56°C	0.939	AF012125 [1]	[11]
	GACAACGGAACCTCTCGTTA					

[1] GenBank (http://www.ncbi.nlm.nih.gov/).

[2] Atlantic salmon Gene Index (http://compbio.dfci.harvard.edu/tgi/).

* geNorm average stability (M value) of reference genes=0.514 [59].

For RT-qPCR, 2 μg of column-purified total RNA per sample was reverse transcribed into cDNA using the High-Capacity cDNA RT kit (Applied Biosystems, Paisley, U.K.), following manufacturer's instructions, but using a mixture of the random primers (1.5 μl as supplied) and anchored oligo-dT (0.5 μl at 400 ng/μl, Eurofins MWG Operon, Ebersberg, Germany). Negative controls (containing no enzyme) were performed to check for genomic DNA contamination. A similar amount of cDNA was pooled from all samples and the remaining cDNA was then diluted 20-fold with water. RT-qPCR analysis used relative quantification with the amplification efficiency of the primer pairs being assessed by serial dilutions of the cDNA pool. Amplifications were carried out in duplicate (Quantica, Techne, Cambridge, U.K.) in a final volume of 20 μl containing 5 μl or 2 μl (for more highly expressed genes) diluted (1/20) cDNA, 0.5 μM of each primer (0.4 μM for lect2) and 10 μl AbsoluteTM QPCR SYBR® Green mix (ABgene). Amplifications were carried out with a systematic negative control (NTC-non template control, containing no cDNA). The RT-qPCR profiles contained an initial activation step at 95°C for 15 min, followed by 30 to 40 cycles: 15 s at 95°C, 15 s at the specific primer pair annealing temperature (Ta; Table 7) and 15 s at 72°C. After the amplification phase, a melt curve of 0.5°C increments from 75°C to 90°C was performed, enabling confirmation of the amplification of a single product in each reaction. Non-occurrence of primer-dimer formation in the NTC was verified. RT-qPCR product sizes and presence of single bands were checked by agarose gel electrophoresis. Additionally, sequencing of amplicons corresponding to new primer designs enabled the confirmation of identities and presence of single sequences for all genes except for trim25, as the sequencing result was of insufficient quality to conclude on the presence of a single gene product, and lrp1, for which results were indicative of quantification of a highly similar, recently duplicated, gene.

Results were analyzed by the ΔΔCt method using the relative expression software tool (REST 2009,http://www.gene-quantification.info/), which employs a pair wise fixed reallocation randomization test (10,000 randomizations) with efficiency correction [56], to determine the statistical significance of expression ratios (or gene expression fold-changes) between two treatments.

Genetic Evaluations of Traits Used In the Salmon Breeding Program

Parental evaluations were confirmed by subsequent analysis of family sibs (at harvest weight, some 1 year after the present study) for a range of traits upon which the breeding program families are under active selection including flesh lipid composition parameters (total lipid) as well as EBVs for weight at harvest, precocious maturation, flesh colour, sealice resistance and resistance to a viral infection (IPN).

AUTHOR'S CONTRIBUTIONS

SM performed laboratory analyses and data analysis. DRG was responsible for family selection. JBT supported the microarray analysis. SM wrote the first draft of the manuscript, followed by contributions from remaining authors. SM, JBT and DRT planned and coordinated the research. DRG, JGB and DRT were project leaders. All authors read and approved the final manuscript.

ACKNOWLEDGEMENTS

This study was funded by the EU FP6 IP "AQUAMAX" (Sustainable Aquafeeds to Maximise the Health Benefits of Farmed Fish for Consumers; 016249–2). SM was supported by Fundação para a Ciência e a Tecnologia, Portugal (SFRH/BPD/64679/2009). Technical assistance from Jacquie Ireland in microarray hybridizations is deeply appreciated.

REFERENCES

1. FAO: The State of World Fisheries and Aquaculture 2008. 2009, Rome: Food and Agriculture Organization of the United Nations
2. Tacon AGJ, Metian M: Global overview on the use of fish meal and fish oil in industrially compounded aquafeeds: Trends and future prospects. Aquaculture. 2008, 285: 146-158. 10.1016/j.aquaculture.2008.08.015.
3. Calder PC, Yaqoob P: Omega-3 polyunsaturated fatty acids and human health outcomes. BioFactors. 2009, 35: 266-272. 10.1002/biof.42.
4. Menoyo D, López-Bote CJ, Obach A, Bautista JM: Effect of dietary fish oil substitution with linseed oil on the performance, tissue fatty acid profile, metabolism, and oxidative stability of Atlantic salmon. J Anim Sci. 2005, 83: 2853-2862.
5. Gjedrem T, Baranski M: Selective Breeding in Aquaculture: an Introduction [Series: Reviews: Methods and Technologies in Fish Biology and Fisheries, Vol. 10]. 2009, London: Springer Science + Business Media B.V
6. Schlechtriem C, Bron JE, Tocher DR: Inter-individual variation in total lipid fatty acid compositions of flesh of Atlantic salmon smolts fed diets containing fish oil or vegetable oil. Aquaculture Res. 2007, 38: 1045-1055. 10.1111/j.1365-2109.2007.01759.x.
7. Leaver MJ, Taggart JB, Villeneuve L, Bron JE, Guy DR, Bishop SC, Houston RD, Matika O, Tocher DR: Heritability and mechanisms of n-3 long chain polyunsaturated fatty acid deposition in the flesh of Atlantic salmon. Comp Biochem Physiol Part D Genomics Proteomics. 2011, 6: 62-69. 10.1016/j.cbd.2010.04.002.
8. Morais S, Pratoomyot J, Torstensen BE, Taggart JB, Guy DR, Bell JG, Tocher DR: Diet × genotype interactions in hepatic cholesterol and lipoprotein metabolism in Atlantic salmon (Salmo salar) in response to replacement of dietary fish oil with vegetable oil. Br J Nutr. 2011, 106: 1457-1469. 10.1017/S0007114511001954.
9. Morais S, Pratoomyot J, Taggart JB, Bron JE, Guy DR, Bell JG, Tocher DR: Genotype-specific responses in Atlantic salmon (Salmo salar) subject to dietary fish oil replacement by vegetable oil: a liver transcriptomic analysis. BMC Genomics. 2011, 12: 255-10.1186/1471-2164-12-255.
10. LeBlanc F, Laflamme M, Gagné N: Genetic markers of the immune response of Atlantic salmon (Salmo salar) to infectious salmon anemia virus (ISAV). Fish Shellfish Immunol. 2010, 29: 217-232. 10.1016/j.fsi.2010.03.007.
11. Leaver MJ, Villeneuve LA, Obach A, Jensen L, Bron JE, Tocher DR, Taggart JB: Functional genomics reveals increases in cholesterol biosynthetic genes and highly unsaturated fatty acid biosynthesis after dietary substitution of fish oil with vegetable oils in Atlantic salmon (Salmo salar). BMC Genomics. 2008, 9: 299-10.1186/1471-2164-9-299.
12. Houston RD, Haley CS, Hamilton A, Guy DR, Tinch AE, Taggart JB, McAndrew BJ, Bishop SC: Major quantitative trait loci affect resistance to infectious pancreatic necrosis in Atlantic salmon (Salmo salar). Genetics. 2008, 178: 1109-1115. 10.1534/genetics.107.082974.
13. Zheng X, Tocher DR, Dickson CA, Bell JG, Teale AJ: Effects of diets containing vegetable oil on expression of genes involved in highly unsaturated fatty acid biosynthesis in liver of Atlantic salmon (Salmo salar). Aquaculture. 2004, 236: 467-483. 10.1016/j.aquaculture.2004.02.003.
14. Zheng X, Torstensen BE, Tocher DR, Dick JR, Henderson RJ, Bell JG: Environmental and dietary influences on highly unsaturated fatty acid biosynthesis and expression of fatty acyl desaturase and elongase genes in liver of Atlantic salmon (Salmo salar). Biochim Biophys Acta. 2005, 1734: 13-24. 10.1016/j.bbalip.2005.01.006.
15. Galli C, Calder PC: Effects of fat and fatty acid intake on inflammatory and immune responses: a critical review. Ann Nutr Metab. 2009, 55: 123-139. 10.1159/000228999.
16. Kolditz CI, Paboeuf G, Borthaire M, Esquerré D, SanCristobal M, Lefèvre F, Médale F: Changes induced by dietary energy intake and divergent selection for muscle fat content in rainbow trout (Oncorhynchus mykiss), assessed by transcriptome and proteome analysis of the liver. BMC Genomics. 2008, 9: 506-10.1186/1471-2164-9-506.
17. Watts JL, Browse J: Genetic dissection of polyunsaturated fatty acid synthesis in Caenorhabditis elegans. Proc Natl Acad Sci U S A. 2002, 99: 5854-5859. 10.1073/pnas.092064799.
18. Sakakura Y, Shimano H, Sone H, Takahashi A, Inoue K, Toyoshima H, Suzuki S, Yamada N: Sterol regulatory element–binding proteins induce an entire pathway of cholesterol synthesis. Biochem Biophys Res Commun. 2001, 286: 176-183. 10.1006/bbrc.2001.5375.
19. Jump DB, Botolin D, Wang Y, Xu J, Christian B, Demeure O: Fatty acid regulation of hepatic gene transcription. J Nutr. 2005, 135: 2503-2506.
20. Sampath H, Ntambi JM: Polyunsaturated fatty acid regulation of genes of lipid metabolism. Annu Rev Nutr. 2005, 25: 317-340. 10.1146/annurev.nutr.25.051804.101917.

21. Flock MR, Green MH, Kris-Etherton PM: Effects of adiposity on plasma lipid response to reductions in dietary saturated fatty acids and cholesterol. Adv Nutr. 2011, 2: 261-274.

22. Morais S, Monroig O, Zheng X, Leaver MJ, Tocher DR: Highly unsaturated fatty acid synthesis in Atlantic salmon: characterization of ELOVL5- and ELOVL2-like elongases. Mar Biotechnol. 2009, 1: 627-639.

23. Kobayashi T, Zadravec D, Jacobsson A: ELOVL2 overexpression enhances triacylglycerol synthesis in 3T3-L1 and F442A cells. FEBS Lett. 2007, 581: 3157-3163. 10.1016/j.febslet.2007.05.081.

24. Zhou Y, Zhang X, Chen L, Wu J, Dang H, Wei M, Fan Y, Zhang Y, Zhu Y, Wang N, Breyer MD, Guan Y: Expression profiling of hepatic genes associated with lipid metabolism in nephrotic rats. Am J Physiol Renal Physiol. 2008, 295: F662-671. 10.1152/ajprenal.00046.2008.

25. Feng D, Witkowski A, Smith S: Down-regulation of mitochondrial acyl carrier protein in mammalian cells compromises protein lipoylation and respiratory complex I and results in cell death. J Biol Chem. 2009, 284: 11436-11445.

26. Jiang Z, Michal JJ, Tobey DJ, Daniels TF, Rule DC, Macneil MD: Significant associations of stearoyl-CoA desaturase (SCD1) gene with fat deposition and composition in skeletal muscle. Int J Biol Sci. 2008, 4: 345-351.

27. Montero D, Grasso V, Izquierdo MS, Ganga R, Real F, Tort L, Caballero MJ, Acosta F: Total substitution of fish oil by vegetable oils in gilthead sea bream (Sparus aurata) diets: effects on hepatic Mx expression and some immune parameters. Fish Shellfish Immunol. 2008, 24: 147-155. 10.1016/j.fsi.2007.08.002.

28. Oxley A, Jolly C, Eide T, Jordal AE, Svardal A, Olsen RE: The combined impact of plant-derived dietary ingredients and acute stress on the intestinal arachidonic acid cascade in Atlantic salmon (Salmo salar). Br J Nutr. 2010, 103: 851-861. 10.1017/S0007114509992467.

29. Estensoro I, Benedito-Palos L, Palenzuela O, Kaushik S, Sitjà-Bobadilla A, Pérez-Sánchez J: The nutritional background of the host alters the disease course in a fish–myxosporean system. Vet Parasitol. 2011, 175: 141-150. 10.1016/j.vetpar.2010.09.015.

30. Yaqoob P: Mechanisms underlying the immunomodulatory effects of n-3 PUFA. Proc Nutr Soc. 2010, 69: 311-315. 10.1017/S0029665110001837.

31. Calder PC: Dietary modification of inflammation with lipids. Proc Nutr Soc. 2002, 61: 345-358. 10.1079/PNS2002166.

32. Shaikh SR, Edidin M: Polyunsaturated fatty acids, membrane organization, T cells, and antigen presentation. Am J Clin Nutr. 2006, 84: 1277-1289.

33. Yaqoob P, Calder PC: Fatty acids and immune function: new insights into mechanisms. Br J Nutr. 2007, 98 (Suppl 1): S41-45.

34. Kmieć Z: Cooperation of liver cells in health and disease. Adv Anat Embryol Cell Biol. 2001, 161: III-XIII. 1–151

35. Knolle PA, Gerken G: Local control of the immune response in the liver. Immunol Rev. 2000, 174: 21-34. 10.1034/j.1600-0528.2002.017408.x.

36. Smith WL: The eicosanoids and their biochemical mechanisms of action. Biochem J. 1989, 259: 315-324.

37. Lee TH, Mencia-Huerta JM, Shih C, Corey EJ, Lewis RA, Austen KF: Effects of exogenous arachidonic, eicosapentaenoic, and docosahexaenoic acids on the generation of 5-lipoxygenase pathway products by ionophore-activated human neutrophils. J Clin Invest. 1984, 74: 1922-1933. 10.1172/JCI111612.

38. Ellis AE: Innate host defense mechanisms of fish against viruses and bacteria. Dev Comp Immunol. 2001, 25: 827-839. 10.1016/S0145-305X(01)00038-6.

39. Stein C, Caccamo M, Laird G, Leptin M: Conservation and divergence of gene families encoding components of innate immune response systems in zebrafish. Genome Biol. 2007, 8: R251-10.1186/gb-2007-8-11-r251.

40. Munir M: TRIM proteins: another class of viral victims. Sci Signal. 2010, 3: jc2-10.1126/scisignal.3118jc2.

41. van der Aa LM, Levraud JP, Yahmi M, Lauret E, Briolat V, Herbomel P, Benmansour A, Boudinot P: A large new subset of TRIM genes highly diversified by duplication and positive selection in teleost fish. BMC Biol. 2009, 7: 7-10.1186/1741-7007-7-7.

42. Uguccioni M, Loetscher P, Forssmann U, Dewald B, Li H, Lima SH, Li Y, Kreider B, Garotta G, Thelen M, Baggiolini M: Monocyte chemotactic protein 4 (MCP-4), a novel structural and functional analogue of MCP-3 and eotaxin. J Exp Med. 1996, 183: 2379-2384. 10.1084/jem.183.5.2379.

43. Li MY, Chen J, Shi YH: Molecular cloning of leucocyte cell-derived chemotaxin-2 gene in croceine croaker (Pseudosciaena crocea). Fish Shellfish Immunol. 2008, 24: 252-256. 10.1016/j.fsi.2007.09.003.

44. Kato H: Regulation of functions of vascular wall cells by tissue factor pathway inhibitor: basic and clinical aspects. Arterioscler Thromb Vasc Biol. 2002, 22: 539-548. 10.1161/01.ATV.0000013904.40673.CC.

45. Asagiri M, Hirai T, Kunigami T, Kamano S, Gober HJ, Okamoto K, Nishikawa K, Latz E, Golenbock DT, Aoki K, Ohya K, Imai Y, Morishita Y, Miyazono K, Kato S, Saftig P, Takayanagi H: Cathepsin K-dependent toll-like receptor 9 signaling revealed in experimental arthritis. Science. 2008, 319: 624-627. 10.1126/science.1150110.

46. Agrawal A, Singh PP, Bottazzi B, Garlanda C, Mantovani A: Pattern recognition by pentraxins. Adv Exp Med Biol. 2009, 653: 98-116. 10.1007/978-1-4419-0901-5_7.

47. Gisladottir B, Gudmundsdottir S, Brown L, Jonsson ZO, Magnadottir B: Isolation of two C-reactive protein homologues from cod (Gadus morhua L.) serum. Fish Shellfish Immunol. 2009, 26: 210-219. 10.1016/j.fsi.2008.03.015.

48. Larkin A, Imperiali B: The expanding horizons of asparagine-linked glycosylation. Biochemistry. 2011, 50: 4411-4426. 10.1021/bi200346n.

49. Cepeda V, Cofre C, González R, MacKenzie S, Vidal R: Identification of genes involved in immune response of

Atlantic salmon (Salmo salar) to IPN virus infection, using expressed sequence tag (EST) analysis. Aquaculture. 2011, 318: 54-60. 10.1016/j.aquaculture.2011.04.045.

50. Alonso MA, Weissman SM: cDNA cloning and sequence of MAL, a hydrophobic protein associated with human T-cell differentiation. Proc Natl Acad Sci U S A. 1987, 84: 1997-2001. 10.1073/pnas.84.7.1997.

51. Folch J, Lees M, Sloane-Stanley GH: A simple method for the isolation and purification of total lipides from animal tissues. J Biol Chem. 1957, 226: 497-509.

52. Christie WW: Lipid analysis. 2003, Bridgewater: The Oily Press

53. Tacchi L, Bron JE, Taggart JB, Secombes CJ, Bickerdike R, Adler MA, Takle H, Martin SAM: 2011 Multiple tissue transcriptomic responses to Piscirickettsia salmonis in Atlantic salmon (Salmo salar). Physiol Genomics. 2011, 43: 1241-1254. 10.1152/physiolgenomics.00086.2011.

54. Krasnov A, Timmerhaus G, Afanasyev S, Jørgensen S-M: Development and assessment of oligonucleotide microarrays for Atlantic salmon (Salmo salar L.). Comp Biochem Physiol Part D Genomics Proteomics. 2011, 6: 31-38. 10.1016/j.cbd.2010.04.006.

55. Brazma A, Hingamp P, Quackenbush J, Sherlock G, Spellman P, Stoeckert C, Aach J, Ansorge W, Ball CA, Causton HC, Gaasterland T, Glenisson P, Holstege FC, Kim IF, Markowitz V, Matese JC, Parkinson H, Robinson A, Sarkans U, Schulze-Kremer S, Stewart J, Taylor R, Vilo J, Vingron M: Minimum information about a microarray experiment (MIAME)-toward standards for microarray data. Nat Genet. 2001, 29: 365-371. 10.1038/ng1201-365.

56. Pfaffl MW, Horgan GW, Dempfle L: Relative expression software tool (REST) for group-wise comparison and statistical analysis of relative expression results in real-time PCR. Nucleic Acids Res. 2002, 30: e36-10.1093/nar/30.9.e36.

57. Haugland O, Torgersen J, Syed M, Evensen O: Expression profiles of inflammatory and immune-related genes in Atlantic salmon (Salmo salar L.) at early time post vaccination. Vaccine. 2005, 23: 5488-5499. 10.1016/j.vaccine.2005.07.034.

58. Todorcević M, Skugor S, Krasnov A, Ruyter B: Gene expression profiles in Atlantic salmon adipose-derived stromo-vascular fraction during differentiation into adipocytes. BMC Genomics. 2010, 11: 39-10.1186/1471-2164-11-39.

59. Vandesompele J, De Preter K, Pattyn F, Poppe B, Van Roy N, De Paepe A, Speleman F: Accurate normalization of real-time quantitative RT-PCR data by geometric averaging of multiple internal control genes. Genome Biology. 2002, 3: 34.1-34.11.

CHAPTER 16

FIRST DETECTION, ISOLATION AND MOLECULAR CHARACTERIZATION OF INFECTIOUS SALMON ANAEMIA VIRUS ASSOCIATED WITH CLINICAL DISEASE IN FARMED ATLANTIC SALMON (SALMO SALAR) IN CHILE

Marcos G Godoy[1], Alejandra Aedo[1], Molly JT Kibenge[2], David B Groman[3], Carmencita V Yason[4], Horts Grothusen[1], Angelica Lisperguer[4], Marlene Calbucura[1], Fernando Avendaño[5], Marcelo Imilán[5], Miguel Jarpa[5] and Frederick SB Kibenge[2]

[1]Biovac SA, Puerto Montt, Chile,

[2]Department of Pathology and Microbiology, OIE Reference Laboratory for ISA, Atlantic Veterinary College, University of Prince Edward Island, Charlottetown, PE, Canada,

[3]Aquatic Diagnostic Services, Atlantic Veterinary College, University of Prince Edward Island, Charlottetown, PE, Canada,

[4]Regional Diagnostic Virology Laboratory, Atlantic Veterinary College, University of Prince Edward Island, Charlottetown, PE, Canada

[5]Marine Harvest S.A., Puerto Montt, Chile

ABSTRACT

Background

Infectious salmon anaemia (ISA) is a viral disease of marine-farmed Atlantic salmon (Salmo salar) caused by ISA virus (ISAV), which belongs to the genus Isavirus, family Orthomyxoviridae. The virus is considered to be carried by marine wild fish and for over 25 years has caused major disease outbreaks in marine-farmed Atlantic salmon in the Northern hemisphere. In the Southern hemisphere, ISAV was first detected in Chile in 1999 in marine-farmed Coho salmon (Oncorhynchus kisutch). In contrast to the classical presentation of ISA in Atlantic salmon, the presence of ISAV in Chile until now has only been associated with a clinical condition called Icterus Syndrome in Coho salmon and virus isolation has not always been possible. During the winter of 2007, unexplained mortalities were registered in market-size Atlantic salmon in a grow-out site located in Chiloé in Region X of Chile. We report here the diagnostic findings of the first significant clinical outbreak of ISA in marine-farmed Atlantic salmon in Chile and the first characterization of the ISAV isolated from the affected fish.

Citation: Marcos G Godoy, Alejandra Aedo, Molly JT Kibenge, David B Groman, Carmencita V Yason, Horts Grothusen, Angelica Lisperguer, Marlene Calbucura, Fernando Avendaño, Marcelo Imilán, Miguel Jarpa and Frederick SB Kibenge* First detection, isolation and molecular characterization of infectious salmon anaemia virus associated with clinical disease in farmed Atlantic salmon (Salmo salar) in Chile, doi:10.1186/1746-6148-4-28.

Results

In mid-June 2007, an Atlantic salmon marine farm site located in central Chiloé Island in Region X of Chile registered a sudden increase in mortality following recovery from an outbreak of Piscirickettsiosis, which rose to a cumulative mortality of 13.6% by harvest time. Based on the clinical signs and lesions in the affected fish, and laboratory tests performed on the fish tissues, a confirmatory diagnosis of ISA was made; the first time ISA in its classical presentation and for the first time affecting farmed Atlantic salmon in Chile. Rapid sequencing of the virus-specific RT-PCR products amplified from the fish tissues identified the virus to belong to the European genotype (Genotype I) of the highly polymorphic region (HPR) group HPR 7b, but with an 11-amino acid insert in the fusion glycoprotein, and ability to cause cytopathic effects (CPE) in CHSE-214 cell line, characteristics which make it distinct from common European Genotype ISAV isolates from Europe and North America.

Conclusion

In conclusion, the present work constitutes the first report of a case of ISA in farmed Atlantic salmon in Chile. The clinical signs and lesions are consistent with the classical descriptions of the disease in marine-farmed Atlantic salmon in the Northern hemisphere. The outbreak was caused by ISAV of European genotype (or Genotype I) of HPR 7b but distinct from common European Genotype ISAV isolates.

BACKGROUND

Infectious salmon anaemia (ISA) is a viral disease of marine-farmed Atlantic salmon (*Salmo salar*) caused by ISA virus (ISAV), which belongs to the genus *Isavirus*, family *Orthomyxoviridae* [1]. In the Northern hemisphere, the first registered outbreak of ISA was in 1984 in Atlantic salmon "parr», on the southwestern coast of Norway [2]. Subsequently the disease was reported in Canada in 1996 [3], in Scotland in 1998 [4], in Faeroe Islands in 1999 [5], and in Maine, USA, in 2000 [6]. The clinical disease in farmed Atlantic salmon is characterized by variable mortality ranging from 0 to 50% with ascites, exophthalmia, petechiation of the visceral adipose tissue, haemorrhagic liver necrosis, renal interstitial haemorrhage and tubular nephrosis, filamental sinus congestion of the gills, splenic congestion with concomitant erythrophagocytosis, and congestion of the lamina propria of the stomach and foregut [2, 7, 8, 9, 10]. ISAV remains an emerging fish pathogen that continues to cause severe economic losses to the salmon-farming industry in an increasing number of countries, although the disease can be successfully eradicated as was managed by Scotland in the late 1990s [11]. In the Southern hemisphere, ISAV was first detected in Chile in 1999 in marine-farmed Coho salmon (*Oncorhynchus kisutch*) and was shown to be of the North American genotype [12]. In contrast to the classical presentation of ISA in Atlantic salmon, the presence of ISAV in Chile up until now has only been associated with a clinical condition called Icterus Syndrome in Coho salmon and virus isolation has not always been possible [13]. Between 2001 and 2003, ISAV of the North American genotype was found in apparently normal Atlantic salmon in Lake Llanquihue (Oscar Gárate, personal communication), which is not in the same zone as where the present disease outbreak occurred. It is not known why the North American variant of ISAV has not caused typical ISA outbreaks as those known to occur in Eastern Canada and Maine, USA, and yet the virus has become widespread in the Atlantic salmon industry in Chile.

ISAV belongs to the family *Orthomyxoviridae*, together with influenza viruses [1]. However, the virus is sufficiently different from influenza viruses to be assigned to its own genus, *Isavirus*. Members of this genus are enveloped particles of 90–140 nm diameter with surface projections consisting of a combined haemagglutinin-esterase (HE) glycoprotein, which is the receptor binding haemagglutinin with receptor destroying enzyme activity demonstrated to be an acetylesterase [14, 15] encoded on segment 6 [14], and a separate fusion (F) glycoprotein encoded on segment 5 [16]. The genome consists of eight segments of linear, single-stranded negative sense RNA ranging in length from 1.0 to 2.4 kb with a total molecular size of approximately 14.3 kb [17]. Sequence analysis of several ISAV isolates on the eight segments consistently reveals two genotypes that are designated with respect to their geographic origin, European and North American, hereafter designated Genotypes I and II, respectively. Sequence variation in a 35-amino acid highly polymorphic region (HPR) on the HE glycoprotein stalk has allowed the separation of ISAV isolates into different HPR groups, with the HPR0 group consisting of the non-cultivable, non-pathogenic viruses detectable only by RT-PCR [18] whereas deletion in the HPR of ≥ 13 amino acids (or if less, with deletion or mutation of the motif at amino acid positions ^{352}FNT^{354}) lead to increased pathogenicity, and ability to replicate in cell culture with production of CPE and consequent virus isolation [19].

This study reports the diagnostic findings of the first occurrence of ISA in the Southern hemisphere, and the first characterization the ISAV isolated from the affected marine-farmed Atlantic salmon in Chile.

RESULTS

Signalment

In mid-June 2007 following recovery from an outbreak of Piscirickettsiosis, a severe increase in mortality was recorded in 2 cages on an Atlantic salmon grow-out site located in central Chiloé in Region X of Chile (Figure. 1A) [see Additional file 1]. The moribund fish were lethargic and had dark integument. By the time the fish were removed from the cages 4 weeks later, the cumulative mortality had reached 70% in one cage and 82% in the second cage. Concomitantly, in other sites of the same area, high mortalities attributed to amoebas and flexibacteriosis were reported, which added to the diagnostic confusion. The total cumulative mortality on the index farm was only 13.6% at the time of harvest due to company decision to cull all remaining cages independently of their clinical status (Figure. 1B).

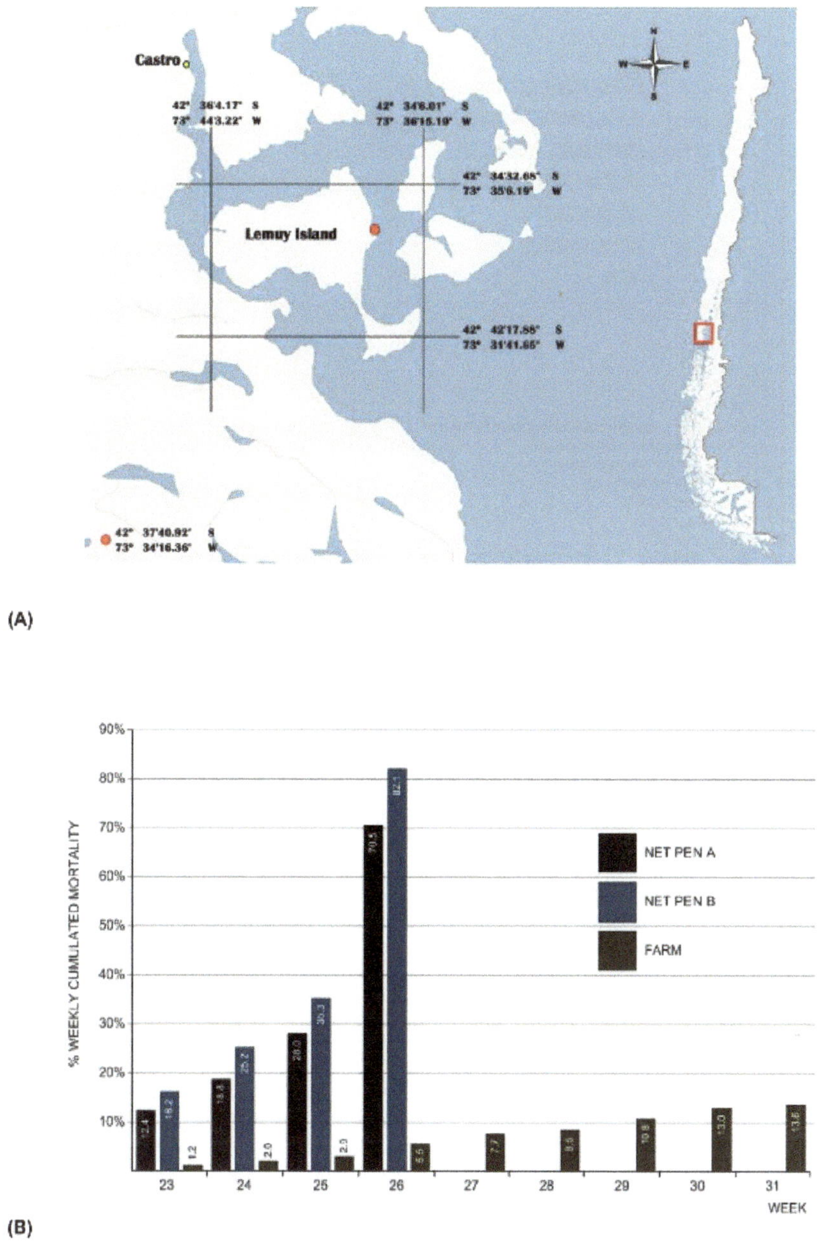

Figure 1 Geographical location and mortality pattern of the first ISA outbreak in Chile. (A)Geographical location of the Atlantic salmon marine farm site: Lemuy Island, Central Chiloé, Region X, Chile. (B)Weekly percent cummulative mortality up to the time of total harvest of the Atlantic salmon marine farm with ISA outbreak. Net Pen A and Net Pen B represent mortality in the 2 cages with ISA. FARM corresponds to the total mortality on the farm up to the time of harvest.

Gross Pathology

The external gross findings noted on Atlantic salmon that were necropsied included exophthalmia, periocular haemorrhage and darkening of the integument. Figure 2A summarizes the frequency of the external lesions noted from 100 affected fish at necropsy. Periocular haemorrhage and paleness of the gills were observed in 38% of the fish, whereas 25% presented with exophthalmia [see Additional file 2] and jaundice in the ventral zone (Fig 2B). The presence of haemorrhages in the central zone was observed in 19% of the fish (Figure. 2B). Of the 100 fish that were examined, 63% were also infested with *Caligus sp.* At necropsy, the internal gross findings included haemorrhaging in the visceral adipose tissue, liver, stomach and gut, with the liver and spleen of some fish appearing dark (Figure. 2B). Some fish had hydropericardium [see Additional file 3], and enlarged spleen and kidney. Figure 2C summarizes the frequency of the internal gross lesions noted from 100 affected fish which were necropsied in this outbreak. The most common lesions were haemorrhagic enteritis in 64%, haemorrhages in pyloric caeca in 43%, and haemorrhages in visceral adipose tissue in 31% of the cases [see Additional files 4 and 5].

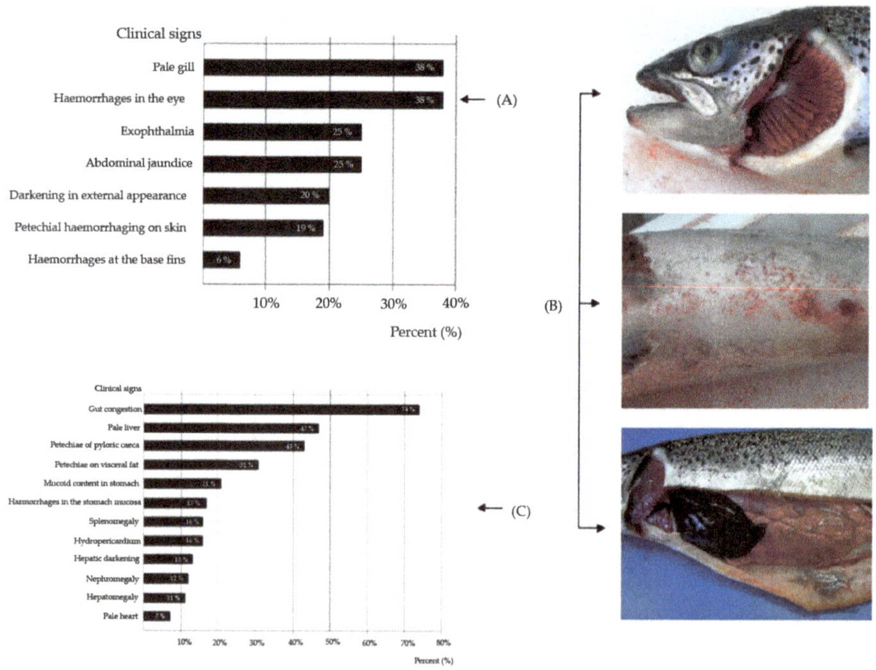

Figure 2 Gross lesions in affected Atlantic salmon from the ISA outbreak. (A) Frequency of the external gross lesions in affected Atlantic salmon from the ISA outbreak. Percentage of fish with a specified lesion among 100 fish necropsied. (B) Common gross lesions seen at necropsy: Top panel – Atlantic salmon with exophthalmia and pale gills. Middle panel – Atlantic salmon with petechial haemorrhages on the abdomen. Bottom panel – Atlantic salmon with very dark liver and haemorrhages on the visceral adipose tissue. (C) Frequency of the internal gross lesions in affected Atlantic salmon from the ISA outbreak. Percentage of fish with a specified lesion among 100 fish necropsied.

Histopathology

Diffuse congestion of lamellar capillaries, with marked infiltration of the filamental subcutis by eosinophillic granular leucocytes was evident in gill tissue. The kidney showed mild diffuse sinusoidal congestion with evidence of increase in circulating granulocytes and intravascular erythrophagia. The intestine and pyloric caeca were the most severely affected, showing marked congestion of the lamina propria vasculature with an associated mixed leucocytic infiltration of submucosa and localized perivascular and intra-lumenal haemorrhage (Figure. 3A). The liver was affected by multifocal to coalescing regions of sinusoidal congestion and peliosis, with mild adjacent hepatocellular necrosis (Figure. 3B). Changes in the spleen included marked sinusoidal congestion and intravascular erythrophagia (Figure. 3C). Mucosal epithelium in the most severely affected regions was denuded. There were no significant morphologic changes noted in the heart, stomach or body wall skeletal muscle. Morphologic diagnoses included: a) necrotizing and congestive hepatitis, subacute, multifocal to bridging, moderate to marked; b) ulcerative mixed leucocytic and congestive enteritis, with marked intra-lumenal haemorrhage, subacute, moderate to marked; and d) congestive splenitis and interstitial nephritis, with intravascular erythrophagia, moderate to marked.

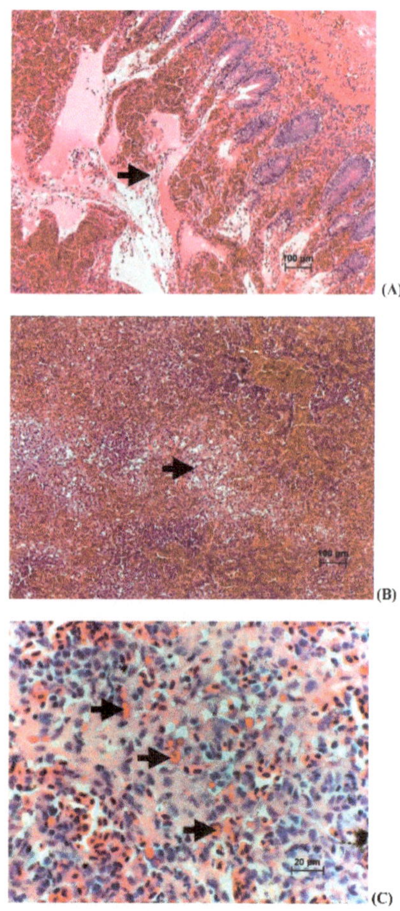

Figure 3 Microscopic lesions in affected Atlantic salmon from the ISA outbreak. (A) Histologic section of intestine of Atlantic salmon.

H&E staining. (→) indicates mucosal ulceration and haemorrhage. (B) Histologic section of liver of Atlantic salmon from the ISA outbreak H&E staining. (→) indicates hepatocellular necrosis, with marked haemorrhage and sinusoidal congestion/peliosis. (C) Histologic section of spleen of Atlantic salmon from the ISA outbreak H&E staining. (→) indicates endothelial erythrophagia.

Virus Antigen Detection

Sections of the kidney, spleen, heart, gills and intestines were subjected to immunohistochemistry staining with anti-ISAV monoclonal antibody P10 (Aquatic Diagnostics Ltd, Stirling, Scotland). ISAV proteins were detected in all the organs tested. Figure 4A shows positive staining of endothelial cells of the heart. Homogenates of pooled spleen and kidney tissues used for virus isolation were also tested in a rapid kit made by Aquatic Diagnostics, based on immunochromatography with anti-ISAV monoclonal antibody. The samples taken from affected fish were positive for ISAV. The infectious haematopoietic necrosis virus (IHNV) ELISA kit and viral haemorrhagic septicaemia virus (VHSV) ELISA kit (Bio-X Diagnostics, Jemelle, Belgium) were used to rule out the presence of IHNV and VHSV in the tissue samples.

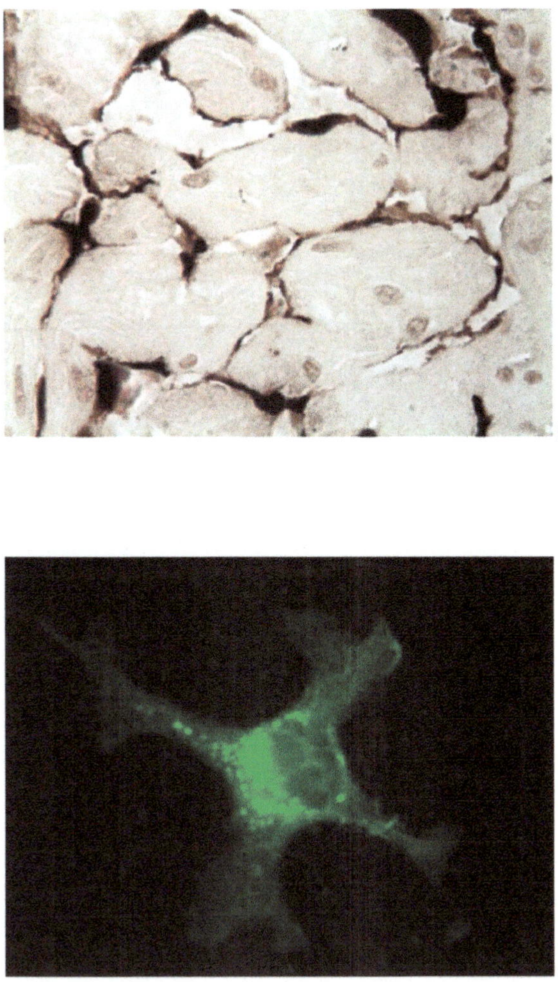

(A)

(B)

Figure 4 ISAV antigen staining of samples from the ISA outbreak. (A) Immunohistochemical staining of histologic section of heart of Atlantic salmon from the ISA outbreak. Dark brown colour indicates positive staining in endothelial cells with anti-ISAV monoclonal antibody. (B)Virus isolation in CHSE-214 cells: ISAV-infected cells (250×) showing fluorescent staining following IFAT with anti-ISAV monoclonal antibody.

Virus Isolation

Homogenates of pooled spleen and kidney tissues inoculated on Chinook salmon embryo (CHSE-214), Salmon head kidney (SHK-1) and *Epithelioma papulosum cyprini* (EPC) cell lines produced CPE between 4 and 7 days post inoculation during primary isolation. The CPE, which was reproduced on subsequent passage on fresh cell line monolayers, was characterized initially by cell vacuolization and rounding and then detachment from the substrate involving the whole cell monolayer. The presence of ISAV in the cell lysates was confirmed using RT-PCR and an indirect immunofluorescent antibody test (IFAT) with anti-ISAV monoclonal antibody P10 (Aquatic Diagnostics Ltd) including appropriate controls. Figure 4B shows positive IFAT of a CHSE-214 cell culture inoculated with a tissue sample from affected fish. Similar fluorescence was seen with EPC cell cultures with CPE (data not shown). Neither CPE nor presence of ISAV was detected by RT-PCR or IFAT in inoculated cultures of Bluegill Fry (BF2) cell line.

RT-PCR

Tissue samples shipped in RNA *later*® (Ambion Inc., Foster City CA) to the Regional Diagnostic Virology Laboratory, AVC, UPEI, Canada, were positive for ISAV by RT-PCR and negative for both IHNV and VHSV by nested RT-PCR.

Sequence Analysis

The nucleotide sequences in this report, which are all of ISAV RNA segment 5, are available through GenBankhttp://www.ncbi.nlm.nih.gov/Genbank, accession nos. EU130923, EU486160, and EU486161. A partial sequence of ISAV RNA segment 6 was generated by the Norwegian Veterinary Institute from duplicate samples of the same ISA outbreak, and appears in GenBank, accession no. AM941715. Tables 1 [see Additional file 6] and 2 [see Additional file 7] show the percent sequence identities between the new Chile ISAV and selected ISAV isolates from Europe and North America. Comparison of the RNA segment 5 sequences of the Chilean ISAV with ISAV strains of Genotype I (European genotype) showed nucleotide sequence identities ≥ 95.9% whereas with ISAV strains of Genotype II (North American genotype), it was ≤ 74.7% [see Additional file 6]. Comparison of the segment 6 sequence of the Chilean ISAV with ISAV strains of Genotype I showed sequence identities ≥ 95.7% whereas with ISAV strains of Genotype II, it was ≤ 79.4% [see Additional file 7]. Thus the new Chile ISAV involved in this outbreak belongs to Genotype I (the European genotype).

Alignment of the amino acid sequences of RNA segment 5, which encodes the F glycoprotein, revealed that the new Chile ISAV has a small insert of 11 amino acids (or 33 nucleotides) relative to ISAV strains from Europe and North America near the putative proteolytic cleavage site of the precursor F_0 protein (Figure. 5). This insert has 100% sequence identity with RNA segment 2 of Genotype I. ISAV RNA segment 2 encodes the PB1 polymerase [22]. Alignment of the amino acid sequences of RNA segment 6, which encodes the HE glycoprotein, revealed that the new Chile ISAV belongs to HPR7b of the European genotype as described by Nylund *et al.* [23] and Plarre *et al.* [24] (Figure. 6) and reviewed by Rimstad *et al.*

[18].

ISAV isolate (GenBank Acc. No.)	Predicted amino acid sequence (250–330)	aa inserted at cleavage site & in-vivo/in-vitro virulence
Norway MR60/01 (AY853944)	RANLANQHGWSKYSFGHSVKKLSN-----------QRA-----------FPGEEFIKCCGFTLGIGGANFQAYLNGEVQGD	8 (IN1)*
Norway MR46/99 (AY853962)	RANLANQHGWSKYSF--------N-----------QRAPGGVLLTETITFPGEEFIKCCGFTLGIGGANFQAYLNGEVQGD	11 (IN2)*
Norway SF70/02 (AY853938)	RASLANQHGWSKYSF--------N-----------QRAIPRTGYVRSA-FPGEEFIKCCGFTLGIGGANFQAYLNGEVQGD	10 (IN3)*
Norway SF57/00 (AY853939)	RASLANQHGWSKYSF--------N-----------QRAIPRTGYVRSA-FPGEEFIKCCGFTLGIGGANFQAYLNGEVQGD	10 (IN3)*
Norway MR62/01 (AY853937)	RASLANQHGWSKYSF--------N-----------QRAIPRTGYVRSA-FPGEEFIKCCGFTLGIGGANFQAYLNGEVQGD	10 (IN3)*
Norway SF71/02 (AY853936)	RASLANQHGWSKYSF--------N-----------HRAIPRTGYVRSA-FPGEEFIKCCGFTLGIGGANFQAYLNGEVQGD	10 (IN3)*
Norway MR61/01 (AY853935)	RASLANQHGWSKYSF--------N-----------HRAIPRTGYVRSA-FPGEEFIKCCGFTLGIGGANFQAYLNGEVQGD	10 (IN3)*
Chile U24636 (EU130923)	RAGLANQHGWSKYNF--------NKGKSANDIISDQRA-----------FPGEEFIKCCGFTLGIGGANFQAYLNGEVQGD	11 (IN4)
Norway SK779/06 (EU118819)	RAGLANQHGWSKYNF--------N-----------QRA-----------FPGEEFIKCCGFTLGIGGANFQAYLNGEVQGD	0 (ISAV HPR0)**
Norway SF83/04 (AY744392)	RAGLANQHGWSKYNF--------N-----------QRA-----------FPGEEFIKCCGFTLGIGGANFQAYLNGEVQGD	0 (low virul. ISAV)***
Can RPC/NB 04-085-1 (EF432567)	RANLANQHGWSKYSF--------Y-----------PRA-----------FPGEEFIKCCGFTLGIGGANFQAYLNGEVQGD	0 (low virul. ISAV)
Norway SF14/95 (AY853925)	RADLANQHGWSKYNF--------N-----------LRA-----------FPGEEFIKCCGFTLGIGGANFQAYLNGEVQGD	0*
Norway N32/98 (AY853921)	RAGLANQHGWSKYNF--------N-----------LRA-----------FPGEEFIKCCGFTLGIGGANFQAYLNGEVQGD	0*
Can NovaScotia NS2003 (AY853919)	RANLANQHGWSKYSF--------N-----------LRA-----------FPGEEFIKCCGFTLGIGGANFQAYLNGEVQGD	0
Norway 810/9/99 (EF217313)	RANLANQHGWSKYSF--------N-----------LRA-----------FPGEEFIKCCGFTLGIGGANFQAYLNGEVQGD	0 (high virul. ISAV)
Scotland 390/98 (AF429988)	RANLANQHGWSKYSF--------N-----------LRA-----------FPGEEFIKCCGFTLGIGGANFQAYLNGEVQGD	0
Norway 485/9/97 (EF217315)	RANLANQHGWSKYSF--------N-----------LRA-----------FPGEEFIKCCGFTLGIGGANFQAYLNGEVQGD	0
Can NovaScotiaU5575-1 (EF217314)	RANLANQHGWSKYSF--------N-----------LRA-----------FPGEEFIKCCGFTLGIGGANFQAYLNGEVQGD	0
Can NBISA01 (DQ465044)	KANFVNKHGWSKYNF--------N-----------LRG-----------FPGEEFIKCCGFTLGVGGANFQAYLNGMVQGD	0 (high virul. ISAV)
Can RPC/NB 98-049-1 (DQ465043)	KANFVNKHGWSKYNF--------N-----------LRG-----------FPGEEFIKCCGFTLGVGGANFQAYLNGMVQGD	0 (high virul. ISAV)
Can RPC/NB 02-0775-14 (DQ465045)	KANFVNKHGWSKYNF--------N-----------LRG-----------FPGEEFIKCCGFTLGVGGANFQAYLNGMVQGD	0
Chile 7833-1 (DQ465046)	KANFVNKHGWSKYNF--------N-----------LRG-----------FPGEEFIKCCGFTLGVGGANFQAYLNGMVQGD	0
Can RPC/NB 02-0775-14 (DQ440507)	KANFVNKHGWSKYNF--------N-----------LRG-----------FPGEEFIKCCGFTLGVGGANFQAYLNGMVQGD	0
Can RPC/NB 01-0593-1 (DQ440508)	KANFVNKHGWSKYNF--------N-----------LRG-----------FPGEKFIKCCGFTLGVGGANFQAYLNGMVQGD	0
Can RPC/NB 01-0973-3 (DQ465047)	KANFVNKHGWSKYNF--------N-----------LRG-----------FPGEEFIKCCGFTLGVGGANFQAYLNGMVQGD	0
Can RPC/NB 98-0280-2 (DQ440506)	KANFVNKHGWSKYNF--------N-----------LRG-----------FPGEEFIKCCGFTLGVGGANFQAYLNGMVQGD	0

Figure 5 Alignment of amino acid sequences in the proteolytic cleavage site of the precursor F_0 protein (modified from Kibenge et al. [19]). The amino acid sequence corresponding to the fusion protein of the Chilean ISAV in this disease outbreak is highlighted in yellow. The designation of amino acid inserts IN1, IN2, and IN3 are as reported by Devold et al. [21]. The unique 11-amino acid insert found in the Chilean ISAV in this disease outbreak is designated IN4. Other sources of information are indicated as *Devold *et al.* [21], **Markussen *et al.* [20], and ***Plarre and Nylund (2004; SF83/04, GenBank Accession No. AY744392). It has been suggested by Markussen *et al.* [20] that ISAV isolate SF83/04 represents a mixed virus infection of HPR0 and another HPR group, which might explain the confusion in the virulence designation.

HPR Group	ISAV isolate (GenBank Acc. No.)	Predicted amino acid sequence (330–370)	aa deleted in HPR	Virus isolated?
HPR0	Norway farmed salmonSK779/06 (EU118820)	SKLQRNITDVKIRVDAIPPQLNQTFNTNQVEQPANSVLSNIFISMGVAGF	0	No*
HPR0	Scotland wild salmon (AJ440971)	SKLQRNITDVKIRVDAIPPQLNQTFNTNQVEQPANSVLSNIFISMGVAGF	0	No**
HPR0	Can Nova Scotia farmed salmon (AY646058)	....RNITDVKIRVDAIPPQLNQTFNTNQVEQPATSVLSNIFISMGV...	0	No CPE in SHK***
HPR0	Can NB farmed salmon (AY646060)	....RNITDVKIRVDAIPPQLNQTFNTNQVEQPATSVLSNIFISMGV...	0	No CPE in SHK***
HPR0	USA Maine farmed salmon (AY973194)	SKLQRNITDVKIRVDAIPPQLNQTFNTNQVEQPSTSVLSNIFISMGVAGF	0	No
	Can RPC/NB 04-085-1 (AY963263)	SKLQRNITDVKIRVDAIPPQLNQTFNTN-------------FISMGVAGF	13	Yes
HPR9	Norway 47/99 SF (AF364888)	SKLQRNITDVKIRVDAIPPQLNQTFNT-----------------MGVAGF	17	Yes
HPR12	Norway N5/89 (AY127882)	SKLQRNITDVKIRVDAIPPQLNQT---------------NIFISMGVAGF	15	Yes
HPR3	Can Nova Scotia U5575-1 (AF294881)	SKLQRNITDVKIRVDAIPPQLNQT-----------------FISMGVAGF	17	Yes
HPR2	Scotland 1490/98 (AF391126)	SKLQRNITDVKIRVDAIPPQLNQT--------------------MGVAGF	20	Yes
	PR2 Can F1-97 (AF427045)	SKLQRNITDVKIRVDAIPPQLNQT-------------L-------GVAGF	20	Yes****
HPR4	Norway T10/93 (AF302801)	SKLQRNITDVKIRVDAIPPQL-----------------SNIFISMGVAGF	17	Yes
HPR5	Norway MR14/95 (AF364873)	SKLQRNITDVKIRVDAIPPQL---------------------ISMGVAGF	21	Yes
HPR8	Norway 48/99 SF (AF364878)	SKLQRNITDVKIRVDAIPPQL------------------------GVAGF	23	Yes
HPR11	Norway 54/00 SF (AF364884)	SKLQRNITDVKIRVDAIPP-------------------RNIFISMGVAGF	19	Yes
	PR12 Norway 97/09/393 (AF427070)	SKLQRNITDVKIRVDAI-------FNTNQVEQPATSVLSNIFISMGVAGF	7	No****
HPR6	Norway ST25/97 (AF364885)	SKLQRNITDVKIRVDAI-----------QVEQPATSVLSNIFISMGVAGF	11	Yes
HPR14	Norway 485/9/97 (AY378181)	SKLQRNITDVKIRVDA-----------NQVEQPATSVLSNIFISMGVAGF	11	Yes
HPR10	Norway MR52/00 (AF364892)	SKLQRNITDVKIK------------------QPATSVLSNIFISMGVAGF	18	Yes
HPR13	Faroe Islands 1173/01/12 (AJ440970)	SKLQRNITDVK-------------------EQPANSVLSNIFISMGVAGF	19	Yes**
HPR1	Norway H1/87 (AF364893)	SKLQRNITDVK---------------------PATSVLSNIFISMGVAGF	21	Yes
	Norway Glesvaer/2/90 (AF283998)	SKLQRNITDVK-------------TCNICVEQ----HL--I-ISMGVAGF	20 (FNT → TCN)	Yes
HPR7	Scotland 390/98 (AF283997)	SKLQRNITDVK-----------------------TSVLSNIFISMGVAGF	23	Yes
HPR7	Chile 1065 (AM941715)	SKLQRNITDVK-----------------------TSVLSNIFISMGVAGF	23	Yes
HPR15	Norway 810/9/99 (AF378180)	SKLQRNITDV--------------------E---TSVLSNIFISMGVAGF	23	Yes
	PR16 Norway 94/09/579 (AF427073)	S-------------DA--------FNTNQVEQPATSVLSNIFISMGVAGF	12 (no ^{333}NIT^{335})	No****
HPR20	Can RPC/NB 98-0280-2 (AF294870)	GKLGRNITDVNNRVDAI-------LGVNQVEQPSTSVPSNIFISMGVAGF	7 (FNT → LGV)	Yes
HPR21	Can NBISA01 (AF283996)	GKLGRNITDVNNRVDAIPPQL-----------------SNIFISMGVAGF	17	Yes
HPR21	Can RPC/NB 01-0593-1 (AY062033)	GKLGRNITDVNNRVDAIPPQL-----------------SNIFISMGVAGF	17	Yes
HPR21	Can RPC/NB 01-0973-3 (AY963266)	GKLGRNITDVNNRVDAIPPQL-----------------SNIFISMGVAGF	17	Yes
HPR21	Can RPC/NB 98-049-1 (AF294876)	GKLGRNITDVNNRVDAIPPQL-----------------SNILISMGVAGF	17	Yes
HPR21	Chile 7833-1 (AF294879)	GKLGRNITDVNNRVDAIPPQL-----------------SNIFISMGVAGF	17	Yes
HPR21	Can RPC/NB 02-0775-14 (AY963265)	GKLGRNITDVNNRVDAIPPQL-----------------SNIFISMGVAGF	17	Yes
HPR21	Can RPC/NB 02-1179-4 (AY963264)	GKLGRNITDVNNRVAAIPPQL-----------------SNIFISMGVAGF	17	Yes
	Can NB1330-2 (AY646063)	....RNITDVNNRVDAIPPQLNQT--------------------MGV...	20	Yes***

HPR

Figure 6 Alignment of amino acid sequences in the highly polymorphic region (HPR) of the HE genes of various strains of ISAV (modified from Kibenge *et al.* [19]). The amino acid sequence corresponding to the HE-HPR of the Chilean ISAV in this disease outbreak is highlighted in yellow, and identifies it as HPR7. Sequences that are not determined are indicated by dots, and amino acid (aa) deletions in the HPR are indicated by dashes. The HPR groups identified are as reported by Nylund *et al.* [23] and Plarre *et al.* [24]; other sources of information are indicated as *Markussen *et al.* [20], **Cunningham *et al.* [25], ***Cook-Versloot *et al.* [26], and ****Mjaaland *et al.* [27].

DISCUSSION

The present work constitutes the first report of a case of ISA in farmed Atlantic salmon in Chile. The outbreak was caused by ISAV of Genotype I (the European genotype). The gross and microscopic lesions observed in this outbreak are consistent with the classical ISA described in Norway [2], Canada [3], Scotland [4], Faeroe Islands [5] and United States [6]. In particular, the liver congestion and necrosis, as well as the renal, splenic and enteric congestion/haemorrhage in this outbreak are morphologically consistent with lesions pathognomonic for ISA. However, where these lesions differed from other presentations of the disease were primarily in the kidney, where in the Canadian situation there is profound interstitial haemorrhage, and in the heart, in the Chilean ISA outbreaks there is hydropericardium and severe myocarditis. Previously, such heart lesions have been shown to be prominent only in experimental infections of rainbow trout with ISAV [28].

The affected fish group in the Chilean ISA outbreak was recovering from Piscirickettsiosis. This particular group had been treated several times against Rickettsia at the farm site, and on this occasion the antibiotic treatment which was administered by injection effectively prevented mortality due to Piscirickettsiosis. Although the pathologic findings in this outbreak were in the range of what is considered as typical ISA, it cannot be categorically stated what contribution, if any, the Piscirickettsiosis disease contributed to the pathology observed.

The virus responsible for the outbreak caused CPE and was recovered using CHSE-214, SHK-1 and EPC cell lines, and confirmed as ISAV by means of RT-PCR and IFA testing. This is the first report of primary isolation of ISAV using EPC cell line, which is a non-salmonid cell line derived from a skin tumour of carp (*Cyprinus carpio*L.) [29] and never before reported to be permissive to ISAV. Most recently, EPC cell line was also reported to be permissive to ISAV of North American genotype [30]. Moreover rapid sequencing of virus-specific RT-PCR products amplified from the fish tissues identified the virus as belonging to the European genotype. Up until now, only some strains of ISAV of the North American genotype [31, 32, 33] but not the European genotype [32] have been known to replicate and cause CPE in the CHSE-214 cell line, a property that is not associated with the antigenic subtype of ISAV nor with the HPR group of the HE gene [34]. Thus, the Chilean ISAV has cell line specificities that are different from those of common European Genotype ISAV isolates from Europe and North America.

Now that we have ISAV isolates of both genotypes from Europe, North America, and South America, we should remove the stigmatizing labeling of ISAV genotypes by geographic reference. We therefore propose to designate the European genotype as Genotype I and the North American genotype as Genotype II. Such designation is consistent with our present knowledge, which is not sufficient to explain the anomalous geographic distribution of the two genotypes of ISAV.

The Chilean ISAV responsible for the ISA outbreak is unique in that it has a 33-nucleotide (or 11-amino acid) insert in the middle of RNA segment 5 which encodes the F glycoprotein. This insert has 100% sequence identity with RNA segment 2 of ISAV of Genotype I, and probably arose through a non-homologous recombination between the F and PB1 genes of the same virus. To date, there have been only 8 other ISAV isolates with inserts in RNA segment 5 [20, 21]. All these isolates are Norwegian ISAV isolates; seven of them were recovered between 1999 and 2002 [21] and one was recovered in 2006 [20]. Seven of these isolates had inserts from different parts of RNA segment 5 while in one isolate, the insert was shown to come from RNA segment 3, which encodes the ISAV nucleoprotein. Therefore, this Chilean ISAV is the first known isolate to have an insert in RNA segment 5 coming from RNA segment 2. Overall, this information indicates that this Chilean ISAV is different from common Genotype I ISAV isolates in Europe. The location of the insert in the Chilean ISAV RNA segment 5 is also unique in that it occurs right in the middle of the ^{265}YP^{266} motif, which was recently shown to be a marker for reduced virulence [19]. The RNA segment 5 of ISAV belonging to HPR0, a non-pathogenic virus, has ^{265}NQ^{266} at this site but without a sequence insertion [20], which would be similar to the Chilean ISAV prior to the non-homologous recombination event between the F and PB1 genes. Such recombination events are well known in avian influenza virus (AIV), involving an insertion in the haemagglutinin (HA) gene of AIV near the cleavage site of the protein and leading to emergence of new virulent strains [35, 36, 37, 38]. For example, the avian influenza disease outbreak that occurred in Chile in 2002 was attributed to the non-homologous recombination between the HA and nucleoprotein genes of H7N3 AIV [38].

The Chilean ISAV described here clearly differs from the 7833-1 isolate (of Genotype II) from coho salmon in Chile [12] in segment 5 and the partial segment 6 amino acid sequences. Thus the 7833-1 virus, which is now wide spread in the Atlantic salmon industry in Chile, is unlikely to have been the source of this new virus. It is interesting that the alignments on RNA segment 6 showed the Chilean ISAV to belong to HPR7b that has previously been found in Scotland and Norway. However, all ISAV isolates of HPR7b sequenced to date on RNA segment 5 do not have the 11-amino acid insert found in the Chilean ISAV. Comparison of the RNA segment 5 sequence of this Chilean ISAV with the HPR0 virus strain SK779/06 [20] showed a nucleotide sequence identity of 98.3% and amino acid sequence identity of 97.1%. In contrast, ISAV strain 390/98, which is also of HPR7 but from Scotland, when compared to the HPR0 virus showed a nucleotide sequence identity of 98.4% and amino acid sequence identity of 99.1% [see Additional file 6]. This suggests that the Chilean ISAV is distinct from common Genotype I ISAV isolates from Europe and North America. The location of the 11-amino acid insert at a motif in the F glycoprotein which is associated with virus virulence, may account for the elusive nature of this Chilean virus. It is therefore necessary to sequence the whole genome of this virus and clarify the epidemiology of this ISA outbreak so as to establish the origin of this virus, its virulence characteristics and the risk factors associated with its presentation and dissemination, in order to institute adequate strategies for the control and prevention of ISA in the Chilean salmon industry.

CONCLUSION

In conclusion, the present work constitutes the first report of a case of ISA in farmed Atlantic salmon in Chile. The clinical signs and lesions are consistent with the classical descriptions of the disease in marine-farmed Atlantic salmon in the Northern hemisphere. The outbreak was caused by ISAV of Genotype I of HPR 7b but distinct from common European ISAV isolates from Europe and North America.

METHODS

Field Sampling

Atlantic salmon with mean weight of 3.9 Kg, held in seawater rearing cages, were necropsied and submitted for diagnostic testing. Moribund fish were submitted for laboratory analysis to the Biovac S.A. laboratory in Puerto Montt, Chile, where a full necropsy was conducted and samples were collected for histological evaluation, virus isolation, and immunohistochemistry and molecular biology analysis.

Histology

Tissue samples for histological analysis were collected in 10% buffered formalin. They were then processed using standard procedures and the sections were stained with Haematoxylin & Eosin (H&E), in order to describe the significant morphological changes. Tissues from 9 fish were additionally submitted to Aquatic Diagnostic Services at the Atlantic Veterinary College in Prince Edward Island, Canada, for histological evaluation.

Virus antigen Detection Assays

The presence of virus antigens in the fish tissues utilized three different types of assays: immunohistochemistry, immunochromatography, and antigen ELISA. The immunohistochemistry analyses were carried out on waterproofed tissue sections of 5 μm thickness, using anti-ISAV monoclonal antibody P10 (Aquatic Diagnostics Ltd, Stirling, Scotland) with the immunohistochemistry kit from Vector Laboratories. Immunochromatography utilized tissue homogenates applied to a rapid kit with anti-ISAV monoclonal antibody according to the manufacturer›s procedures (Aquatic Diagnostics Ltd). The IHNV ELISA kit (Bio-X Diagnostics) and VHSV ELISA kit (Bio-X Diagnostics) were used to rule out IHNV and VHSV in the tissue samples. The ELISA procedures followed the manufacturer›s protocols.

Virus Isolation

Homogenized pools of spleen and kidney were inoculated on monolayers of CHSE-214, EPC, SHK-1 and BF2 cell lines following standard protocols in the OIE Aquatic Manual [39].

RT-PCR and Nucleic acid Sequencing

Total RNA was extracted from the same homogenates of pooled spleen and kidney tissues used for virus isolation, using EZNA® kits (Omega Biotech) or the RNeasy® kit (Qiagen) according to the manufacturer›s recommended protocol. For tissues preserved in RNA *later*® (Ambion Inc), the tissues were first washed three times with phosphate buffered saline (PBS) and then homogenized using a tissue homogenizer (Brinkman, Ontario, Canada) prior to total RNA extraction using the RNeasy® kit (Qiagen).

For the detection of ISAV in tissue homogenates, ISAV-specific primers and conditions described by Devold *et al.* [40] and Mjaaland *et al.* [41] for RNA segment 8, and the primers described in the OIE Aquatic Manual [39] for RNA segment 6 were used. Briefly, amplification was performed using 50 µl reaction mixture utilizing EzrTth RT-PCR kit (Applied Biosystems, Montreal, Quebec) as follows: EzrTth DNA polymerase – 1 µl, ISA forward primer-0.4 uM, ISA reverse primer- 0.4 uM, Mn-2.5 mM, dNTPs-200 uM each, 5× PCR buffer-10 µl, ddH$_2$O-19 µl and 5 µl RNA template. The amplification was performed in Perkin Elmer Gene Amp PCR System 2400 using the following conditions: 1 cycle at 53°C for 45 min, one cycle at 94°C for 2 min, 35 cycles at 94°C for 15 s, 60°C for 15 s and 72°C for 15 s and 1 cycle at 72°C for 7 min before holding the reactions at 4°C.

Samples were tested for IHNV and VHSV using nested RT-PCR with primers and conditions described by Batts and Winton [42] and [43], respectively, with minor modifications. The RT-PCR was performed using 50 µl reaction mixture utilizing EzrTth RT-PCR kit (Applied Biosystems) as follows: EzrTth DNA Polymerase-2.5 units, forward primer-0.4 uM, reverse primer -0.4 uM, Mn-2.5 mM, dNTPs -200 uM each, 5× PCR buffer-10 µl, ddH$_2$O-19 µl and 5 µl RNA template. For IHNV, the primers consisted of IHNV 1F primer and IHNV 1R primer whereas for VHSV they were VHSV 1F primer and VHSV 1R primer. The amplification was performed in Perkin Elmer Gene Amp PCR System 2400 using the following conditions: 1 cycle at 50°C for 15 min, 1 cycle at 95°C for 2 min, 25 cycles at 95°C for 30 s, 50°C for 30 s and 72°C for 60 s, 1 cycle at 72°C for 1 min and 1 cycle at 72°C for 7 min before soaking at 4°C. The second amplification was done using AmpliTaq Gold (Applied Biosystems) utilizing 50 µl mixture as follows: AmpliTaq Gold Polymerase-2.5 units, IHNV 2F primer – 0.4 uM, IHNV 2R primer-0.4 uM for IHNV or VHSV 2F primer – 0.4 uM, VHSV 2R primer-0.4 uM for VHSV, Mg-2 mM, dNTPs -200 mM each, 10× PCR buffer – 5 µl, ddH$_2$O- 33.5 µl and 2 µl RT-PCR mixture. The following cycling conditions were used: 1 cycle of 95°C for 2 min, 25 cycles at 95°C for 30 s, 50°C for 30 s and 72°C for 60 s and 1 cycle at 72°C for 7 min before holding the reactions at 4°C.

For DNA sequencing of ISAV RT-PCR products, total RNA was extracted from 300 µl of tissue homogenate using Trizol LS® (Invitrogen). The RT-PCR was carried out using the Titan® One Tube RT-PCR (Roche Diagnostic), in a PTC-200 DNA Engine Peltier thermal cycler (MJ Research, Inc.) using oligonucleotide primers and cycling conditions as previously described [19, 44]. The PCR products were then either directly sequenced or they were cloned into the pCRII vector using a TOPO TA cloning kit (Invitrogen) in preparation for nucleotide sequencing. Plasmid DNA for sequencing was prepared as described before [45]. DNA sequencing was performed as previously described [19] by ACGT Corporation (Toronto, Ontario, Canada). DNA Sequencing was done either directly on RT-PCR products or on plasmid DNA containing the cloned RT-PCR products obtained from reactions using total RNA from tissue samples. Sequence analysis used the BLAST programs [46], the Sequence Manipulation suite version 2 [47], and the FASTA program package for personal computers [48].

ACKNOWLEDGEMENTS

Funding for this study was provided by Marine Harvest S.A., Puerto Montt, Chile, Biovac S.A., Puerto Montt, Chile, and OIE Reference Laboratory for ISA, Atlantic Veterinary College, University of Prince Edward Island, Charlottetown, PE, Canada.

AUTHORS' CONTRIBUTIONS

MGG made the veterinary investigation of the outbreak, performed the necropsy and histological analysis, coordinated the laboratory investigation, and helped to write the manuscript. AA coordinated the laboratory investigation and performed the immunohistochemistry analysis. MJTK performed the RT-PCR and sequence analysis of RNA segment 5 and edited the manuscript. DBG performed the histological evaluation and edited the manuscript. CVY performed the RT-PCR for ISAV, IHNV and VHSV, provided total RNA of U24636 and U24637 to the OIE Ref Lab, and edited the manuscript. HG performed RT-PCR on tissue homogenates and cell culture lysates. AL and MC isolated the virus in cell culture. FA coordinated all the MH laboratory procedures. MI made the veterinary investigation of the increased mortality in the outbreak. MJ coordinated the MH laboratory investigation and wrote the manuscript. FSBK coordinated the viral analysis and DNA sequence analysis of RNA segment 5, and helped to write the manuscript. All authors read and approved the final manuscript.

REFERENCES

1. Kawaoka Y, Cox NJ, Haller O, Hongo S, Kaverin N, Klenk H-D, Lamb RA, McCauley J, Palese P, Rimstad E, Webster RG: Infectious Salmon Anaemia Virus. Virus Taxonomy – Eight Report of the International Committee on Taxonomy Viruses. Edited by: Fauquet CM, Mayo MA, Maniloff J, Desselberger U, Ball LA. Elsevier Academic Press: New York; 2005:681-693.
2. Thorud KE, Djupvik HO: Infectious salmon anaemia in Atlantic salmon (*Salmo salar* L). Bull Eur Assoc Fish Pathol. 1988, 8: 109-111.

3. Mullins JE, Groman D, Wadowska D: Infectious salmon anaemia in salt water Atlantic salmon (*Salmo salar* L) in New Brunswick, Canada. Bull Eur Assoc Fish Pathol. 1998, 18: 110-114.

4. Rodger HD, Turnbull T, Muir F, Millar S, Richards R: Infectious salmon anaemia (ISA) in United Kingdom. Bull Eur Assoc Fish Pathol. 1998, 18: 115-116.

5. Lyngøy C: Infectious salmon anaemia in Norway and the Faeroe Islands: An industrial approach. International response to infectious salmon anaemia: prevention, control, and eradication: proceedings of a symposium. Tech Bull 1902. Edited by: Miller O, Cipriano RC. Washington, DC: U.S. Department of Agriculture, Animal and Plant Health Inspection Service; U.S. Department of the Interior, U.S. Geological Survey; U.S. Department of Commerce, National Marine Fisheries Service, 2003:97-109.

6. Bouchard DA, Brockway K, Giray C, Keleher W, Merrill PL: First report of infectious salmon anaemia (ISA) in the United States. Bull Eur Assoc Fish Pathol. 2001, 21: 86-88.

7. Evensen O, Thorud KE, Olsen YA: A morphological study of the gross and microscopic lesions of infectious salmon anaemia in Atlantic salmon (*Salmo salar*). Res Vet Sci. 1991, 51: 215-222.

8. Jones SRM, MacKinnon AM: Virulence and pathogenicity of infectious salmon anaemia virus isolated from farmed salmon in Atlantic Canada. J Aquat Anim Health. 1999, 11: 400-405. 10.1577/1548-8667(1999)011<0400:VAPOIS>2.0.CO;2.

9. Simko E, Brown LL, MacKinnon AM, Byrne PJ, Ostland VE, Ferguson HW: Experimental infection of Atlantic salmon,*Salmo salar* L., with infectious salmon anaemia virus: a histological study. J Fish Dis. 2000, 23: 27-32. 10.1046/j.1365-2761.2000.00203.x.

10. Gattuso A, Mazza R, Imbrogno S, Sverdrup A, Tota B, Nylund A: Cardiac performance in *Salmo salar* with infectious salmon anaemia (ISA): putative role of nitric oxide. Dis Aquat Org. 2002, 52: 11-20. 10.3354/dao052011.

11. Stagg RM, Bruno DW, Cunningham CO, Raynard RS, Munro PD, Murray AG, Allan CET, Smail DA, McVicar AH, Hastings TS: Epizootiological investigations into an outbreak of infectious salmon anaemia (ISA) in Scotland. Report No 13/01, FRS Marine Laboratory, Aberdeen. 2001.

12. Kibenge FSB, Garate ON, Johnson G, Arriagada R, Kibenge MJT, Wadowska D: Isolation and identification of infectious salmon anaemia virus (ISAV) from Coho salmon in Chile. Dis Aquat Org. 2001, 45: 9-18. 10.3354/dao045009.

13. Smith PA, Larenas J, Contreras J, Cassigoli J, Venegas C, Rojas ME, Guajardo A, Pérez S, Díaz S: Infectious haemolytic anaemia causes jaundice outbreaks in seawater-cultured coho salmon, *Oncorhynchus kisutch* (Walbaum), in Chile. J Fish Dis. 2006, 29: 709-715. 10.1111/j.1365-2761.2006.00766.x.

14. Falk K, Aspehaug V, Vlasak R, Endresen C: Identification and characterization of viral structural proteins of infectious salmon anaemia virus. J Virol. 2004, 78: 3063-3071. 10.1128/JVI.78.6.3063-3071.2004.

15. Hellebø A, Vilas U, Falk K, Vlasak R: Infectious salmon anaemia virus specifically binds to and hydrolyzes 4-O-Acetylated sialic acids. J Virol. 2004, 78: 3055-3062. 10.1128/JVI.78.6.3055-3062.2004.

16. spehaug VT, Mikalsen AB, Snow M, Biering E, Villoing S: Characterization of the infectious salmon anaemia virus fusion protein. J Virol. 2005, 79: 12544-12553. 10.1128/JVI.79.19.12544-12553.2005.

17. Clouthier SC, Rector T, Brown NEC, Anderson ED: Genomic organization of infectious salmon anaemia virus. J Gen Virol. 2002, 83: 421-428.

18. Rimstad E, Biering E, Brun E, Falk K, Kibenge FSB, Mjaaland S, Snow M, Winton J: Which risk factors relating to spread of infectious salmon anaemia (ISA) require development of management strategies?. Opinion of the Panel on Animal Health and Welfare of the Norwegian Scientific Committee for Food Safety, ad hoc group. 2006, [http://www.vkm.no/eway/default.aspx?pid=0;oid=-2trg=__new__new=-2:17005]

19. Kibenge FSB, Kibenge MJ, Wang Y, Qian B, Hariharan S, McGeachy S: Mapping of putative virulence motifs on infectious salmon anaemia virus surface glycoprotein genes. J Gen Virol. 2007, 88: 3100-3111. 10.1099/vir.0.83097-0.

20. Markussen T, Jonassen CM, Numanovic S, Braaen S, Hjortaas M, Nilsen H, Mjaaland S: Evolutionary mechanisms involved in the virulence of infectious salmon anaemia virus (ISAV), a piscine orthomyxovirus. Virology. 2008, 374: 515-527. 10.1016/j.virol.2008.01.019.

21. Devold M, Karlsen M, Nylund A: Sequence analysis of the fusion protein gene from infectious salmon anaemia virus isolates: evidence of recombination and reassortment. J Gen Virol. 2006, 87: 2031-2040. 10.1099/vir.0.81687-0.

22. Krossøy B, Hordvik I, Nilsen F, Nylund A, Endresen C: The putative polymerase sequence of infectious salmon anaemia virus suggests a new genus within the Orthomyxoviridae. J Virol. 1999, 73: 2136-2142.

23. Nylund A, Devold M, Plarre H, Isdal E, Aarseth M: Emergence and maintenance of Infectious Salmon Anaemia Virus (ISAV) in Europe: a new hypothesis. Dis Aquatic Org. 2003, 56: 11-24. 10.3354/dao056011.

24. Plarre H, Devold M, Snow M, Nylund A: Prevalence of infectious salmon anaemia virus (ISAV) in wild salmonids in Western Norway. Dis Aquatic Org. 2005, 66: 71-79. 10.3354/dao066071.

25. Cunningham CO, Gregory A, Black J, Simpson I, Raynard RS: A novel variant of infectious salmon anaemia virus (ISAV) haemagglutinin gene suggests mechanisms for virus diversity. Bull Eur Ass Fish Pathol. 2002, 22: 366-374.

26. Cook-Versloot M, Griffiths S, Cusack R, McGeachy S, Ritchie R: Identification and characterization of infectious salmon anaemia virus (ISAV) hemagglutinin gene highly polymorphic region (HPR) type 0 in North America. Bull Eur Assoc Fish Path. 2004, 24: 203-208.

27. Mjaaland S, Hungnes O, Teig A, Dannevig BH, Thorud K, Rimstad E: Polymorphism in the infectious salmon anaemia virus haemagglutinin gene: importance and possible implications for evolution and ecology of infectious salmon anaemia disease. Virology. 2002, 304: 379-391. 10.1006/viro.2002.1658.

28. MacWilliams C, Johnson GR, Groman D, Kibenge FSB: Morphologic description of infectious salmon anaemia virus (ISAV)-induced lesions in rainbow trout, *Oncorhynchus mykiss* (Walbaum). Dis Aquatic Org. 2007, 78: 1-12. 10.3354/dao01866.

29. Fijan N, Sulimanovic D, Bearzotti M, Muzinic D, Zwillenberg LO, Chilmonczyk S: Some properties of the Epithelioma papulosum cyprini (EPC) cell line from carp *Cyprinus carpio*. Annu Virol Inst Pasteur. 1983, 134E: 207-220. 10.1016/S0769-2617(83)80060-4.

30. Goíc B, Bustamante J, Miquel A, Alvarez M, Ines Vera M, Valenzuela PDT, Burzio LO: The nucleoprotein and the viral RNA of infectious salmon anaemia virus (ISAV) are localized in the nucleolus of infected cells. Virology. 2008.

31. Bouchard D, Keleher W, Opitz HM, Blake S, Edwards KC, Nicholson BL: Isolation of infectious salmon anaemia virus (ISAV) from Atlantic salmon in New Brunswick, Canada. Dis Aquatic Org. 1999, 35: 131-137. 10.3354/dao035131.

32. Kibenge FS, Lyaku JR, Rainnie D, Hammell KL: Growth of infectious salmon anaemia virus in CHSE-214 cells and evidence for phenotypic differences between virus strains. J Gen Virol. 2000, 81: 143-150.

33. Griffiths S, Cook M, Mallory B, Ritchie R: Characterization of ISAV proteins from cell culture. Dis Aquat Org. 2001, 45: 19-24. 10.3354/dao045019.

34. Kibenge FSB, Kibenge MJT, McKenna PK, Stothard P, Marshall R, Cusack RR, McGeachy S: Antigenic variation among isolates of infectious salmon anaemia virus correlates with genetic variation of the viral haemagglutinin gene. J Gen Virol. 2001, 82: 2869-2879.

35. Khatchikian D, Orlich M, Rott R: Increased viral pathogenicity after insertion of a 28S ribosomal RNA sequence into the haemagglutinin gene of an influenza virus. Nature. 1989, 340: 156-157. 10.1038/340156a0.

36. Orlich M, Gottwald H, Rott R: Nonhomologous recombination between the haemagglutinin gene and the nucleoprotein gene of an influenza virus. Virology. 1994, 204: 462-465. 10.1006/viro.1994.1555.

37. Hirst M, Astell CR, Griffith M, Coughlin SM, Moksa M, Zeng T, Smailus DE, Holt RA, Jones S, Marra MA, Petric M, Krajden M, Lawrence D, Mak A, Chow R, Skowronski DM, Tweed SA, Goh S, Brunham RC, Robinson J, Bowes V, Sojonky K, Byrne SK, Li Y, Kobasa D, Booth T, Paetzel M: Novel avian influenza strain outbreak, British Columbia. Emerg Infect Dis. 2004, 10: 2192-2195.

38. Suarez DL, Senne DA, Banks J, Brown IH, Essen SC, Chang-Won Lee, Manvell RJ, Mathieu-Benson C, Mareno V, Pedersen J, Panigrahy B, Rojas H, Spackman E, Alexander DJ: A virulence shift in the influenza A subtype H7N3 virus responsible for a natural outbreak of avian influenza in Chile appears to be the result of recombination. Emerg Infect Dis. 2004, 10: 693-699.

39. OIE Manual of Diagnostic Tests for Aquatic Animals. 5th edition 2006 World Organization for Animal Health, Paris, [http://www.oie.int/eng/normes/fmanual/A_summry.htm]Fifth

40. Devold M, Krossøy B, Aspehaug V, Nylund A: Use of RT-PCR for diagnosis of infectious salmon anaemia virus (ISAV) in carrier sea trout *Salmo trutta* after experimental infection. Dis Aquat Organ. 2000, 24: 9-18. 10.3354/dao040009.

41. Mjaaland S, Rimstad E, Cunningham CO: Molecular diagnosis of infectious salmon anaemia. Molecular Diagnosis of Salmonid Diseases. Edited by: Cunningham CO. Kluwer Academic Publishers, Dordrecht, The Netherlands; 2002:1-22.

42. Batts WN, Winston JR: Infectious Hematopoietic Necrosis Virus RT-nPCR Assay. Nucleic Acid Amplification Assays for Diagnosis of Animal Diseases. Edited by: Lloyd H Lauerman. American Association of Veterinary Laboratory Diagnostician; 1998:118-120.

43. Batts WN, Winston JR: Viral Hemorrhagic Septicemia Virus RT-nPCR Assay. Nucleic Acid Amplification Assays for Diagnosis of Animal Diseases. Edited by: Lloyd H Lauerman. American Association of Veterinary Laboratory Diagnostician; 1998: 132-133.

44. Kibenge FSB, Xu H, Kibenge MJT, Qian B, Joseph T: Characterization of gene expression on genomic segment 7 of infectious salmon anaemia virus. Virol J. 2007, 4: 34-10.1186/1743-422X-4-34.

45. Kibenge FSB, Dybing JK, McKenna PK: Rapid procedure for large-scale isolation of plasmid DNA. Biotechniques. 1991, 11: 65-67.

46. Altschul SF, Gish W, Miller W, Myers EW, Lipman DJ: Basic local alignment search tool. J Mol Biol. 1990, 215: 403-410.

47. Sequence Manipulation Suite version 2. [http://www.ualberta.ca/~stothard/javascript/index.html]

48. Pearson WR, Lipman DJ: Improved tools for biological sequence comparison. Proc Nat Acad Sci USA. 1988, 85: 2444-2448. 10.1073/pnas.85.8.2444.

CHAPTER 17

PISCINE ORTHOREOVIRUS CAN INFECT AND SHED THROUGH THE INTESTINE IN EXPERIMENTALLY CHALLENGED ATLANTIC SALMON (SALMO SALAR L.)

Helena Hauge, Maria Dahle, Torfinn Moldal, Even Thoen, Anne-Gerd Gjevre, Simon Weli, Marta Alarcón and Søren Grove

Norwegian Veterinary Institute, Sentrum, 0106 Oslo, Norway

ABSTRACT

Piscine orthoreovirus (PRV) is a ubiquitous virus in Norwegian salmon farms associated with the disease heart and skeletal muscle inflammation (HSMI). Experimental challenge has shown that the virus replicates in circulating red blood cells of Atlantic salmon prior to infecting heart myocytes. The infection route from water to blood is however still unknown. The related mammalian orthoreovirus primarily infects the lungs and gastrointestinal (GI) tract and is proposed to spread mainly through the faecal–oral route. To investigate the role of the salmonid GI tract in PRV-infection, oral and anal administration of virus was compared to intraperitoneal (i.p.) injection. When administered anally, PRV was transferred to blood 4 days post challenge (dpc) and levels peaked at 42 dpc, similar to i.p. injected fish. PRV was detected in heart and faeces with corresponding kinetics, and inflammatory heart lesions consistent with HSMI were observed from 49 dpc. The orally intubated group showed slower virus kinetics in both blood and heart, and no signs of HSMI. Compared to the oral and i.p. administration routes, leakage of virus inoculate by anal intubation was minor and challenge was restricted to the mid- and distal intestine. These findings show that anal intubation is an efficacious method for PRV delivery to the GI tract and demonstrates that PRV can establish infection through the intestine with the potential for transmission via faeces.

INTRODUCTION

In Norway, *Piscine orthoreovirus* (PRV) is associated with heart and skeletal muscle inflammation (HSMI) in Atlantic salmon (*Salmo salar* L.) [1, 2, 3]. HSMI is one of the most common diseases in Norwegian salmon farming [4] and is characterized by epicardial and myocardial inflammation, usually combined with inflammation in the red skeletal muscle and signs of circulatory disturbances [1]. The relationship between PRV infection and development of HSMI remains unclear as PRV infection is not always associated with disease. In Norway, HSMI typically occurs in seawater staged farmed Atlantic salmon although PRV is ubiquitous in seawater and freshwater salmon farms [5] and is occasionally detected in wild salmon [6, 7] and marine species [8]. PRV has also been commonly detected in salmon and trout species in Chile and western North America without the occurrence of HSMI [9, 10, 11]. Furthermore, despite replicating to high viral loads in Atlantic salmon, PRV from western North America failed to cause HSMI or other disease [12, 13]. Whether this is due to differential virulence of PRV or multifactorial disease triggers is not known.

Citation: Helena Hauge, Maria Dahle, Torfinn Moldal, Even Thoen, Anne-Gerd Gjevre, Simon Weli, Marta Alarcón and Søren Grove, Piscine orthoreovirus can infect and shed through the intestine in experimentally challenged Atlantic salmon (Salmo salar L.), DOI: 10.1186/s13567-016-0343-z

PRV is a non-enveloped virus with a genome consisting of ten segments of double stranded RNA. The virus is related to the well characterized mammalian orthoreovirus (MRV) [14]. MRV primarily infects the lungs and GI tract and is so common in humans that most people are exposed early in life and develop immunity without any signs of disease [15]. Several isotypes of MRV have been characterized, and some are associated with gastroenteritis in young children. In murine models, MRV has also been found associated with pathology in the liver and heart [16]. Although the relative sequence homology is low, potential homologues of all main MRV protein-encoding genes have been detected in the PRV genome [17]. Studies of PRV by electron microscopy have so far revealed that the virus shape, size and viral factory formation in the cytoplasm of erythrocytes shares similarities to MRV [18]. Accordingly, it was of interest to determine the role of the GI tract in PRV infections.

The epithelial surfaces in e.g. gills, skin and intestine are major sites of viral entry in fish [19]. The intestine of Atlantic salmon posterior to the stomach has been defined into five different regions; the pyloric caeca, the first segment of the mid-intestine with pyloric caeca, the first segment of the mid-intestine posterior to pyloric caeca (in the following called mid-intestine), the second segment of the mid-intestine (in the following called distal intestine) and the posterior segment (Figure 1B) [20]. The distal intestine is characterized by a larger diameter than the mid-intestine and by pronounced complex folds in addition to simple folds. In contrast to mammals where mesenteric lymph nodes and distinct lymphoid follicles are harboured in the intestinal mucosa, teleost fish like Atlantic salmon have their immune competent cells including antigen presenting cells and T and B lymphocytes, more diffusely spread in the intestinal tissue [21]. Higher transcript levels of immune-related genes are observed in the distal intestine compared to the proximal regions [22], and the presence of cells that resemble mammalian M cells, important in stimulating mucosal immunity, has also been demonstrated in this segment [23].

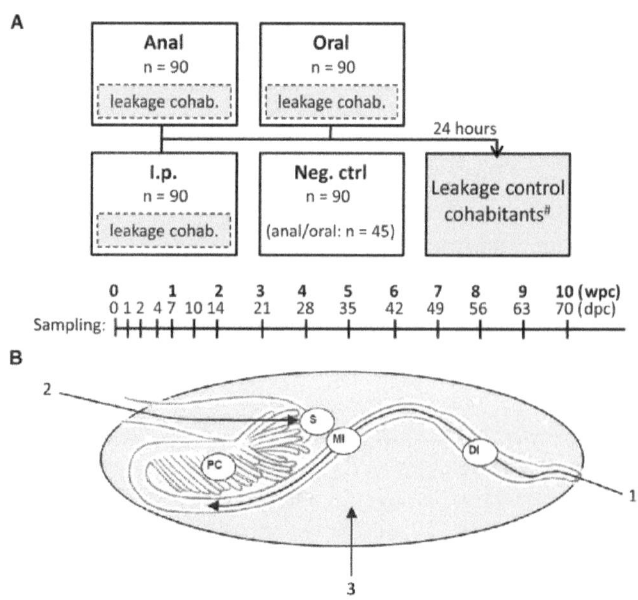

Figure 1 Study design (A) and administration routes (B). A Treatment groups (n = 90) were challenged as described in the "Materials and methods" section. To assess potential leakage of challenge inoculate, naïve fish were introduced to challenged groups after two rounds of water exchange, and then after 24 h transferred to a separate tank. Timeline specifies sampling schedule days post challenge (dpc) and weeks post challenge (wpc). #Leakage control cohabitants were only sampled at 56 dpc. **B** Arrows indicate where challenge material was deposited and circles indicate where gastrointestinal samples were taken (stomach (S), pyloric caeca (PC), mid-intestine (MI) and the distal intestine (DI)). **B** modified with courtesy of Inge Bjerkås, Oslo, Norway.

PRV has been shown to transmit through cohabitation with infected fish, and controlled challenge trials have demonstrated that the virus infects and replicates in circulating red blood cells prior to infection of heart myocytes [18, 24, 25]. The infection route from water to blood is still unknown.

Studying the early events of infection presents obvious difficulties when using established experimental models such as i.p. injection and cohabitation. Ideally, fish should be challenged simultaneously with identical virus doses targeting the same infection site. Since MRV is detected in faeces and proposed to spread mainly through the faecal–oral route, we set out to study if a similar transmission route is relevant for PRV in Atlantic salmon, using an experimental model of oral and anal intubation of virus inoculates.

The current study shows that Atlantic salmon are efficiently infected by PRV after anal intubation, leading to transfer of virus from the intestinal lumen to blood, then heart and subsequently to development of HSMI. Furthermore, PRV was found to shed into faeces, indicating a potential virus transmission route. In contrast to established experimental models, the anal administration challenge route limits virus exposure mainly to the intestine and hence allows studies of viral infections over the epithelium at the very early time points.

MATERIALS AND METHODS

Fish and Rearing Conditions

The experimental study was carried out in accordance with the recommendations in the current animal welfare regulations: FOR-1996-01-15-23 (Norway), and the protocols were approved by the Norwegian Animal Research Authority. All fish handling was performed under anesthesia with Aqui-S vet. (Scan Aqua, Årnes, Norway) at a concentration of 10 mg/L, and the fish were monitored daily for mortality and signs of disease.

Unvaccinated Atlantic salmon (*Salmo salar* L.) pre-smolts (indigenous to the river Drammenselva, Buskerud County, Norway), 20–40 g (33.7 g average), originating from Hellefoss cultivation station, were transported to the research aquarium facility at the Norwegian Veterinary Institute. The fish were here reared in freshwater in standard 160 L fibreglass tanks with flow-through water supplied from the municipal water works. The water was passed through a carbon filter column and aerated mechanically in the aquarium facility, before entering the fish tanks. Water temperature and oxygen saturation was monitored daily and ranges were 8.3–12.1 °C and 86–99%, respectively, during the study period. The fish were hand-fed a 2 mm pelleted commercial diet (Skretting, Stavanger, Norway) at a rate of 2% of calculated biomass/tank/day. The fish were acclimatised for 3 weeks. Prior to the experimental start, histology was performed on 12 fish to rule out any health issues in the fish population, and three fish were tested for common salmonid viruses; PRV, salmonid alphavirus (SAV), piscine myocarditis virus (PMCV) and infectious pancreatic necrosis virus (IPNV) by real time RT-PCR using primers and probes as described elsewhere [7, 26, 27, 28].

Preparation of Virus Homogenate

The PRV inoculum originated from a Norwegian field HSMI outbreak in 2012 and had previously been passaged through two controlled PRV challenge trials also resulting in HSMI where the challenge inoculum was prepared from homogenized PRV infected blood cells. Here, blood collected on heparinized tubes from PRV infected fish with HSMI was centrifuged, plasma removed, and the blood cell pellets stored frozen at −80 °C until use. At the day of infection, the pellets were diluted 1:2 in sterile phosphate-buffered saline (0.01 M, pH 7.4) (PBS) and combined, freeze-thawed in three rapid cycles until cell lysis, further diluted fivefold in PBS and used as challenge material. RNA corresponding to 25 μL of virus inoculate analysed by one-step real time RT-PCR targeting the PRV-segment λ1, resulted in a Ct value of 17. An inoculum dose of 0.1 mL, equivalent to Ct 15, was administered to the individual fish.

Challenge Experiment

Fish were kept in freshwater and starved for 6 days prior to challenge, partly to facilitate intubation and absorption of challenge material and partly to reduce potential leakage and rapid excretion of the inoculate e.g. via faeces. Treatment groups, comprising 90 fish each, were administered as follows under anesthesia with benzocaine (Figure 1):

Anal intubation: The challenge inoculate was administered by anal intubation using a sterile syringe fitted with a disposable mouse feeding needle (7202K, Fuchigami, Akirunoshi, Japan). After introduction via the anus, the feeding needle was inserted to a length where the 0.1 mL inoculate would become deposited in the mid-intestine (Figure 1B). Then the feeding needle was slowly withdrawn to minimize leakage.

Oral intubation: Fish were intubated using the same type of equipment and challenge material and the catheter was introduced through the mouth down to the stomach (bend) where challenge material was deposited (Figure 1B).

Intraperitoneal injection (i.p.): Fish were injected i.p. with 0.1 mL inoculate.

Negative controls: Fish were intubated via the anal- or oral route respectively with 0.1 mL PBS.

Leakage control: In order to minimize adverse effects of inoculate leakage, i.e. infection via non-GI routes, the following procedures were undertaken. After challenge, fish were first dipped into a tank with fresh water and then left in a temporary tank with maximum water-flow (12 L/min) until all fish in the group had been treated. The water was then changed completely twice within the first 12 h before the fish were finally transferred to a permanent tank for the rest of the experiment. To further reduce potential excretion of infective material via intestinal waste, fish were not fed the first 24 h after intubation. To assess potential leakage of challenge material from challenged fish and the consequential risk of virus transmission via undesirable routes, 10 naïve fish ("leakage control cohabitants") were introduced to the tanks of each of the three challenged groups after the two changes of water. These leakage control fish were left in the tank for 24 h, and then transferred to separate tanks where they were kept until 56 dpc when they were sacrificed and sampled.

Fish were monitored for a total period of 70 dpc where upon all remaining fish were sampled.

Sampling

Six fish per group were euthanized with an overdose of Finquel vet. (Scan Aqua) and sampled at 15 individual time points after challenge, i.e. at 6 h post-challenge (in the following denoted 0.25 dpc) as well as 1, 2, 4, 7, 10, 14, 21, 28, 35, 42, 49, 56, 63 and 70 dpc.

Weight-, stomach- and intestinal content registration did not reveal any significant differences between experimental groups. Neither did haematocrit measurements.

Detection of Viral RNA

For virus analyses, samples of heart (tip), spleen, head kidney, stomach, pyloric caeca, mid- and distal intestine, and faeces were collected on RNAlater and stored at −80 °C until further processing. Total RNA was isolated manually (heart, stomach, faeces) or automatically (QIAsymphony, QIAGEN, Hilden, Germany) (pyloric caeca, mid-intestine, distal intestine, spleen, head kidney) with the RNeasy Mini Kit (QIAGEN) according to the manufacturer's recommendations. 50 μL of blood was transferred to NucliSENS easyMag lysis buffer (280134, bioMérieux, Marcy l'Etoile, France) before isolation of nucleic acids (easyMag, bioMérieux). The RNA-concentration was measured using a NanoDrop™ 2000 spectrophotometer (Thermo Scientific, Wilmington, DE, USA), and the samples were stored in RNase-free water at −80 °C. Real time RT-PCR was performed on 500–1000 ng total RNA using OneStep RT-PCR kit (QIAGEN) and primers and probes targeting the PRV-segment λ1 as described elsewhere [7] and according to the manufacturer's instructions. PCR was run for 45 cycles, and all virus negative samples were given an arbitrary Ct-value of 45 allowing them to be included in statistical analyses. Validation of assays and data handling was according to the MxPro Manual, and baseline and cycle threshold (CT) were set manually. Analyses were based on the mean Ct-value of six fish per group and time point.

Histology

For histological examination, samples of gills, pseudobranch, skin, muscle, heart, liver, spleen, kidney, stomach, pyloric caeca, mid- and distal intestine were fixed in 10% buffered formalin. After 24–48 h of fixation, the samples were transferred to 70% ethanol until processed and embedded in paraffin wax. Samples from a total of 171 fish were sectioned, stained with haematoxylin & eosin (H&E) and examined by light microscopy as follows: all anally intubated fish, all negative controls until 63 dpc, oral group fish at 49 and 56 dpc, and i.p. group fish at 63 and 70 dpc. Throughout the study, histology showed that most of the fish examined suffered sparse gill inflammation. No consistent lesions were noted in muscle, pseudobranch, gastrointestinal tract, liver, kidney or spleen. A total of nine fish died; three in the anal group (14,

41 and 48 dpc), 1 (63 dpc) in the oral group, and 5 (2, 24, 24, 26, 28 dpc) in the control group. Mortalities displayed similar macroscopical and microscopical lesions; mild to moderate skin lesions related to *Saprolegnia*-infection and mild to moderate bacteria gill inflammation.

Immunohistochemistry

A polyclonal antibody against the putative PRV outer capsid protein σ1 (anti-σ1) (kindly provided by Øystein Wessel, Norwegian University of Life Sciences, Oslo, Norway) was used to investigate the presence of viral proteins in hearts of anally infected fish. We used a previously described protocol for immunohistochemical detection of PRV with minor modifications [24]. In short, heart sections mounted on poly l-lysine coated glass slides were heated at 60 °C for 20 min, deparaffinized in xylene and rehydrated through graded alcohol baths. Antigen retrieval was done by microwave treatment for 2 × 6 min in citrate buffer (0.1 M, pH 6.0). To block non-specific binding sites, sections were incubated 20 min in goat serum diluted 1:50 in 5% skimmed dry milk in TBS (pH 7.6, 0.05 M Tris/HCl, 0.15 M NaCl). Primary antibody (anti-σ1) diluted 1:2500 in 1% skimmed milk in TBS was then added and the sections incubated in a humidity chamber at 4 °C overnight. The UltraVision ONE detection system kit (Thermo Scientific) was used for detection of bound antibody according to the manufacturer's instructions, employing Fast Red (1 mg/mL) as substrate. Finally the sections were counterstained with haematoxylin and mounted (Aquamount, Thermo Scientific).

Statistical Analyses

Statistical analyses were performed using the JMP™ software (SAS Institute Inc., Cary, NC USA). Non-parametric Wilcoxon was used to compare groups and non-parametric Kendall's τ and Spearman's ρ was used for testing correlation.

RESULTS

PRV levels in Blood during the Course of Infection

PRV was detected by real time RT-PCR in blood samples in all the challenged groups but not in the control group (Figure 2A). In the i.p. group, PRV was detected in all sampled fish already 0.25 dpc, after which the mean level increased steadily until peaking at around 42 dpc (mean Ct 19). In the anally challenged group, only one single fish was positive at 0.25, 1 and 2 dpc, and then the number of positive fish increased until all fish were PRV positive from day 14. Peak levels were reached at 42 dpc (mean Ct 22.7), as observed for the i.p. injected fish. In contrast, PRV was only detected consistently in orally injected fish from 28 dpc and onward and even then never in all sampled fish. From 28 dpc, the mean level of PRV increased in orally challenged fish until 63 dpc when 5 out of 6 oral fish had PRV Ct values comparable to the other challenge groups.

A

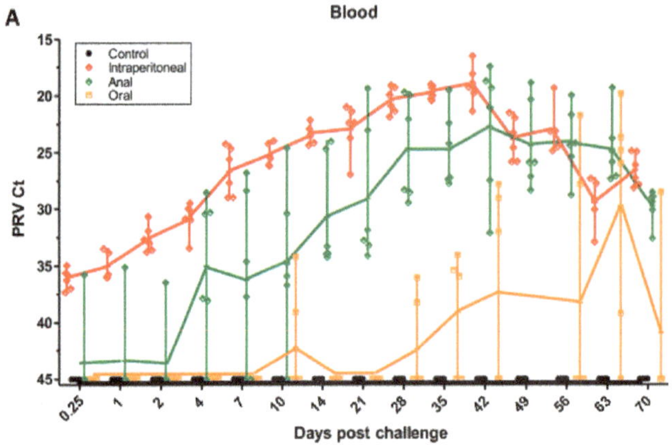

	0.25	1	2	4	7	10	14	21	28	35	42	49	56	63	70
Anal	1	1	1	5	4	5	6	6	6	6	6	6	6	6	6
Oral	0	0	0	0	0	2	0	0	2	3	3	-	2	5	1
i.p.	6	6	6	6	6	6	6	6	6	6	6	6	6	6	6

B

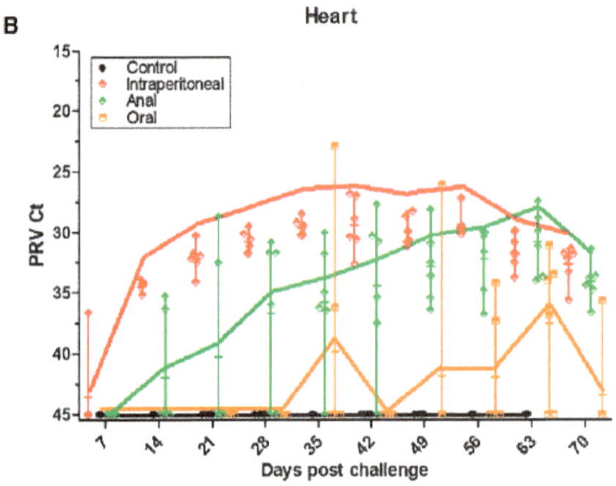

	7	14	21	28	35	42	49	56	63	70
Anal	0	2	2	4	5	5	6	6	6	6
Oral	0	0	0	0	2	0	1	2	4	1
i.p.	1	6	6	6	6	6	6	6	6	6

Figure 2 PRV infection dynamics in blood and heart according to the different modes of challenge. PRV levels (Ct values) in blood (**A**) and heart (**B**) of i.p. (red), oral (orange), anal (green) and control (black) group fish. Markers indicate individual fish, horizontal bars group means and vertical bars the range. Graph for control group and parts of graph for oral group has been slightly parallel offset from x = 45 for improved intelligibility. Table indicates number of positive fish per group ($n = 6$ for the challenge groups except for 70 dpc, $n = 4$ for control group).

Temporal Development in PRV Levels in Heart Tissue

Also for heart tissue, PRV was detected by real time RT-PCR in all three challenged groups (Figure 2B). The data for heart were overall similar to the findings in blood, with the i.p. group exhibiting the earliest detection (one out of six fish at 7 dpc) and highest initial PRV levels. In the anal challenge group, the number of PRV positive fish increased from two out of six fish at 14 dpc until all fish were positive from 49 dpc and onward, with PRV-levels peaking at 63 dpc (mean Ct 30.8). The mean PRV level of the anal challenge group was accordingly significantly lower ($p < 0.05$) than in the i.p. group between 14 and 49 dpc, after which similar levels were reached. PRV was detected much later in the oral infection group and was never seen in every fish at any sampling point, peaking with four out of six fish positive at 63 dpc (mean Ct 37.6).

Histology and PRV-immunohistochemistry of Heart Tissue

Inflammatory heart lesions consistent with HSMI were observed in 5 out of 12 fish in the i.p. group (63–70 dpc) and 4 out of 24 fish in the anal group (49–63 dpc) (Figure 3A and Table 1). Minor myocarditis was seen at early samplings in a few individuals. The fish in the oral and control groups showed no HSMI lesions.

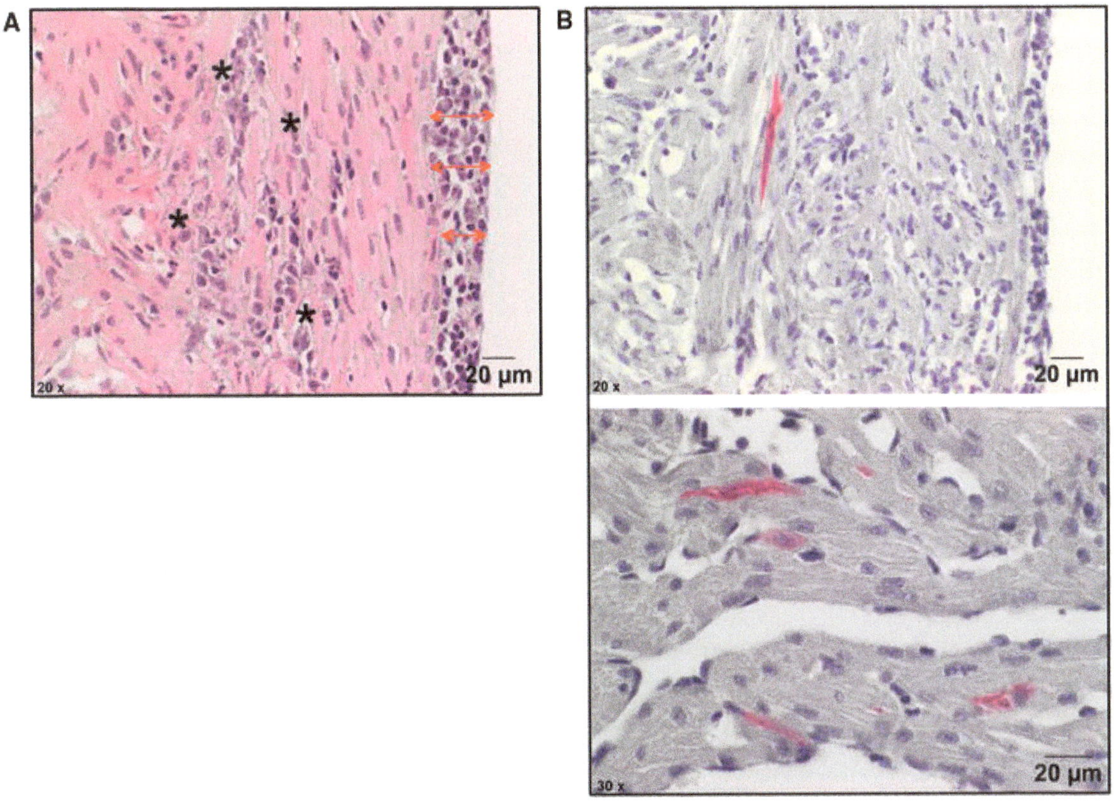

Figure 3 Histology (A) and PRV immunohistochemistry (B) of hearts from anally intubated fish. A Cardiac ventricle showing epicarditis (red arrows) and myocarditis in the underlying compact myocardium (asterisk) in the anal group at 9 wpc. **B** Immunostaining of heart tissue in the anal group at 9 wpc. PRV antigen detected with σ1 antibodies in cardiomyocytes in the cardiac ventricle (red colour).

Table 1 Histology and immunohistochemistry for heart tissue

wpc dpc	0–4 0.25–28	5 35	6 42	7 49	8 56	9 63	10 70
Anal (IHCM)	0/54 (0/10)	0/6 (1/2)	0/6 (2/2)	1/6 (2/6)	1/6 (2/6)	2/6 (1/1)	0/6 (0/1)
Oral (IHCM)	0/6	–	–	0/1 (0/1)	0/2 (0/2)	–	–
i.p.	–	–	–	–	–	4/6	1/6
Neg. ctrl.	0/26	0/2	0/3	0/3	0/3	0/3	–

Number of fish with HSMI in the anal, oral, i.p. and negative control group at different time points. Positive fish are expressed out of the total analysed fish per group. In parentheses the number of positive fish detected in situ by immunohistochemistry.

Hearts from fish with high PRV-levels in blood (Ct < 29) in the anally intubated group were further analysed by immunohistochemistry, and positive staining was detected in 8 out of 28 analysed fish (35–63 dpc) (Table 1). At 35 and 42 dpc, PRV positive cells were mainly circulating blood cells located in the cardiac lumen (suspected erythrocytes and leukocyte-like cells). At later time points PRV was found mainly in cardiomyocytes (Figure 3B). Positive staining was not observed at earlier (0.25–28 dpc) or at the last sampling point (70 dpc).

Distribution Dynamics of PRV in Gastrointestinal Tract of Challenged Fish

To investigate the distribution of PRV during the early phase post challenge, four segments of the GI tract, from the stomach to the distal intestine, (Figure 1B) were analysed by PRV real time RT-PCR from 0.25 to 10 dpc (Figure 4). The data showed that both tissue distribution and distribution dynamics differed between the challenge groups in accordance with the different administration routes. For the i.p. group, PRV levels were comparable between the four GI segments at all sampling times, indicating a uniform distribution of virus in the peritoneum.

From a medium high level (mean Ct 32.4–35.4) at 0.25 dpc, the PRV levels gradually decreased until a distinct minimum was reached by 4 dpc, after which the PRV levels increased in all four segments, in line with virus replication in blood. Compared to the i.p. group, the anal and oral challenge groups exhibited clear differences in the distribution of virus in the four GI segments. At early time points, PRV levels in the oral challenge group were distinctly higher in stomach where the inoculum was deposited, compared to the other segments. From high levels in stomach at 0.25 dpc, PRV amounts gradually decreased during 1 and 2 dpc until becoming undetectable at 4 dpc. In pyloric caeca, three of six oral fish were slightly PRV positive (Ct > 40) at both 1 and 2 dpc, after which PRV became undetectable at 7 dpc. In the mid- and distal intestine, PRV was only detected at low levels in 5 out of 36 fish from 0.25 to 10 dpc (Ct > 35). In the anal challenge group, PRV levels in stomach and pyloric caeca were consistently low with only a few detections, except for in pyloric caeca at 4 dpc when 4 out of 6 fish were positive (Ct > 34). In line with the deposition of the virus inoculum, PRV was detected in mid- and distal intestine at relatively high levels at 0.25 dpc (mean Ct 34.1 and 30.0, respectively), gradually decreasing until 10 dpc when PRV was still detectable in 2 out of 6 fish. In contrast to the i.p. and oral challenge groups, the anal challenge group did not exhibit the same marked minimum in PRV detection at 4 dpc.

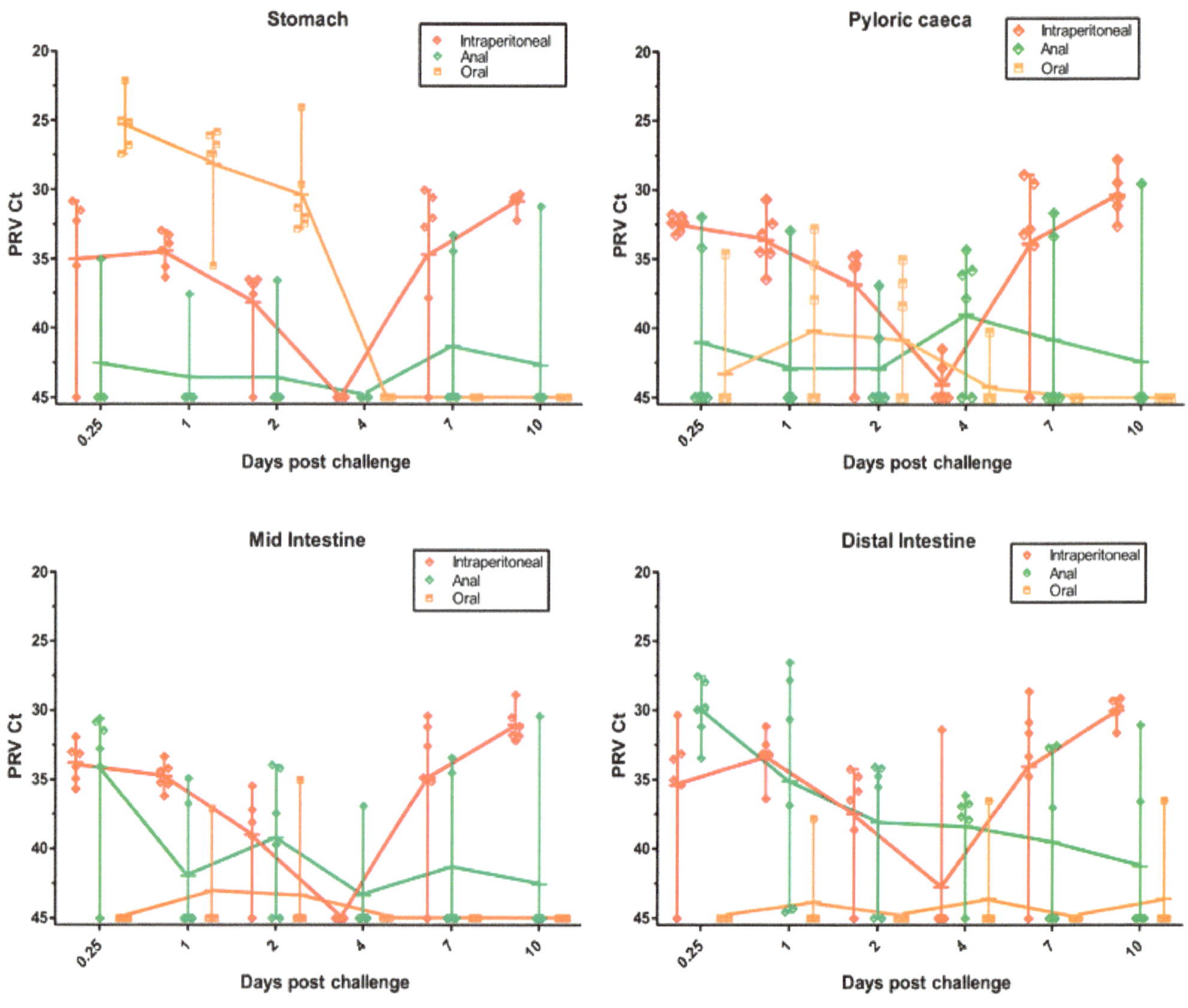

Figure 4 Distribution dynamics of PRV in the GI tract of challenged fish. Viral loads (Ct values) in four segments of the gastrointestinal tract from 0.25 to 10 dpc by i.p. (red), oral (orange), anal (green) administration routes. Markers indicate individual fish, horizontal bars group means and vertical bars the range. Graph for control group and parts of graph for oral group has been slightly parallel offset from x = 45 for improved intelligibility.

PRV in Faecal Contents

Since inoculated virus was detectable in the intestine of anally challenged fish as late as 10 dpc, only faeces samples collected from the intestine of the i.p. and anal group fish from 14 dpc and later were analysed for viral RNA. PRV was detected in faeces of both groups, reaching a maximum at 42 dpc (mean Ct 28.3) in the i.p. fish and at 63 dpc (mean Ct 34.1) in the anal challenge group, corresponding to the virus levels in blood (Figure 5).

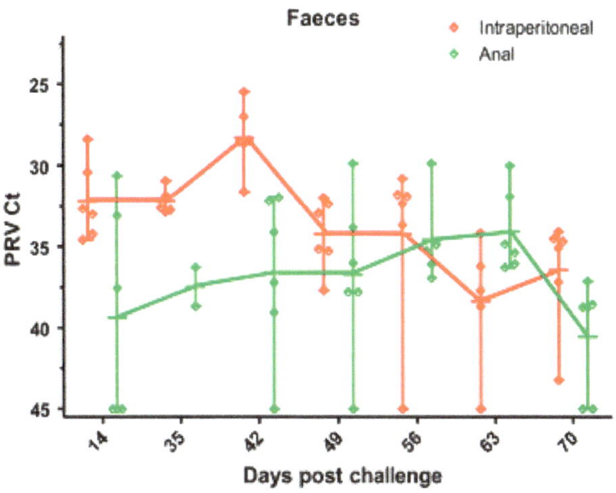

Figure 5 PRV in faecal contents. PRV levels (Ct values) in faeces samples from i.p. (red) anal (green) groups. Markers indicate individual fish, horizontal bars group means and vertical bars the range.

Leakage of PRV from challenged fish

To assess potential leakage of inoculated virus from challenged fish, PRV real time RT-PCR was applied on samples of pooled heart-, spleen- and head kidney tissue from leakage control cohabitants (leakage) at 56 dpc and compared to the corresponding samples from challenged fish. The results demonstrated notable differences between the leakage groups (Figure 6). While most fish were positive for PRV in the oral leakage group and quite surprisingly also in the i.p. leakage group, the majority of fish in the anal leakage group were negative. The mean PRV Ct values in the oral leakage group was significantly lower ($p < 0.01$) than that of the anal leakage group, indicating a significantly higher viral load in the former group. Between the i.p. leakage and anal leakage groups, a similar difference was apparent but not significant ($p = 0.09$).

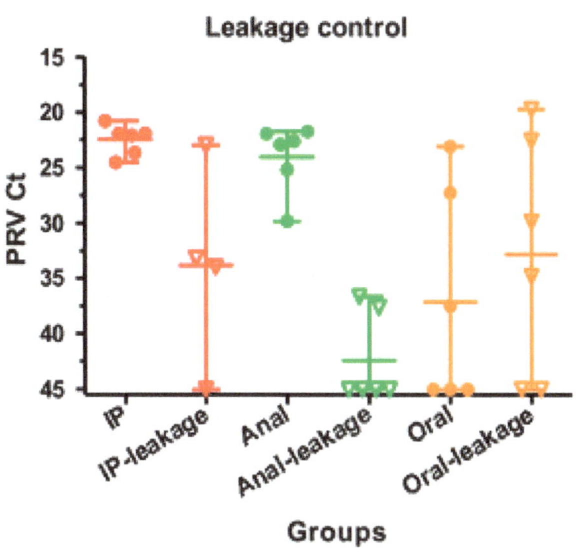

Figure 6 Viral loads (Ct-values) in leakage control cohabitants and challenge groups at 56 dpc. Pooled samples of heart, spleen and head kidney tissue from leakage groups (i.p.-leakage, anal-leakage and oral-leakage) were analysed by PRV qPCR and compared to the corresponding samples from challenge groups (i.p. anal and oral). Markers indicate individual fish, horizontal bars group means and vertical bars the range.

Integrated Analyses

To further characterize the kinetics of the PRV-infection, PRV levels in GI segments, heart, faeces and blood were analysed for possible correlations in the i.p. and anal groups. Comparable levels of correlation were observed for anal and i.p. challenge groups but due to a generally lower variation in Ct values in the i.p., only data from this group are presented in Additional files 1, 2 and 3. Comparison of PRV levels in the four GI segments versus blood from 4 to 14 dpc, showed a strongly significant correlation for all four GI tissues ($p < 0.0001$ by non-parametric Spearman's ρ testing) (Additional file 1). A strongly significant correlation was also found between PRV levels in heart and blood ($p < 0.0001$ by non-parametric Spearman's ρ testing). When sampling time was taken into account, it became clear that a certain spatiotemporal development took place during the infection (Additional file 2). For all sampling time points, the PRV Ct values for blood and heart grouped into relative narrow clusters in the "blood-heart space". The centre of these clusters further exhibited an interesting development with time. Between 7 and 21 dpc, heart seemed to have a comparatively stronger increase in PRV levels which was then succeeded by a more equal increase in both tissues until the peak was reached at 42 dpc. After 42 dpc, PRV levels decreased slowly in heart but more rapidly in blood. PRV levels in blood and faeces were strongly correlated ($p < 0.0001$ by non-parametric Spearman's ρ testing) and again the PRV Ct values in these samples formed sampling time-dependent clusters (Additional file 3). In contrast to blood-heart, the temporal development of these blood-faeces clusters followed a relative linear trajectory, in particular after 42 dpc where both sample types experienced a comparable decrease in PRV levels.

DISCUSSION

Understanding how pathogenic organisms enter their host and spread in the environment is crucial for the management of disease. For PRV, the mechanisms of infection and transmission are yet unknown. In this study, we investigated if the GI tract could serve as a putative site of infection and potential source of shedding for PRV as shown for the related mammalian orthoreovirus [15].

A cohabitation challenge model is often preferred in viral infection studies because the transfer of virus to the cohabitants mimics natural infection. However, in order to study site-specific virus–host interactions at the first encounter it is essential to control when and where the challenge takes place, as well as standardizing the viral dose. Whereas injecting virus into the peritoneum of the host is assumed to ensure these factors, the intrinsic defense mechanisms of the mucus, skin, gills and gut of the fish are bypassed, and the infection cannot be considered natural. We therefore set out to establish an experimental model for studying PRV-infection with particular focus on the GI tract at early time points post infection. The GI tract is responsible for uptake of water and nutrients but also serves as a protective barrier towards the environment. Intestinal challenge may therefore mimic the conditions of a natural infection. The intestinal epithelium is very likely to be an entry port for pathogens transmitting through the fecal-oral route, which include orthoreoviruses in mammals and possibly also PRV. In this study, a PRV inoculum was administered to fish via either oral or anal intubation or by i.p. injection, the latter serving as a basis for comparison. Analysis of the PRV levels in blood and heart showed that anal, but not oral intubation of virus resulted in infection kinetics comparable to i.p. injected fish. Histology further demonstrated that fish infected through the anal route develop HSMI. The real time RT-PCR analysis of GI samples showed that the challenge inoculum is largely restricted to stomach and pyloric caeca in the orally intubated group and to the mid- and distal intestine in the anally intubated group. The low levels of virus RNA in mid- and distal intestine of orally challenged fish suggest that only limited amounts of virus reached this compartment, which could be due to the apparent leakage of virus to the water column that was likely caused by regurgitation. The relative delayed and milder infection kinetics of the oral group may hence correspond to a lesser amount of infective material reaching the intestinal segments that likely serve as an infection site for PRV. Oral and anal vaccination against enteric redmouth disease in rainbow trout revealed that much of the orally fed antigen is digested in the stomach before it reaches the second segment of the intestine where it can be taken up as an immunogenic antigen [29], and this may also in part provide an explanation for the reduced PRV load observed in our oral treatment group.

A feature of mammalian reovirus is that it depends on proteolytic activity for optimal infection. Specific proteolysis of the outer capsid to form an infective subviral particle (ISVP) triggers uptake of the virus, and endosomal proteolysis and low pH further induces virus transfer into the cell cytoplasm [30]. Analysis of the PRV µ1 protein sequence has revealed the presence of a central proteolytic cleavage site, indicating that similar mechanisms may apply for PRV [17]. MRV given orally is demonstrated to be readily transferred to the intestine (within 10–30 min) as ISVPs, most likely due to intestinal protease activity and acidic conditions, indicating that oral uptake should be well suited for reoviruses [31]. Chymotrypsin activity (<10 µg/mL) can trigger MRV ISPV formation and increase infectivity ex vivo, but high levels (>100 µg/mL) may have an inhibitory effect [32]. Compared to mammals, the salmonid stomach is less acidic [33], and starvation has been demonstrated to increase the chymotrypsin levels in the pyloric caeca [34], factors that may affect PRV infectivity. In contrast to PRV, a recent report on IPNV reported slightly higher antigen uptake by oral than anal administration [35]. IPNV has been shown to resist harsh environments well [36, 37] but as PRV is also a non-enveloped virus, this difference may reflect separate infection strategies. It should also be noted that fish in the IPNV-study were starved only 24 h prior to intubation as compared to 6 days in the present study that in our experience was necessary to empty the intestine and enable challenge of the mid-intestine. Under natural circumstances, it is more likely that virus is introduced orally than anally. It could still be that infective virus is able to pass through the stomach and reach the intestine if embedded in fat-containing biological material, i.e. faeces or feed, which could protect from acids and digestive enzymes in the stomach. If the trial period had been extended or the virus inoculum absorbed to a matrix, results might have shown that a PRV-infection can be established also through the oral route.

To control for infection via non-GI routes, naïve fish were cohabitated with challenged fish for 24 h and then held in the same unit until 56 dpc. Although this could allow for individual infected fish to shed virus and further infect the cohort, analysis revealed interesting differences between the various challenge groups. While the relative higher virus levels in the oral leakage group could be explained by leakage of inoculate through regurgitation, the data suggest that the inoculate delivered anally remain relatively confined at the site of inoculation, which is somewhat surprising given the quite direct contact with the water through the anal opening. Although not statistically significant, results indicate that virus can leak through the narrow puncture in the peritoneal wall of i.p. injected fish. In cohabitation challenge experiments, commonly performed with 30–50% of i.p. injected shedders, cohabitant infection efficacies often vary and may be influenced by virus leakage to water from the i.p. injected fish. Moreover, the i.p. challenge method is considered particularly targeted, but leakage short-circuiting the intended i.p. route could add a potentially significant mucosal/immersion component to this challenge type. In comparison, the low leakage observed in the anal challenge, suggest this route to be relatively targeted.

Regarding the kinetics of the PRV-infection, the observed correlation between PRV levels in blood and heart is complex, illustrated by the non-straightforward spatiotemporal development (Additional file 2). These data suggest that heart has an "eigenvalue" in the pathogenesis of PRV in which heart tissue contributes independently to overall virus load, e.g. by local replication. In contrast the relation between PRV in blood and faeces is simpler, almost linear (Additional file 3). This suggests that PRV levels in faeces can be explained by blood levels alone, and that intestine hence is less likely to contribute to faeces levels by local replication. The strong correlation between PRV levels in blood and the four GI segments (Additional file 1) further corroborates the determining role of blood for the PRV findings. Hence, the demonstration of relatively high levels of PRV RNA in faeces samples suggests that faeces is a likely route for transmission of PRV, and that PRV reaches the intestinal lumen via active/passive transport from blood across the intestinal wall. A study on SAV similarly showed that virus was detected in faeces and mucus following challenge [38]. These observations suggest that analysis of fish droppings collected from the net pens or freshwater tanks could be used as a diagnostic tool for early detection of virus without sacrificing fish.

In conclusion, this study demonstrates that anal intubation is an efficacious and well suited model for studying PRV-infection. Furthermore, the results show that PRV can establish an infection through the intestine and can likely shed to the environment through faecal contents. To our knowledge, this is the first report to document that the intestine can be a port of entry for PRV under experimental conditions. The more specific PRV infection mechanisms should be a subject for future studies.

AUTHORS' CONTRIBUTIONS

All authors participated in the design and execution of the study, and writing of the manuscript. Coordination of the study and virus analyses: HH, fish and rearing conditions: ET, histological examination and immunohistochemistry: MA, figures and statistical analyses: SG. All authors read and approved the final manuscript.

ACKNOWLEDGEMENTS

This study was funded by the Norwegian Veterinary Institute. We greatly thank Øystein Wessel and Espen Rimstad, Norwegian University of Life Sciences, for providing anti-PRV σ1 (Anti-σ1) antibody and to Miroslava Hansen, Terje Markussen, Elin Trettenes, Karen Bækken-Soleim, Inger H. Austrheim and Randi Faller, Norwegian Veterinary Institute, for technical assistance. Figure 1B modified with courtesy of Inge Bjerkås.

REFERENCES

1. Kongtorp RT, Kjerstad A, Taksdal T, Guttvik A, Falk K (2004) Heart and skeletal muscle inflammation in Atlantic salmon,*Salmo salar* L: a new infectious disease. J Fish Dis 27:351–358
2. Kongtorp RT, Taksdal T (2009) Studies with experimental transmission of heart and skeletal muscle inflammation in Atlantic salmon, *Salmo salar* L. J Fish Dis 32:253–262
3. Palacios G, Løvoll M, Tengs T, Hornig M, Hutchison S, Hui J, Kongtorp RT, Savji N, Bussetti AV, Solovyov A, Kristoffersen AB, Celone C, Street C, Trifonov V, Hirschberg DL, Rabadan R, Egholm M, Rimstad E, Lipkin WI (2010) Heart and skeletal muscle inflammation of farmed salmon is associated with infection with a novel reovirus. PLoS One 5:e11487
4. Bornø G, Lie Linaker M (eds.) (2015) The health situation in Norwegian aquaculture 2014.http://www.vetinst.no/eng/Publications/Fish-Health-Report/Fish-Health-Report-2014. Norwegian Veterinary Institute
5. Løvoll M, Alarcon M, Jensen BB, Taksdal T, Kristoffersen AB, Tengs T (2012) Quantification of piscine reovirus (PRV) at different stages of Atlantic salmon *Salmo salar* production. Dis Aquat Organ 99:7–12
6. Garseth ÅH, Biering E, Aunsmo A (2013) Associations between piscine reovirus infection and life history traits in wild-caught Atlantic salmon *Salmo salar* L. in Norway. Prev Vet Med 112:138–146
7. Garseth Å, Fritsvold C, Opheim M, Skjerve E, Biering E (2013) Piscine reovirus (PRV) in wild Atlantic salmon, *Salmo salar*L., and sea-trout, *Salmo trutta* L., in Norway. J Fish Dis 36:483–493
8. Wiik-Nielsen CR, Løvoll M, Sandlund N, Faller R, Wiik-Nielsen J, Jensen BB (2012) First detection of piscine reovirus (PRV) in marine fish species. Dis Aquat Organ 97:255–258
9. Kibenge MJ, Iwamoto T, Wang Y, Morton A, Godoy MG, Kibenge FS (2013) Whole-genome analysis of piscine reovirus (PRV) shows PRV represents a new genus in family *Reoviridae* and its genome segment S1 sequences group it into two separate sub-genotypes. Virol J 10:230
10. Siah A, Morrison DB, Fringuelli E, Savage P, Richmond Z, Johns R, Purcell MK, Johnson SC, Saksida SM (2015) Piscine reovirus: genomic and molecular phylogenetic analysis from farmed and wild salmonids collected on the Canada/US Pacific coast. PLoS One 10:e0141475
11. Marty GD, Morrison DB, Bidulka J, Joseph T, Siah A (2015) Piscine reovirus in wild and farmed salmonids in British Columbia, Canada: 1974–2013. J Fish Dis 38:713–728
12. Garver KA, Johnson SC, Polinski MP, Bradshaw JC, Marty GD, Snyman HN, Morrison DB, Richard J (2016) Piscine orthoreovirus from Western North America is transmissible to Atlantic salmon and Sockeye salmon but fails to cause heart and skeletal muscle inflammation. PLoS One 11:e0146229
13. Garver KA, Marty GD, Cockburn SN, Richard J, Hawley LM, Müller A, Thompson RL, Purcell MK, Saksida S (2016) Piscine reovirus, but not jaundice syndrome, was transmissible to Chinook salmon, *Oncorhynchus tshawytscha* (Walbaum), Sockeye salmon, *Oncorhynchus nerka* (Walbaum), and Atlantic salmon, *Salmo salar* L. J Fish Dis 39:117–128
14. Nibert ML, Duncan R (2013) Bioinformatics of recent aqua- and orthoreovirus isolates from fish: evolutionary gain or loss of FAST and fiber proteins and taxonomic implications. PLoS One 8:e68607
15. Virgin HW, Tyler KL, Dermody TS (1997) Reoviruses. In: Nathanson N (ed) Viral pathogenesis. Lippincott-Raven, New York
16. Stanley N, Leak P, Walters M, Joske R (1964) Murine infections with reovirus. II. The chronic disease following reovirus type 3 infection. Br J Exp Pathol 45:142–149
17. Markussen T, Dahle MK, Tengs T, Løvoll M, Finstad Ø, Wiik-Nielsen CR, Grove S, Lauksund S, Robertsen B, Rimstad E (2013) Sequence analysis of the genome of piscine orthoreovirus (PRV) associated with heart and skeletal muscle inflammation (HSMI) in Atlantic salmon (*Salmo salar*). PLoS One 8:e70075
18. Finstad ØW, Dahle MK, Lindholm TH, Nyman IB, Løvoll M, Wallace C, Olsen CM, Storset AK, Rimstad E (2014) Piscine orthoreovirus (PRV) infects Atlantic salmon erythrocytes. Vet Res 45:35
19. Secombes CJ, Ellis AE (2012) The immunology of teleosts. In: Roberts RJ (ed) Fish pathology, 4th edn. Blackwell Publishing Ltd, Oxford
20. Løkka G, Austbø L, Falk K, Bjerkås I, Koppang EO (2013) Intestinal morphology of the wild Atlantic salmon (*Salmo salar*). J Morphol 274:859–876
21. Rombout J, Abelli L, Picchietti S, Scapigliati G, Kiron V (2011) Teleost intestinal immunology. Fish Shellfish Immunol 31:616–626
22. Løkka G, Austbø L, Falk K, Bromage E, Fjelldal PG, Hansen T, Hordvik I, Koppang EO (2014) Immune parameters in the intestine of wild and reared unvaccinated and vaccinated Atlantic salmon (*Salmo salar* L.). Dev Comp Immunol 47:6–16
23. Fuglem B, Jirillo E, Bjerkås I, Kiyono H, Nochi T, Yuki Y, Raida M, Fischer U, Koppang EO (2010) Antigen-sampling cells in the salmonid intestinal epithelium. Dev Comp Immunol 34:768–774
24. Finstad ØW, Falk K, Løvoll M, Evensen Ø, Rimstad E (2012) Immunohistochemical detection of piscine reovirus (PRV) in hearts of Atlantic salmon coincide with the course of heart and skeletal muscle inflammation (HSMI). Vet Res 43:11

25. Wessel Ø, Olsen CM, Rimstad E, Dahle MK (2015) Piscine orthoreovirus (PRV) replicates in Atlantic salmon (*Salmo salar*L.) erythrocytes ex vivo. Vet Res 46:26

26. Hodneland K, Endresen C (2006) Sensitive and specific detection of salmonid alphavirus using real-time PCR (TaqMan (R)). J Virol Methods 131:184–192

27. Løvoll M, Wiik-Nielsen J, Grove S, Wiik-Nielsen CR, Kristoffersen AB, Faller R, Poppe T, Jung J, Pedamallu CS, Nederbragt AJ, Meyerson M, Rimstad E, Tengs T (2010) A novel totivirus and piscine reovirus (PRV) in Atlantic salmon (*Salmo salar*) with cardiomyopathy syndrome (CMS). Virol J 7:309

28. Ørpetveit I, Mikalsen AB, Sindre H, Evensen Ø, Dannevig BH, Midtlyng PJ (2010) Detection of infectious pancreatic necrosis virus in subclinically infected Atlantic salmon by virus isolation in cell culture or real-time reverse transcription polymerase chain reaction: influence of sample preservation and storage. J Vet Diagn Invest 22:886–895

29. Villumsen KR, Neumann L, Ohtani M, Strøm HK, Raida MK (2014) Oral and anal vaccination confers full protection against enteric redmouth disease (ERM) in rainbow trout. PLoS One 9:e93845

30. Guglielmi KM, Johnson EM, Stehle T, Dermody TS (2006) Attachment and cell entry of mammalian orthoreovirus. Curr Top Microbiol Immunol 309:1–38

31. Sturzenbecker LJ, Nibert M, Furlong D, Fields BN (1987) Intracellular digestion of reovirus particles requires a low pH and is an essential step in the viral infectious cycle. J Virol 61:2351–

32. Madren JA, Sarkar P, Danthi P (2012) Cell entry-associated conformational changes in reovirus particles are controlled by host protease activity. J Virol 86:3466–3473

33. Krogdahl Å, Sundby A, Holm H (2015) Characteristics of digestive processes in Atlantic salmon (*Salmo salar*). Enzyme pH optima, chyme pH, and enzyme activities. Aquaculture 449:27–36

34. Rungruangsak-Torrissen K, Moss R, Andresen LH, Berg A, Waagbø R (2006) Different expressions of trypsin and chymotrypsin in relation to growth in Atlantic salmon (*Salmo salar* L.). Fish Physiol Biochem 32:7–23

35. Chen LH, Evensen Ø, Mutoloki S (2015) IPNV antigen uptake and distribution in Atlantic salmon following oral administration. Viruses 7:2507–2517

36. Myrmel M, Modahl I, Nygaard H, Lie KM (2014) Infectious pancreatic necrosis virus in fish by-products is inactivated with inorganic acid (pH 1) and base (pH 12). J Fish Dis 37:349–355

37. Nygaard H, Modahl I, Myrmel M (2012) Thermal inactivation of infectious pancreatic necrosis virus in a peptone-salt medium mimicking the water-soluble phase of hydrolyzed fish by-products. Appl Environ Microbiol 78:2446–2448

38. Graham DA, Brown A, Savage P, Frost P (2012) Detection of salmon pancreas disease virus in the faeces and mucus of Atlantic salmon, *Salmo salar* L., by real-time RT-PCR and cell culture following experimental challenge. J Fish Dis 35:949–951

CHAPTER 18

EFFECTS OF FUNCTIONAL FEEDS ON THE LIPID COMPOSITION, TRANSCRIPTOMIC RESPONSES AND PATHOLOGY IN HEART OF ATLANTIC SALMON (SALMO SALAR L.) BEFORE AND AFTER EXPERIMENTAL CHALLENGE WITH PISCINE MYOCARDITIS VIRUS (PMCV)

Laura Martinez-Rubio[1], Øystein Evensen[2], Aleksei Krasnov[3], Sven Martin Jørgensen[3], Simon Wadsworth[4], Kari Ruohonen[4], Jose LG Vecino[4] and Douglas R Tocher[1]

[1]Institute of Aquaculture, School of Natural Sciences, University of Stirling, Stirling FK9 4LA, Scotland, UK.

[2]Norwegian Life Science University, Faculty of Veterinary Medicine, Dep, Oslo N-0033, Norway.

[3] Nofima AS, Ås N-1431, Norway.

[4]EWOS Innovation AS, Dirdal N-4335, Norway

ABSTRACT

Background

Cardiomyopathy syndrome (CMS) is a severe cardiac disease of Atlantic salmon (Salmo salar) **recently associated with a double-stranded RNA virus, Piscine Myocarditis Virus (PMCV). The disease has been diagnosed in 75-85 farms in Norway each year over the last decade resulting in annual economic losses estimated at up to €9 million. Recently, we demonstrated that functional feeds led to a milder inflammatory response and reduced severity of heart lesions in salmon experimentally infected with Atlantic salmon reovirus, the causal agent of heart and skeletal muscle inflammation (HSMI). In the present study we employed a similar strategy to investigate the effects of functional feeds, with reduced lipid content and increased eicosapentaenoic acid levels, in controlling CMS in salmon after experimental infection with PMCV.**

Citation: Martinez-Rubio et al.: Effects of functional feeds on the lipid composition, transcriptomic responses and pathology in heart of Atlantic salmon (Salmo salar L.) before and after experimental challenge with Piscine Myocarditis Virus (PMCV). BMC Genomics 2014 15:462. doi:10.1186/1471-2164-15-462

Rseults

Hepatic steatosis associated with CMS was significantly reduced over the time course of the infection in fish fed the functional feeds. Significant differences in immune and inflammatory responses and pathology in heart tissue were found in fish fed the different dietary treatments over the course of the infection. Specifically, fish fed the functional feeds showed a milder and delayed inflammatory response and, consequently, less severity of heart lesions at earlier and later stages after infection with PMCV. Decreasing levels of phosphatidylinositol in cell membranes combined with the increased expression of genes related with T-cell signalling pathways revealed new interactions between dietary lipid composition and the immune response in fish during viral infection. Dietary histidine supplementation did not significantly affect immune responses or levels of heart lesions.

Conclusions

Combined with the previous findings on HSMI, the results of the present study highlight the potential role of clinical nutrition in controlling inflammatory diseases in Atlantic salmon. In particular, dietary lipid content and fatty acid composition may have important immune-modulatory effects in Atlantic salmon that could be potentially beneficial in fish balancing the immune and tissue responses to viral infections.

BACKGROUND

Cardiomyopathy syndrome (CMS) is a severe cardiac disease of Atlantic salmon (Salmo salar) recently associated with a double-stranded RNA virus termed piscine myocarditis virus (PMCV) [1]. The disease was first described and diagnosed in Norway in 1985 [2], and has since also been diagnosed in the Faroe Islands [3], Scotland [4] and, possibly, Canada [5]. In 2012, the Norwegian Veterinary Institute diagnosed the presence of CMS at 89 salmon farming sites located in the mid-coast of Norway [6]. Early stages of the disease have been reported in adult Atlantic salmon close to harvest, around 14 to 18 months after transfer to seawater [7], and so the economic impact can be considerable [8].

Histopathologically, CMS is characterized by severe inflammation and necrosis of the spongy myocardium of the atrium and ventricle [9], but liver may also be affected due to the circulatory disturbance associated with the heart lesions. Mortality is usually moderate although morbidity can be very high with the associated chronic inflammatory response lasting for several months. The pathogenesis of experimentally induced CMS was previously assessed by transcriptomic profiling using oligonucleotide microarrays in association with pathology [10, 11]. In these studies, cardiac pathology and viral load were correlated, with an up regulation of T cell response genes. It appeared that the cellular effector response mediated by CD8+ T cells contributed to successful clearance of the virus infection, although it was also correlated with an up regulation of apoptotic genes, which could contribute to tissue damage.

The lack of commercial vaccines to CMS currently makes the use of alternative therapies crucial. Factors that modulate the inflammatory process might be key to mitigate the clinical symptoms and improve the performance of affected fish. The concept of clinical nutrition is well known in humans [12], and is becoming of increasing interest in aquaculture. Functional feeds are defined as high-quality feeds that, beyond their nutritional composition, are formulated with health promoting features that could be beneficial in supporting disease resistance and mitigation of clinical disease symptoms. Thus a clinical nutrition approach using functional feeds potentially enables a shift away from chemotherapeutic and antibiotic treatments, lowering the costs of disease treatment and management [13]. There are several studies in fish linking nutrition and immunology, recently reviewed by Kiron (14). The inclusion in aquaculture feeds of additives such as prebiotics, probiotics, immunostimulants, vitamins and nucleotides, is reported to increase growth and feed conversion efficiency, as well as having positive effects on the immune system and protection against bacterial infections [13]. Moreover, macronutrients like proteins and lipids are reported to play key roles in the regulation of pathways of the immune system and have been widely studied due to the necessity of the aquafeed industry to replace dietary fishmeal (FM) and fish oil (FO) [14].

Polyunsaturated fatty acids (PUFA) are involved in the regulation of both innate and adaptive immunity, and the inflammatory response through four potential mechanisms including gene expression, eicosanoid metabolism, cellular signalling and membrane organization [15]. Specific transcription factors are activated by a number of PUFA, long-chain (LC)-PUFA and eicosanoid ligands [16], and subsequently regulate the expression of genes related with inflammatory, B-cell and T-cell responses, which play important roles in viral infections. The roles of PUFA in cell membrane organization and signalling pathway mechanisms have been widely studied in humans [15]. Inhibition of T-cell signalling pathways by PUFA is mainly linked to the suppression of the elevation of cytoplasmic calcium concentration through the phosphatidylinositol (PI) signalling system, which is a key event in T-cell activation [17]. Eicosanoids are key mediators of inflammation and the regulation of T and B lymphocyte functions. Arachidonic acid (20:4n-6; ARA) and eicosapentaenoic acid (20:5n-3; EPA), released from membrane phospholipids through the action of phospholipases, are converted to pro- or anti-inflammatory eicosanoids, respectively and, although the enzymes of eicosanoid metabolism have a preference for ARA, increased levels of EPA in the membranes of immune cells inhibit the production of pro-inflammatory eicosanoids [18]. Studies have demonstrated that LC-PUFA have similar roles in immune modulation in fish. Substitution of dietary FO by vegetable oil (VO), with consequent reduction in the ratio of n-3/n-6 LC-PUFA, promotes the synthesis of pro-inflammatory eicosanoids [19, 20], and alters humoral immunity and the expression of pro-inflammatory cytokine genes [21]. Most studies have been performed in vitro[22–24], and so little is known about the roles of LC-PUFA during viral infections in fish.

In a previous study, the use of functional feeds led to a milder inflammatory response and reduced severity of heart lesions in salmon experimentally infected by Atlantic salmon reovirus (ASRV), the causal agent of heart and skeletal muscle inflammation (HSMI) [25]. As both HSMI and CMS have similar symptoms, we have employed a similar strategy in the present study to investigate the effects of functional feeds in controlling CMS. The functional feeds contained increased EPA levels, from 3.6% in a reference feed to 14%, and reduced dietary lipid, from 31% to 18% in the functional feeds. Supplementation of histidine was also assessed in one of the functional feeds as this amino acid and related compounds, such as N-α-acetyl-histidine (NAH), have important roles in muscle pH buffering [26, 27] and tissue antioxidant systems [28, 29]. In particular, histidine has been associated with a cardio-protective role in human studies, being potentially beneficial in the alleviation of oxidative stress associated with viral diseases [30]. Thus, we hypothesized that dietary supplementation of this amino acid could have a potentially beneficial effect in fish suffering CMS.

Fish were fed the experimental feeds for 8 weeks prior to infection by PMCV and throughout the study post-infection. The incorporation of LC-PUFA into membrane phospholipids of heart, the main organ affected by the disease, was assessed, and both heart and liver tissues were subjected to histological evaluation. The inflammatory and immune status in heart tissue was assessed after infection by determining gene expression through oligonucleotide microarray analysis.

RESULTS

Lipid and Fatty Acid Composition of Heart Tissue

The total lipid content of heart tissue was unaffected by dietary treatment and did not show any significant changes over the course of infection (Table 1). Proportions of triacylglycerol (TAG) were lower and those of total PL and the major PL classes, phosphatidylcholine (PC) and phosphatidylethanolamine (PE), significantly higher in heart tissue of fish fed the functional feeds compared with fish fed the REF diet (2-way ANOVA p-value diet, < 0.05) (Figure 1). These differences were particularly apparent at the end of the pre-feeding phase and, after PMCV infection, PL levels generally decreased, significantly for phosphatidylinositol (PI) and phosphatidylserine (PS), over the course of the infection.

Table 1 Fatty acid compositions (percentage of total fatty acids) of total phospholipids of heart tissue from fish fed the reference (REF) and functional (CMS1 and CMS2) feeds at different times before (PreCh) and after (6, 8, 10, 12 and 14 weeks) infection with PMCV

	PreCh			6 wpc			8 wpc			10 wpc			12 wpc			14 wpc			TWO WAY ANOVA *P-value*		
	REF	**CMS1**	**CMS2**	**REF**	**CMS1**	**CMS2**	**REF**	**CMS1**	**CMS2**	**REF**	**CMS1**	**CMS2**	**REF**	**CMS1**	**CMS2**	**REF**	**CMS1**	**CMS2**	**Diet**	**Week**	**Diet*Week**
Saturated	25.9±0.8	31.2±0.9	31.9±0.8	27.6±0.8	29.7±0.7	30.1±0.6	27.2±0.6	29.9±0.7	29.2±0.5	27.1±0.5	29.5±0.3	29.9±0.5	27.6±1.5	30.6±1.4	31.2±1.0	27.9±0.5	31.3±0.4	30.4±0.7	0.000	0.005	ns
18:1n-9	13.2±0.6	7.4±0.6	6.8±0.3	9.5±0.4	6.8±0.2	7.2±0.4	10.1±0.3	7.2±0.2	8.0±0.4	10.2±0.8	7.5±0.6	7.5±0.7	11.0±1.5	7.3±0.1	7.6±0.3	11.5±0.6	7.5±0.2	7.3±0.5	0.000	0.003	0.003
Monounsat	24.6±2.6	15.4±0.9	15.3±0.7	18.0±1.0	14.7±0.8	15.1±1.1	19.5±0.7	14.9±0.7	16.8±0.9	18.9±1.6	15.6±1.3	15.8±1.9	20.4±3.0	15.5±0.4	15.8±0.7	21.5±1.4	15.7±0.6	15.5±1.0	0.000	0.033	0.065
18:2n-6	4.3±0.4	1.9±0.1	1.7±0.1	3.3±0.1	1.7±0.1	1.7±0.1	3.4±0.1	1.8±0.1	2.1±0.2	3.3±0.2	1.8±0.2	1.8±0.1	3.4±0.4	1.7±0.0	1.7±0.1	3.5±0.2	1.8±0.1	1.8±0.1	0.000	0.008	0.021
20:3n-6	0.6±0.0	0.1±0.0	0.2±0.0	0.6±0.0	0.2±0.0	0.2±0.0	0.6±0.0	0.3±0.0	0.3±0.0	0.5±0.0	0.3±0.0	0.3±0.0	0.4±0.0	0.3±0.0	0.3±0.0	0.4±0.0	0.2±0.0	0.2±0.0	0.000	0.000	0.000
20:4n-6	1.4±0.0	1.7±0.1	1.6±0.1	1.5±0.1	1.7±0.1	1.7±0.1	1.5±0.0	1.7±0.1	1.6±0.0	1.4±0.1	1.7±0.0	1.7±0.2	1.3±0.1	1.8±0.1	1.7±0.1	1.4±0.1	1.8±0.1	1.8±0.1	0.000	0.056	0.002
n-6 PUFA	7.3±0.4	4.7±0.2	4.5±0.1	6.2±0.1	4.5±0.1	4.5±0.1	6.2±0.1	4.7±0.0	4.8±0.2	6.1±0.2	4.8±0.2	4.7±0.1	6.0±0.4	4.6±0.1	4.6±0.1	6.4±0.2	5.0±0.1	4.9±0.1	0.000	0.000	0.000
18:3n-3	1.5±0.1	0.5±0.1	0.4±0.0	1.1±0.0	0.4±0.0	0.4±0.0	1.2±0.1	0.5±0.0	0.6±0.1	1.3±0.1	0.5±0.0	0.5±0.0	1.3±0.2	0.4±0.0	0.4±0.0	1.4±0.0	0.5±0.0	0.5±0.0	0.000	0.000	0.001
20:5n-3	6.3±0.3	9.7±0.3	9.6±0.4	7.7±0.3	10.6±0.3	10.4±0.3	8.0±0.3	10.5±0.2	10.2±0.3	8.4±0.4	11.3±0.4	10.8±0.4	8.0±0.1	10.9±0.6	10.4±0.2	7.7±0.3	10.3±0.4	11.0±0.6	0.000	0.000	0.038
22:6n-3	31.5±1.4	33.8±1.3	33.9±1.5	35.8±1.4	35.3±0.9	34.8±1.5	34.6±0.7	34.8±0.5	33.6±0.7	34.5±0.8	33.2±1.9	33.6±1.5	33.4±2.7	32.9±1.2	32.8±1.4	31.7±1.5	32.4±0.5	32.7±1.0	ns	0.000	ns
n-3 PUFA	42.0±1.4	48.3±1.7	48.0±1.8	47.4±1.2	50.7±0.8	50.0±1.6	46.8±0.8	50.3±0.8	48.8±0.8	47.3±1.2	49.8±1.7	49.4±1.8	45.7±2.5	48.9±1.8	48.1±1.5	43.8±1.6	47.5±0.5	48.8±1.6	0.000	0.000	ns
PUFA	49.3±0.8	53.0±0.9	52.6±0.7	53.7±1.1	55.2±0.8	54.5±1.6	53.0±0.8	55.0±0.8	53.6±0.6	53.4±1.0	54.6±1.5	54.0±1.8	51.7±2.2	53.5±1.7	52.6±1.5	50.2±1.5	52.5±0.6	53.7±1.5	0.000	0.000	ns
EPA/ARA	4.6±0.2	5.8±0.4	5.8±0.3	5.2±0.2	6.3±0.2	6.3±0.3	5.5±0.2	6.1±0.0	6.5±0.3	6.0±0.2	6.6±0.1	6.6±0.4	6.1±0.5	6.2±0.3	6.3±0.1	5.7±0.1	5.7±0.4	6.0±0.3	0.000	0.000	0.001
n-3/n-6	5.7±0.2	10.3±0.5	10.6±0.3	7.6±0.3	11.2±0.3	11.2±0.5	7.5±0.2	10.6±0.2	10.1±0.5	7.7±0.4	10.4±0.8	10.6±0.3	7.7±1.0	10.6±0.6	10.5±0.3	6.9±0.4	9.6±0.1	10.1±0.5	0.000	0.000	0.066
Total lipid	2.8±0.3	3.5±0.2	3.6±0.2	3.32±0.2	3.34±0.2	3.49±0.2	3.8±0.4	3.34±0.2	3.63±0.6	3.84±0.3	3.55±0.2	3.58±0.3	3.85±0.2	3.56±0.1	3.46±0.1	3.81±0.3	3.44±0.5	3.46±0.4	ns	ns	ns

ARA, arachidonic acid; DHA, docosahexaenoic acid (DHA); EPA, eicosapentaenoic acid; PUFA, polyunsaturated fatty acid.

Diet*Week indicates the p value of the interaction of those two factors in the 2-way ANOVA statistical analysis.

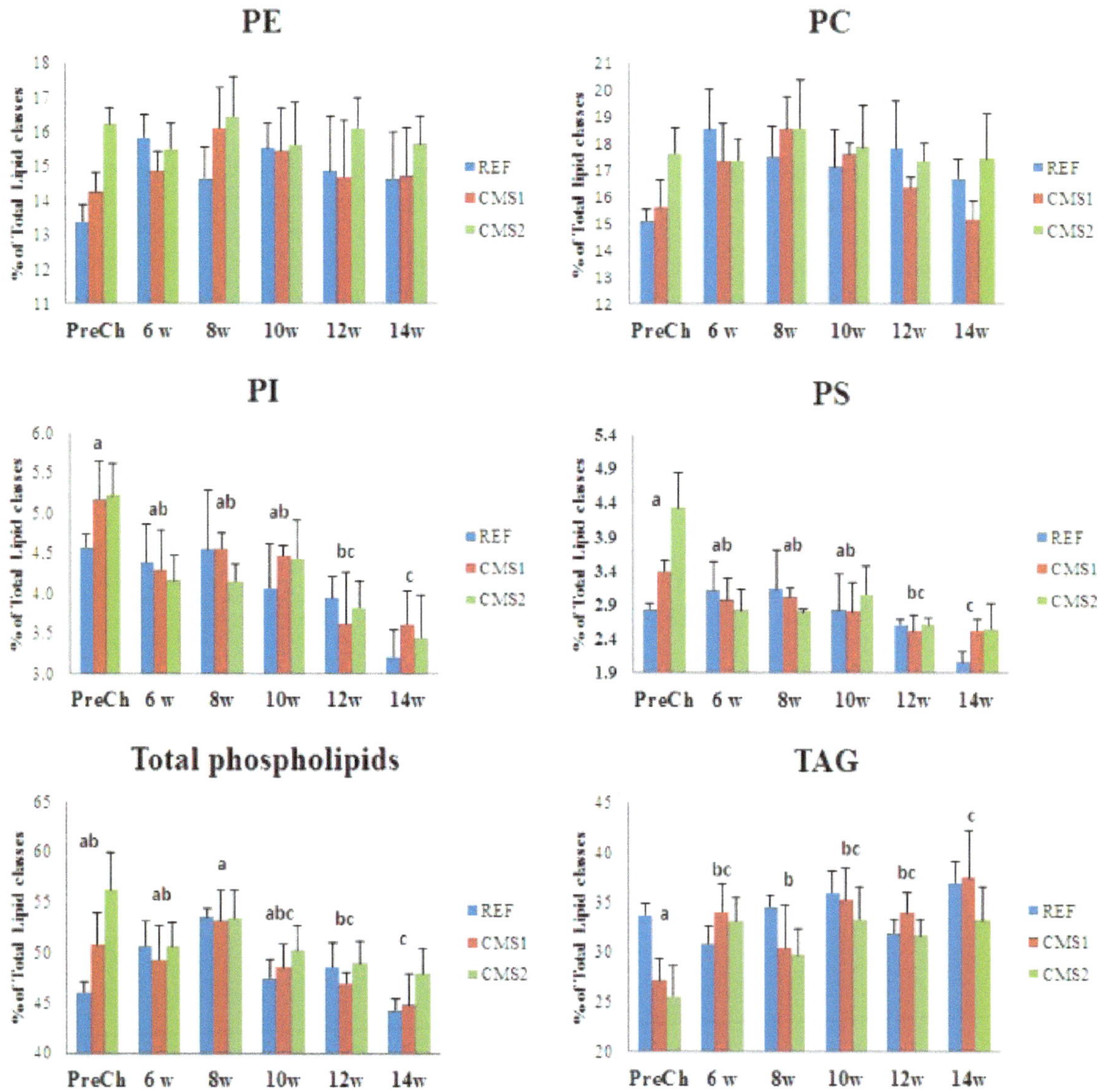

Figure 1 Proportions of lipid classes in total lipid of heart tissue from fish fed the reference (REF) and functional (CMS1 and CMS2) feeds at different times before (PreCh) and after (6, 8, 10, 12 and 14 weeks) infection with PMCV.Data were analysed by 2-way ANOVA with "time-course" and "diet" as the two factors. Different letters above time-points indicate significant differences between time-points over the time-course of the infection. The factor diet showed significant differences in PC, PE, total phospholipids and TAG (not indicated). PC, phosphatidycholine; PE, phosphatidylethanolamine; PI, phosphatidylinositol; PS, phosphatidylserine; TAG, triacylglycerol.

The fatty acid compositions of total PL of heart tissue of fish fed the functional feeds reflected the composition of the diets showing different proportions of the LC-PUFA that could potentially influence the immune response [31] (Table 1). Thus, the proportions of EPA, ARA and the EPA/ARA ratio were significantly higher in fish fed the functional feeds compared to fish fed the REF diet (P value diet, <0.05). However, levels of docosahexaenoic acid (22:6n-3; DHA) were similar between the different dietary groups, which was consistent with our previous study using similar feeds in which tissue levels of DHA were more conserved despite differences in the composition of the diets [25]. The proportions of the inflammation-related LC-PUFA in heart total PL changed during the time course of the infection but were similar in fish fed the three dietary treatments (Table 1). The changes in ARA post-infection were not significant, but the proportions of EPA and DHA were significantly higher at 6-wpc compared with pre-challenge. After 6-wpc, levels of DHA progressively decreased during the time-course of the infection but proportions of EPA were significantly higher at 10-wpc. As a consequence of the latter, the EPA/ARA ratio in heart total PL was also higher in all dietary treatments at 10-wpc. There were no differences in the EPA/ARA ratio in heart PL between fish fed the different diets in the later stages of the infection.

Heart and Liver Tissue Histopathology

Histological changes were consistent with CMS and first observed in the atrium at 6-wpc (Figure 2), characterized by focal infiltration of inflammatory cells dominated by lymphocytes. A degenerative process associated with the inflammatory changes was also observed in cardiomyocytes of the atrium. The most marked increase in histopathological scores was found in the ventricle of fish from all the dietary groups from 6- to 8-wpc. These changes were typified by multifocal infiltration of inflammatory cells, dominated by lymphocytes and macrophages, and concomitant degeneration and necrosis of myocytes. Another typical finding was hyperplasia of endothelial cells in inflamed areas. Inflammatory changes were greatest in the atrium from 6- to 8-wpc and significantly higher for all groups at both time points. By 10-wpc, the inflammatory changes and myocyte necrosis had levelled off in all groups, and was not different from 8-wpc. At 12- and 14-wpc there was a moderate decline in inflammatory scores (Figure 2). Differences between dietary groups were found mainly in the early stages of the disease, at 6- and 8-wpc. Fish fed both functional feeds showed significantly lower histoscores in the atrium at both 6- and 8-wpc compared to fish fed the REF diet (Figure 2). Lesions in the ventricle were also lower in fish fed the functional feeds, being significant in both dietary groups at 8-wpc but only for the fish fed with the CMS2 diet at 6-wpc. There were no differences in inflammation in either part of the heart between dietary groups at 10- and 12-wpc. At the end of the trial, 14-wpc, lesions in the atrium were not significantly different between groups, although fish fed with CMS2 had lower histoscores compared with the other groups. In ventricle, fish fed both functional feeds showed lower histoscores although they were not significantly different (Figure 2).The liver histology scores, based on the degree of steatosis were interpreted as vacuole-formation in hepatocyte cytoplasm characterized by both micro- and macro-vesicular lesions. There were clear differences between the fish fed the functional feeds and fish fed the REF diet at all samplings points post-challenge, with the latter group presenting higher micro- and macro-vesicular steatosis. Severity of the liver histopathology was significantly higher at the beginning of the infection and greater at the end of the challenge (12- and 14-wpc) in salmon fed the REF diet compared to fish fed the two functional feeds (Figure 3).

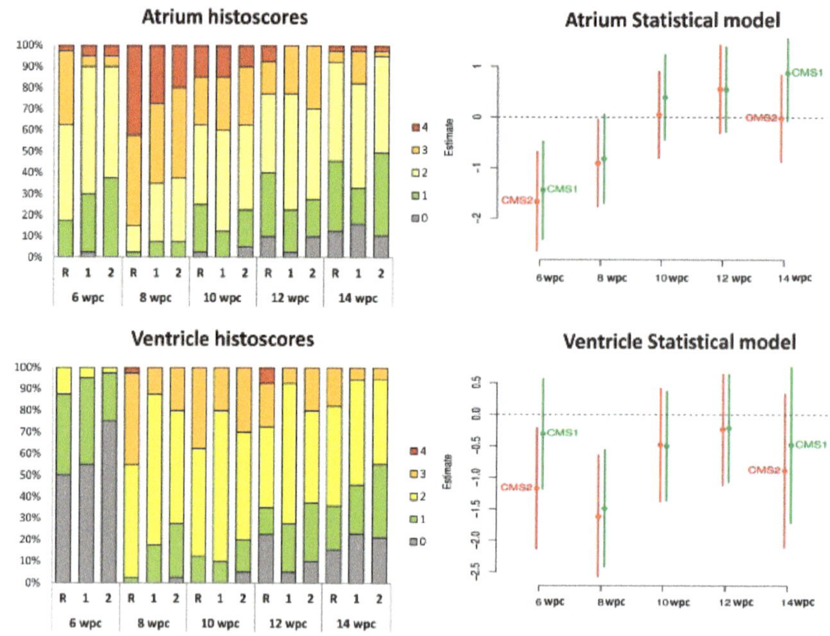

Figure 2 Histopathology results in heart tissue. Figures on the left are representing incidence (percentage of fish sampled) and severity of histopathology (based on the criteria described in methods) in both parts of the heart, atrium and ventriculum at 6-, 8-, 10-, 12-, 14-weeks post-challenge in fish fed the Reference diet (R), and the two functional feeds CMS1 (1) and CMS2 (2). Figures on the right are representing the statistical analysis of the atrium and ventriculum histoscores. Estimated effects of CMS1 and CMS2 diets in comparison to the REF diet by sampling weeks. Negative estimates mean there are lower scores and positive that there are higher scores than for the REF dietary group. Error bars denote approximate 95% confidence limits.

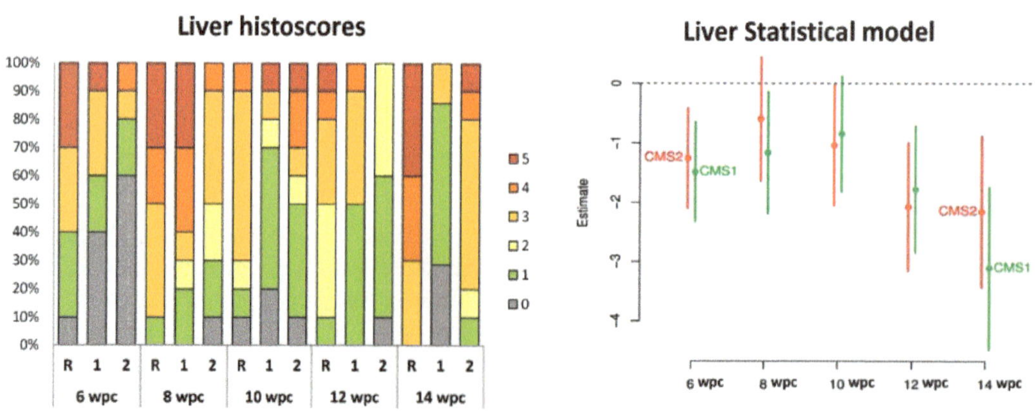

Figure 3 Histopathology results in liver tissue. Figure on the left is representing incidence (percentage of fish sampled) and severity of histopathology (based on the criteria described in methods) at 6-, 8-, 10-, 12-, 14-weeks post-challenge in fish fed the Reference diet (R), and the two functional feeds CMS1 (1) and CMS2 (2). Figure on the right is representing the statistical analysis of liver histoscores. Estimated effects of CMS1 and CMS2 diets in comparison to the REF diet by sampling weeks. Negative estimates mean there are lower scores and positive that there are higher scores than for the REF dietary group. Error bars denote approximate 95% confidence limits.

Viral Load

Viral load was similar for the three dietary groups at 6-wpc, however by 8-wpc, although individual variation in viral RNA was observed, heart viral load was higher in fish fed the REF diet compared with fish fed the functional feeds, significantly so in the case of fish fed CMS2 (Figure 4).

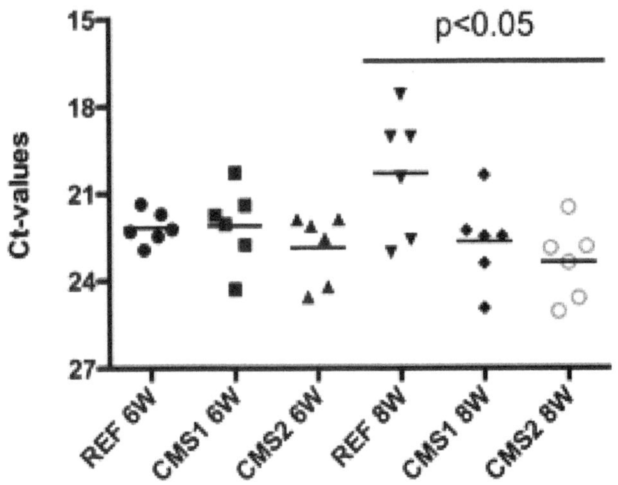

Figure 4 Viral load in heart tissue of the different dietary groups at 6- and 8-weeks post-challenge. Viral load was determined by quantitative real-time PCR analysis of Piscine Myocarditis Virus. Results are presented as CT values (normalized) as a basis for showing the relative level of virus expression (n=6) Levels of significant differences (t-test) were calculated using the relative expression software tool (REST 2009).

CARDIAC TRANSCRIPTOMIC RESPONSES

Time-course of Disease

Comparison of gene expression changes in fish fed the REF diet over the course of infection (6-, 8- and 14-wpc) relative to pre-challenge levels showed highly co-ordinated up-regulation of genes involved in immune pathways (Figure 5); data for representative genes included in Figure 5 are in Table 2. The antiviral and IFN response (62 features), mainly represented by previously annotated virus-responsive genes [32], showed highest expression at 6- and 8-wpc that levelled off by 14-wpc. The largest group of genes associated with T cell responses (99 features) reached peak levels by 8-wpc coinciding with the peak of cardiac histoscores. A similar profile was observed for genes involved in antigen presentation via major histocompatibility complexes (MHC class I and II), however their induced expression remained stable to 14-wpc. The B cell response, represented by various immunoglobulins (26 features), showed increasing expression from 6- and 8-wpc until peak levels at 14-wpc. Recovery in cardiac histopathology by 14-wpc was consistent with significantly reduced expression of several inflammatory markers, such as granzyme A, serum amyloid A, TNF decoy receptor (tnfrsf6b) and neutrophil cytosolic factor 4. It is also noteworthy that negative immune regulators such as suppressor of cytokine signaling 3 and lymphocyte antigen 75/cd205 were induced by 14-wpc.

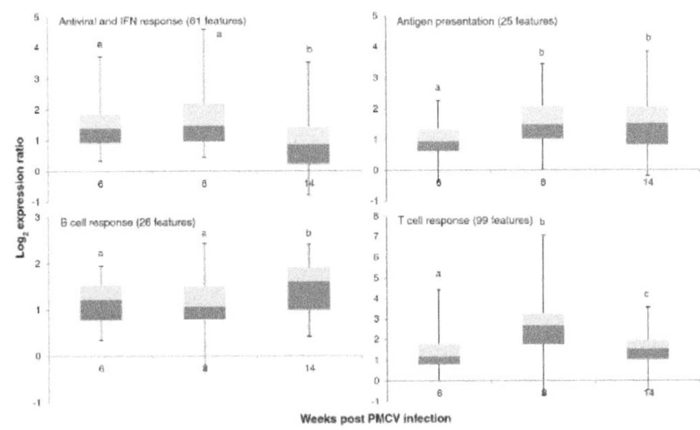

Figure 5 Time-course of four immune pathways to PMCV infection examined with microarray in salmon fed with REF diet. Boxes show median $\log_2$ gene expression ratio (total number of features for each pathway indicated in panel header) between pre-challenge and time points post infection, with the 25th (dark grey) and 75th (light gray) percentile, and whiskers indicate minimum and maximum values. Letters indicate significant differences between time points ($p<0.05$, Student t-test).

Table 2 Examples of immune genes induced in CMS challenged salmon. Data are challenge to pre-challenged expression ratios (fold changes) on the fish fed with the REF diet

Genbank	Gene, symbol	6 wpc	8 wpc	14 wpc
	Antigen presentation and VRG			
U20945	MHCII beta chain	4.80	10.91	8.10
DY713447	CD40	1.73	3.06	3.26
AF508864	MHC class I antigen	3.60	6.17	10.80
DW538400	IFN-induced protein 44, ifit44	6.93	24.09	6.77
117530907	Barrier-to-autointegration factor, bainf	13.20	14.39	2.96
EG795847	GTPase IMAP family member 7, gimap7	3.65	9.05	3.44
	T cells			
AJ841811	Interferon gamma, ifng	3.60	8.03	1.29
DY710412	Granzyme A, graa	21.78	131.75	6.62
NM_0011235	CD8 alpha	3.13	12.06	6.44

213070237	CD45	7.11	10.20	11.83
EG819885	Cytotoxic T-lymphocyte protein 4, ctla	5.18	14.95	3.04
	Inflammatory markers, regulators			
CX032230	TRAF interacting protein, traip	1.26	2.32	3.27
117427898	Chemokine CCL-C24	2.98	10.93	16.14
AJ517803	TNFR associated factor 3, traf3	2.70	1.29	3.15
CU652893	Serum amyloid A5, saa	3.11	1.76	1.53
EG881931	TNF decoy receptor, tnfrsf6b	3.54	10.64	1.41
EG904168	IL1b	2.48	2.66	2.84
117496064	Neutrophil cytosolic factor 4, ncl4	1.43	2.64	0.94
213077403	TNF alpha-induced protein 2, tnfaip2	3.39	5.28	3.95
S48406409	Lymphocyte antigen 75, Ly75/CD205	2.21	2.64	9.20
213081931	Helicase lymphoid-specific, hells	1.03	6.06	6.52
DW578491	Suppressor of cytokine signaling 3b, socs 3b	1.34	1.42	3.41

Effects of functional feeds pre- and post-challenge

Effects on immune responses

A large fraction of differentially expressed genes encoded proteins of the immune system and genes from several functional groups showed similar expression profiles over the course of the infection (Figure 6; Additional file 1: Table S1); results for selected genes are given in Table 3. Overall, the functional feeds resulted in down-regulated levels of immune genes, suggestive of immune suppressive actions (Figure 7). At pre-challenge this was most significant for antiviral and IFN responses, which showed reduced expression in fish fed CMS1 and CMS2 compared to fish fed the REF diet (Figure 7). This down-regulation was also significant at 8-wpc, although with lower magnitude, and this coincided with the most significant reduction in ventricle histoscore in fish fed the functional feeds (Figure 2). Among antiviral and IFN genes, the largest expression differences were found in genes known for potent responses to viruses, such as very large inducible GTPase 1, IFN inducible MX protein and receptor-transporting protein 3. Notable exceptions were two genes encoding IFN regulatory factors – IRF4 and IRF7 – that had from 2.6 to 4.4-fold higher expression before challenge in salmon fed with the functional feeds compared to the REF diet. The expression profiles of immunoglobulin (Ig) genes most likely reflected the influx and amount of B cells in the heart. Fish fed the functional feeds had slightly lower abundance of 28 Ig transcripts before challenge that were further reduced by 6-wpc, followed by normalized levels from 8- to 14-wpc. The most striking immunosuppressive effect of functional feeds was observed for T cell response genes, as suggested by significantly reduced expression of 64 transcripts at pre-challenge, 6- and 8-wpc, and moderately lower reduction at 14-wpc. Down-regulation of CD8 beta, granzyme A/K and IFN gamma could imply that a large fraction of this population was represented by cytotoxic T cells. Granzyme A and TNF decoy receptor tnfrsf6b, which were also down-regulated by the functional feeds, showed strongest correlation with heart histopathology level in a previous CMS study [11].

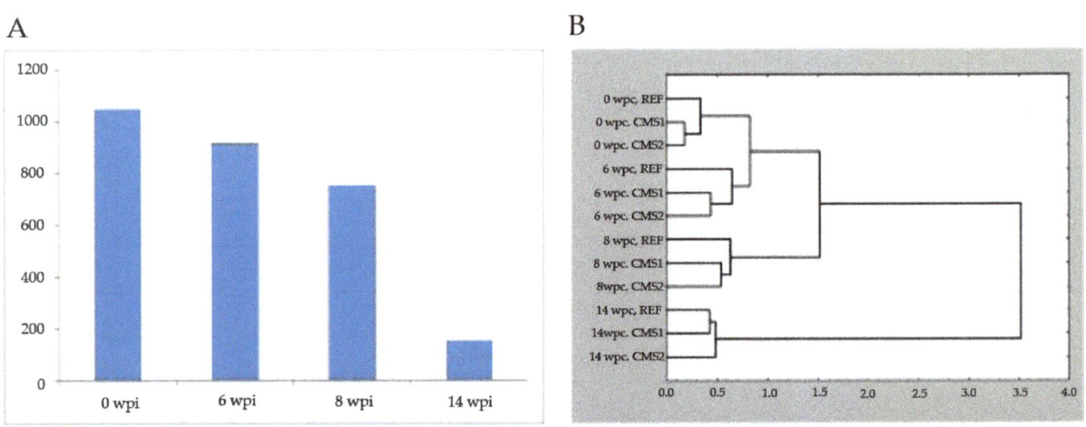

Figure 6 Effect of functional feeds on gene expression in heart tissue of Atlantic salmon examined with microarray. A: Number of differential expressed genes (y-axis) per time point post infection (x-axis). B: Hierarchical clustering by 2419 features (Pearson r, Ward's method)

Table 3 Examples of immune genes that responded to the functional feeds

Genbank Or Probename	Gene, Symbol	0 wpc		6 wpc		8 wpc		14 wpc	
		CMS1	CMS2	CMS1	CMS2	CMS1	CMS2	CMS1	CMS2
	Antiviral responses								
BQ035726	Very large inducible GTPase 1-3, vlig	−3.71	−1.87	−1.30	−2.75	−3.53	−4.80	−1.51	−3.71
?	IFN inducible mx protein	−5.37	−2.61	1.23	0.78	−1.77	−4.04	−1.09	−1.08
DY741158	STAT2	−2.70	−4.19	1.26	1.03	1.06	−1.67	1.20	1.08
CB500614	Pyrin, bty	−1.38	−2.76	1.12	−1.03	−2.39	−3.50	−1.79	−1.14
BX909789	IRF4	2.82	2.64	3.28	1.46	−1.73	−2.14	1.31	−1.06
213061919	IRF7	3.33	4.43	−1.01	−1.19	1.54	−1.18	1.27	1.45
DW538275	Receptor-transporting protein 3	−1.90	−2.31	−0.09	−0.43	−0.85	−1.17	0.45	−0.13
	T cells								
Ssa#TC108524	MHC class I antigen	−2.33	−1.57	−0.36	−1.02	−0.05	−0.34	−0.33	−0.52
?	IFNg	−1.66	−1.09	−2.18	−1.98	−2.44	−1.99	−1.32	−1.91
AY693394	CD8 beta	1.19	−1.35	−2.07	−1.76	−2.02	−1.43	−1.60	−1.74
CA368982	Kidney injury molecule 1, havcr1	1.36	−1.31	−1.46	−1.64	−4.27	−2.62	−1.36	1.19
DY710412	Granzyme A-1, graa	−2.24	1.30	−2.65	−2.18	−3.04	−2.94	−1.04	−2.07
?	Granzyme A-2, graa	−1.48	−1.11	−2.76	−3.27	−3.56	−2.50	−1.19	−2.61
CR369847	NF activated T-cells calcineurin-dependent 1, nfatc1	−1.10	2.27	−1.63	−1.03	4.62	3.09	−1.88	−1.35
	Inflammatory markers, regulators								
213077403	TNF alpha-induced protein 2, tnfaip2	2.53	2.48	−3.51	−1.35	−2.53	−2.29	2.03	1.11
EG871595	c3a anaphylatoxin chemotactic receptor, c3ar1	−1.02	1.53	1.12	1.13	−2.53	−1.84	−1.17	−1.77
DW576911	Cytochrome P450 family 2 subfamily V, cyp2u1	−1.90	1.13	−5.38	−2.73	−1.74	−2.67	−2.35	−1.54
213081824	FcRgamma-like protein, fcer1g	−6.40	−3.06	−2.44	1.16	−3.18	−2.05	1.20	−2.59
EG881931	TNF decoy receptor, tnfrsf6b	1.36	1.13	−1.73	−1.69	−2.96	−3.18	1.21	−1.10
213081931	Helicase lymphoid-specific, hells	2.24	4.47	1.34	2.20	−2.25	−1.98	−1.06	1.41
CK888744	fetuin B	1.16	1.39	0.87	1.19	1.42	0.58	1.51	−0.56
S48406409	Ly75/CD205	2.79	2.75	−1.13	−1.31	2.06	1.55	−1.27	−1.34

Data are expression ratios (fold changes) of fish fed the functional feeds (CMS1 and CMS2) relative to the REF diet during the course of the infection.

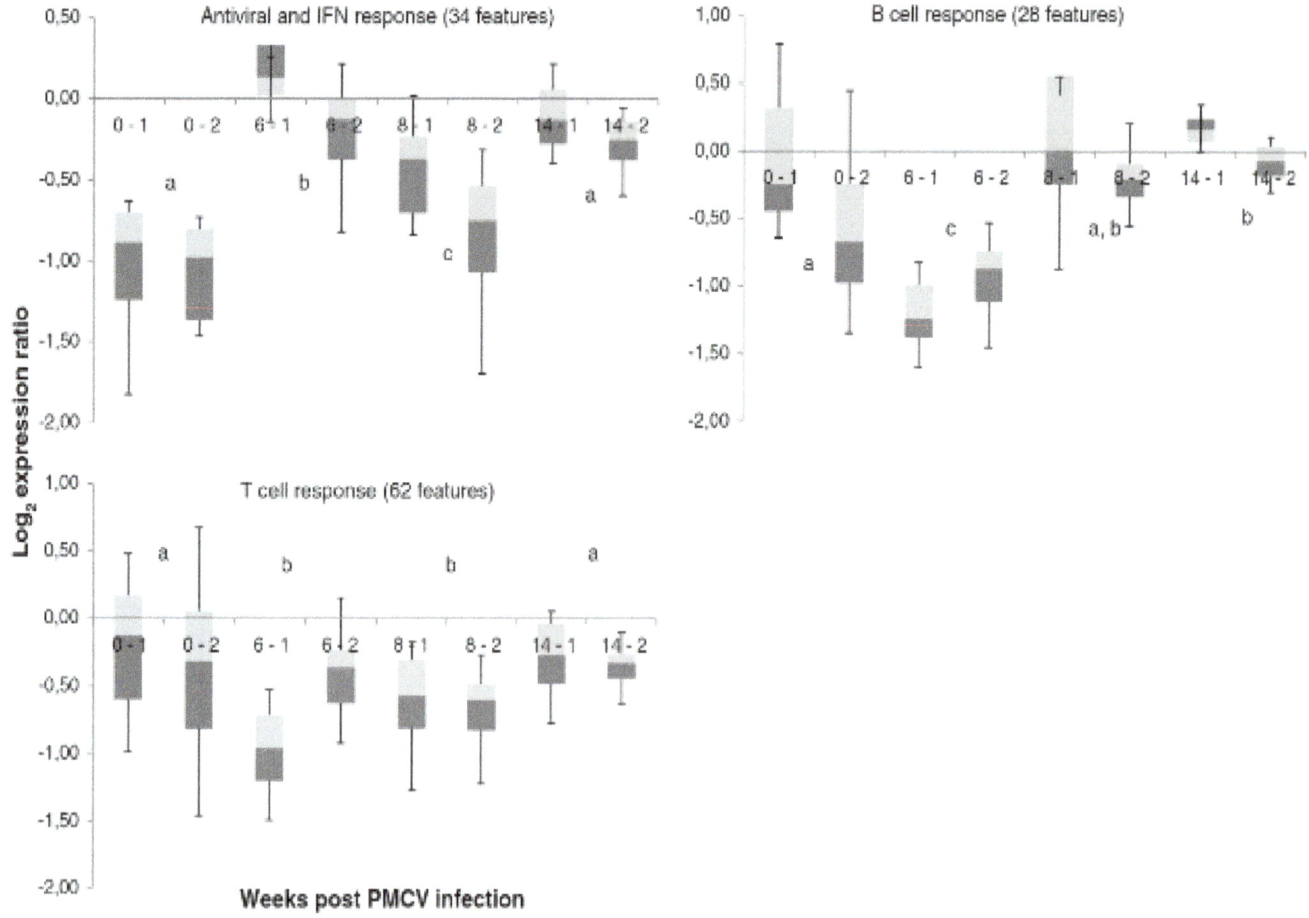

Figure 7 Effect of the two functional feeds on immune gene expression in heart tissue of Atlantic salmon pre (week 0) and post (weeks 6, 8, 14) PMCV infection, as examined with microarray. Data are $\log_2$ expression ratios between functional feeds (CMS1 and 2, noted respectively as -1 and -2 after each time points) and reference feed (REF). Boxes show median values (total number of features for each pathway indicated in panel header) with the 25th (dark grey) and 75th (light gray) percentile, and whiskers indicating minimum and maximum values. Letters indicate significant differences between time points ($p<0.05$, Student t-test).

Effects on lipid metabolism

Differences in expression of genes involved in lipid metabolism that could potentially modulate the immune response were also observed over the course of the infection (Additional file 1: Table S1). Among the significantly relevant genes were those related to the PI signalling system (Figure 8) and the biosynthesis of LC-PUFA. In general, the expression of the genes related to the PI signalling pathway was not correlated with tissue pathology or the expression of immune-related genes, as there was higher activation of those genes at all sampling points after PMCV infection in fish fed the functional feeds compared with fish fed the REF diet. Of the lipid metabolism genes (Additional file 1: Table S1), the most significant effect was detected with probes for the delta-6 fatty acyl desaturase (Fadsd6) gene (Figure 9A). Differences between the three dietary groups were most prominent at 6-wpc, when expression of this gene was higher in the fish fed with the functional feeds compared with fish fed the REF diet. However, these differences appeared to correlate more with heart tissue pathology rather than diet as the expression of this gene significantly decreased over the time course of the infection in all dietary groups (Figure 9B).

		PreCh		6 wpc		8 wpc		14 wpc	
BlastxHit	Probe name	CMS1-REF	CMS2-REF	CMS1-REF	CMS2-REF	CMS1-REF	CMS2-REF	CMS1-REF	CMS2-REF
Diacylglycerol O-acyltransferase homolog 1 (Mouse)	Ssa#DY731573	0.55	0.07	2.60	0.58	0.63	2.03	0.33	0.15
Phosphatidylinositol-4-phosphate 5-kinase_ type I_ alpha	Omy#S34308961	-0.69	-1.51	1.09	0.76	0.68	0.50	0.20	0.59
SH2 domain containing 5	Ssa#DW178537	0.99	0.79	1.37	1.56	1.43	2.05	-0.94	[illegible]
Inositol 1_4_5-trisphosphate receptor-interacting protein	Ssa#TC93931	1.39	1.27	-0.28	1.36	-0.25	-0.75	-0.14	-0.35
Phosphatidylinositol phosphate 5-kinase type III isoform d	Ssa#S30258921	0.06	0.16	0.48	0.59	0.91	0.88	-0.45	0.08
Tyrosine-protein kinase receptor	Omy#S15340917	0.70	0.43	[illegible]	[illegible]	0.94	1.67	0.85	0.62
Tyrosine-protein phosphatase non-receptor type 12	Ssa#CL279Ctg1	-0.35	0.51	0.90	1.11	0.44	1.69	0.58	1.73
Protein kinase C	Omy#TC170368	0.68	1.69	-0.07	0.22	-0.76	0.78	0.08	0.32

Figure 8 Expression between functional feeds (CMS1 and CMS2) and reference diet (REF) of genes related with phosphatidyl inositol signalling pathway. At cut off log2-ER=0.8 (1.75-fold). Red/orange colour intensity indicates higher expression and green/blue colour intensity indicates lower expression.

A

	P-value time	PreCh CMS1-REF	PreCh CMS2-REF	6 wpc CMS1-REF	6 wpc CMS2-REF	8 wpc CMS1-REF	8 wpc CMS2-REF	14 wpc CMS1-REF	14 wpc CMS2-REF
Salmo salar delta-6 fatty acyl desaturase (Fadsd6)	8.41E-09	0.66	0.33	1.27	1.73	0.05	0.37	-1.29	0.82
Salmo salar delta-6 fatty acyl desaturase (Fadsd6) gene	4.44E-09	1.01	-0.45	2.65	2.54	0.52	0.16	-0.94	-0.09
Salmo salar delta-6 fatty acyl desaturase (Fadsd6) gene	4.57E-10	0.47	0.05	1.81	1.74	-0.08	1.27	-1.30	0.36
Delta-6 fatty acyl desaturase D6fad_a	2.00E-04	1.00	-1.90	1.53	2.06	1.16	-0.36	-0.61	0.10
Salmo salar delta-6 fatty acyl desaturase (Fadsd6) gene	2.18E-06	-1.12	-0.82	0.88	1.98	-0.33	-0.77	-2.13	-2.09

B

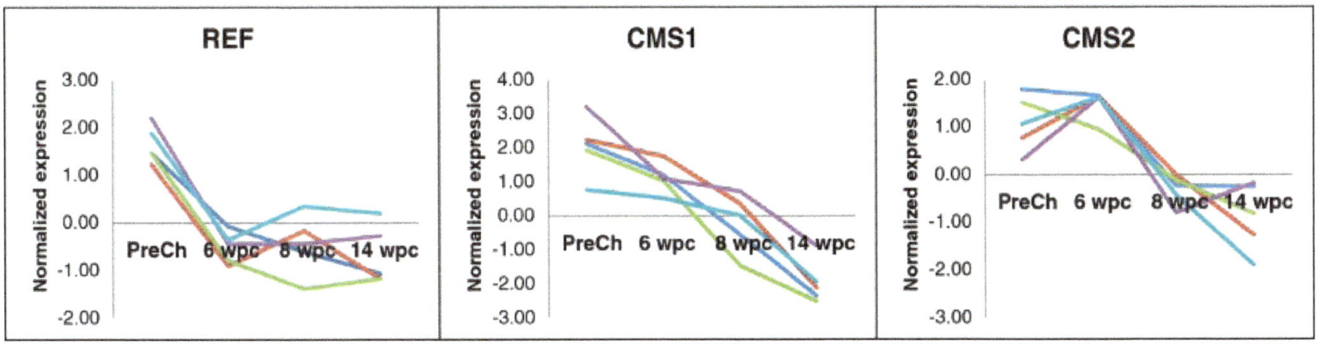

Figure 9 Expression of fatty acyl desaturase (Fadsd6) gene. A) Normalized expression of different probes of the oligoarray from delta-6 fatty acyl desaturase (Fadsd6) gene over the time course of the PMCV infection on fish fed with the REF diet and the functional feeds (CMS1 and CMS2). **B)** Expression ratios (fold changes) of different probes of the oligoarray from delta-6 fatty acyl desaturase (Fadsd6) gene. Data are fish fed the functional feeds (CMS1 and CMS2) relative to the REF diet during the course of the infection. Red/orange colour intensity indicates higher expression and green/blue colour intensity indicates lower expression.

DISCUSSION

The present study demonstrated the potential of dietary immunomodulation for reducing the pathological outcome of virus-associated heart diseases in salmon. Specifically, increased dietary levels of EPA and reduced lipid content were associated with altered expression of genes related with the immune response after an infection with PMCV, significantly reduced pathology in heart and liver tissue, and reduced viral loads at 8 wpc when there was a peak in heart pathology. In contrast, the addition of histidine did not appear to improve the performance of the fish as the level of heart lesions and the expression of genes related with the immune response were not further reduced in fish fed the histidine supplemented diet.

The specific mechanism explaining the potential role(s) of reduced dietary lipid content in the effects of the functional feeds is not clearly established [25, 33]. Previous studies on HSMI showed that there was reduced lipid deposition (steatosis) in livers of fish fed the functional feeds with lower lipid content at initial stages of the disease and the steatosis was usually more frequent when the severity of the heart lesions was high [33]. Thus, there was some association between liver lipid metabolism and on-going viral infection, although earlier studies had found no correlation between the severity of heart and liver lesions during different stages of HSMI disease, associating the liver lesions to the circulatory disturbances as a consequence of heart pathology [34]. Therefore, although the precise mechanism of the reduced liver steatosis index at initial stages in fish fed the functional feeds was not identified in detail, lower dietary lipid was a likely contributing factor [33]. The functional feeds used in the present study were beneficial for the prevention of liver pathology associated with CMS. Fish fed functional feeds showed lower liver histoscores over the whole course of the infection, significantly so at 6-, 12- and 14-wpc, possibly due to the lower dietary lipid content that could reduce liver lipid metabolism when circulatory disturbances are potentially affecting this organ.

Levels of EPA and ARA, and the EPA/ARA ratio, in heart tissue PL were significantly altered by the functional feeds, and hence one of the main strategies of the experimental design was achieved. Specifically, the potential bioavailability of EPA and its proportion relative to ARA was always higher over the time-course of the infection in fish fed the functional feeds compared with fish fed the REF diet. Interestingly, the use of Southern hemisphere FO in the functional feeds in the present trial also increased the level of ARA in heart tissue PL and thus the EPA/ARA ratio was actually lower than in heart tissue of fish in the previous trial on HSMI [25]. The relative proportions of ARA and EPA remained generally constant during the time-course of the PMCV infection although, as described above for DHA, there were increased levels of both ARA and EPA when, according to gene expression analysis, there was activation of the immune response. However, this was only significant for EPA, possibly reflecting the higher absolute levels of EPA generated by the functional feeds in the present trial.

As previously described in humans [35] and fish [21, 36, 37], the fatty acid compositions of membrane PL play a critical role in the regulation of the innate and adaptive immune response. Fatty acids such as ARA, EPA and DHA, released from membrane PL through the action of phospholipases, modulate the immune response not only through eicosanoids, but also by modifying signalling pathways and stimulating immune-related nuclear transcription factors (PPARs and NFkβ) and the production of cytokines [15]. The n-6 LC-PUFA, ARA, is associated with pro-inflammatory responses as the precursor of pro-inflammatory eicosanoids, and EPA, as well as being the precursor of anti-inflammatory eicosanoids, can also decrease the production of ARA-derived eicosanoids through competition for the eicosanoid-synthesising enzymes. DHA has also an anti-inflammatory role in humans, particularly relevant at the recovery phase of an inflammatory process, as it is a precursor of immune-resolving resolvins and protectins [31].

Although macrophages, primarily produced in the head kidney of fish, are a major source of eicosanoids, production of eicosanoids at the site of the infection (e.g. heart in CMS) is also highly relevant due to the short life of these LC-PUFA derivatives [38]. As previously reported in unchallenged [36] and challenged fish [25], DHA levels in heart tissue PL did not reflect dietary levels, being generally similar in fish fed the three diets in the present study. This may reflect the fundamental role of DHA in the maintenance of cellular membrane structure and fluidity [39]. In our previous study investigating the effects of similar diets on HSMI, levels of DHA were not significantly affected during the time-course of the infection after challenge with ASRV [25]. In contrast, the levels of DHA in heart tissue PL changed significantly over the time-course of the infection after PMCV challenge. Levels of DHA were higher when there was enhanced immune response and the heart lesions were more prominent. Whether DHA levels were related with the activation of immune pathways involved in the control of the inflammatory process requires further investigation in fish. However, the present results suggest that changes in the levels of DHA may be associated with PMCV infection and therefore this fatty acid could have a role in the immune response in fish as has been described in humans [31]. The associations between levels of ARA, EPA and DHA in heart PL, the extent of heart histological lesions, and cardiac expression of genes related to the immune response, indicated that dietary effects on incorporation of LC-PUFA into cell membranes of affected tissues could have important immunomodulatory roles in viral infections.

The relevance of dietary supplementation with EPA was supported when evaluating the expression of a key enzyme of LC-PUFA biosynthesis, delta-6 fatty acyl desaturase over the course of the infection. Higher expression of this enzyme and potentially higher biosynthesis of n-3 LC-PUFA was reported previously in liver of fish fed similar functional feeds compared with fish fed a reference diet [25], later linked to dietary lipid/energy levels [40]. However, the present study is the first to report a generally decreased expression of this enzyme in all dietary groups over the course of PMCV infection after 6-wpc, possibly indicating a negative effect of the viral infection on LC-PUFA biosynthesis. This highlights the importance of dietary supplementation of anti-inflammatory n-3 LC-PUFA when fish are suffering viral heart disease although further studies are required.

In addition to fatty acid composition, the PL class composition was also affected during viral infection, with decreased proportions of PI and PS as the infection progressed. Phosphorylated derivatives of PI such as phosphatidylinositol 4,5-bisphosphate (PIP_2) are involved in the production of the intracellular second messengers diacylglycerol (DAG) and inositol 1,4,5-triphosphate (IP_3) through the action of the phospholipase C [17]. DAG and IP_3 activate calcium channels, increasing Ca^{+2} concentration in the cytosol, with this being an essential step for the survival of some human viruses [41]. PS has also been related with this signalling pathway as it is an important activator of protein kinase C [42, 43]. The difference in expression of genes related with PI signalling between the dietary groups along with decreased proportions of PI and PS, suggested relevance of this pathway during the infection, and could reflect utilization of these PL classes associated with an immunomodulatory role. This is a novel finding and the composition/content of specific PL classes could be an interesting area for future studies on clinical nutrition in fish.

Diet alone had marked effects on the cardiac expression of immune genes prior to challenge with functional feeds suppressing genes related with innate antiviral responses and even those specific for lymphocytes. Broadly similar effects were recently reported in a trial investigating substitution of dietary FO with VO in salmon [44]. In that study, expression levels of immune-related genes in liver of fish fed a FO diet, with a fatty acid composition similar to the functional feeds, were generally lower than in fish fed a VO diet, with a composition similar to the present REF diet. However, in the earlier study, genes related with the T-cell response were expressed at a lower level in fish fed the VO diet [44]. This may be explained by the higher levels of EPA present in the functional feeds used in the present study compared with the FO diet used in the earlier study, as the influence of EPA in controlling T-cell signalling pathways has been documented previously [15, 18]. Expression of genes related with non-specific immune responses were also higher in grouper (Epinephelus malabaricus) fed a VO/FO blend compared with fish fed FO alone [37]. So it appears that, even without any apparent infection, inclusion of FO could dampen both innate and adaptive immune responses in fish. Thus, the generally higher immune status of the fish fed the REF diet could be a factor in the earlier development of inflammation and heart lesions associated with CMS in these fish compared with the fish fed the functional feeds. Put another way, the functional feeds generally lowered the immune/inflammatory status prior to viral challenge and, in doing so, delayed and moderated the response to the infection. This was supported by the recent finding that showed low-responder fish to CMS, which developed high infection levels in the absence of cardiac pathology, had ablated expression of adaptive immunity genes and genes involved in T cell responses [11]. However, it must be emphasised that immune and inflammatory responses are essential components of the host defence against infections [45]. Therefore, strategies modulating the inflammatory response must be a balance between promoting resolution and mitigating negative effects of chronic inflammation that can lead to damaging fibrosis without preventing the initial onset of the response [31].

Interestingly, IFN regulatory factors 4 and 7 were among the few innate antiviral genes that were induced by the functional feeds prior to infection. While IRF7 presumably is an important regulator of the antiviral IFN response in fish, as in mammals [46], the suggested role of IRF4 as a negative regulator of TLR signalling and pro-inflammatory cytokines in mice peritoneal macrophages [47] is interesting in view of the down-regulated antiviral state pre-challenge of fish fed the functional feeds and hence the possible implications for the positive effects post-infection. Whether IRF4 has a similar function in fish has yet to be confirmed.

The time-courses for the development and resolution of histopathology and immune gene responses were similar to those observed previously [10, 11]. In addition, there were correlations between the expression of genes related with immune responses, viral load and heart lesions, in agreement with Timmerhaus et al. [10]. Importantly, however, the temporal differences observed between the different dietary treatments in the present study strengthened these correlations. Thus, the changes in the progression of CMS heart pathology, gene expression and viral load, in fish fed the REF diet were temporally similar to the development of the infection described previously [10], whereas there was a clear delay in the appearance of those changes in fish fed the functional feeds. Although viral load was not monitored at later stages of the PMCV infection, the aforementioned correlation between histopathology and gene expression in the REF group allow us to hypothesize that viral load in fish fed the functional feeds would be higher at the later stages. Therefore, the lower viral load in heart of fish fed the functional feeds at early stages of PMCV infection was probably a direct effect of diet on initial viral replication as reported previously after ASRV infection [25]. Furthermore, in addition to a delay in the appearance of heart pathology in fish fed the functional feeds, a slightly improved performance of fish fed the CMS2 diet was observed, which could be associated with reduced expression of genes associated with the innate immune response.

At 6-wpc, gene expression and heart tissue histopathology of fish fed the REF diet were in agreement with those reported previously [10]. Thus, atrium was affected by the infection at 6-wpc with higher histoscores in fish fed the REF diet compared to fish fed the functional feeds, whereas ventricle was only slightly affected. A peak in the expression of genes related with the complement response, B and T-cell responses, and apoptosis was expected at this time post-infection [10], and many of the genes related to these pathways showed lower expression in fish fed the functional feeds compared to fish fed the REF diet. In addition, genes related with antiviral and interferon responses often showed higher expression in fish fed the functional feeds compared with fish fed the REF diet. The peak in expression of the latter genes was reported to be at 2-4 wpc in the earlier study [10], supporting the conclusion that there was a delayed response in fish fed the functional feeds. A similar delayed response to viral infection of around 2 weeks was observed previously after ASRV infection in salmon fed functional feeds [25].

As the CMS disease progressed the extent of lesions in the heart increased in both atrium and ventricle with the differences between the fish fed the functional feeds and fish fed the REF diet being most pronounced at 8-wpc. At this time-point, fish fed the functional feeds presented lower histoscores in both atrium and ventricle compared with fish fed the REF diet, which could be due to the up-regulation of genes related with host immune response observed at 6-wpc in fish fed the REF diet. At 8-wpc, following the hypothesis of delayed development of CMS in fish fed the functional feeds, higher expression of genes involved in antiviral and innate immune pathways could have been expected in these fish compared with fish fed the REF diet but, in contrast, these groups showed lower activation of these pathways. Dietary modulation of the inflammatory response has, of course, been described previously in fish. Higher incorporation of n-3 LC-PUFA in biological membranes of immune cells led to a lower expression of pro-inflammatory cytokines in gilthead seabream (Sparus aurata) [21], decreased antibody production and macrophage killing activity in rainbow trout [48], and lower production of pro-inflammatory eicosanoids in Atlantic salmon [19, 20]. Furthermore, dietary EPA suppressed production and release of TNF, interleukins and IFN in humans [49]. Dietary EPA also suppressed key components of the antiviral response in gilthead sea bream, and dietary VO inclusion modulated the expression of Mx proteins, interferon-induced mediators of innate resistance to RNA virus [50]. According to these previous studies, the increased proportions of dietary EPA in the functional feeds used in the present study could be a factor in preventing an uncontrolled immune response that could be more harmful to the fish.

As expected, many genes related with the T-cell response showed lower expression in fish fed the functional feeds compared with fish fed the REF diet [10]. Although gene expression was not assessed at 10- and 12-wpc, increased transcription of genes involved in virus clearance may be predicted based on histopathology of heart lesions at these time-points. By 14-wpc, there was a clear remission in the lesions in both parts of the heart in all dietary groups and therefore no significant differences in the pathology were observed between them. Even though the list of genes differentially expressed at 14-wpc in fish fed the different diets was shorter than the ones from previous sampling points, some pathways represented in the list of genes differentially expressed at 14-wpc in fish fed the different diets could still indicate a more effective remediation of the inflammatory response by the group of fish fed with the functional feeds. Thus, important markers of the T-cell response such as T cell receptor alpha, CD8 beta and alpha and CD82 antigen showed lower expression in fish fed the functional feeds, but expression of genes related with PI signalling was higher in those fish compared with fish fed the REF diet. As mentioned above, the PI signalling pathway is intimately involved in the regulation of the T-cell response [17], and dependent on the levels of membrane PL classes and their fatty acid compositions. As the lesions in heart tissue in CMS were associated with the action of CD8+ cytotoxic T-cells [10], factors controlling this response, such as anti-inflammatory n-3 LC-PUFA, could be key to the general better performance observed in the fish fed the functional feeds. Furthermore, results from the previous study evaluating the use of functional feeds in salmon infected with HSMI were consistent with this hypothesis. Thus, milder expression of genes related with inflammatory response and virus clearance, including those related with T-cell response, were reported in salmon fed the functional feeds leading to better performance of those fish over the course of the HSMI infection [25].

CONCLUSIONS

The present study is the first to describe the effects of functional feeds on the expression of genes related with the immune response after infection of Atlantic salmon with PMCV. Significant differences in immune and inflammatory responses and pathology in heart tissue were found in fish fed the different dietary treatments over the course of the infection, highlighting the potential immune modulatory role of dietary fatty acid composition and lipid content in viral infections in salmon. The strategy was effective in significantly reducing heart pathology and also decreasing the time-course of the viral infection. Furthermore, liver pathology was significantly reduced during the time course of the infection in fish fed the functional feeds. Molecular signatures showed lower expression of immune-inflammatory genes, especially those reported to be correlated with heart lesions in the clinical phase (i.e. T cells). Moreover, transcriptome analysis revealed a novel pathway of the fish immune response, the PI signalling pathway, that functional feeds could modulate, as well as potential suppression of LC-PUFA biosynthesis as the infection progresses, stressing the importance of dietary supplementation with n-3 LC-PUFA.

METHODS

Experimental Feeds and Fish

Three fishmeal-based diets, based on a previous study investigating HSMI [25], were formulated and manufactured by EWOS Innovation (Dirdal, Norway) (Table 4). The reference diet (REF) was essentially a standard, commercial formulation with 31% lipid with the added oil being a blend of Northern hemisphere FO and rapeseed oil. The two functional feeds (CMS1 and CMS2) both contained a lower level of lipid (18%) that was balanced by increased protein, provided by fishmeal and krill meal. Therefore, as a major factor differing between the REF and functional feeds was dietary lipid level, the feeds were not isolipidic or isoproteic. The added oil in the functional feeds was provided by a Southern hemisphere FO. As a result the CMS1 and CMS2 feeds had a similar fatty acid profile containing higher proportions of EPA (almost 14%) and a n-3/n-6 PUFA ratio around 4, in comparison to the REF feed (<4% EPA and an n-3/n-6 ratio of 1.4) (Table 5). The only major difference between the functional feeds was that CMS1 was supplemented with additional histidine for reasons described above [30].

Table 4 Formulation (g/Kg) and proximate composition (percentage) of the reference (REF) and functional (CMS1 and CMS2) feeds

Component (g/100 g)[1]	REF	CMS1	CMS2
Fish meal and hydrolysates	42.1	53.0	53.0
Plant protein concentrates[2]	21.5	18.0	18.0
Northern fish oil	13.0	0.0	0.0
Southern fish oil	0.0	10.0	10.0
Rapeseed oil	10.7	0.0	0.0
Carbohydrate-based binders[3]	11.9	11.9	12.1
Micro premixes[4]	0.8	1.7	1.9
Krill meal[5]	0.0	5.0	5.0
Histidine (synthetic)	0.0	0.4	0.0
Total	100	100	100
Proximate composition			
Moisture	6.5	6.5	6.5
Fat	31.0	18.0	18.0
Protein	42.2	53.4	53.4

1) All ingredients sourced from EWOS stocks unless otherwise stated.

2) Includes soy protein concentrate, pea protein concentrate, wheat gluten and sunflower meal.

3) Includes wheat grain.

4) Includes vitamins, minerals, crystalline amino acids, ammonium phosphate.

5) Aker Biomarine AS.

Table 5 Fatty acid compositions (percentage of total fatty acids) of the reference (REF) and functional (CMS1 and CMS2) feeds

Fatty acid	REF	CMS1	CMS2
Saturated	19.96	26.62	25.83
Mounsaturated	56.64	36.42	36.08
18:2n-6	8.38	4.94	5.55
20:3n-6	0.15	0.30	0.30
20:4n-6	0.18	0.64	0.66
n-6 PUFA	9.50	6.91	7.16
18:3n-3	3.11	1.27	1.31
20:5n-3	3.51	13.66	13.38
22:6n-3	4.77	8.41	8.45
n-3 PUFA	13.28	27.55	27.79
PUFA	23.40	36.97	38.09
n-3/n-6	1.40	3.98	3.88
EPA/ARA	19.35	21.40	20.20
% Lipid	25.60	17.30	16.46

ARA, Arachidonic acid; EPA, eicosapentaenoic acid; PUFA, polyunsaturated fatty acids.

A total of 675 Atlantic salmon (Salmo salar L.), SalmoBreed IPN strain (average weight ca.150 g), were distributed into nine tanks (1 m³) at the VESO facility, Vikan, Norway and fed one of the three feeds (3 tanks of 75 fish each per dietary treatment) for a period of 8 weeks prior to being transferred to the challenge tanks. Feeding ration was a maximum of 2%. Seawater/brackish water (ca. 25 ‰-35 ‰) delivery was flow-through sufficient to maintain oxygen-satiation in effluent water 70%. Water temperature was maintained at 12 ± 1°C, and a photoperiod 24:0 h light/dark regime was followed. Stocking density of the tanks was a maximum of 60 Kg/m³. Feeding and infection challenge trials were approved by The National Animal Research Authority (http://www.fdu.no) according to the 'European Convention for the Protection of Vertebrate Animals used for Experimental and other Scientific Purposes' (EST 123).

Growth Performance

Growth performance of fish was assessed using repeated measurements of tank mean weights determined at the start of the trial (day 0), challenge-day (day 92) and then at 6-, 8-, 10-, 12- and 14-weeks post-challenge. A linear mixed-effects (multilevel) model was fitted between the response (weight) and predictor (diet) by allowing the intercepts and slopes of the time variable (day) to vary to account for tank level correlations and variability in growth trajectories. The model was estimated with the lmer function in the lme4 package of the R language (R Development Core Team, 2008). All treatment effects were based on posterior simulation (n=2,500) with 95% credible intervals that were interpreted as statistical significant (p=0.05%) when the interval did not overlap the reference value in question. There were no significant differences in overall growth between dietary treatments at the end of the trial (i.e. after 192 days) (Additional file 2: Figure S1). However, after 92 days of feeding (pre-challenge) fish fed diet CMS2 had higher weights than those fed REF and CMS1 diets.

PMCV Challenge

After the pre-feeding period, two fish from each pre-challenge tank, 18 in total (n=6 per dietary treatment) were collected, as described below, prior to challenge (0 time-point) and the remaining fish were transferred to challenge tanks (1 m³). The fish were unvaccinated and tested for PMCV, ASRV, infectious pancreatic necrosis virus (IPNV) and salmon pancreas disease virus (SPDV) and confirmed as free from CMS, HSMI, IPN and PD. Of the available salmon, a total of 600 fish, pre-challenge average weight 479 ± 13 g for the fish fed with the REF diet; 475 ± 13 g for the fish fed with the CMS1 diet and 502 ± 14 g for the fish fed with the CMS2 diet; were distributed into 12 tanks (4 tanks per dietary treatment with 50 fish/tank). The fish were acclimated for 2 weeks prior to challenge and were fed with the same diets during the acclimation period and throughout the period of the challenge (14 weeks) that they were fed prior to transfer. No previous diseases were described. For challenge, the fish were sedated using Aqui-S at a final concentration of 5 mg/L of isoeugenol, followed by anaesthesia in benzocaine (20%, Benzoak®) using a final concentration of 30 ml/L. All 576 fish were challenged by intramuscular injection of 0.1 ml PMCV inoculum on each side of the fish in the lateral muscle tissue beneath the dorsal fin (0.2 ml in total per fish). Production of the PMCV inoculum was described previously [1]. Briefly, the virus was originally isolated from heart tissue homogenate collected from a clinical outbreak of CMS and filtered through a 0.22-μm filter. Challenge of Atlantic salmon was performed using supernatant of GF-1 cell culture in which the virus was inoculated.

Sampling

Ten fish from each tank (total of 40 per dietary treatment) were sampled at 6-, 8-, 10-, 12- and 14-weeks post-challenge. Fish were anaesthetized as above and killed by a blow to the head and heart tissue collected for analyses. Liver tissue and a portion of each heart were transferred to 10% buffered formalin for histological analyses. Half of the remaining portions of heart tissue were immediately transferred to RNAlater® (following manufacturers protocol) and later stored at -20°C prior to molecular analyses (i.e. 5 samples of heart tissue per tank, 20 per dietary treatment). The remaining portions of heart tissue were frozen in liquid N_2 and stored at -80oC prior to lipid analyses (five per tank, 20 per treatment).

Histology

The histological changes in heart were assessed as described in detail previously [1]. Briefly, the inflammatory changes were evaluated separately for atrium and ventricle, and to the extent present, inflammatory scores of the epicardium were also recorded. In addition to atrium and ventricle in the same sample, bulbus arteriosus was also included but not assessed for histopathological changes. Samples were fixed in 10% buffered formalin, embedded in paraffin wax and sectioned at 3 - 4 micron and stained with hematoxylin and eosin according to standard methods. All sections were evaluated randomly and without knowledge to which feeding group they belonged (double-blind).

Liver steatosis scores were ranked according to a non-continuous score grade from 0 to 4 (Table 6). Briefly, a score of 0-1 indicated lesions in the cytoplasm involving less than 10% of the hepatocytes and including less than 25% of the area of the individual hepatocytes, and low levels of leukocyte infiltration. A score of 4 indicated lesions in the cytoplasm involving more than 75% of the hepatocytes and including more than 80% of the area of the individual hepatocytes, and severe leukocyte infiltration. The initial "continuous" scores were converted to a fewer number of discrete scores (3 - 6 classes depending on the response). Scores were an ordinal response and a multilevel ordinal regression was used to analyse the effects of feeds on the scores. This was achieved with the ordinal package Sweave le processed by LATEX of the R language (http://www.R-project.org). Each of the scores was used as the response variable in separate analyses that had feed type (REF, CMS1, CMS2) as the fixed effect term. Since multiple fish were examined from each replicate tank, a random effect of tank was added to the model.

Table 6 Scoring system for liver steatosis in individual sections

Score	Type of changes
0	No pathological findings or negligible number of inflammatory cells present.
0.1-0.9	Increasing number of inflammatory cells per foci and moderate myocyte necrosis per foci.
1	One or a few distinct focal lesions with cardiomyocyte necrosis, and distinct infiltration of leukocytes (1-5 per foci)
1.1-1.9	Increasing number of lesion individually increasing in size per focus
2	Several distinct lesions and increased number of leukocytes
2.1-2.9	Multifocal myonecrosis with increasing number of inflammatory cells
3	Multifocal to confluent lesions and moderate to severe increase in number of leukocytes
3.1-3.9	Increasing number of foci towards diffuse lesions. Inflammation increases as do endothelial hyperplasia.
4	Severe confluent lesions comprising >75% of the tissue and massive leukocyte infiltration

The changes observed by histological examination have been indicated on a visual analog scale (0-4) based on the criteria given above.

Lipid Analyses

Lipid and fatty acid analyses were performed on heart tissue from all sampling points. Three pre-challenge heart samples per dietary treatment were randomly chosen for lipid analysis. Heart tissue from the post-challenge samples was pooled per tank so that heart tissue of five fish became a pool resulting in 4 pools/treatment/sampling point. Total lipid from each pool was extracted according to Folch et al. [51] by homogenization in chloroform/methanol (2:1, v/v) containing 0.1% butylated hydroxytoluent (BHT) and lipid content determined gravimetrically. Total lipid extracts were re-suspended in chloroform/methanol at a concentration of 10 mg lipid/ml and stored at -70^{oC} until analysed. Phospholipid (PL) fractions were prepared from 0.5 mg of total lipid by thin-layer chromatography as described by Bell et al. [52]. FAME were separated and quantified by GLC (Fisons 8160; Carlo Erba, Milan, Italy) using a 60 m × 0.32 mm × 0.25 μm film thickness capillary column (ZB-WAX; Phenomenex, Macclesfield, Cheshire, UK) as also described previously [52]. Tissue and diet lipid class compositions were determined by single-dimension double-development high-performance thin-layer chromatography (HPTLC) and densitometry as described previously [22]. Significance of differences due to diet and time were determined by two-way ANOVA ($p<0.05$) using the SPSS 19.0 statistical package (SPSS Inc., Chicago IL, USA).

Microarray Experimental Design

Heart tissue, as the main organ affected during a CMS outbreak, was selected for transcriptomic analysis. The choice of appropriate time-points for analysis was based on the description of expression of genes of the immune response after a PMCV infection described by Timmerhaus et al. [10]. Thus samples from fish at pre-challenge, 6-, 8-, and 14- weeks post-challenge (wpc) were analyzed. The pre-challenge sampling point was selected to evaluate changes due to dietary compositions, and 6- and 8-wpc sampling points were selected to evaluate dietary effects on the expression of genes from immune response pathways at the peak of infection. As a plateau phase in the expression of immune related genes at 10- and 12-wpc was described previously [10], the 14-wpc sampling point was selected to evaluate gene expression during the recovery phase of the disease. Within each of these sampling points, the available samples were arranged by individual histoscore and 6 samples per dietary group were randomly selected from the interquartile range (25 – 75%).

RNA Extraction and Purification

Heart tissue of the selected samples was homogenized in TRI Reagent (Ambion, Applied Biosystems, Warrington, U.K.) using an Ultra-Turrax homogenizer (Fisher Scientific, Loughborough, U.K.). Total RNA was isolated following manufacturer's instructions, and RNA quality and quantity assessed by gel electrophoresis and spectrophotometry (NanoDrop ND-1000, Thermo Scientific, Wilmington, U.S.A.), respectively.

Microarray Hybridisations and Analysis

The transcriptomic experiment used an Atlantic salmon custom-made oligoarray with 44 k features per array on a four-array-per-slide format, with each feature printed singly (Agilent Technologies UK Ltd., Wokingham, UK). The probes were co-designed by researchers at the Institute of Aquaculture, University of Stirling, U.K. and Nofima (Ås, Norway), and array design is available on request. A dual-labelled experimental design was employed for the microarray hybridizations. Each experimental sample was competitively hybridized against a common pooled-reference sample, which comprised equal amounts of each of the replicates used in the study. This design permits valid statistical comparisons across all treatments to be made. The entire experiment comprised 72 hybridizations; 4 time-points (pre-challenge, 6-, 8- and 14-wpc) × 3 diets (REF, CMS1 and CMS2) × 6 biological replicates.

Indirect labelling methodology was employed in preparing the microarray targets. Amplified antisense RNA (aRNA) was produced from each RNA sample using the TargetAmp™ 1-RoundAminoallyl-aRNA Amplification Kit (Ambion, Applied Biosystems, Warrington, UK), following the manufacturer's methodology, followed by Cy3 or Cy5 fluor incorporation through a dye-coupling reaction. Briefly, 250 ng of total RNA per sample were amplified and column-purified according to manufacturer's instructions. Resultant aRNA was quantified and quality assessed as above. Subsequently, Cy dye suspensions (Cy3 & Cy5) in sufficient quantity for all labelling reactions were prepared by adding 38 μL high purity dimethyl sulphoxide (Stratagene, Agilent Technologies UK Ltd.) to each tube of Cy dye (PA23001 or PA25001, GE Healthcare, Chalfont St. Giles, UK). To attach the Cy dyes, 3 μg each aRNA sample was suspended in 10 μL nuclease-free H_2O and heated to 70°C for 2 min. When cooled to room temperature, 3 μL of coupling buffer (0.5 M $NaHCO_3$; pH 9.2) and 2 μL of Cy3 dye suspension stock was added and then incubated for 1 h at 25°C in the dark. To label the common pooled reference sample with Cy5, a scaled-up batch reaction was similarly performed. Unincorporated dye was removed by column purification (Illustra AutoSeq G-50 spin columns; GE Healthcare). Dye incorporation and aRNA yield were quantified by spectrophotometry (NanoDrop) and further quality controlled by separating 0.4 μL of the sample on a thin mini-agarose gel and visualizing products on a fluorescence scanner (Typhoon Trio, GE Healthcare).

Hybridization of a total of 6 slides (24 arrays) was performed in a single day, with sample order semi-randomized, using SureHyb hybridisation chambers in a DNA Microarray Hybridization Oven (Agilent Technologies) as described previously [25]. Scanning was performed at 5 μm resolution using an Axon GenePix 4200AL Scanner (MDS Analytical Technologies, Wokingham, Berkshire, U.K.). Laser power was kept constant (80%) and the "auto PMT" function within the acquisition software (v.4) was enabled to adjust PMT for each channel such that less than 0.1% of features were saturated and that the mean intensity ratio of the Cy3 and Cy5 signals was close to one. Agilent Feature Extraction Software (v 9.5) was used to identify features and extract fluorescence intensity values from the resultant TIF images. The remaining analysis was then performed in the GeneSpring GX version 12 analysis platform (Agilent Technologies). After removing control features and filtration of low quality spots (saturated, non-uniform, population outliers and spots non-significantly different from background), lowess normalization of log2-expression ratios (ER) was performed.

As the primary objective of the present study was to evaluate the effects of functional feeding (i.e. comparison between diets), differentially expressed genes were selected based on expression changes (cut off log2-ER = 0.8, 1.75-fold change, $p < 0.05$, one-way ANOVA) between the functional feed groups (CMS1 and CMS2) and the fish fed the REF diet, or between the two functional feed groups at each sampling point (pre-challenge-, 6-, 8- and 14-wpc). Annotation of selected genes was performed using STARS software developed at Nofima (Ås, Norway) [53], KEGG analysis, and information from literature and public databases such as Gene Expression Omnibus (GEO).

Validation of the Microarray

Expression of 11 selected genes showing a significant diet × time interaction in the microarray analysis at 2 different time points of the infection and related to relevant immune-related pathways was studied by reverse transcription quantitative real-time PCR (qPCR) as previously described [25]. The qPCR primer sequences, annealing temperature (Tm) and size of amplicon are given in Additional file 3: Table S2. The sequences were obtained either by literature searches or designed from EST sequences corresponding to microarray clones or candidate genes of interest using Primer3 software (http://biotools.umassmed.edu/bioapps/primer3_www.cgi). Cofilin-2 and elf-1α were used as reference genes as they were sufficiently stable across treatments for normalization and had been identified in previous salmon cDNA microarray and qPCR studies as suitable reference genes on the basis of constant expression between different feeds and time points [44, 54]. Data were analysed using the relative expression software tool (REST 2009, http://www.gene-quantification.info/), which employs a pair-wise fixed reallocation randomization test (10,000 randomizations) with efficiency correction, to determine the statistical significance of expression ratios (or gene expression fold-changes) between two treatments [55]. There was good agreement between microarray results and RT-qPCR quantification, with most of the comparisons showing identical regulation and similar fold differences (Additional file 4: Table S3). Analysis indicated that the gene expression data obtained using the oligoarray were reliable and robust, which enabled interpretations to be made with confidence, especially with immune response-related genes.

Assessment of Viral Load

Relative quantification of PMCV by qPCR was used to determine differences in the viral load between the dietary treatments at the two critical points in the time-course of the infection (6- and 8- weeks post-challenge) when differential expression of genes related with the inflammatory process was greatest between fish fed the REF diet and fish fed the functional feeds. The qPCR primer sequences (Additional file 3: Table S2) previously described by Timmerhaus et al. [11] were employed following the same procedure and from the same RNA samples used for the validation of the microarray (as described above).

Ethical Statement

All experimentation performed by the Institute of Aquaculture (IoA) is subjected to a thorough Ethical Review Process prior to any work being approved. This involves all projects, irrespective of where they are carried out, to be submitted to the IoA Ethical Committee for approval using detailed Ethical Approval forms that require all aspects of the experimentation to be described including conditions for the human experimenters as well as animal subjects. This procedure ensures all ethical issues are addressed before an experiment can be initiated. The present research was assessed by the IoA Ethical Review Committee and passed the Ethical Review Process of the University of Stirling.

Availability of Supporting Data

The data set supporting the results of this article is included within the article (Additional file 1: Table S1) and microarray design is available in ArrayExpress repository, A-MEXP-2400.

ACKNOWLEDGEMENTS

We wish to thank Dr John Taggart for advice and direction with the microarray and bioinformatic analyses, and Jacquie Ireland for technical assistance with microarray analyses.

REFERENCES

1. Haugland O, Mikalsen AB, Nilsen P, Lindmo K, Thu BJ, Eliassen TM, Roos N, Rode M, Evensen Ø: Cardiomyopathy syndrome of Atlantic salmon (Salmo salar L.) is caused by a double-stranded RNA virus of the Totiviridae family. J Virol. 2011, 85: 5275-5286.
2. Amin A, Trasti J: Endomyocarditis in Atlantic salmon in Norwegian seafarms. Bull Eur Assoc Fish Pathol. 1988, 8: 70-71.
3. Poppe TT, Sande RD: Cardiomyopathy in farmed Atlantic salmon: a review, introducing an ultrasound technique for clinical examination. 1994, Oslo: Norwegian School of Veterinary Science
4. Rodger H, Turnbull T: Cardiomyopathy syndrome in farmed Scottish salmon. Vet Rec. 2000, 146: 500-501.
5. Brocklebank J, Raverty S: Sudden mortality caused by cardiac deformities following seining of preharvest farmed Atlantic salmon (Salmo salar) and by cardiomyopathy of postintraperitoneally vaccinated Atlantic salmon parr in British Columbia. Can Vet J. 2002, 43: 129-130.Central
6. Johansen R: The Health Situation in Norwegian Aquaculture, Fish Health Report 2012. 2013, Oslo: Norwegian Veterinary Institute
7. Fritsvold C, Kongtorp RT, Taksdal T, Orpetveit I, Heum M, Poppe TT: Experimental transmission of cardiomyopathy syndrome (CMS) in Atlantic salmon Salmo salar. Dis Aquat Organ. 2009, 87: 225-234.
8. Brun E, Poppe T, Skrudland A, Jarp J: Cardiomyopathy syndrome in farmed Atlantic salmon Salmo salar:oCcurrence and direct financial losses for Norwegian aquaculture. Dis Aquat Organ. 2003, 56: 241-247.
9. Ferguson HW, Poppe T, Speare DJ: Cardiomyopathy in farmed Norwegian salmon. Dis Aquat Organ. 1990, 8: 225-231.
10. Timmerhaus G, Krasnov A, Nilsen P, Alarcon M, Afanasyev S, Rode M, Takle H, Jørgensen SM: Transcriptome profiling of immune responses to cardiomyopathy syndrome (CMS) in Atlantic salmon. BMC Genomics. 2011, 12: 459-
11. Timmerhaus G, Krasnov A, Takle H, Afanasyev S, Nilsen P, Rode M, Jørgensen SM: Comparison of Atlantic salmon individuals with different outcomes of cardiomyopathy syndrome (CMS). BMC Genomics. 2012, 13: 205-
12. Roberfroid MB: Functional foods: concepts and application to inulin and oligofructose. Br J Nutr. 2002, 87: S139-S143.
13. Tacchi L, Bickerdike R, Douglas A, Secombes CJ, Martin SAM: Transcriptomic responses to functional feeds in Atlantic salmon (Salmo salar). Fish Shellfish Immunol. 2011, 31: 704-715.
14. Tacon AGJ, Metian M: Global overview on the use of fishmeal and fish oil in industrially compounded aquafeeds: trends and future prospects. Aquaculture. 2008, 285: 146-158.
15. Shaikh SR, Edidin M: Polyunsaturated fatty acids, membrane organization, T cells, and antigen presentation. Am J Clin Nutr. 2006, 84: 1277-1289.
16. Moraes LA, Piqueras L, Bishop-Bailey D: Peroxisome proliferator-activated receptors and inflammation. Pharmacol Therapeutics. 2006, 110: 371-385.
17. Stulnig TM, Zeyda M: Immunomodulation by polyunsaturated fatty acids: impact on T-cell signaling. Lipids. 2004, 39: 1171-1175.
18. Calder PC: The relationship between the fatty acid composition of immune cells and their function. Prostaglandins Leukot Essent Fatty Acids. 2008, 79: 101-108.
19. Bell JG, Ashton I, Secombes CJ, Weitzel BR, Dick JR, Sargent JR: Dietary lipid affects phospholipid fatty acid compositions, eicosanoid production and immune function in Atlantic salmon (Salmo salar). Prostaglandins Leukot Essent Fatty Acids. 1996, 54: 173-182.
20. Bell J, Farndale B, Dick J, Sargent J: Modification of membrane fatty acid composition, eicosanoid production, and phospholipase A activity in Atlantic salmon (Salmo salar) gill and kidney by dietary lipid. Lipids. 1996, 31: 1163-1171.
21. Montero D, Mathlouthi F, Tort L, Afonso JM, Torrecillas S, Fernandez-Vaquero A, Negrin D, Izquierdo MS: Replacement of dietary fish oil by vegetable oils affects humoral immunity and expression of pro-inflammatory

cytokines genes in gilthead sea breamSparus aurata. Fish Shellfish Immunol. 2010, 29: 1073-1081.

22. Bell JG, Dick JR, McVicar AH, Sargent JR, Thompson KD: Dietary sunflower, linseed and fish oils affect phospholipid fatty acid composition, development of cardiac lesions, phospholipase activity and eicosanoid production in Atlantic salmon (Salmo salar). Prostaglandins Leukot Essent Fatty Acids. 1993, 49: 665-673.

23. Bell JG, Castell JD, Tocher DR, MacDonald F, Sargent JR: Effects of different dietary arachidonic acid - docosahexaenoic acid ratios on phospholipid fatty-acid compositions and prostaglandin production in juvenile turbot (Scophthalmus maximus). Fish Physiol Biochem. 1995, 14: 139-151.

24. Gjøen T, Kleveland EJ, Moya-Falcon C, Froystad MK, Vegusdal A, Hvattum E, Berge RK, Ruyter B: Effect of dietary thia fatty acids on lipid composition, morphology and macrophage function of Atlantic salmon (Salmo salar L.) kidney. Comp Biochem Physiol. 2007, 148: 103-111.

25. Martinez-Rubio L, Morais S, Evensen O, Wadsworth S, Ruohonen K, Vecino JLG, Bell JG: Functional feeds reduce heart inflammation and pathology in Atlantic salmon (Salmo salarL.) following experimental challenge with Atlantic salmon reovirus (ASRV). Plos One. 2012, 7: e40266-

26. Ogata HY: Muscle buffering capacity of yellowtail fed diets supplemented with crystalline histidine. J Fish Biol. 2002, 61: 1504-1512.

27. Suzuki T, Hirano T, Suyama M: Free imidazole compounds in white and dark muscles of migratory marine fish. Comp Biochem Physiol. 1987, 87B: 615-619.

28. Kim H, Park SY, Choi YH: Determination of the antioxidant anserine in muscle of juvenile and mature rainbow trout, Oncorhynchus mykiss. Arch Ryb Fish. 1999, 7: 65-72.

29. Wade MA, Tucker HN: Antioxidant characteristics of L-histidine. J Nutr Biochem. 1998, 9: 308-315.

30. Hasnain SE, Begum R, Ramaiah KVA, Sahdev S, Shajil EM, Taneja TK, Mohan M, Athar M, Sah NK, Krishnaveni M: Host-pathogen interactions during apoptosis. J Biosci. 2003, 28: 349-358.

31. Calder PC, Albers R, Antoine JM, Blum S, Bourdet-Sicard R, Ferns GA, Folkerts G, Friedmann PS, Frost GS, Guarner F, Løvik M, Macfarlane S, Meyer PD, M'Rabet L, Serafini M, van Eden W, van Loo J, Vas Dias W, Vidry S, Winklhofer-Roob BM, Zhao J: Inflammatory disease processes and interactions with nutrition. Br J Nutr. 2009, 101: S1-S45.

32. Krasnov A, Timmerhaus G, Schiotz BL, Torgersen J, Alanasyev S, Iliev D, Jørgensen J, Takle H, Jørgensen SM: Genomic survey of early responses to viruses in Atlantic salmon,Salmo salar L. Mol Immunol. 2011, 49: 163-174.

33. Martinez-Rubio L, Morais S, Evensen O, Wadsworth S, Vecino JG, Ruohonen K, Bell JG, Tocher DR: Effect of functional feeds on fatty acid and eicosanoid metabolism in liver and head kidney of Atlantic salmon (Salmo salar L.) with experimentally induced heart and skeletal muscle inflammation. Fish Shellfish Immunol. 2013, 34: 1533-1545.

34. Kongtorp R, Halse M, Taksdal T, Falk K: Longitudinal study of a natural outbreak of heart and skeletal muscle inflammation in Atlantic salmon, Salmo salar L. J Fish Dis. 2006, 29: 233-244.

35. Calder PC: N-3 polyunsaturated fatty acids and cytokine production in health and disease. Annal Nutr MetaB. 1997, 41: 203-234.

36. Bell J, McVicar A, Park M, Sargent JR: High dietary linoleic-acid affects the fatty-acid compositions of individual phospholipids from tissues of Atlantic salmon (Salmo salar) - association with stress susceptibility and cardiac lesion. J Nutr. 1991, 121: 1163-1172.

37. Lin Y, Shiau S: Effects of dietary blend of fish oil with corn oil on growth and non-specific immune responses of grouper, Epinephelus malabaricus. Aquacult NutR. 2007, 13: 137-144.

38. Tocher DR: Metabolism and functions of lipids and fatty acids in teleost fish. Rev Fisheries Sci. 2003, 11: 107-184.

39. Stillwell W, Wassall S: Docosahexaenoic acid: membrane properties of a unique fatty acid. Chem Phys Lipids. 2003, 126: 1-27.

40. Martinez-Rubio L, Wadsworth S, Vecino JLG, Bell JG, Tocher DR: Effect of dietary digestible energy content on expression of genes of lipid metabolism and LC-PUFA biosynthesis in liver of Atlantic salmon (Salmo salar L.). Aquaculture. 2013, 384: 94-103.

41. Chami M, Oules B, Paterlini-Brechot P: Cytobiological consequences of calcium-signaling alterations induced by human viral proteins. Biochim Biophys Acta. 2006, 1763: 1344-1362.

42. Bell MV, Sargent JR: Protein-kinase-C activity in the spleen of trout (Salmo gairdneri) and the rectal gland of dogfish (Scyliorhinus canicula), and the effects of phosphatidylserine and diacylglycerol containing (n-3) polyunsaturated fatty acids. Comp Biochem Physiol. 1987, 87B: 875-880.

43. Tocher DR, Bendiksen EA, Campbell PJ, Bell JG: The role of phospholipids in nutrition and metabolism of teleost fish. Aquaculture. 2008, 280: 21-34.

44. Morais S, Pratoomyot J, Taggart JB, Bron JE, Guy DR, Bell JG, Tocher DR: Genotype-specific responses in Atlantic salmon (Salmo salar) subject to dietary fish oil replacement by vegetable oil: a liver transcriptomic analysis. BMC Genomics. 2011, 12: 255-

45. Kiron V: Fish immune system and its nutritional modulation for preventive health care. Anim Feed Sci Technol. 2012, 173: 111-133.

46. Holland JW, Bird S, Williamson B, Woudstra C, Mustafa A, Wang T, Zou J, Blaney SC, Collet B, Secombes CJ: Molecular characterization of IRF3 and IRF7 in rainbow trout,Oncorhynchus mykiss: Functional analysis and transcriptional modulation. Mol Immunol. 2008, 46: 269-285.

47. Negishi H, Ohba Y, Yanai H, Takaoka A, Honma K, Yui K, Matsuyama T, Taniguchi T, Honda K: Negative regulation of toll-like-receptor signaling by IRF-4. Proc Natl Acad Sci U S A. 2005, 102: 15989-15994.

48. Kiron V, Fukuda H, Takeuchi T, Watanabe T: Essential fatty acid nutrition and defense mechanisms in rainbow trout Oncorhynchus mykiss. Comp Biochem Physiol. 1995, 111A: 361-367.

49. Das UN: Long-chain polyunsaturated fatty acids in the growth and development of the brain and memory. Nutrition. 2003, 19: 62-65.

50. Montero D, Grasso V, Izquierdo M, Ganga R, Real F, Tort L, Caballero MJ, Acosta F: Total substitution of fish oil by vegetable oils in gilthead sea bream (Sparus aurata) diets: effects on hepatic Mx expression and some immune parameters. Fish Shellfish Immunol. 2008, 24: 147-155.

51. Folch J, Lees M, Sloane-Stanley GH: A simple method for the isolation and purification of total lipids from animal tissues. J Biol Chem. 1957, 226: 497-509.

52. Bell JG, Miller D, MacDonald DJ, MacKinlay EE, Dick JR, Cheseldine S, Boyle RM, Graham C, O'Hare AE: The fatty acid compositions of erythrocyte and plasma polar lipids in children with autism, developmental delay or typically developing controls and the effect of fish oil intake. Br J Nutr. 2010, 103: 1160-1167.

53. Krasnov A, Timmerhaus G, Afanasyev S, Jørgensen SM: Development and assessment of oligonucleotide microarrays for Atlantic salmon (Salmo salar L.). Comp Biochem Physiol. 2011, 6D: 31-38.

54. Jørgensen SM, Kleveland EJ, Grimholt U, Gjoen T: Validation of reference genes for real-time polymerase chain reaction studies in Atlantic salmon. Mar Biotechnol. 2006, 8: 398-408.

55. Pfaffl MW, Horgan GW, Dempfle L: Relative expression software tool (REST) for group-wise comparison and statistical analysis of relative expression results in real-time PCR. Nucleic Acids Res. 2002, 30: e36-

CHAPTER 19

ANTI-INFLAMMATORY EFFECTS OF TETRADECYLTHIOACETIC ACID (TTA) IN MACROPHAGE-LIKE CELLS FROM ATLANTIC SALMON (SALMO SALAR L.)

Fabian Grammes[1,2] and Harald Takle[2,3]

[1] Institute of Animal and Aquaculture Sciences, Norwegian University of Life Sciences, N-1432 As-UMB, Norway.

[2] NOFIMA, N-1432 s, Norway. 3 AVS Chile SA, Casilla 300, Puerto Varas, Chile

ABSTRACT

Background

Commercial Atlantic salmon is fed diets with high fat levels to promote fast and cost-effective growth. To avoid negative impact of obesity, food additives that stimulate fat metabolism and immune function are of high interest. TTA, tetradecylthioacetic acid, is a synthetic fatty acid that stimulates mitochondrial β-oxidation most likely by activation of peroxysome proliferator-activated receptors (PPARs). PPARs are important transcription factors regulating multiple functions including fat metabolism and immune responses. Atlantic salmon experiments have shown that TTA supplemented diets significantly reduce mortality during natural outbreaks of viral diseases, suggesting a modulatory role of the immune system.

Results

To gain new insights into TTA effects on the Atlantic salmon immune system, a factorial, high-throughput microarray experiment was conducted using a 44K oligo nucleotide salmon microarray SIQ2.0 and the Atlantic salmon macrophage-like cell line ASK. The experiment was used to determine the transcriptional effects of TTA, the effects of TTA in poly(I:C) elicited cells and the effects of pretreating the cells with TTA. The expression patterns revealed that a large proportion of genes regulated by TTA were related to lipid metabolism and increased mitochondrial β-oxidation. In addition we found that for a subset of genes TTA antagonized the transcriptional effects of poly(I:C). This, together with the results from qRT-PCR showing an increased transcription of anti-inflammatory IL10 by TTA, indicates anti-inflammatory effects.

Citation: Fabian Grammes and Harald Takle Anti-inflammatory effects of tetradecylthioacetic acid (TTA) in macrophage-like cells from Atlantic salmon (Salmo salar L.), doi: 10.1186/1471-2172-12-41.

Conclusions

We demonstrate that TTA has significant effects on macrophage-like salmon cells that are challenged by the artificial dsRNA poly(I:C). The immune stimulatory effect of TTA in macrophages involves increased lipid metabolism and suppressed inflammatory status. Thus, suggesting that TTA directs the macrophage-like cells towards alternative, anti-inflammatory, activation. This has positive implications for TTA as a feed additive.

BACKGROUND

In commercial Atlantic salmon (Salmo salar L.) aquaculture, diets usually contain high levels of fat as it provides an inexpensive source of energy and promotes rapid growth. However, high levels of dietary fat cause excess fat to be deposited in the tissues, most notably in muscle, liver and as visceral fat [1]. Thereby the diet may affect the general health of the Atlantic salmon and also the quality of the filet as the final product [2]. In order to reduce the negative side effects of high fat diets, several studies have experimented with food additives promoting fat catabolism. Tetradecylthioacetic acid (TTA) is one of these additives, considered to be highly interesting.

TTA is a synthetic fatty acid with a sulfur substitution at the 3rd position in the carbon chain. The chemical properties of TTA resemble those of a normal fatty acid of similar length. What makes TTA interesting is that the sulfur atom makes it resistant towards further catabolism via β-oxidation and that TTA serves as activator for all peroxysome proliferator-activated receptor (PPAR) subtypes [3, 4]. PPARs are important transcription factors regulating multiple functions, most notably fat metabolism and immune responses (reviewed in [5]). TTA has been reported to induce pleiotropic effects in mammals. These effects include increased mitochondrial β-oxidation in muscle and liver, decreased plasma lipid levels as well as antioxidant and anti-inflammatory effects (reviewed in [6, 7]). It appears that most of the metabolic effects of TTA are mediated through activation of PPARs.

Studies addressing metabolic effects of TTA in fish are scarce. Nevertheless, the studies that have been conducted with members of the salmonidae family indicate similar metabolic effects of TTA in fish by showing increased lipid metabolism in liver [8, 9] and muscle [10]. In contrast to mammals, high levels of dietary TTA have been reported to induce mortality in Atlantic salmon [9, 11] but not in Rainbow trout (Oncorhynchus mykiss Walbaum) [8]. It has been hypothesized that the mortality may be related to accumulation of TTA-metabolites in the kidney at low temperatures [12].

To date only one study investigated immune modulatory functions of TTA in Atlantic salmon, showing that TTA reduces the production of leukotriene B4 (LTB_4) and prostaglandin E_2 (PGE_2) in headkidney macrophages isolated from TTA fed fish [12]. It should be noted that the reduction was found only to be significant for cells cultured at 5°C, but not in those maintained at 12°C [12]. Intriguingly, two studies, both reporting significant reduced mortality in Atlantic salmon pre fed with a low-level TTA supplemented diet during natural outbreaks of viral diseases in sea [10, 13], indicate important immune modulatory effects of TTA. Hence, TTA may be of interest as a feed additive in Atlantic salmon aquaculture.

To address immune modulatory functions of TTA in Atlantic salmon we conducted a "genome wide" microarray approach in a factorial designed experiment to identify transcriptional effects of TTA using the established, macrophage-like cell line ASK [14]. The cell line was chosen because macrophages pose a key role in the innate immunity, which is considered to be of primary importance to fight off pathogens since it can take weeks to mount an acquired immune response in poikilothermic fish [15]. Macrophages posses a remarkable plasticity that allows them to change their phenotype according to environmental signals. In vitrotwo major macrophage phenotypes have been characterized. Classically activated macrophages displaying a pro-inflammatory profile and alternative activated macrophages showing anti-inflammatory properties [16]. Macrophages recognize invading pathogens via pattern recognition receptors that bind conserved pathogen-associated molecular patterns (PAMPs). Activation of macrophages via PAMPs induces the classical activated phenotype. Because most viral pathogens produce dsRNA during one part of their replication [17], we decided to use the double stranded RNA analogue poly(I:C), a frequently used PAMP to simulate infection with a viral pathogen in our experiment.

A large proportion of the genes regulated by TTA was related to increased lipid metabolism and were also activated by TTA in poly(I:C) elicited ASK cells. The main effect seen for TTA in poly(I:C) elicited cells was to antagonize the induction of a subset of poly(I:C) dependent genes. Further we found increased expression of anti-inflammatory IL10 and Krueppel-like factor 11 (KLF11). Thus the results indicate that TTA polarizes the macrophage-like cells towards the alternative activated state.

RESULTS

We used the permanent Atlantic salmon macrophage-like cell line ASK in a factorial designed experiment to study transcriptional effects of TTA. An overview of the experimental design is given in Table 1. RNA from two biological replicates was used in the microarray analysis with the Atlantic salmon SIQ2.0 microarray [18] in a one-color setup, resulting in totally 16 arrays. One array did not pass the quality control analysis and was subsequently excluded from the final data set (Table 1, indicated by*). To identify differentially expressed (DE) probes, a linear model (Equation 2), containing the main effects was fitted to each probe set. Subsequently we extracted specific contrasts (Table 2), using the Bioconductor package "limma" [19].

Table 1 Experimental design

Pretreatment	Treatment	biol. repl.[a]	array[b]
Control	Control	4	2
	TTA	4	2
	TTA+poly(I:C)	4	2
	poly(I:C)	4	2
TTA	Control	4	2
	TTA	4	2
	TTA+poly(I:C)	4	2
	poly(I:C)	4	2*

[a] Replicates, using different passages of the cells

[b] Number of replicates used in the array experiment

* One of the arrays did not pass the quality control and was excluded from further analysis

Table 2 Specific contrasts

Contrast	H_0 to test	Interpretation when H_0 is rejected and the corresponding $\log_2$FC passes the cut-off
preTTA vs. CTR	$H_0 : P_{TTA} = P_{contr}$	DE in: Samples pretreated with TTA vs. pretreatment Control
TTA vs. CTR	$H_0 : T_{TTA} = T_{contr}$	DE in: Samples treated with TTA vs. Control treated
TTAIC vs. CTR	$H_0 : T_{TTA+polyI:C} = T_{contr}$	DE in: Samples treated with TTA+poly(I:C) vs. Control treated
IC vs. CTR	$H_0 : T_{polyI:C} = T_{contr}$	DE in: Samples treated with poly(I:C) vs. Control treated
IC vs. TTAIC	$H_0 : T_{TTA+polyI:C} = T_{polyI:C}$	DE in: Samples treated with TTA+poly(I:C) vs. poly(I:C) treated

Contrasts, extracted for each probe after the full linear model (Equation 2) was fitted to the data.

Pretreatment with TTA (preTTA vs. CTR) significantly altered the expression of 121 probes (108 genes), while treatment with TTA (TTA vs. CTR) affected 49 probes (41 genes) (Table 3). Stimulating the cells with poly(I:C) showed the strongest effect with regard to number of DE probes and $\log_2$FC with 3636 DE probes (2785 genes) in the contrast IC vs. CTR. We observed significant alteration of 3487 probes (2697 genes) when treating the cells with TTA and poly(I:C) simultaneously (TTAIC vs. CTR). 92 probes (70 genes) were DE when comparing the concordant TTA and poly(I:C) treatment vs. poly(I:C) treatment (IC vs. TTAIC), indicating that TTA exerts effects on transcription in poly(I:C) challenged cells.

Table 3 Summary of significant probes identified by the different contrasts

Contrast	UP/DOWN	Probes	Probes (lfc>1)	Probes (lfc>2)
Pretreatment				
preTTA vs. CTR	UP	119	4	0
	DOWN	2	0	0
Treatment				
TTA vs. CTR	UP	27	6	0
	DOWN	22	1	0
TTAIC vs. CTR	UP	1043	476	250
	DOWN	2444	764	72
IC vs. CTR	UP	1491	594	227
	DOWN	2145	787	80
IC vs. TTAIC	UP	22	4	0
	DOWN	70	9	0

In this study we were primarily interested in transcriptional effects related to TTA. Thus, we will focus for the rest of the paper on the three TTA related contrasts: preTTA vs. CTR, TTA vs. CTR and IC vs. TTAIC. However, a comprehensive list of all DE probes for each contrast can be found in Additional file 1, Table S1.

The number of DE probes and their overlap between the different contrasts is shown in Figure 1.

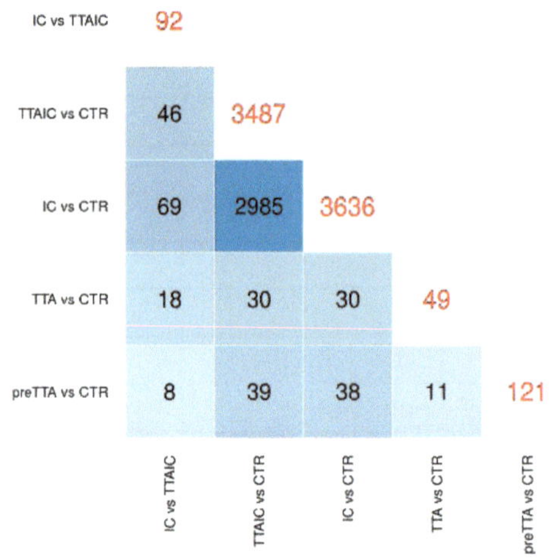

Figure 1 Overlap matrix. Overlap of the DE probes between the different contrasts. The numbers within cells display the number of common probes, while the red numbers in the diagonal display the total number of DE probes in that contrast.

Transcriptional effects in IC vs. TTAIC

92 probes were DE when comparing ASK cells treated with poly(I:C) alone vs. cells that received concordant TTA and poly(I:C) treatment (Figure 1 and 2). Considering that poly(I:C) treatment alone significantly altered the expression of 3636 probes (Figure 1), the effects of TTA under poly(I:C) elicitation may be considered relatively small.

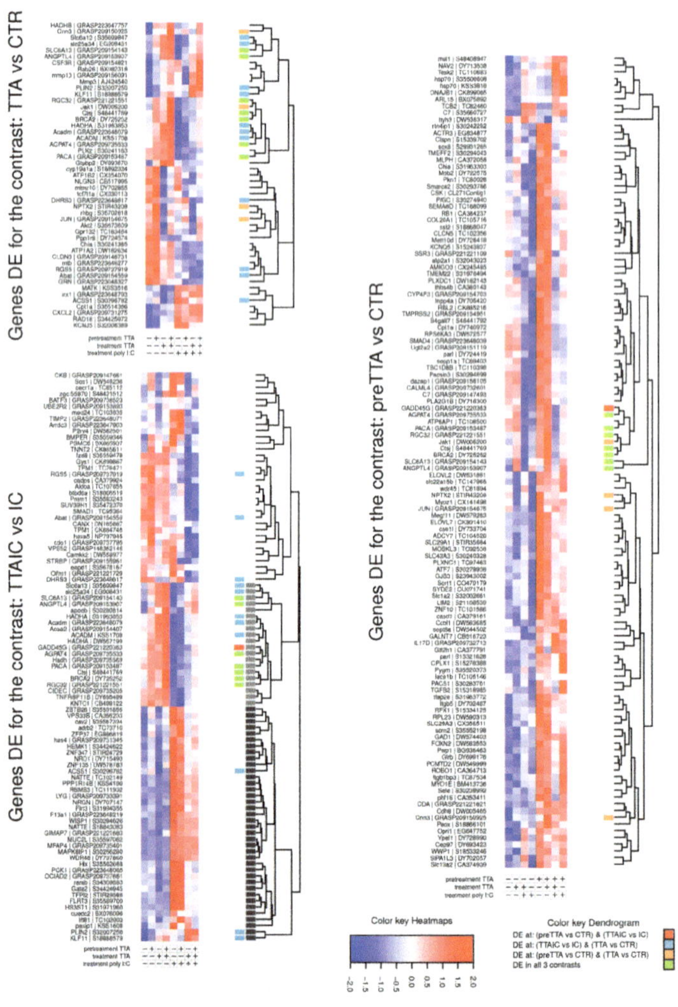

Figure 2 Heatmaps. Heatmaps displaying the relative expression of the probes significant in the three TTA related contrasts (IC vs. TTA-IC, TTA vs. CTR and preTTA vs. CTR. The probes were clustered hierarchical using euclidean distance as indicated by the dendrogram on the

side. The color code attached to the dendrogram indicates overlap for the corresponding gene between the 3 contrasts. In the dendrogram of the heatmap for contrast IC vs. TTAIC, the three major clusters are highlighted additionally (lightgrey - cluster.1; grey - cluster.2; darkgrey - cluster.3).

We observed that after hierarchical clustering of the DE probes for the IC vs. TTAIC contrast the genes could be separated into three major clusters (Figure 2). First, probes showing mostly a moderate down-regulation in response to poly(I:C), but a generally stronger down-regulation upon concordant TTA poly(I:C) treatment (cluster.1). Second, probes that were up-regulated by TTA, but down-regulated by poly(I:C) treatment (cluster.2). And third, probes that were less up-regulated when treated with TTA in combination with poly(I:C) (cluster.3). Especially for the second and the third cluster of probes we observed profound antagonistic effects for TTA in poly(I:C) elicited macrophages.

We found that cluster.2 contained mostly genes that were related to fat metabolism, and that most of them were shared with the TTA vs CTR contrast (see next section). Genes of particular interest were: ANGPTL4, as the highest ranking gene, Acadm, AGPAT4, THIM, HADHA, PLIN2 and KLF11. Angiopoetin-related protein 4 (ANGPTL4) has been reported, beside having a role in regulating angiogenesis, to act as a important stimulator of lipid metabolism [20]. Acetyl-CoA acyltransferase THIM and the mitochondrial trifunctional enzyme HADHA are enzymes of the mitochondrial β-oxidation. The gene AGPAT4 encodes an acyltransferase, participating in de-novo phospholipid synthesis. The gene remains yet to be fully characterized, however, other isoforms have been reported to be regulated by PPAR-α[21]. The protein Perilipin 2 (PLIN2) is a lipid droplet associated protein, recently suggested to stimulate lipolysis in macrophages [22]. Moreover, KLF11 an inhibitor of cellular PGE_2 synthesis [23] and APOEB, the teleost homolog to apoE, was found to be strongly up-regulated. PGE_2 is an important mediator of inflammation and has been previously reported to be down-regulated in response to TTA treatment [12, 24]. ApoE secreted by macrophages participates in the regulation of cholesterol efflux, reducing the cholesterol-ester accumulation in macrophages attached to the arterial wall [25]. In addition, ApoE in macrophages has been reported to act as suppressor of pro-inflammatory signals and vice versa [26], and is reported to respond to PPAR-γ agonists in macrophages [27]. Hence, these results suggest that TTA induces lipid metabolism and also affects lipid mediators of inflammation in poly(I:C) elicited macrophages.

The genes in cluster.1 and cluster.3, down-regulated for the IC vs. TTAIC contrast, contained several genes with reported functions in immune modulation. The genes encoding Nattectin (NATTE) and Lysozyme G (LYG) both showed a strong increase in transcription in response to poly(I:C), which was attenuated when TTA was added. Nattectin has been identified as C-type lectin in fish [28]. C-type lectins function as pattern recognition receptors, recognizing pathogens and may subsequently opsonize them for phagocytosis. C-type lectins are induced following infection with bacterial [29] and viral [30] pathogens in fish, hence, suggesting host defense properties. Lysozyme is primarily associated with the defense against bacterial pathogens through hydrolyzing β 1,4 glycoside bounds of peptidoglucans in the bacterial cell wall [31]. Expression of the anti-inflammatory marker Pituitary adenylate cyclase-activating polypeptide (PACA) [32], was strongly decreased in the IC vs. CTR contrast, but showed increased expression in the IC vs. TTAIC contrast. A similar pattern was observed for Os-teoprotegerin (TNFRSF11B). Osteoprotegerin is a member of the tumor necrosis receptor superfamily. Initially Osteoprotegrin was characterized for its ability to suppress osteoclast formation [33]. However Osteoprotegrin has also been reported to have a role in controlling inflammation. Exposure of mammalian endothelial cells to inflammatory cytokines has been reported to increase Osteoprotegrin expression, which in turn promotes leucocyte adhesion in vitro and in vivo[34].

Transcriptional effects in TTA vs. CTR

49 probes were DE in response to treatment with TTA (Figure 1, Figure 2). From these 49 probes we found 27 to be up- and 22 probes to be down-regulated. The transcriptional changes in general were modest, with 6 probes showing an absolute $\log_2 FC \geq 1$ (Table 3). As shown in Figure 1 (and indicated in the dendrogram of Figure 2) the overlap to the previously described IC vs. TTAIC contrast was 37%, demonstrating that TTA affects a slightly different subsets of genes when applied alone to ASK cells.

Treatment with TTA generally increased the expression of genes involved in lipid metabolism. The gene showing the highest $\log_2 FC$ in this contrast was Angiopoetin-related protein 4 (ANGPTL4). Further we found a couple of genes encoding crucial enzymes for mitochondrial β-oxidation. We found carnitine palmitoyltransferase 1A (Cpt1a), the acyl-CoA dehydrogenase (ACADM) and mitochondrial trifunctional enzyme (HADHB) to be up-regulated. Also we observed increased expression of the gene encoding the lipid particle associated protein PLIN2 and short-chain dehydrogenase/reductase DHRS3. Thus, indicating increased lipid metabolism in ASK cells following TTA treatment.

We observed TTA regulation of a number of genes associated to anti-inflammatory effects. The second highest ranking gene up-regulated was the PGE_2 inhibitor KLF11[23]. Moreover we observed attenuated transcription of Grannulins (GRN)/ progrannulin and the neuropeptide pituitary adenylate cyclase-activating polypeptide (PACA). Both genes are reported to exhibit anti-inflammatory effects. Macrophages isolated from mice deficient for the Grannulin gene showed higher transcription levels of pro-inflammatory cytokines [35], indicating an anti-inflammatory role for progrannulin. PACA has been identified to exert anti-inflammatory effects [32]. Increased expression of the matrix metalloproteinases MMP3, MMP13 and chemokine ligand 2(CXCL2) was found in TTA treated ASK cells. Matrix metalloproteinases (MMPs) possess pleiotropic functions such as acting as matrix degrading enzymes, immunomodulary functions and have been suggested to be required for the migration of inflammatory cells [36]. CXCL2 has been implicated to act as chemotactic or activating factor for monocytes [37] and was further found to be highly induced by poly(I:C) treatment. Thus suggesting an increased potential for cell migration induced by TTA.

Further, megakaryocyte-associated tyrosine kinase (MATK) and the transcription factor AP-1 (JUN) proofed to be down-regulated. MATK is a nonreceptor tyrosine kinase that negatively regulates the activity of the Src family protein kinases - key regulators in signal transduction. A study showed higher levels of interferon-γ in MATK K.O mice challenged with antigens, thus suggesting a function for MATK in the immunological response [38]. JUN is an important transcription factor responding to cytokines and stress and has been reported to be up-regulated in response to viral challenges in Atlantic salmon [39, 40]. Further we found a decreased expression of NPTX2, a member of the pentraxin superfamily that are known to act as pattern recognition receptors and can facilitate pathogen recognition and removal [15].

Transcriptional effects in preTTA vs. CTR

In the two experiments showing lower mortality in Atlantic salmon fed with TTA upon natural (viral) disease outbreak, the TTA feeding period ended before the onset of disease related mortality [10, 13]. So we speculated that TTA might have long-term effects on the immunity in Atlantic salmon. To identify these effects we included the factor pretreatment in the experimental design. We found that pretreatment with TTA caused DE of 121 probes, which with the exception of two probes, all showed increased transcription (Table 3). The list of DE probes showed relatively little overlap with the two previously described TTA contrasts (7% and 9% to the contrasts IC vs. TTAIC and TTA vs. CTR respectively, Figure 2). Hence, suggesting that TTA pretreatment affects a different subset of genes. However, 38 of the probes in this contrast were also DE in the contrast IC vs. CTR (Figure 1). It should be noted that, similar to the effects observed in IC vs. TTAIC, the gene regulation in response to TTA pretreatment was mostly antagonistic to the effect found in IC vs. CTR (Figure 3).

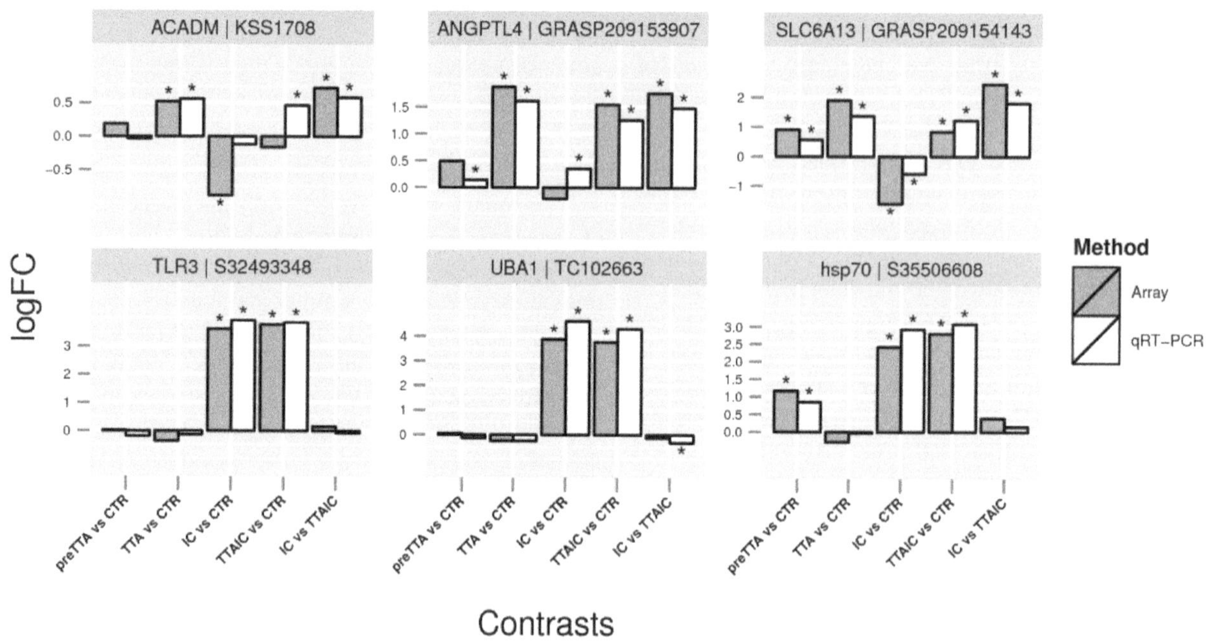

Figure 3 qRT PCR verification. Comparison of the log_2 fold changes and the estimated significances between qRT-PCR (n = 4) and array (n = 2). Significances for the different contrasts were estimated by linear models for the array (as described in Methods) and 1-way ANOVA (Equation.1) followed by a Tukey test for qRT-PCR. Before calculating the fold changes the qRT-PCR expression values were normalized to the expression of the housekeeping gene Elongation factor 1α (EF1α). Contrasts found to be significant are marked with an asterisk. **ACDAM**: acyl-Coenzyme A dehydrogenase, medium-chain; **ANGPTL4**: Angiopoetin-like protein 4; **hsp70**: Heat shock protein 70; **SLC6A13**: Solute carrier family 6 member 13/sodium- and chloride-dependent GABA transporter 2; **TLR3**: Toll-like receptor 3;**UBA1**: ubiquitin-like modifier activating enzyme 1.

Fatty acid metabolism and anti-inflammatory gene expression were less affected by pretreatment compared to treatment with TTA. The two fat metabolism related genes ANGPTL4 and elongation of very long chain fatty acids protein 2 (ELOV2) were both up-regulated. Further we observed increased transcription of Interleukin 17 D (IL17D). The IL 17 family possesses an essential role in protecting the host from infection by pathogens [41]. The anti-inflammatory gene PACA and the pattern recognition receptor NPTX2, both described in the previous contrasts showed increased gene expression. It should be noted that the gene regulation of NPTX2in preTTA vs. CTR was opposite to that found in TTA vs CTR. Two transmembrane proteins, Integrin β 5 (ITB5) and thrombospondin-4 (thbs4b), both contributing to cell adhesion and migration showed increased transcription following pre-treatment with TTA. In addition we found increased expression of the transmembrane receptors roundabout-1 (ROBO1) and plexin-C1 (PLXNC1). Both receptors are considered to be repulsive receptors, contributing to the release of cell adhesion upon ligand activation [42, 43]. These results may indicate that pretreatement with TTA supports increased cell migration in ASK cells, similar to the effects found in TTA vs. CTR.

Approximately 10% of the genes DE following TTA pre-treatment were transporters/ion channels. The genes included the GABA transporter (SLC6A13), Na^+/dicarboxylate co-transporter (SLC13A2), SLC22A15b, SLC43A3, Cl^- transporter (CLCN5), v-type proton ATPase subunit s1 (ATP6AP1), ATP2A1, K-channel (KCNQ5), the neurotransmitter-gated ion channel GLRB and the Ca^{2+} activated chloride channel TTYH3 as the only one showing down-regulation. CLCN5 and ATP6AP1 are responsible for the acidification of the endosomal lumen. Hence, the increased expression suggests a higher capacity for acidification of the endosome, which may promote phagocytosis [44].

qRT-PCR

To validate the microarray results, six genes were randomly chosen from the list of DE genes for all contrasts. The genes were analyzed by qRT-PCR (Figure 4). In four cases we were able to detect significance by qRT-PCR, but not by array. However, in these cases the $\log_2$FC proofed to be lower than the cut-off of 0.5, that we used to identify significant genes in the array experiment. In one case we detected significance by array, but could not confirm this by qRT-PCR. Overall the results indicate that the parameters chosen to account for multiple hypothesis testing were appropriate. The results of the $\log_2$FCs for qRT-PCR matched those obtained by array. When comparing the $\log_2$FCs obtained for all genes and contrasts between qRT-PCR and array we found a highly significant correlation (pearson correlation of $r = 0.96$, $p = 2.2e^{-16}$ and an $R^2 = 0.92$).

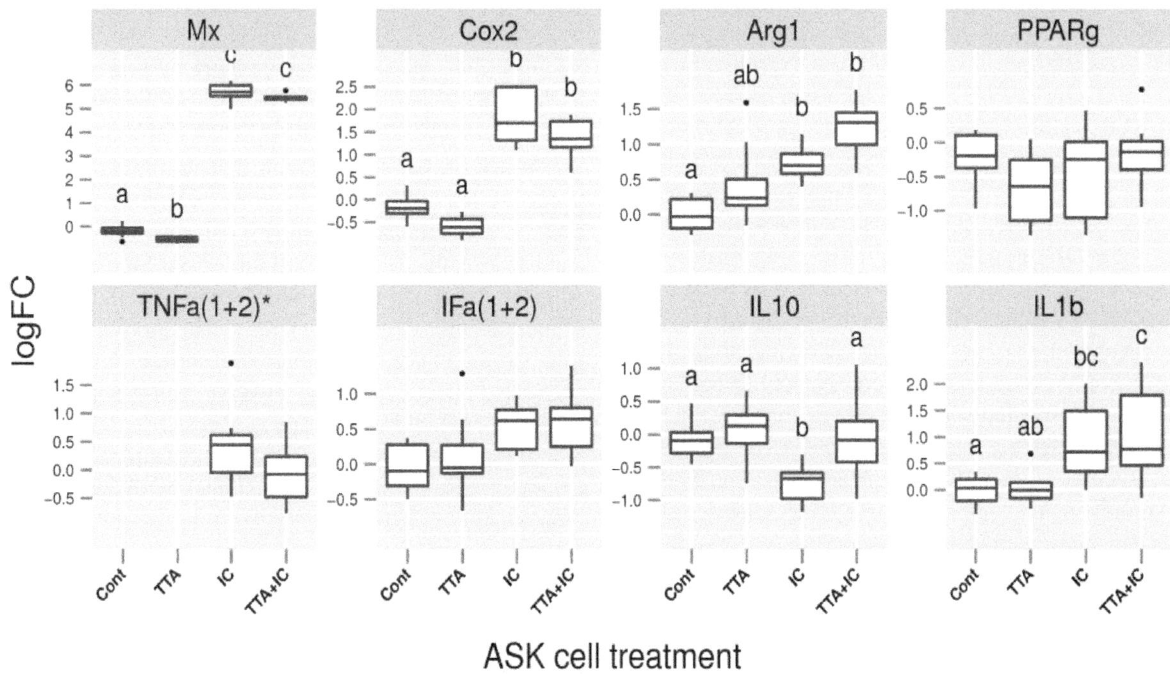

Figure 4 qRT-PCR. Box plots showing $\log_2$ gene expression relative to the mean expression of the samples that received control pretreatment and control treatment. Gene expression was normalized to the expression of the housekeeping gene EF1α. Significance was assessed using 1-way ANOVA (Equation.1). None of the tested genes was significant affected by TTA pretreatment. Different letters between the box plots indicate significant difference between the treatments, as determined by Tukey test (p-value ≤ 0.05, $n = 4$). **Mx**: myxovirus resistance 1; **Cox2**: Cyclooxygease 2; **Arg1**: Arginase 1; **PPARg**: peroxysome proliferator-activated receptor γ ; **TNFa**: Tumor necrosis factor α (1+2); **IFa1+2**: Interferon α (1+2); **IL10**: Interleukin 10; **IL1b**: Interleukin 1β* Reliable gene expression could be only measured for samples that received poly IC treatment. Expression for TNFα is displayed relative to the samples that received control pretreatment and TTA + poly(I:C) treatment.

To further study immune modulatory effects of TTA in ASK cells, we used qRT-PCR to measure the expression of selected genes previously reported to respond directly to TTA or other PPAR agonists (Cox2, IL10, PPARγ,TNFα, Arg1) and/or are important mediators of inflammation (IL1β, IFα(1+2), Mx) (Figure 4). None of the analyzed genes was found to respond significantly to the effect of pretreatment with TTA. Several genes, however, responded significantly to the four different treatments.

For Interleukin 10 (IL10), which is generally considered to have anti-inflammatory properties, we found significant lower expression in the group that received poly(I:C), compared to all other treatment groups. The two pro-inflammatory cytokines TNFα and Interferon α showed no DE in response to the different treatments. It should be noted that the expression of TNFα could only be detected in poly(I:C) elicited samples and that concordant application of TTA and poly(I:C) reduced the mean expression by 50%. PPARγ showed no altered expression in response to the different treatments. Arginase 1 was significantly higher expressed in all samples that received poly(I:C) treatment, additionally we observed that TTA also increased the Arginase 1expression. The mean expression of the samples receiving TTA and poly(I:C) was 67% higher compared to the poly(I:C) samples. Even though not statistical significant, we observed a reduction in the expression of COX2upon TTA treatment. Interleukin 1β is considered to be pro-inflammatory. We found a significant increase in transcription upon poly(I:C) elicitation, while concordant treatment with TTA and poly(I:C) elevated the mean expression (not significant). Mx is considered to be an antiviral protein [45], TTA treatment caused a significant reduction of Mx gene expression. Treatment with poly(I:C) increased the expression strongly.

DISCUSSION

In the present study we characterized the transcriptional responses to the PPAR activator TTA in the Atlantic salmon macrophage-like ASK cell line. Considering the strong effects documented for various PPAR agonists in different tissues like intestine or fat tissue (reviewed [46]) we found relatively few genes responding with altered transcription to TTA treatment. However, based on the results of the Microarray and qRT-PCR analysis TTA was found to stimulate the capacity for fat metabolism and have anti-inflammtory effects in the ASK cell line.

The innate immune response recognizes invading pathogens by pattern recognition receptors (PRR) through recognition of pathogen-associated molecular patterns. Toll-like receptors are a family of PRRs where each receptor displays specificity for it's lig-and [47]. Viral pathogens produce double stranded (ds) RNA during one part of their reproduction cycle. DsRNA is recognized by TLR3, triggering the production and release of pro-inflammatory cytokines [48]. The results from our microarray study showed that indeed a high number of genes were DE in poly(I:C) elicited macrophages. By comparing the group that received concordant treatment with TTA and poly(I:C) against the group that received poly(I:C) alone we identified a subset of genes showing antagonistic regulation. Several of these genes were sparsely characterized in the literature, especially in relation to immune function. However, the fact that we found those genes to be induced by poly(I:C) suggests a function in the Atlantic salmon immune response. Indicating that application of TTA to Atlantic salmon ASK cells has immune modulatory properties.

Expression of anti-inflammatory markers by macrophages is a hallmark for the alternative activated state. Anti-inflammatory effects of TTA have been mainly ascribed to increased expression of anti-inflammatory IL10. Indeed, we found a significant 98% increased transcription of IL10. Hence, indicating that TTA increases expression of IL10 in poly(I:C) activated macrophages, similar to the effects reported from human PBMCs stimulated with TTA [49, 50].

Further we identified the PGE_2 inhibitor KLF11[23] as up-regulated in response to TTA. PGE_2 is a mediator of inflammation, produced from PGH_2 which is mainly produced by COX2. We observed that in poly(I:C) elicited ASK cells, TTA caused a mean reduction of 24% in COX2 gene expression. Arguably, these results indicate that TTA attenuates the capacity for production of pro-inflammatory PGE_2, as it has been shown in primary macrophages from TTA fed fish reared at 5°C, but not for those kept at 12°C [12].

Arg1 competes with iNOS, an enzyme controlling the production of Nitric oxide for it's substrate. Arg1 is considered to be a key marker for alternative activated macrophages [51]. Recent studies in mice have shown that the expression of Arg1 in alternative activated macrophages is essential for suppressing inflammation [52]. Thus, our observation that TTA increased the expression of Arg1 by 67%, supports anti-inflammatory effects of TTA, and is also in accordance with a described regulation of Arg1 expression by PPAR-γ and β/δ[51]. Studies conducted on murine macrophages have shown that an up-regulation in fatty acid oxidation is crucial to direct them towards an alternative activated state [53]. In agreement with this we found that TTA induced the expression of genes that are related to lipid metabolism in ASK cells. Namely we identified increased expression of Angiopoetin-like protein 4 (ANGPTL4), carnitine palmitoyltransferase 1 (Cpt1a), acyl-CoA dehydrogenase (Acdam), enoyl-Coenzyme A hydratase (HADH) and the lipid droplet associated protein PLIN2. These effects of TTA are also consistent with reports of increased fatty acid oxidation upon TTA treatment [6,8] and also with effects associated to PPAR activation [5]. Taken together, the results provided by this study indicate that TTA directs the macrophage-like ASK cells towards an anti-inflammational, alternative activated state. Studies conducted in mice have reported that macrophages in the alternative activated state, via macrophage-parenchymal cell crosstalk, affect fat catabolism in liver and adipose tissues [54, 55]. This effect was especially found to be important in animals under high-fat diet. Thus, we may speculate that TTA directed macrophage activation in Atlantic salmon fed high fat diets could contribute to an improved metabolic and inflammatory state.

Many studies have shown that the immune response is necessary in order to evict an invading pathogen, however, an excessive inflammatory response can cause unintended tissue damage, finally causing mortality [56]. Studies challenging Atlantic salmon with the viral pathogen ISAV (infectious salmon anemia virus) have shown that survival is not related to a strong innate inflammatory response, but rather to an attenuated innate and a stronger cell-mediated immune response [40]. Hence, it appears possible that the reduced mortality as reported from TTA fed Atlantic salmons during natural disease outbreaks [10, 13] may partly result from anti-inflammatory effects of TTA.

TTA has been reported to act as an agonist for all PPAR subtypes in vitro[3, 4], and it is assumed that most physiological effects of TTA are mediated via activation of PPARs. In our experiment we found neither significant transcriptional regulation of PPAR-α, β/δ (represented on the Array, data not shown) nor PPAR-γ, (Figure 4). However, the fact that a large proportion of the genes identified in our study have been reported to be regulated by PPARs and that the list of DE genes shows considerable overlap with the results from PPAR stimulated murine macrophages K.O for PPAR-γ or β/δ[57] support our assumption that the ASK polarization towards an alternative activated state may be induced via activation of PPAR-γ and β/δ by TTA.

CONCLUSIONS

The results presented in this study suggests that TTA directs the macrophage-like cells towards an alternative activated state which is indicated by a stimulated lipid metabolism and attenuated inflammatory response. Further the pattern of TTA regulated genes indicate that the stimulation occurs via PPAR activation. The effects confirm a function of TTA in Atlantic salmon macrophages similar to those documented in mammals. These findings supports that TTA modulates the immune system of Atlantic salmon.

METHODS

Cell Culture

The Atlantic salmon kidney cell line (ASK: ATCC: CRL-2747) [14] was cultured and propagated at 20°C according to standard cell culture procedures. Compared to mammalian cell lines, fish cell lines thrive slowly [58], so the ASK cells were studied for the respectively long period of 7 days. The cells were seeded out at day 1 in 6 well format (density 5 × 10^4 cells/ml). On day 2-4 the cells were pretreated with either TTA or control medium. Day 5-7: cells were treated with either Control, TTA, TTA+poly(I:C) or poly(I:C) medium.

Growth medium was L-15 (Invitrogen, MD, USA) supplemented with 10% FBS (Invitrogen), 40μ M β-Mercaptoethanol and 50μ g/ml Gentamycin. TTA medium was prepared by dissolving TTA (ThiaMed-ica, Norway) in 0.1 M NaOH at 80°C to a concentration of 100 mM, this stock-solution was diluted to a final concentration of 100μ M in growth-medium [59]. Control-medium was obtained by adding the same volume of 0.1 M NaOH as used in the TTA medium to the growth medium. Poly I:C was prepared by adding poly(I:C) (invitrogen) to a final concentration of (50 μ g/ml) to Control-medium. TTA+poly(I:C) medium was obtained by adding poly(I:C) to a final concentration of (50 μ g/ml) to the TTA-medium.

The cells were harvested on day 7 for RNA extraction. An overview of the experimental design is given in Table1. The experiment was repeated four times with four different passages (passages 6-9) from the cells.

RNA Extraction

Total RNA was extracted using Qiazol (Qiagen, Halle, Germany) and purified from the cells using column purification (Qiagen) according to the manufacturer's instructions. Traces of genomic DNA in the samples were eliminated by on-column-DNase (Qiagen) digestion. RNA concentrations were measured for all samples using a NanoDrop 1000 Spectrophotometer. RNA quality for samples later used in the Microarray was determined using a Agilent 2100 Bioanalyzer (RNA 6000 NanoLabChip, Agilent, CA, USA). All samples had RIN values ranging from 9.6 to 10, indicating high quality of the RNA. Only individual RNA samples were used for subsequent qRT-PCR and microarray experiments.

Quantitative RT-PCR

Single strand cDNA was reverse transcribed from 500 ng of total, DNase treated RNA using oligo dT primers and the Taq Man reverse transcription Kit (Applied Biosystems, Tx, USA). qRT-PCR was performed in 96-well optical plates on a Light-Cycler 480 (Roche, Switzerland). For the PCR reaction 2× SYBR green I master Mix (Roche), 0.41 nM of each primer and the cDNA template were mixed in a total reaction volume of 10μ l. Primer sequences are listed in Table 4. A three step PCR protocol with 45 cycles (15 s 95°C, 15 s 60°C, 15 s 72°C) was used. To verify specific amplification a melting curve analysis step was done at the end of the program. All samples were analyzed in duplicates and for each measured gene a standard curve was produced using a serial dilution from a pool of all cDNA samples.

Table 4 qRT-PCR primer sequences

ID/Gene.Acc.Nr.	Gene	Forward primer	Reverse primer
AF321836	EF1α	CACCACCGGCCATCTGATCTACAA	TCAGCAGCCTCCTTCTCGAACTTC
KSS1708	ACADM	GCCATCTCAGCCAACAGGAA	GACAAGACCAGACCGCCAGT
GRASP209147493	S6A13	CCGTATGGGGGATGATGCTAA	GGTAGTATGCTGACGACTGACACCT
S35506608	HSP70	AAGGCAAGATCAGCGAGGAG	TGCCTGATCTCCACAGCAAC
GRASP209153907	ANGL4	AGGAGATGCAGCAGGAGAGG	TTCTGAGCTTCCAGCATCCA
S32493348	TLR3	TCCAAGCCTGGTGACTCTCA	CAGGTGGAGGACTCGTTAGGG
TC102663	UBA1	GCACTGTTTCCCCTCTGACC	AGCAGTCTCGTGTTCCCACA
AF321836	TNFα (1+2)	GCTTGTCTCTTGTTGCCACCA	TGTGTGGGATGAGGATTTGGTT
AF088999	COX2	ACCTATGGGAACCGCAGGA	GTGAAGTGCTGGGCGAAGA
EG929369	ARG1	AGCCATGCGTATCAGCCAA	AAGGCGATCCACCTCAGTCA
AJ416951	PPARγ	CATTGTCAGCCTGTCCAGAC	TTGCAGCCCTCACAGACATG
EF165028	IL10	ATGAGGCTAATGACGAGCTGGAGA	GGTGTAGAATGCCTTCGTCCAACA
AY216594+AY216595	IFα (1+2)	TGCAGTATGCAGAGCGTGTG	TCTCCTCCCATCTGGTCCAG
SSU66477	MX	TGATCGATAAAGTGACTGCATTCA	TGAGACGAACTCCGCTTTTCA
AY617117	IL1β	AGCAGGGTTCAGCAGTACAT	CTCCATAGCCTCACTCATCA

The expression levels were calculated using the standard curve method (Applied Biosystems User Bulletin 2). Expression levels were standardized to the expression of the housekeeping gene Elongation factor 1α (EF1α, [60]), and log_2FCs were calculated relative to the average standardized sample that received control pretreatment and control treatment. Significance was tested by conducting a analysis of variance (ANOVA) test on the log_2FCs for each gene using the model:

$$y_{ijk} = \mu + P_j + T_k + \varepsilon_{ijk} \quad (1)$$

Where y_{ijk} is the log_2FC of the ith sample that received pretreatment j and treatment k. P_j is the effect of pretreatment j = {contr, TTA}, T_k is the effect of treatment k = {contr, TTA, TTA+polyI:C, polyI:C}, μ is a constant and ε is the error. A Tukey test was used subsequently to identify differences within significant effects.

Microarray Hybridization

The customized 4 × 44 K oligo (60-mer) Atlantic salmon microarrays SIQ2.0 (Agilent) [18] was used in the experiment. Unless otherwise stated all equipment and reagents were from Agilent. Amplification and labeling of 500 ng of total RNA and spike-ins for each sample was performed using the Quick Amp Labeling Kit (One-Color). After fragmentation of the Cy3 labeled cRNA, the samples were hybridized to the microarray at 65°C for 17 h. After hybridization the arrays were washed with Gene Expression wash Buffer 1 and 2 and scanned using a Agilent Microarray Scanner. All steps were conducted according to the Agilent protocol (One-Color Quick Amp Labeling, Version 5.7). Analysis of the mi-croarray images was done in Agilent Feature Extraction (Version 10.7.1.1) Software using the one-color (GE1_107_Sep09) protocoll. Array quality was assessed through Agilent control features and spike-in controls.

Data Analysis

One array did not pass the quality checks and was excluded from subsequent analysis. Normalization and analysis of the data was done in R/Bioconductor [61] using the Bioconductor package "limma" [19]. To generate the data-set, background corrected spot intensity signals (gProcessedSignal) were filtered according to the following criteria provided by Feature Extraction: gIsFound, gIsPosAndSignif, gIsWellAboveBG and ControlType (a description of the feature results can be found in the Feature Extraction Software Reference Guide). All Control spots were removed from the data set. The mean of the intensity signal for the duplicated features was calculated and the data was subsequently normalized by quantile normalization in order to adjust the scale of intensities across arrays [62]. The data was then log_2 transformed and probe sets showing more than two missing values were discarded from the data set, classifying 12496 (59%) probe sets as present. Processed and raw microarray data is publicly available at NCBI's GEO repository (http://www.ncbi.nlm.nih.gov/geo/, Acc.Nr: **GSE25328**). A linear model (using zero intercept parameterization) was fitted to each probe set, where y_{ijkl} is the log_2 intensity of the ith sample that received pretreatment j and treatment k. P_j is the effect of pretreatment j = {contr, TTA}, T_k is the effect of treatment k = {contr, TTA, TTA+polyI:C, polyI:C}, B_l is the batch effect l = {1,2}, μ is a constant and ε is the error.

$$y_{ijkl} = P_j + T_k + B_l + \varepsilon_{ijkl} \quad (2)$$

Subsequent to fitting the model to each probe set, a number of specific contrasts of interest (listed in Table 2) were extracted for each probe set. For a graphical representation of the samples that were compared by the contrasts see Additional file 2, Figure S1. Improved variance estimates for the contrasts were computed using empirical Bayes moderated t-statistics from the "limma" package [63]. Before accounting for multiple hypothesis testing, probe sets without annotation were removed, reducing the data set to 10627 (50%) probe sets. Multiple hypothesis testing was accounted for by controlling the false discovery rate (FDR) [64] across all probes and contrasts simultaneously (using the method: global in the decideTests function from the 'limma" package). Probes were declared significant when the q-value was ≤ 0.15 and the $log_2FC \geq 0.5$.

Annotations for the probes were obtained using the function topBlast from the program Blast2GO [65]. Full length cDNA sequences of the probes have been compared in a BlastX search against the Swissprot protein database using an E-value cut-off of 10^{-6} and the default program settings. All plots were generated using the R package "ggplot2" [66] and heatmaps were produced using «lattice» [67].

ACKNOWLEDGEMENTS

The work was funded by the Research Council of Norway (NRC project no: 179481/I30). We would like to thank Dr. Torben Luders from the Institutt for klinisk epidemiologi og molekylrbiologi (EpiGen) for his support in scanning the microarrays. Also we would like to thank Dr. Aleksei Krasnov (Nofima) for providing the microarray platform and Carlos Lozano (AFGC) for his help with the analysis.

REFERENCES

1. Sargent JR, Tocher DR, Bell JG: The Lipids. Fish Nutrition. Edited by: Halver JE, Hardy RW. 2002, London: Academic Press, 181-257. 3
2. Herme G, Sandnes K: Effect of dietary lipid level on muscle composition in Atlantic salmon Salmo salar. Aquaculture Nutrition. 1999, 5: 9-16. 10.1046/j.1365-2095.1999.00081.x.
3. Berge K, Tronstad KJ, Flindt EN, Rasmussen TH, Madsen L, Kristiansen K, Berge RK: Tetradecylthioacetic acid inhibits growth of rat glioma cells ex vivo and in vivo via PPAR-dependent and PPAR-independent pathways. Carcinogenesis. 2001, 22 (11): 1747-55. 10.1093/carcin/22.11.1747.
4. Westergaard M, Henningsen J, Svendsen ML, Johansen C, Jensen UB, Schrøder HD, Kratchmarova I, Berge RK, Iversen L, Bolund L, Kragballe K, Kristiansen K: Modulation of keratinocyte gene expression and differentiation by PPAR-selective ligands and tetradecylthioacetic acid. J Invest Dermatol. 2001, 116 (5): 702-12. 10.1046/j.1523-1747.2001.01329.x.
5. Chinetti G, Fruchart JC, Staels B: Peroxisome proliferator-activated receptors (PPARs): nuclear

receptors at the crossroads between lipid metabolism and inflammation. Inflamm Res. 2000, 49 (10): 497-505. 10.1007/s000110050622.

6. Berge RK, Skorve J, Tronstad KJ, Berge K, Gudbrandsen OA, Grav H: Metabolic effects of thia fatty acids. Curr Opin Lipidol. 2002, 13 (3): 295-304. 10.1097/00041433-200206000-00010.

7. Skrede S, Sørensen HN, Larsen LN, Steineger HH, Høvik K, Spydevold OS, Horn R, Bremer J: Thia fatty acids, metabolism and metabolic effects. Biochim Biophys Acta. 1997, 1344 (2): 115-31.

8. Kennedy S, Bickerdike R, Berge R, Dick J, Tocher D: Influence of conjugated linoleic acid (CLA) or tetradecylthioacetic acid (TTA) on growth, lipid composition, fatty acid metabolism and lipid gene expression of rainbow trout (Oncorhynchus mykiss L.). Aquaculture. 2007, 272 (1-4): 489-501. 10.1016/j.aquaculture.2007.06.033.

9. Kleveland EJ, Ruyter B, Vegusdal A, Sundvold H, Berge RK, Gjøen T: Effects of 3-thia fatty acids on expression of some lipid related genes in Atlantic salmon (Salmo salar L.). Comp Biochem Physiol B, Biochem Mol Biol. 2006, 145 (2): 239-48. 10.1016/j.cbpb.2006.07.012.

10. Rørvik KA, Alne H, Gaarder M, Ruyter B, Måseide NP, Jakobsen JV, Berge RK, Sigholt T, Thomassen MS: Does the capacity for energy utilization affect the survival of post-smolt Atlantic salmon, Salmo salar L., during natural outbreaks of infectious pancreatic necrosis?. Journal of Fish Diseases. 2007, 30 (7): 399-409. 10.1111/j.1365-2761.2007.00823.x.

11. Moya-Falcón C, Hvattum E, Dyrøy E, Skorve J, Stefansson SO, Thomassen MS, Jakobsen JV, Berge RK, Ruyter B: Effects of 3-thia fatty acids on feed intake, growth, tissue fatty acid composition, beta-oxidation and Na+, K+-ATPase activity in Atlantic salmon. Comp Biochem Physiol B, Biochem Mol Biol. 2004, 139 (4): 657-68. 10.1016/j.cbpc.2004.08.009.

12. Gjøen T, Kleveland EJ, Moya-Falcón C, Frøystad MK, Vegusdal A, Hvattum E, Berge RK, Ruyter B: Effects of dietary thia fatty acids on lipid composition, morphology and macrophage function of Atlantic salmon (Salmo salar L.) kidney. Comp Biochem Physiol B, Biochem Mol Biol. 2007, 148: 103-11. 10.1016/j.cbpb.2007.04.021.

13. Alne H, Thomassen MS, Takle H, Terjesen BF, Grammes F, Oehme M, Refstie S, Sigholt T, Berge RK, Rørvik KA: Increased survival by feeding tetradecylthioacetic acid during a natural outbreak of heart and skeletal muscle inflammation in So Atlantic salmon, Salmo salar L. Journal of Fish Diseases. 2009, 32 (11): 953-61. 10.1111/j.1365-2761.2009.01078.x.

14. Devold M, Krossøy B, Aspehaug V, Nylund A: Use of RT-PCR for diagnosis of infectious salmon anaemia virus (ISAV) in carrier sea trout Salmo trutta after experimental infection. Dis Aquat Org. 2000, 40: 9-18.

15. Ellis AE: Innate host defense mechanisms of fish against viruses and bacteria. Dev Comp Immunol. 2001, 25 (8-9): 827-39. 10.1016/S0145-305X(01)00038-6.

16. Gordon S: Alternative activation of macrophages. Nat Rev Immunol. 2003, 3: 23-35. 10.1038/nri978.

17. Weber F, Wagner V, Rasmussen SB, Hartmann R, Paludan SR: Double-stranded RNA is produced by positive-strand RNA viruses and DNA viruses but not in detectable amounts by negative-strand RNA viruses. J Virol. 2006, 80 (10): 5059-64. 10.1128/JVI.80.10.5059-5064.2006.

18. Krasnov A, Timmerhaus G, Afanasyev S, Jørgensen SM: Development and assessment of oligonucleotide microarrays for Atlantic salmon (Salmo salar L.). Comparative biochemistry and physiology Part D, Genomics & proteomics. 2010

19. Smyth GK: Limma: linear models for microarray data. Bioinformatics and Computational Biology Solutions using R and Bioconductor. Edited by: Gentleman R, Carey V, Dudoit S, R Irizarry WH. 2005, New York: Springer, 397-420.

20. Mandard S, Zandbergen F, van Straten E, Wahli W, Kuipers F, Müller M, Kersten S: The fasting-induced adipose factor/angiopoietin-like protein 4 is physically associated with lipoproteins and governs plasma lipid levels and adiposity. The Journal of Biological Chemistry. 2006, 281 (2): 934-44.

21. Lu B, Jiang YJ, Zhou Y, Xu FY, Hatch GM, Choy PC: Cloning and characterization of murine 1-acyl-sn-glycerol 3-phosphate acyltransferases and their regulation by PPARalpha in murine heart. Biochem J. 2005, 385 (Pt 2): 469-77.

22. Persson J, Degerman E, Nilsson J, Lindholm MW: Perilipin and adipophilin expression in lipid loaded macrophages. Biochemical and Biophysical Research Communications. 2007, 363 (4): 1020-6. 10.1016/j.bbrc.2007.09.074.

23. Buttar NS, DeMars CJ, Lomberk G, Rizvi S, Bonilla-Velez J, Achra S, Rashtak S, Wang KK, Fernandez-Zapico ME, Urrutia R: Distinct role of Kruppellike factor 11 in the regulation of prostaglandin E2 biosynthesis. The Journal of Biological Chemistry. 2010, 285 (15): 11433-44. 10.1074/jbc.M109.077065.

24. Bivol LM, Hultström M, Gudbrandsen OA, Berge RK, Iversen BM: Tetradecylthioacetic acid downregulates cyclooxygenase 2 in the renal cortex of two-kidney, one-clip hypertensive rats. Am J Physiol Regul Integr Comp Physiol. 2008, 295 (6): R1866-73. 10.1152/ajpregu.00850.2007.

25. Fazio S, Babaev VR, Murray AB, Hasty AH, Carter KJ, Gleaves LA, Atkinson JB, Linton MF: Increased atherosclerosis in mice reconstituted with apolipoprotein E null macrophages. Proc Natl Acad Sci USA. 1997, 94 (9): 4647-52. 10.1073/pnas.94.9.4647.

26. liang Zhang H, Wu J, Zhu J: The role of apolipoprotein E in Guillain-Barré syndrome and experimental autoimmune neuritis. J Biomed Biotechnol. 2010, 2010: 357412-

27. Akiyama TE, Sakai S, Lambert G, Nicol CJ, Matsusue K, Pimprale S, Lee YH, Ricote M, Glass CK, Brewer HB, Gonzalez FJ: Conditional disruption of the peroxisome proliferator-activated receptor gamma gene in mice results in lowered expression of ABCA1, ABCG1, and apoE in macrophages and reduced cholesterol efflux.

Molecular and Cellular Biology. 2002, 22 (8): 2607-19. 10.1128/MCB.22.8.2607-2619.2002.

28. Gourdine JP, Markiv A, Smith-Ravin J: The three-dimensional structure of codakine and related marine C-type lectins. Fish & Shellfish Immunology. 2007, 23 (4): 831-9. 10.1016/j.fsi.2007.03.009.

29. Martin SAM, Blaney SC, Houlihan DF, Secombes CJ: Transcriptome response following administration of a live bacterial vaccine in Atlantic salmon (Salmo salar). Mol Immunol. 2006, 43 (11): 1900-11. 10.1016/j.molimm.2005.10.007.

30. O'Farrell C, Vaghefi N, Cantonnet M, Buteau B, Boudinot P, Benmansour A: Survey of transcript expression in rainbow trout leukocytes reveals a major contribution of interferon-responsive genes in the early response to a rhabdovirus infection. J Virol. 2002, 76 (16): 8040-9. 10.1128/JVI.76.16.8040-8049.2002.

31. Magnadóttir B: Innate immunity of fish (overview). Fish & Shellfish Immunology. 2006, 20 (2): 137-51. 10.1016/j.fsi.2004.09.006.

32. Delgado M, Abad C, Martinez C, Juarranz MG, Leceta J, Ganea D, Gomariz RP: PACAP in immunity and inflammation. Ann N Y Acad Sci. 2003, 992: 141-57. 10.1111/j.1749-6632.2003.tb03145.x.

33. Simonet WS, Lacey DL, Dunstan CR, Kelley M, Chang MS, Lüthy R, Nguyen HQ, Wooden S, Bennett L, Boone T, Shimamoto G, DeRose M, Elliott R, Colombero A, Tan HL, Trail G, Sullivan J, Davy E, Bucay N, Renshaw-Gegg L, Hughes TM, Hill D, Pattison W, Campbell P, Sander S, Van G, Tarpley J, Derby P, Lee R, Boyle WJ: Osteoprotegerin: a novel secreted protein involved in the regulation of bone density. Cell. 1997, 89 (2): 309-19. 10.1016/S0092-8674(00)80209-3.

34. Zauli G, Corallini F, Bossi F, Fischetti F, Durigutto P, Celeghini C, Tedesco F, Secchiero P: Osteoprotegerin increases leukocyte adhesion to endothelial cells both in vitro and in vivo. Blood. 2007, 110 (2): 536-43. 10.1182/blood-2007-01-068395.

35. Yin F, Banerjee R, Thomas B, Zhou P, Qian L, Jia T, Ma X, Ma Y, Iadecola C, Beal MF, Nathan C, Ding A: Exaggerated inflammation, impaired host defense, and neuropathology in progranulin-deficient mice. J Exp Med. 2010, 207: 117-28. 10.1084/jem.20091568.

36. Elkington PTG, O'Kane CM, Friedland JS: The paradox of matrix metalloproteinases in infectious disease. Clin Exp Immunol. 2005, 142: 12-20. 10.1111/j.1365-2249.2005.02840.x.

37. Luster AD, Rothenberg ME: Role of the monocyte chemoattractant protein and eotaxin subfamily of chemokines in allergic inflammation. J Leukoc Biol. 1997, 62 (5): 620-33.

38. Lee BC, Avraham S, Imamoto A, Avraham HK: Identification of the nonreceptor tyrosine kinase MATK/CHK as an essential regulator of immune cells using Matk/CHK-deficient mice. Blood. 2006, 108 (3): 904-7. 10.1182/blood-2005-12-4885.

39. Schiøtz BL, Jørgensen SM, Rexroad C, Gjøen T, Krasnov A: Transcriptomic analysis of responses to infectious salmon anemia virus infection in macrophage-like cells. Virus Res. 2008, 136 (1-2): 65-74. 10.1016/j.virusres.2008.04.019.

40. Jørgensen SM, Afanasyev S, Krasnov A: Gene expression analyses in Atlantic salmon challenged with infectious salmon anemia virus reveal differences between individuals with early, intermediate and late mortality. BMC Genomics. 2008, 9: 179-10.1186/1471-2164-9-179.

41. Onishi RM, Gaffen SL: Interleukin-17 and its target genes: mechanisms of interleukin-17 function in disease. Immunology. 2010, 129 (3): 311-21. 10.1111/j.1365-2567.2009.03240.x.

42. Prasad A, Qamri Z, Wu J, Ganju RK: Slit-2/Robo-1 modulates the CXCL12/CXCR4-induced chemotaxis of T cells. J Leukoc Biol. 2007, 82 (3): 465-76. 10.1189/jlb.1106678.

43. Walzer T, Galibert L, Smedt TD: Dendritic cell function in mice lacking Plexin C1. Int Immunol. 2005, 17 (7): 943-50. 10.1093/intimm/dxh274.

44. Christensen EI, Devuyst O, Dom G, Nielsen R, der Smissen PV, Verroust P, Leruth M, Guggino WB, Courtoy PJ: Loss of chloride channel ClC-5 impairs endocytosis by defective trafficking of megalin and cubilin in kidney proximal tubules. Proc Natl Acad Sci USA. 2003, 100 (14): 8472-7. 10.1073/pnas.1432873100.

45. Haller O, Stertz S, Kochs G: The Mx GTPase family of interferon-induced antiviral proteins. Microbes Infect. 2007, 9 (14-15): 1636-43. 10.1016/j.micinf.2007.09.010.

46. Bunger M, Hooiveld G, Kersten S, Muller M: Exploration of PPAR functions by microarray technology--A paradigm for nutrigenomics. Biochimica et Biophysica Acta (BBA) - Molecular and Cell Biology of Lipids. 2007, 1771 (8): 1046-1064. 10.1016/j.bbalip.2007.05.004.

47. Akira S, Uematsu S, Takeuchi O: Pathogen recognition and innate immunity. Cell. 2006, 124 (4): 783-801. 10.1016/j.cell.2006.02.015.

48. Alexopoulou L, Holt AC, Medzhitov R, Flavell RA: Recognition of double-stranded RNA and activation of NF-kappaB by Toll-like receptor 3. Nature. 2001, 413 (6857): 732-8. 10.1038/35099560.

49. Pettersen RJ, Muna ZA, Kuiper KK, Svendsen E, Müller F, Aukrust P, Berge RK, Nordrehaug JE: Sustained retention of tetradecylthioacetic acid after local delivery reduces angioplasty-induced coronary stenosis in the minipig. Cardiovascular Research. 2001, 52 (2): 306-13. 10.1016/S0008-6363(01)00404-7.

50. Aukrust P, Wergedahl H, Müller F, Ueland T, Dyrøy E, Damås JK, Frøland SS, Berge RK: Immunomodulating effects of 3-thia fatty acids in activated peripheral blood mononuclear cells. Eur J Clin Invest. 2003, 33 (5): 426-33. 10.1046/j.1365-2362.2003.01140.x.

51. Odegaard JI, Ricardo-Gonzalez RR, Goforth MH, Morel CR, Subramanian V, Mukundan L, Eagle AR, Vats D, Brombacher F, Ferrante AW, Chawla A: Macrophage-specific PPARgamma controls alternative activation and improves insulin resistance. Nature. 2007, 447 (7148): 1116-20. 10.1038/nature05894.

52. Pesce JT, Ramalingam TR, Mentink-Kane MM, Wilson MS, Kasmi KCE, Smith AM, Thompson RW, Cheever AW, Murray PJ, Wynn TA: Arginase-1-expressing macrophages suppress Th2 cytokine-driven inflammation and fibrosis. PLoS Pathog. 2009, 5 (4): e1000371-10.1371/journal.ppat.1000371.

53. Vats D, Mukundan L, Odegaard JI, Zhang L, Smith KL, Morel CR, Wagner RA, Greaves DR, Murray PJ, Chawla A: Oxidative metabolism and PGC-1beta attenuate macrophage-mediated inflammation. Cell Metabolism. 2006, 4: 13-24. 10.1016/j.cmet.2006.05.011.

54. Odegaard JI, Ricardo-Gonzalez RR, Eagle AR, Vats D, Morel CR, Goforth MH, Subramanian V, Mukundan L, Ferrante AW, Chawla A: Alternative M2 activation of Kupffer cells by PPARdelta ameliorates obesity-induced insulin resistance. Cell Metabolism. 2008, 7 (6): 496-507. 10.1016/j.cmet.2008.04.003.

55. Kang K, Reilly SM, Karabacak V, Gangl MR, Fitzgerald K, Hatano B, Lee CH: Adipocyte-derived Th2 cytokines and myeloid PPARdelta regulate macrophage polarization and insulin sensitivity. Cell Metabolism. 2008, 7 (6): 485-95. 10.1016/j.cmet.2008.04.002.

56. Medzhitov R: Inflammation 2010: new adventures of an old flame. Cell. 2010, 140 (6): 771-6. 10.1016/j.cell.2010.03.006.

57. Welch JS, Ricote M, Akiyama TE, Gonzalez FJ, Glass CK: PPARgamma and PPARdelta negatively regulate specidic subsets of lipopolysaccharide and IFN-gamma target genes in macrophages. Proc Natl Acad Sci USA. 2003, 100 (11): 6712-7. 10.1073/pnas.1031789100.

58. Fryer JL, Lannan CN: Three decades of fish cell culture: A current listing of cell lines derived from fishes. Journal of Tissue Culture Methods. 1994, 16: 87-94. 10.1007/BF01404816.

59. Tronstad KJ, Berge K, Flindt EN, Kristiansen K, Berge RK: Optimization of methods and treatment conditions for studying effects of fatty acids on cell growth. Lipids. 2001, 36 (3): 305-13. 10.1007/s11745-001-0722-8.

60. Olsvik PA, Lie KK, Jordal AEO, Nilsen TO, Hordvik I: Evaluation of potential reference genes in real-time RT-PCR studies of Atlantic salmon. BMC Mol Biol. 2005, 6: 21-10.1186/1471-2199-6-21.

61. Gentleman RC, Carey VJ, Bates DM, et al: Bioconductor: Open software development for computational biology and bioinformatics. Genome Biology. 2004, 5: R80-10.1186/gb-2004-5-10-r80.

62. Bolstad BM, Irizarry RA, Astrand M, Speed TP: A comparison of normalization methods for high density oligonucleotide array data based on variance and bias. Bioinformatics. 2003, 19 (2): 185-93. 10.1093/bioinformatics/19.2.185.

63. Smyth GK: Linear models and empirical bayes methods for assessing differential expression in microarray experiments. Stat Appl Genet Mol Biol. 2004, 3: Article3-

64. Benjamini Y, Yosef H: Controlling the false discovery rate: a practical and powerful approach to multiple testing. Journal of the Royal Statistical Society, Series B (Methodological). 1995, 57: 289-300.

65. Conesa A, Götz S, García-Gómez JM, Terol J, Talón M, Robles M: Blast2GO: a universal tool for annotation, visualization and analysis in functional genomics research. Bioinformatics. 2005, 21 (18): 3674-6. 10.1093/bioinformatics/bti610.

66. Wickham H: Ggplot: Elegant graphics for data analysis. 2009, New York: Springer

67. Sarkar D: Lattice: Multivariate Data Visualization with R. 2008, New York: Springer

INDEX

M

N

P

Q

R

S

T

U

V

W